DIGITAL TERRAIN ANALYSIS IN SOIL SCIENCE AND GEOLOGY

DIGITAL TERRAIN ANALYSIS IN SOIL SCIENCE AND GEOLOGY

Igor V. Florinsky
Institute of Mathematical Problems of Biology
Russian Academy of Sciences
Pushchino, Russia

AMSTERDAM • BOSTON • HEIDELBERG • LONDON • NEW YORK • OXFORD
PARIS • SAN DIEGO • SAN FRANCISCO • SINGAPORE • SYDNEY • TOKYO

Academic Press is an imprint of Elsevier

Academic Press is an imprint of Elsevier
The Boulevard, Langford Lane, Kidlington, Oxford OX5 1GB, UK
Radarweg 29, PO Box 211, 1000 AE Amsterdam, The Netherlands
225 Wyman Street, Waltham, MA 02451, USA
525 B Street, Suite 1900, San Diego, CA 92101-4495, USA

First edition 2012

Copyright © 2012 Elsevier Inc. All rights reserved

No part of this publication may be reproduced, stored in a retrieval system or transmitted in any form or by any means electronic, mechanical, photocopying, recording or otherwise without the prior written permission of the publisher Permissions may be sought directly from Elsevier's Science & Technology Rights Department in Oxford, UK: phone (+44) (0) 1865 843830; fax (+44) (0) 1865 853333; email: permissions@elsevier.com. Alternatively you can submit your request online by visiting the Elsevier web site at http://elsevier.com/locate/permissions, and selecting Obtaining permission to use Elsevier material

Notice
No responsibility is assumed by the publisher for any injury and/or damage to persons or property as a matter of products liability, negligence or otherwise, or from any use or operation of any methods, products, instructions or ideas contained in the material herein. Because of rapid advances in the medical sciences, in particular, independent verification of diagnoses and drug dosages should be made

Library of Congress Cataloging-in-Publication Data
Florinsky, Igor V.
  Digital terrain analysis in soil science and geology / Igor V. Florinsky.
    p. cm.
  Includes bibliographical references and index.
  ISBN 978-0-12-385036-2 (hardback)
1. Geology—Data processing. 2. Soils—Analysis—Data processing. 3. Geographic information systems. 4. Digital mapping. I. Title.
  QE48.8.F58 2011
  551.0285—dc23

British Library Cataloguing-in-Publication Data
A catalogue record for this book is available from the British Library

For information on all Academic Press publications
visit our web site at www.elsevierdirect.com

Printed and bound in USA

12 13 14 15   10 9 8 7 6 5 4 3 2 1

**Working together to grow
libraries in developing countries**

www.elsevier.com | www.bookaid.org | www.sabre.org

ELSEVIER   BOOK AID International   Sabre Foundation

On the cover:
The cover shows the painting *Koktebel* by Maximilian Voloshin, 1932 (watercolor on paper) from the collection of the Chelyabinsk Museum of Arts, Chelyabinsk, Russia

Contents

Preface ix
Acknowledgments xv
Abbreviations and acronyms xvii

1. Digital Terrain Modeling: A Brief Historical Overview 1

I
PRINCIPLES AND METHODS OF DIGITAL TERRAIN MODELING 5

2. Morphometric Variables

   2.1 Topographic Surface 7
   2.2 Local Morphometric Variables 9
   2.3 Nonlocal Morphometric Variables 16
   2.4 Structural Lines 19
   2.5 Solar Morphometric Variables 21
   2.6 Combined Morphometric Variables 23
   2.7 Landform Classifications 23

3. Digital Elevation Models

   3.1 DEM Generation 31
   3.2 DEM Grid Types 36
   3.3 DEM Resolution 38
   3.4 DEM Interpolation 40

4. Calculation Methods

   4.1 The Evans–Young Method 43
   4.2 Calculation of Local Morphometric Variables on a Plane Square Grid 45
   4.3 Calculation of Local Morphometric Variables on a Spheroidal Equal Angular Grid 54
   4.4 Calculation of Nonlocal Morphometric Variables 59
   4.5 Calculation of Structural Lines 61

5. Errors and Accuracy

5.1 Sources of DEM Errors   66
5.2 Estimation of DEM Accuracy   70
5.3 Calculation Accuracy of Local Morphometric Variables   71
5.4 Ignoring of the Sampling Theorem   81
5.5 The Gibbs Phenomenon   88
5.6 Grid Displacement   93
5.7 Linear Artifacts   98

6. Filtering

6.1 Tasks of DTM Filtering   103
6.2 Methods of DTM Filtering   109
6.3 Two-dimensional Singular Spectrum Analysis   122

7. Mapping and Visualization

7.1 Peculiarities of Morphometric Mapping   133
7.2 Combined Visualization of Morphometric Variables   135
7.3 Cross Sections   135
7.4 Three-dimensional Topographic Modeling   136
7.5 Combining Hill-shading Maps with Soil and Geological Data   141

II
DIGITAL TERRAIN MODELING IN SOIL SCIENCE

8. Influence of Topography on Soil Properties

8.1 Introduction   145
8.2 Local Morphometric Variables and Soil   146
8.3 Nonlocal Morphometric Variables and Soil   148
8.4 Discussion   149

9. Adequate Resolution of Models

9.1 Motivation   151
9.2 Theory   153
9.3 Field Study   157

10. Predictive Soil Mapping

10.1 The Dokuchaev Hypothesis as a Central Idea of Soil Predictions   167
10.2 Early Models   170
10.3 Current Predictive Methods   172
10.4 Topographic Multivariable Approach   187

11 Analyzing Relationships in the "Topography–Soil" System

11.1 Motivation   191
11.2 Study Sites   192
11.3 Materials and Methods   195
11.4 Results and Discussion   214

III
DIGITAL TERRAIN MODELING IN GEOLOGY

12. Folds and Folding

12.1 Introduction   223
12.2 Fold Geometry and Fold Classification   223
12.3 Predicting the Degree of Fold Deformation and Fracturing   225
12.4 Folding Models and the *Theorema Egregium*   226

13. Lineaments and Faults

13.1 Motivation   231
13.2 Theory   235
13.3 Method Validation   237
13.4 Two Case Studies   241

14. Accumulation Zones and Fault Intersections

14.1 Motivation   255
14.2 Study Area   257
14.3 Materials and Methods   258
14.4 Results and Discussion   261

15 Global Topography and Tectonic Structures

15.1 Motivation   263
15.2 Materials and Data Processing   266
15.3 Results and Discussion   269

16. Synthesis   285

Appendix A: The Mathematical Basis of Local Morphometric Variables, by Peter A. Shary

    A.1 Gradient, Flow Lines, and Special Points   289
    A.2 Aspect and Insolation   294
    A.3 Curvatures   297
    A.4 Generating Function   312

Appendix B: LandLord—A Brief Description of the Software   315

References   317
Index   367
Color Plate Section

Preface

This book is the first attempt to synthesize knowledge on theory, methods, and applications of digital terrain analysis in the context of multiscale problems of soil science and geology. The content of the book is based on the author's long-standing, interdisciplinary research.

The book is addressed to geomorphometrists, soil scientists, geologists, geoscientists, geomorphologists, geographers, and GIS scientists (at scholar, lecturer, and postgraduate student levels, with mathematical skills). This book is also intended for GIS professionals in industry and research laboratories focusing on geoscientific and soil research.

The book is divided into three parts. Part I represents the main concepts, principles, and methods of digital terrain modeling. Part II discusses various aspects of the use of digital terrain analysis in soil science. Part III looks at applications of digital terrain modeling in geology.

Chapter 1 presents a brief historical overview of the progress of geomorphometry and digital terrain analysis in the context of soil and geological studies. It is demonstrated that there are four main research trends in soil- and geology-oriented digital terrain modeling: (1) analysis and modeling of relationships between soil properties and topographic characteristics; (2) use of the resulting data and knowledge in predictive mapping of soil properties; (3) analysis of forms of geological features, such as folds and domes; and (4) revelation and analysis of lineaments and faults, as well as their relationships with other components of geosystems.

Part I comprises six chapters, Chapters 2–7.

Chapter 2 discusses the basic notions of digital terrain modeling. First, we introduce the concept of the topographic surface and its restrictions, as well as the notion of a morphometric variable. Then, we present definitions, formulas, and interpretations for five main groups of morphometric variables, such as local, nonlocal, solar, and combined attributes, as well as structural lines. Finally, we discuss three key types of landform classifications: the Gaussian classification, the Efremov–Krcho classification, and the Shary classification, based on signs of several local morphometric variables. The concept of relative accumulation, transit, and dissipation zones is also presented.

In Chapter 3, first, we present a brief review of techniques to produce digital elevation models (DEMs). Specifically, we mention conventional topographic surveys, kinematic GPS surveys, photogrammetric approaches, radar techniques, laser surveys, shipboard echo sounding, satellite radar altimetry, three-dimensional (3D) seismic survey, and digitizing of topographic and geological contours. Second, we describe main types of DEM grids: plane irregular and regular grids, as well as spheroidal regular grids. Third, we discuss issues of DEM resolution, in particular, the fundamental sampling theorem and its three sequences. Finally, there is a brief review of interpolation approaches used in digital terrain analysis.

Chapter 4 deals with calculation methods of digital terrain modeling. First, we describe the Evans–Young method, a conventional technique to compute local topographic attributes. Second, we present two methods to derive models of local morphometric variables from DEMs based on plane square grids and spheroidal equal angular grids. The method for plane square grids uses the approximation of the third-order polynomial to the 5×5 moving window by the least-squares approach. The method can be employed in soil and geological studies at a field, watershed, and regional scales. The method for spheroidal equal angular grids is based on the approximation of the second-order polynomial to the 3×3 moving window by the least-squares approach. The method is intended for geological and soil studies at regional, continental, and global scales. Finally, we briefly review techniques to compute nonlocal morphometric variables on plane square grids, as well as approaches to detect structural lines including calculation of the generating function.

Chapter 5 investigates the problem of errors and accuracy of digital terrain models (DTMs). First, we look at sources and types of DEM errors, as well as methods of their detection and analysis. Second, we describe a method to estimate calculation accuracy of digital models of local morphometric attributes using the criterion of root mean square error of a function of measured variables. Third, we discuss DTM errors responsible for artificial landforms and, hence, critical for DTM-based geological studies. Such artifacts are caused by: (1) the ignoring of the sampling theorem; (2) the Gibbs phenomenon; and (3) discretization errors due to grid displacement. Finally, we explore intrinsic properties of local morphometric variables as a possible cause of orthogonal and diagonal linear artifacts.

Chapter 6 considers various aspects of DTM filtering. There are three main tasks of DTM filtering: (1) DTM decomposition, that is, separation of high- and low-frequency components of the topographic surface to study its structure and elements of different scales; (2) DTM denoising; and (3) DTM generalization. We briefly describe main methods of DTM filtering: trend-surface analysis, the Filosofov method, two-dimensional

(2D) discrete Fourier transform, 2D discrete wavelet transform, smoothing, row and column elimination from DTMs, and the cutting method. The algorithm of the 2D singular spectrum analysis is described in detail and is exemplified by a portion of the Northern Andes.

Chapter 7 examines peculiarities of DTM visualization. For the correct perception of morphometric maps, it is reasonable: (1) to apply layer tinting; (2) to subdivide values of morphometric variables into intervals relative to the zero value; (3) to select contrast color schemes; and (4) to employ a logarithmic transform. We briefly describe some approaches of DTM visualization: combined displaying of several morphometric variables on a single map, construction of cross sections and 3D models, as well as superimposition of geological and soil information on hill-hading maps.

Part II consists of four chapters, Chapters 8−11.

It is well-known that topography is one of the soil-forming factors. Chapter 8 discusses the main regularities in the influence of topography on spatial distribution of soil properties exemplified by soil moisture content. In particular, we explain the role of slope gradient and aspect, horizontal, vertical, and mean curvatures, as well as the catchment area and topographic index.

The central problem of combined analysis of DTMs and soil data is the selection of the correct value for the DTM grid spacing. Chapter 9 describes a method to determine adequate grid spacing based on the concept of representative elementary volume. The method includes the following steps: (1) deriving a set of DTMs using different values of grid spacing; (2) analyzing correlations between a soil property and a morphometric variable calculated with different values of the grid spacing; (3) plotting a graph of correlation coefficients between the soil property and the morphometric variable versus the grid spacing values; and (4) finding smooth portions of the graph indicating intervals of the adequate grid spacing. The method is exemplified by the study of relationships between topography and soil moisture.

Chapter 10 looks at predictive soil mapping, a growing branch of soil science. It is demonstrated that Vasily Dokuchaev explicitly defined its central idea and statement of the problem in 1886. In predictive soil mapping, it is common to use morphometric variables as key predictors of soil properties. We briefly review predictive soil-mapping methods developed in the pre-computer era. Then we present a classification and a brief review of predictive soil-mapping methods based on digital terrain modeling and various mathematical approaches (i.e., multiple regression analysis, hybrid geostatistical approaches, fuzzy logic, discriminant analysis, artificial neural networks, decision trees, etc.). We also discuss problems of small-scale soil prediction and upscaling of relationships in the system "topography–soil". Finally, we describe our approach to map spatial distribution of physical, chemical, and biological properties of soil based on digital terrain modeling and multiple regression analysis.

Chapter 11 presents two case studies on DTM-based analysis of relationships between topography and soil. Study sites were located in the Canadian prairies. We studied: (1) temporal variability in the influence of topography on soil properties; (2) changes in the influence of topography on soil dynamic properties depending on the soil layer depth; and (3) the effect of topography on the activity of denitrifiers under different soil moisture conditions. We demonstrate that (a) spatial distribution of soil dynamic properties depends on topographic variables only if soil moisture content is higher than some threshold value; (b) dependence of soil dynamic properties on morphometric attributes may both decrease and increase as the soil layer depth increases; and (c) there exists temporal variability in relationships between spatial distribution of soil dynamic properties and morphometric attributes.

Part III includes four chapters, Chapters 12–15.

Folds of various scales are the most abundant and studied geological features. Chapter 12 probes into applications of data on curvatures of the land and stratigraphic surfaces in research of folds and folding processes. In particular, such data are used (1) to describe the geometry of folds and to classify folds; (2) to predict the degree of fold deformation and strain, as well as fracture orientation and fracture density of folded strata; and (3) to estimate the degree of plasticity of folds and to reconstruct paleotopography using Gauss's *Theorema Egregium*.

Lineaments are usually associated with faults and linear zones of fracturing, bending deformation, and increased permeability of the crust. Chapter 13 describes a method to reveal and classify topographically expressed lineaments. The method is based on the derivation of horizontal and vertical curvatures from DEMs of the land or stratigraphic surfaces. Lineaments revealed on horizontal curvature maps are associated with faults formed mostly by horizontal tectonic movements (i.e., strike-slip faults). Lineaments recognized by mapping of vertical curvature relate to faults formed mainly by vertical motions (i.e., dip-slip and reverse faults) and thrusting. Lineaments recorded on maps of both horizontal and vertical curvatures indicate, as a rule, oblique-slip and gaping faults. The method is validated using an artificial DEM with modeled faults. Finally, we present results of the method's implementation for a seismically active, mountainous terrain (the Crimean Peninsula) and a platform plain region (the Kursk nuclear power plant area).

Topography controls in many ways gravity-driven overland and intrasoil transport of water. At the same time, valley networks, determining principal routes of overland flows, can often be connected with fault networks, which serve as pathways for upward transport of groundwaters. For tectonic terrains, it was established that topographically expressed zones of flow accumulation, as a rule, coincide with sites of fault intersections due to increased rock fracturing. Chapter 14 looks at relationships between zones of flow accumulation and natural phenomena *a priori* associated with fault intersections (i.e., sites of intensive rock fracturing and sites of springs/boreholes with abnormally high discharges). The analysis was performed for the Crimean Peninsula. It was found that the phenomena under study are spatially correlated with the location of accumulation zones. This testifies that in accumulation zones, soil moisture depends on both upward transport of deep groundwaters and accumulation of overland lateral water flows. In other words, topographically expressed accumulation zones/fault intersections are areas of contact and interaction between overland and deep substance flows.

In the past few decades, there have been proposals suggesting that hidden global linear (helical) structures exist, which are tectonically and topographically expressed. Chapter 15 examines this hypothesis using spheroidal digital terrain modeling and global DEMs of the Earth, Mars, Venus, and the Moon. Local and nonlocal topographic variables were calculated and mapped for the entire surface of the celestial bodies. Digital terrain analysis provided support for the existence of global lineaments: on catchment area maps of the Earth, it was possible to detect five mutually symmetrical pairs of helical structures encircling the planet from pole to pole. On catchment area maps of Mars and Venus, it was also possible to detect several helical structures. All these structures are apparently associated with traces of the planetary rotational stresses.

Chapter 16 concludes the book summarizing its main themes.

Appendix A presents a mathematical proof for the formulas of local morphometric variables used in this book. First, we discuss slope lines, flow lines, special and nonspecial points of the topographic surface. Second, we develop equations for gradient, aspect, insolation, plan curvature, horizontal curvature, vertical curvature, unsphericity curvature, rotor, difference curvature, total ring curvature, total accumulation curvature, and generating function. Third, we probe the physical meaning of local morphometric variables and interrelationships between curvatures.

Appendix B briefly describes the software LandLord intended for digital terrain analysis.

I would appreciate receiving readers' suggestions and pointing out of errors.

Igor V. Florinsky
Pushchino−Kiev, April 2011

Acknowledgments

This book is mostly based on my research conducted from 1989 to 2010 in two institutes of the Russian Academy of Sciences: the Institute of Soil Science and Photosynthesis, and the Institute of Mathematical Problems of Biology (Pushchino, Moscow Region), as well as in two Canadian organizations: the Manitoba Land Resource Unit, Agriculture and Agri-Food Canada, and the Department of Soil Science, University of Manitoba (Winnipeg, MB).

For over 20 years, I have discussed various aspects of this research with many scholars. The most fruitful discussions were with the late I. N. Stepanov, P. A. Shary, G. A. Kuryakova, A. M. Molchanov, J. L. Meshalkina, and V. G. Trifonov.

Section 6.3 is based on my collaborative work with N. E. Golyandina and K. D. Usevich. Experimental sections of Chapter 9 are based on the field data collected together with G. A. Kuryakova and P. A. Shary. Chapter 11 includes some results of my collaborative studies with R. G. Eilers, G. R. Manning, D. L. Burton, and S. K. McMahon. Appendix A is written by P. A. Shary.

I am grateful to all my colleagues for their help, criticism, and enthusiasm.

Finally, I thank Liza for her patience.

Abbreviations and Acronyms

ASTER	Advanced Spaceborne Thermal Emission and Reflection Radiometer
AVHRR	Advanced Very High Resolution Radiometer
DEM	digital elevation model
DSM	digital soil mapping
DTM	digital terrain model
ERS	European Remote-Sensing Satellite
ET	eigentriple
GIS	geographical information system
GPS	global positioning system
IBCAO	International Bathymetric Chart of the Arctic Ocean
InSAR	interferometric synthetic aperture radar
K-S	Kolmogorov–Smirnov
LiDAR	light detection and ranging
LOLA	Lunar Orbiter Laser Altimeter
MOLA	Mars Orbiter Laser Altimeter
MSS	Multispectral Scanner
R^2	coefficient of determination
RMSE	root mean square error
SAR	synthetic aperture radar
SPOT	Satellite Pour l'Observation de la Terre
SRTM	Shuttle Radar Topography Mission
SSA	singular spectrum analysis
TIN	triangulated irregular network
1D	one-dimensional
2D	two-dimensional
2D-SSA	two-dimensional singular spectrum analysis
3D	three-dimensional
4D	four-dimensional
A	slope aspect
CA	catchment area
DA	dispersive area
E	difference curvature
G	slope gradient
H	mean curvature
$I(\theta,\psi)$	insolation
IS	shape index
K	Gaussian curvature

K_a	accumulation curvature
k_h	horizontal curvature
k_{he}	horizontal excess curvature
k_{max}	maximal curvature
k_{min}	minimal curvature
k_p	plan curvature
K_r	ring curvature
k_v	vertical curvature
k_{ve}	vertical excess curvature
M	unsphericity curvature
$Moist$	soil moisture
SCA	specific catchment area
SDA	specific dispersive area
SI	stream power index
T	generating function
TI	topographic index
w	square grid spacing
z	elevation

CHAPTER

1

Digital Terrain Modeling
A Brief Historical Overview

Topography is one of the main factors controlling processes taking place in the near-surface layer of the planet (Huggett and Cheesman, 2002). In particular, topography is one of the soil-forming factors (Dokuchaev, 1883; Zakharov, 1913; Neustruev, 1927; Jenny, 1941; Huggett, 1975; Fridland, 1976; Gerrard, 1981; Schaetzl and Anderson, 2005) since it influences: (1) climatic and meteorological characteristics, which controls hydrological and thermal regimes of soils (Geiger, 1927; Romanova, 1977; Kondratyev et al., 1978; Raupach and Finnigan, 1997; Böhner and Antonić, 2009); (2) prerequisites for gravity-driven overland and intrasoil lateral transport of water and other substances (Kirkby and Chorley, 1967; Young, 1972; Speight, 1980); and (3) spatial distribution of vegetation cover (Yaroshenko, 1961; Franklin, 1995). At the same time, being a result of the interaction of endogenous and exogenous processes of different scales, topography can reflect the geological structure of a terrain (Penck, 1924; Gerasimov, 1959; Meshcheryakov, 1965; Ollier, 1981; Ufimtsev, 1988; Burbank and Anderson, 2001; Scheidegger, 2004; Lopatin, 2008; Brocklehurst, 2010). In this connection, qualitative and quantitative topographic information is widely used in the geosciences.

Before the 1990s, topographic maps were the main source of quantitative information on topography. They were analyzed using geomorphometric[1] techniques to calculate manually morphometric variables (e.g., slope gradient, drainage density, horizontal curvature, etc.) and produce morphometric maps (Vakhtin, 1930; Weinberg, 1934a; Chentsov, 1940; Horton, 1945; Volkov, 1950; Strahler, 1957; Clarke, 1966; Devdariani, 1967;

---

[1]Pike (2000, p. 1) defines geomorphometry as "the quantitative study of topography."

Pannekoek, 1967; Mark, 1975b; Gardiner and Park, 1978; Stepanov et al., 1984; Lastochkin, 1987). Conventional geomorphometric techniques have received wide acceptance in geological studies (Berlyant, 1966): to represent graphically the shape of deposits (Sobolevsky, 1932), to explore oil and gas bearing and ore controlling structures (Levorsen, 1927; Filosofov, 1960; Murray, 1968; Volchanskaya, 1981; Guberman et al., 1997), to analyze block structure of the Earth's crust (Orlova, 1975; Glasko and Rantsman, 1996), to study seismicity (Gelfand et al., 1972), and so on. In soil science, conventional geomorphometric techniques have been used to investigate relationships between soil cover and topography (Dokuchaev, 1891; Ototzky, 1901); to predict quantitative soil properties (Romanova, 1963, 1970, 1971); to produce soil maps (Anisimov et al., 1977; Stepanov et al., 1987, 1998; Stepanov, 1989; Stepanov and Loshakova, 1998); and to study regularities in the structure of the soil cover and its relations with geological features (Filatov, 1927; Stepanov and Sabitova, 1983; Kuryakova and Florinsky, 1991; Stepanov, 1996).

In the mid-1950s, a new research field—digital terrain modeling—emerged in photogrammetry (Rosenberg, 1955). Within its framework, digital elevation models (DEMs), two-dimensional discrete functions of elevation, became the main source of information on topography. DEMs were used to calculate digital terrain models (DTMs), two-dimensional discrete functions of morphometric variables. Initially, digital terrain modeling has mainly been applied to produce raised-relief maps using computer-controlled milling machines (Spooner et al., 1957; Lyubkov and Martynenko, 1963), and to design highways and railways (Miller and Leflamme, 1958; Konovalov, 1960).

Subsequent advances in computer, space, and geophysical technologies were responsible for the transition from conventional geomorphometry to digital terrain modeling[2] (Evans, 1972; Koshkarev, 1982; Burrough, 1986; Dikau, 1988; Serbenyuk, 1990). This was supported by the development of the physical and mathematical theory of the topographic surface[3] in gravity (Krcho, 1973; Evans, 1979; Shary, 1991, 1995; Koenderink and van Doorn, 1994; Rudy, 1999; Shary et al., 2002b). Currently, digital terrain modeling is widely used to solve various multiscale problems of geomorphology, hydrology, remote sensing, soil science, geology, geophysics, geobotany, glaciology, oceanology, climatology, planetology, and other disciplines—see reviews (McCullagh, 1988; Moore et al., 1991; Shary et al., 1991; Weibel and Heller, 1991;

---

[2]Pike (2000, p. 1) noted that geomorphometry "is known variously as terrain analysis or quantitative geomorphology, although the newer term digital terrain modelling increasingly seems preferred."

[3]For the definition of the term *topographic surface*, see Section 2.1.

Band, 1993; Franklin, 1995; Florinsky, 1998b; Pike, 1995, 2000, 2001; Deng, 2007; Brocklehurst, 2010), books (Felicísimo, 1994a; Wilson and Gallant, 2000; El-Sheimy et al., 2005; Li et al., 2005; Hengl and Reuter, 2009), and bibliography (Pike, 2002). Digital terrain modeling evolved into the science of quantitative modeling and analysis of the topographic surface and relationships between topography and other natural and artificial components of geosystems.[4]

As early as the 1960s, soil science and geology have begun to use methods of digital terrain modeling. Two DTM-based research avenues have arisen in that time:

- Analysis of the influence of topography on the formation of soil properties. In that period, the first attempts were made to model soil properties with digital topographic data (Troeh, 1964; Walker et al., 1968).
- Study of geological structures using DEMs of both the land surface and stratigraphic surfaces (Muñoz-Espinoza, 1968; Abelsky and Lastochkin, 1969; Robinson et al., 1969). One of the pioneering works was conducted by Belonin and Zhukov (1968) who studied an uplift evolution of an elevated block using digital models of the Gaussian, mean, and principal curvatures of geological surfaces.

Although the first effective methods to calculate morphometric variables were developed in the 1970–1980s (Young, 1978; Evans, 1979; Zevenbergen and Thorne, 1987; Jenson and Domingue, 1988; Martz and de Jong, 1988), digital terrain modeling was still relatively uncommon in soil and geological studies of this period. In the 1980s, however, two mentioned DTM-based research trends have been further developed in both soil science (Sinai et al., 1981; Burt and Butcher, 1985; Kachanoski et al., 1985a, 1985b; Pennock et al., 1987) and geology (Moore and Simpson, 1983; Schowengerdt and Glass, 1983; Onorati et al., 1987; Zeilik et al., 1989). In the 1990s, the widespread use of personal computers was responsible for the mass transition from conventional geomorphometric techniques to digital terrain analysis in both soil science and geology. In the first decade of the twenty-first century, advances in aerial, space, and geophysical technologies have opened new horizons for digital terrain modeling. First, large-scale and detailed DEMs of the land surface became available due to the progress in kinematic GPS survey (Ghilani and Wolf, 2008) and LiDAR aerial survey (French, 2003). Second, global DEMs marked by a relatively high resolution and accuracy were produced using satellite surveys (Farr et al., 2007; Hato et al., 2009). Public access to these materials via Internet extended the

---

[4]Detailed historical overview of digital terrain modeling can be found elsewhere (Pike et al., 2009).

capabilities to conduct DTM-based regional geological and soil studies. Finally, the advances in three-dimensional seismic survey (Chopra and Marfurt, 2005, 2007a) enhanced the production of DEMs of geological surfaces.

Currently, there are four main research trends in soil- and geology-oriented digital terrain modeling:

1. Analysis and modeling of relations between soil properties and topographic characteristics (Chapters 8, 9, and 11).
2. Use of the resulting data and knowledge in predictive mapping of soil properties (Chapters 10 and 11).
3. Analysis of forms of geological features, such as folds and domes (Chapter 12).
4. Revealing and analysis of lineaments and faults, as well as their relations with other components of geosystems (Chapters 13–15).

Methods of digital terrain modeling are also used in three research fields, which, from the formal point of view, are associated with geology. However, they have a closer connection with adjacent research areas, and hence they are not discussed in this book:

1. Study of geodynamics as a factor of terrain evolution (reviews can be found elsewhere—Codilean et al., 2006; Brocklehurst, 2010). This is the subject of tectonic geomorphology.
2. Analysis of microtopography of geological samples (Pollard et al., 2004). To carry out such works, researchers use superdetailed DTMs with resolution of about 0.2 mm. This scientific field is close to industrial surface metrology (Pike, 2001).
3. Study of geophysical fields using approaches of digital terrain modeling (e.g., Rybakov et al., 2003). This research trend is, in fact, a modification of the well-known geophysical method of second derivatives (Elkins, 1951).

PART I

PRINCIPLES AND METHODS OF DIGITAL TERRAIN MODELING

CHAPTER 2

Morphometric Variables

OUTLINE

2.1 Topographic Surface	7
2.2 Local Morphometric Variables	9
2.3 Nonlocal Morphometric Variables	16
2.4 Structural Lines	19
2.5 Solar Morphometric Variables	21
2.6 Combined Morphometric Variables	23
2.7 Landform Classifications	23
2.7.1 The Gaussian Classification	23
2.7.2 The Concept of Accumulation Zones. The Efremov–Krcho Classification	25
2.7.3 The Shary Classification	28

2.1 TOPOGRAPHIC SURFACE

The Earth's surface is too complex for rigorous mathematical treatment because it is not smooth and regular (Shary, 2008). However, for many practically important problems, it is sufficient to approximate the Earth's surface by the topographic surface. Here, we define the topographic surface as a closed, oriented, continuously differentiable, two-dimensional manifold S in the three-dimensional Euclidean space E^3. Three key restrictions are true for the topographic surface (Evans, 1979; Mark, 1979; Shary, 1991, 1995):

1. The topographic surface is uniquely defined by a continuous, single-valued function $z = f(x, y)$, where z is elevation and x and y are the Cartesian coordinates. In particular, this condition means that caves, grottos, and similar landforms are ignored.

2. The planar size of the topographic surface is essentially less than the Earth's radius. It is generally assumed that the curvature of the planet may be ignored if the size of the surface portion is less than 0.1 of the average radius of the planet.
3. The topographic surface is characterized by a uniform gravity field. This condition is important because the Earth's surface is the surface in gravity. Its description by measuring elevations (Section 3.1) implicitly considers the direction of the gravitational acceleration vector.

We also assume that topography is a scale-dependent phenomenon (Clarke, 1988). In such a case, a fractal component of topography can be considered as a high-frequency noise. Fractal topographic models (Clarke, 1988; Xu et al., 1993; Ivanov, 1994; McClean and Evans, 2000) are not discussed in this book.

The notion of the topographic surface can be easily applied to describe not only the Earth's surface, but also other similar surfaces studied in soil science and geology. In this book, the term *topographic surface* means the following:

- The Earth's surface, that is, a contact surface between (a) the lithosphere and atmosphere, (b) the lithosphere and hydrosphere, and (c) the ice sheet and the atmosphere.
- The surface of a celestial body, that is, a contact surface between its lithosphere and atmosphere (for celestial bodies with atmospheres), or its lithosphere and space (for celestial bodies without atmospheres).
- The surface of a soil horizon, stratigraphic horizon, geological structure (e.g., a fold), and so on, that is, a contact surface between (a) adjacent geological horizons, (b) adjacent soil horizons, and (c) the lithosphere and the ice sheet (for the Earth and some celestial bodies).

A morphometric (or topographic) variable or attribute is a two-dimensional function describing the topographic surface. In this book, we mainly discuss fundamental topographic variables associated with concepts of differential geometry and theory of the topographic surface (Shary, 1991, 1995; Shary et al., 2002b). Such an approach is connected with the concept of general geomorphometry, which Evans (1972, p. 18) defined as

> the measurement and analysis of those characteristics of landform which are applicable to any continuous rough surface. This is distinguished from "specific geomorphometry", the measurement and analysis of specific landforms such as cirques, drumlins and stream channels ... General geomorphometry as a whole provides a basis for the quantitative comparison even of qualitatively different

landscapes, and it can adapt methods of surface analysis used outside geomorphology. Specific geomorphometry is more limited; it involves more arbitrary decisions, and leaves more room for subjectivity in the quantification of its concepts.

There are five main groups of morphometric variables[1]:

1. Local variables (Section 2.2)[2]
2. Nonlocal variables (Section 2.3)
3. Structural lines (Section 2.4)
4. Solar variables (Section 2.5)
5. Combined variables (Section 2.6)

Being a morphometric variable, elevation is outside the groups. All topographic attributes are derived from elevation data.

2.2 LOCAL MORPHOMETRIC VARIABLES

Local morphometric variables describe the geometry of the topographic surface in the vicinity of each point of the surface (Speight, 1974), along directions determined by two pairs of mutually perpendicular normal sections[3] (Fig. 2.1a). The first pair includes principal sections AA' and BB,' well known in differential geometry (Pogorelov, 1957, § 3; Nikolsky, 1977a, § 7.24). These are normal sections with extreme (maximal and minimal) bending at the given point of the surface. The second pair of normal sections CC' and DD' is defined on the topographic surface by gravity: the normal section CC' includes the gravitational acceleration vector at the given point.

Two classes of local morphometric variables—form attributes and flow attributes—are related to the two pairs of normal sections (Shary, 1995; Shary et al., 2002b). Form attributes are associated with principal sections. They are gravity field invariants; that is, they do not depend on the direction of the gravitational acceleration vector. Among these

---

[1] Alternative classifications of morphometric variables can be found elsewhere (Shary et al., 2002; Evans and Minár, 2011).

[2] For morphometric variables, the terms *local* and *nonlocal* are used regardless of the study scale or DTM resolution. These terms are associated with the mathematical sense of a particular variable. A local morphometric attribute describes the surface geometry in the vicinity of the given point. A nonlocal topographic variable characterizes a relative position of the given point on the surface.

[3] A normal section is a curve formed by the intersection of a surface with a plane containing the normal to the surface at the given point.

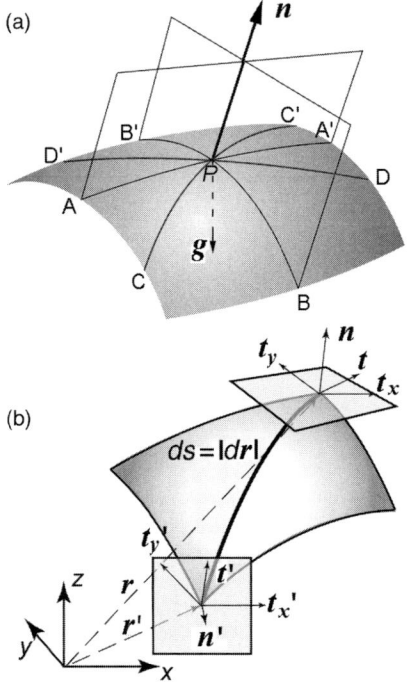

FIGURE 2.1 Illustrations for some notions of geomorphometry: (a) Four normal sections of the topographic surface at the point P: AA′ and BB′ are principal sections, CC′ and DD′ are normal sections setting by gravity; *n* is the external normal to the surface at the point P, *g* is the gravitational acceleration vector at the point P. Modified from (Lisle et al., 2010, Fig. 2), reproduced with permission. (b) The curvature of a surface can be defined as the rate of change of the tangent vector *t* moved along the curve *ds* on the surface (thick line). The vector *n* is the unit normal to the surface at the initial point defined by the vector *r*; t_x and t_y are tangent vectors defining the tangent plane at the initial point. The primed symbols indicate corresponding vectors at the end point. From (Bergbauer and Pollard, 2003, Fig. 1), reproduced with permission.

are minimal curvature[4] (k_{min}), maximal curvature (k_{max}),[5] mean curvature (H), the Gaussian curvature (K), and unsphericity curvature (M). Flow attributes are gravity field-specific variables because they depend on the direction of the gravitational acceleration vector. Among these are slope gradient (G), slope aspect (A), horizontal curvature (k_h),[6] vertical curvature (k_v),[7] difference curvature (E), accumulation curvature (K_a), ring curvature (K_r), vertical excess curvature (k_{ve}), horizontal excess curvature (k_{he}), generating function (T), and some others.

---

[4]In this book, we systematically use terms and concepts associated with the notion of surface curvature. For explanation, see Fig. 2.1b and Section A.3. It is assumed that the curvature is positive for convex landforms (e.g., hills and ridges) and the curvature is negative for concave ones (e.g., depressions and valleys).

[5]k_{min} and k_{max} are principal curvatures (Pogorelov, 1957, § 3; Nikolsky, 1977a, § 7.24).

[6]Do not confuse this with plan curvature (Section A.3.1).

[7]k_h is sometimes called *plan curvature* or *tangential curvature*, while k_v is called *profile curvature* (e.g., Shary, 2006). Roberts (2001) used the terms *strike curvature, dip curvature, dip angle,* and *azimuth* for k_h, k_v, G, and A, correspondingly. Such terminological "innovations" introduce confusion.

2.2 LOCAL MORPHOMETRIC VARIABLES

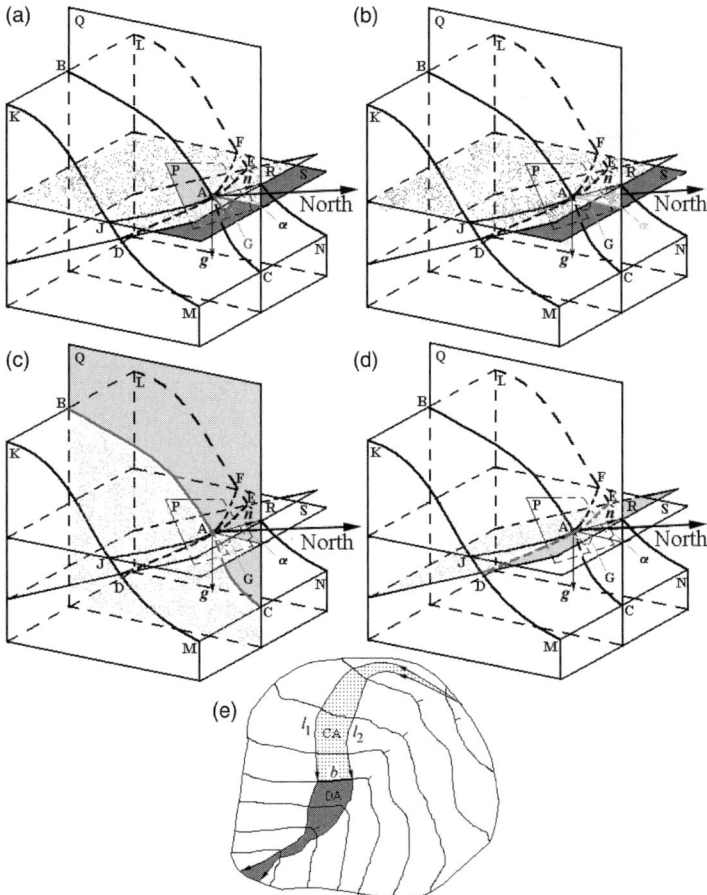

FIGURE 2.2 Illustrations for the definition of some morphometric variables: (a) slope gradient, (b) slope aspect, (c) vertical curvature, (d) horizontal curvature, (e) catchment and dispersive areas. For an explanation, see Table 2.1. *From (Florinsky, 2010, Fig. 1.2).* (See page 1 in Color Plate Section)

Table 2.1 presents definitions and interpretations of local morphometric variables.

Local topographic variables are functions of partial derivatives of elevation

$$g = \frac{\partial^3 z}{\partial x^3}, \ h = \frac{\partial^3 z}{\partial y^3}, \ k = \frac{\partial^3 z}{\partial x^2 \partial y}, \ m = \frac{\partial^3 z}{\partial x \partial y^2}, \ r = \frac{\partial^2 z}{\partial x^2},$$
$$t = \frac{\partial^2 z}{\partial y^2}, \ s = \frac{\partial^2 z}{\partial x \partial y}, \ p = \frac{\partial z}{\partial x}, \ q = \frac{\partial z}{\partial y}. \tag{2.1}$$

TABLE 2.1 Definitions and Interpretations of Some Morphometric Variables

Variable and Unit	Definition and Interpretation
Local Morphometric Variables	
Form Attributes	
Minimal curvature, m^{-1}	A curvature of a normal section with the lowest value of curvature among all normal sections of the topographic surface at the given point (Gauss, 1828) (Fig. 2.1a). $k_{min} > 0$ corresponds to hills, while $k_{min} < 0$ relates to valleys (Section A.3.5).
Maximal curvature, m^{-1}	A curvature of a normal section with the highest value of curvature among all normal sections of the topographic surface at the given point (Gauss, 1828) (Fig. 2.1a). $k_{max} > 0$ corresponds to ridges, while $k_{max} < 0$ relates to closed depressions (Section A.3.5).
Mean curvature, m^{-1}	A half-sum of curvatures of two orthogonal normal sections of the topographic surface at the given point (Young, 1805). H presents convergence and relative deceleration of flows with equal weights (Section A.3.5).
Gaussian curvature, m^{-2}	A product of maximal and minimal curvatures. According to *Teorema egregium*, K retains values in each point of the topographic surface after its bending without breaking, stretching, and compressing (Gauss, 1828) (Section A.3.5).
Unsphericity curvature, m^{-1}	A half-difference of maximal and minimal curvatures (Shary, 1995). $M = 0$ on a sphere; thus values of M show the extent to which the shape of the topographic surface is non-spherical at the given point (Section A.3.6).
Flow Attributes	
Slope gradient, °	An angle G between the tangent plane P and the horizontal plane S at the given point A of the topographic surface (Fig. 2.2a) (Lehmann, 1816). G determines the velocity of gravity-driven flows (Section A.1.1).
Slope aspect, °	A clockwise angle α from north to a projection of the external normal n to the horizontal plane S at the given point A of the topographic surface (Fig. 2.2b). A is a measure of the direction of gravity-driven flows (Section A.2.1).
Vertical curvature, m^{-1}	A curvature of the normal section BAC formed by the intersection of the topographic surface with the plane Q, which contains the gravitational acceleration vector g at the given point A of the surface (Fig. 2.2c). k_v is a measure of relative deceleration and acceleration of gravity-driven flows (Aandahl, 1948; Speight, 1974; Shary, 1991). Overland and intrasoil lateral flows are decelerated when $k_v < 0$, and they are accelerated when $k_v > 0$ (Section A.3.3).

(Continued)

TABLE 2.1 (Continued)

Variable and Unit	Definition and Interpretation
Horizontal curvature, m^{-1}	A curvature of the normal section DAE formed by the intersection of the topographic surface with the plane R, which is orthogonal to the normal vertical section BAC at the given point A of the surface (Fig. 2.2d). k_h is a measure of flow convergence and divergence (Sobolevsky, 1932; Kirkby and Chorley, 1967; Shary, 1991). Gravity-driven overland and intrasoil lateral flows are converged when $k_h < 0$, and they are diverged when $k_h > 0$ (Sections A.3.1 and A.3.2).
Difference curvature, m^{-1}	A half-difference of vertical and horizontal curvatures (Shary, 1995). E shows to what extent k_v is larger than k_h at the given point of the topographic surface (Section A.3.8).
Horizontal excess curvature, m^{-1}	A difference of horizontal and minimal curvatures (Shary, 1995). k_{he} describes to what extent k_h is larger than k_{min} at the given point of the topographic surface (Section A.3.9).
Vertical excess curvature, m^{-1}	A difference of vertical and minimal curvatures (Shary, 1995). k_{ve} describes to what extent k_v is larger than k_{min} at the given point of the topographic surface (Section A.3.9).
Accumulation curvature, m^{-2}	A product of vertical and horizontal curvatures (Shary, 1995). K_a is a measure of the extent of flow accumulation at the given point of the topographic surface (Section A.3.11).
Ring curvature, m^{-2}	A product of horizontal excess and vertical excess curvatures (Shary, 1995). $K_r = 0$ for any point of a radially symmetrical landform with a vertical axis of symmetry (Section A.3.10).

Nonlocal Morphometric Variables

Catchment area, m^2	An area of the closed figure CA formed by the contour portion b at the given point of the topographic surface and two flow lines l_1 and l_2 coming from upslope to the ends of the contour portion (Fig. 2.2e) (Speight, 1974). A measure of the contributing area.
Specific catchment area, m^2/m	A ratio of CA to the length of the contour portion b (Fig. 2.2e) (Speight, 1974).
Dispersive area, m^2	An area of the closed figure DA formed by the contour portion b at the given point of the topographic surface and two flow lines l_1 and l_2 going down slope from the ends of the contour portion (Fig. 2.2e) (Speight, 1974). A measure of a downslope area potentially exposed by flows passing through the given point of the topographic surface.
Specific dispersive area, m^2/m	A ratio of DA to the length of the contour portion b (Fig. 2.2e) (Speight, 1974).

(Continued)

TABLE 2.1 (Continued)

Variable and Unit	Definition and Interpretation
Drainage density, m/m^2	The total length of the permanent and seasonal streams and rivers for the unit area (Horton, 1945). A measure of landscape dissection.
Combined Morphometric Variables	
Topographic index	The logarithm of the ratio of CA to tan G at the given point of the topographic surface. A measure of the extent of flow accumulation (Beven and Kirkby, 1979).
Stream power index	The logarithm of the product of CA and tan G at the given point of the topographic surface. SI can be used to describe potential flow erosion and related landscape processes (Moore et al., 1991).

They are calculated using the following formulas (Shary, 1991, 1995, 2006):

$$G = \arctan\sqrt{p^2 + q^2}, \tag{2.2}$$

$$A = -90\left[1 - \operatorname{sign}(q)\right](1 - |\operatorname{sign}(p)|) + 180\left[1 + \operatorname{sign}(p)\right] - \frac{180}{\pi}\operatorname{sign}(p)\arccos\left(\frac{-q}{\sqrt{p^2 + q^2}}\right), \tag{2.3}$$

$$k_h = -\frac{q^2 r - 2pqs + p^2 t}{(p^2 + q^2)\sqrt{1 + p^2 + q^2}}, \tag{2.4}$$

$$k_v = -\frac{p^2 r + 2pqs + q^2 t}{(p^2 + q^2)\sqrt{(1 + p^2 + q^2)^3}}, \tag{2.5}$$

$$K = k_{min}k_{max} = \frac{rt - s^2}{(1 + p^2 + q^2)^2}, \tag{2.6}$$

$$H = \frac{1}{2}(k_{min} + k_{max}) = \frac{1}{2}(k_h + k_v) = -\frac{(1 + q^2)r - 2pqs + (1 + p^2)t}{2\sqrt{(1 + p^2 + q^2)^3}} \tag{2.7}$$

$$E = \frac{1}{2}(k_v - k_h) = \frac{q^2 r - 2pqs + p^2 t}{(p^2 + q^2)\sqrt{1 + p^2 + q^2}} - \frac{(1 + q^2)r - 2pqs + (1 + p^2)t}{2\sqrt{(1 + p^2 + q^2)^3}}, \tag{2.8}$$

2.2 LOCAL MORPHOMETRIC VARIABLES

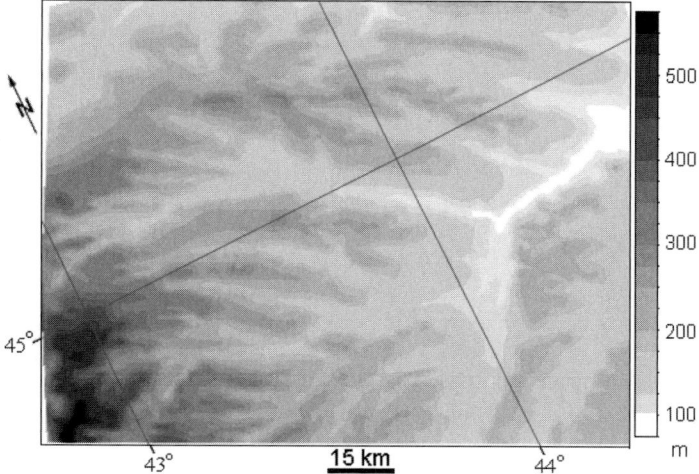

FIGURE 2.3 The Stavropol Upland: elevation. For the DEM description, see Section 4.2.3.1. *From (Florinsky, 2009b, Fig. 2).*

$$K_a = k_h k_v = \frac{(q^2 r - 2pqs + p^2 t)(p^2 r + 2pqs + q^2 t)}{[(p^2 + q^2)(1 + p^2 + q^2)]^2}, \quad (2.9)$$

$$M = \frac{1}{2}(k_{max} - k_{min}) = \sqrt{H^2 - K}, \quad (2.10)$$

$$K_r = k_{he} k_{ve} = M^2 - E^2 = \left[\frac{(p^2 - q^2)s - pq(r - t)}{(p^2 + q^2)(1 + p^2 + q^2)}\right]^2, \quad (2.11)$$

$$k_{he} = k_h - k_{min} = M - E, \quad (2.12)$$

$$k_{ve} = k_v - k_{min} = M + E, \quad (2.13)$$

$$k_{min} = H - M, \quad (2.14)$$

$$k_{max} = H + M. \quad (2.15)$$

For the development of these formulas, see Appendix A.

Notice that not all local morphometric variables can be determined at special points of the topographic surface (for details, see Appendix A). The following expression is true for the locus of special points: $p^2 + q^2 = 0$ (Nikolsky, 1977a, § 7.23 and 7.24; Shary, 1991). Among these are isolated points (local extrema, such as hill summits and pit bottoms), saddles, as well as horizontal planes (Section A.1.2).

Methods for calculating local topographic attributes are discussed in Sections 4.1–4.3. Figures 2.3–2.5 illustrate their calculation.

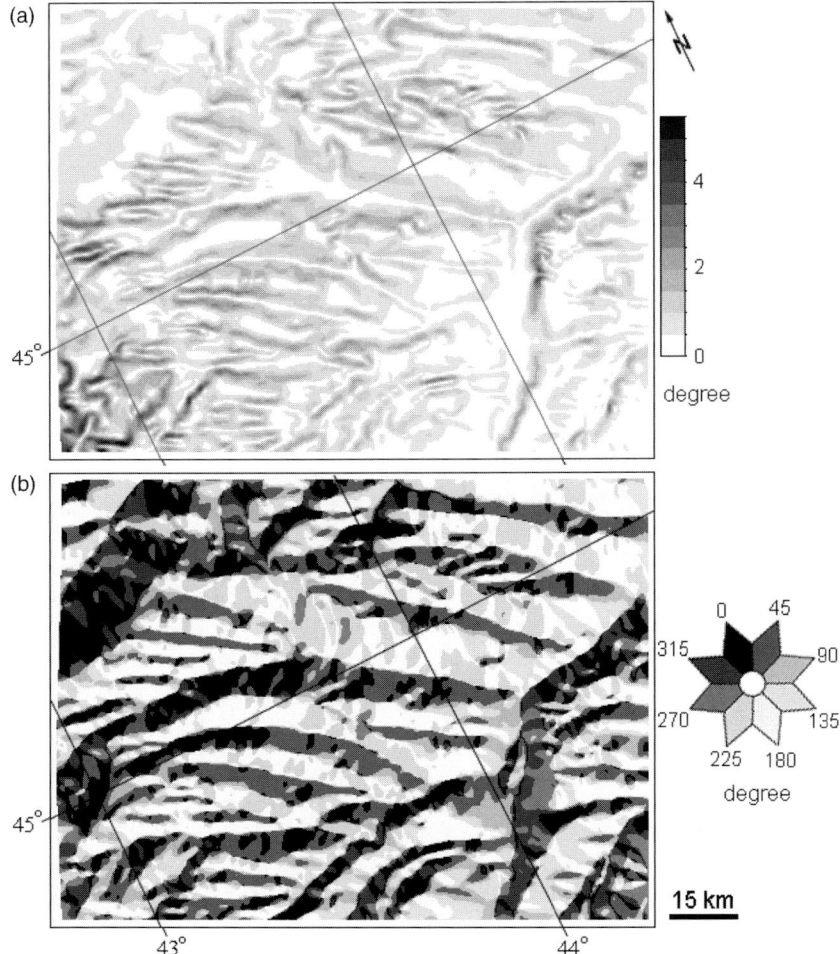

FIGURE 2.4 The Stavropol Upland: (a) slope gradient, (b) slope aspect. For details of calculation, see Section 4.2.3.1; for the elevation map, see Fig. 2.3. *From (Florinsky, 2010, Fig. 1.4).*

2.3 NONLOCAL MORPHOMETRIC VARIABLES

Nonlocal morphometric variables describe a relative position of a point on the surface (Speight, 1974). To determine such attributes, one should analyze a large territory with boundaries located far away from the given point (e.g., an entire upslope portion of a watershed). Among

2.3 NONLOCAL MORPHOMETRIC VARIABLES

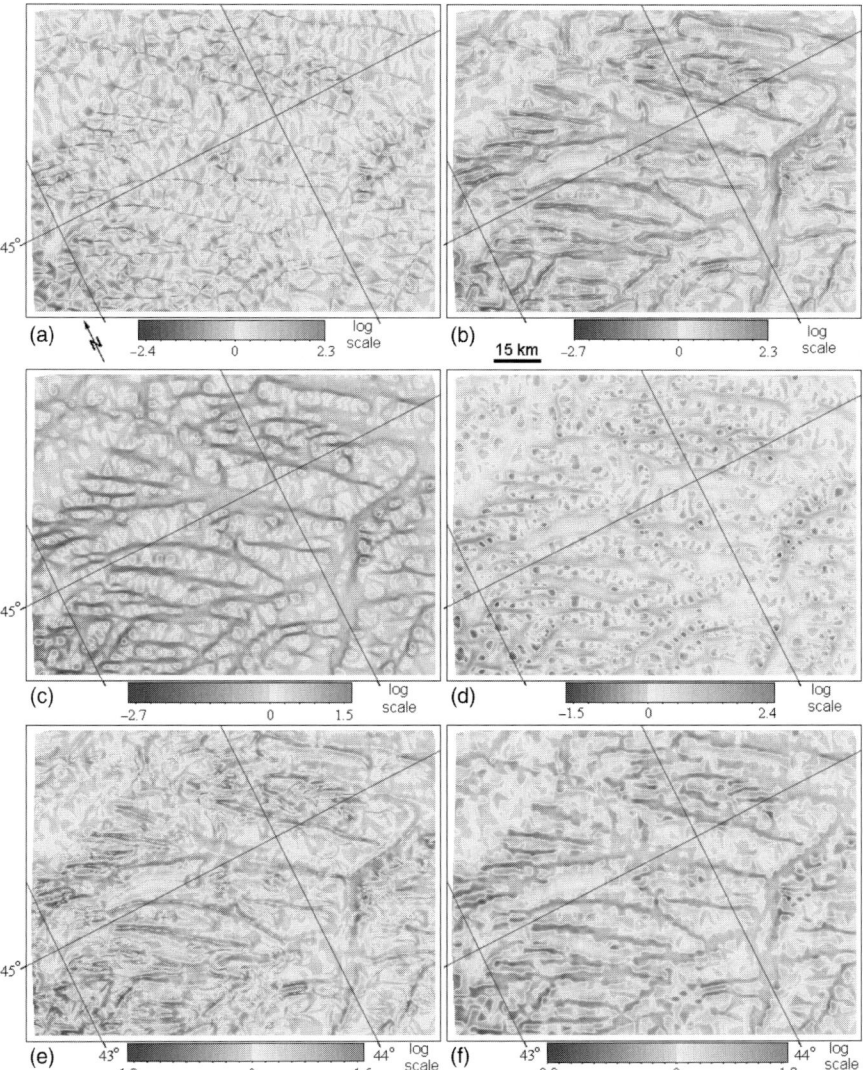

FIGURE 2.5 The Stavropol Upland: (a) horizontal curvature, (b) vertical curvature, (c) minimal curvature, (d) maximal curvature, (e) difference curvature, (f) mean curvature, (g) horizontal excess curvature, (h) vertical excess curvature, (i) the Gaussian curvature, (j) accumulation curvature, (k) ring curvature, and (l) unsphericity curvature. For details of calculation, see Section 4.2.3.1; for the elevation map, see Fig. 2.3. *From (Florinsky, 2009a, Fig. 2).* (See pages 2 and 3 in Color Plate Section)

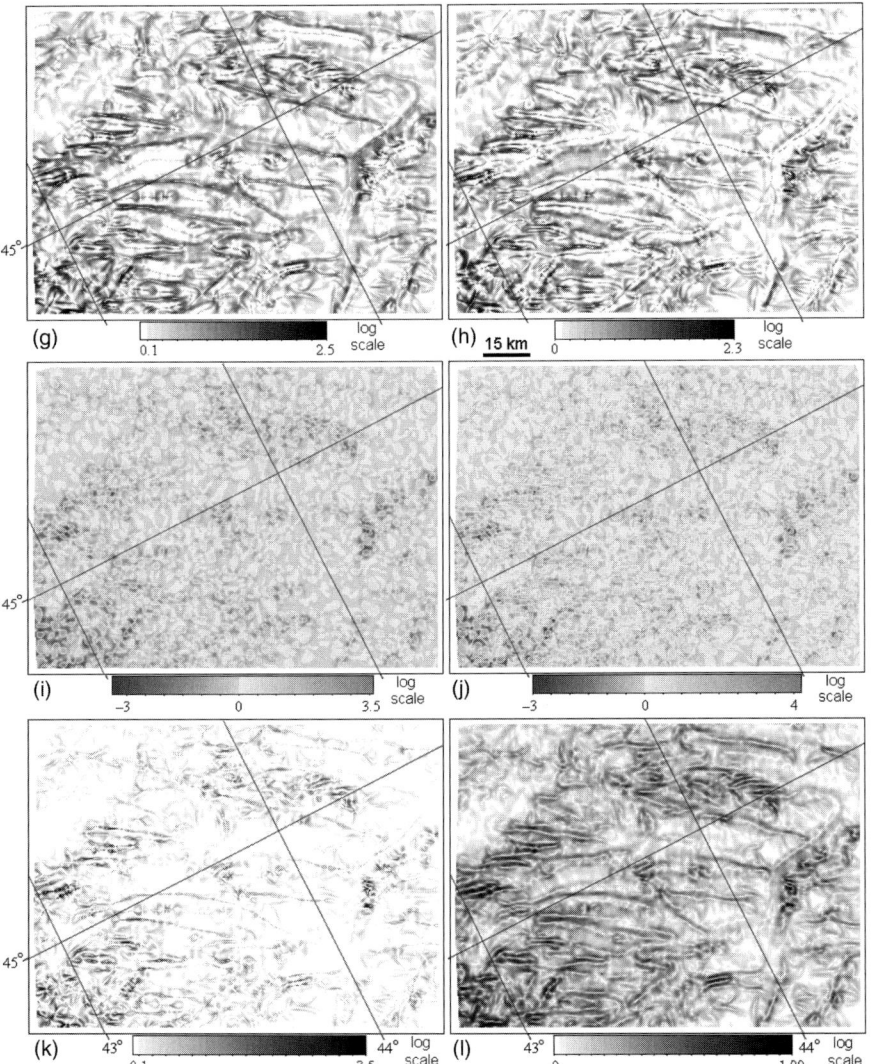

FIGURE 2.5 (Continued).

nonlocal topographic variables are catchment area[8] (*CA*), specific catchment area (*SCA*), dispersive area (*DA*), specific dispersive area (*SDA*), drainage density (D_d), and some others.

---

[8]Definitions and interpretations of some morphometric variables use such terms and notions as flow line, catchment, and flow. To avoid misunderstanding, we should stress that topographic attributes describe a surface in gravity regardless of the surface origin and the existence of water or other liquids on the surface.

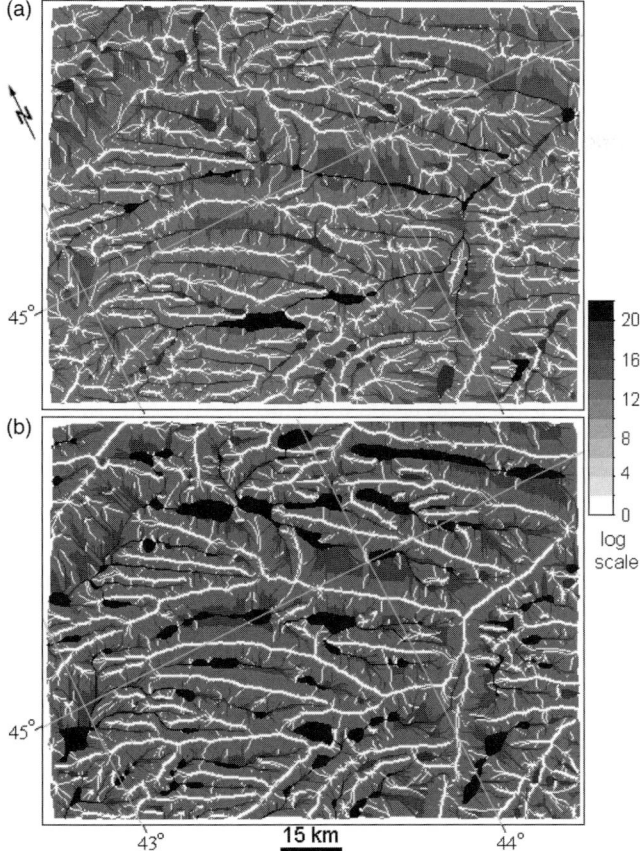

FIGURE 2.6 The Stavropol Upland: (a) catchment area, (b) dispersive area. Calculation was carried out by the Martz–de Jong method (Section 4.4). For the elevation map, see Fig. 2.3. *From (Florinsky, 2010, Fig. 1.7).*

For definitions and interpretations of some nonlocal morphometric attributes, see Table 2.1. Methods for calculating CA and DA are discussed in Section 4.4. Examples of CA and DA calculation can be seen in Fig. 2.6.

2.4 STRUCTURAL LINES

One can distinguish two families of spatial curves on the topographic surface: contour lines and slope lines (Cayley, 1859) (Section A.1.3). A contour is a locus of intersection of a horizontal plane with the topographic surface. For any point of a slope line, the direction of a tangent vector to the curve coincides with the direction of a tangential component of the gravity vector. Slope lines are not defined at special points of the surface, such as local maxima, minima, saddles, and flat areas. Slope lines and contours are mutually perpendicular at their intersections.

Considering these families of spatial curves, one can distinguish two groups of loci of extreme curvature of the topographic surface: (1) a locus of extreme curvature of contours and (2) a locus of extreme curvature of slope lines. Obviously, extreme curvature varies in sign: one can set the positive sign to convex areas and negative—to concave ones. Loci of extreme curvature may *partially* describe four types of structural lines:

1. Ridge lines, or crests—the locus of positive extreme curvature of contours (Weinberg, 1934a)
2. Valley lines, or thalwegs—the locus of negative extreme curvature of contours (Weinberg, 1934a)
3. Convex break lines—the locus of positive extreme curvature of slope lines (Shary and Stepanov, 1991)
4. Concave break lines—the locus of negative extreme curvature of slope lines (Shary and Stepanov, 1991)

Note that structural "lines" are formed not only by the loci of extreme curvature, but also by loci of special points (Section A.1.2).

At the same time, one can consider ridge and valley lines as two topologically connected treelike hierarchical structures. Maxwell (1870) defined a ridge as a slope line connecting a sequence of local maximal and saddle points, and a thalweg as a slope line connecting a sequence of local minimal and saddle points.

Methods for digital revealing of structural lines are discussed in Section 4.5. The revealing of crests and thalwegs is illustrated in Fig. 2.7.

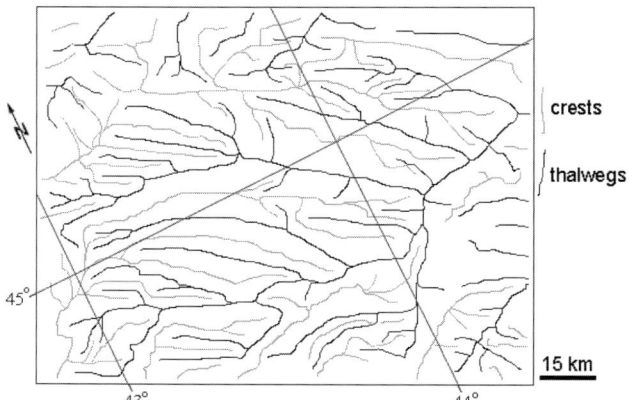

FIGURE 2.7 The Stavropol Upland: ridge and thalweg lines. For details of calculation, see Section 4.5.1; for the elevation map, see Fig. 2.3. *From (Florinsky, 2009b, Fig. 8a).*

2.5 SOLAR MORPHOMETRIC VARIABLES

Solar morphometric variables describe relations between the topographic surface and solar irradiation. Among these variables are reflectance, insolation ($I(\theta,\psi)$), and some related functions. Reflectance and insolation can be estimated with several models of light reflectance from a surface: Lambertian, Lommel–Seeliger, and so on (Horn, 1981).

Insolation is a measure of the topographic surface illumination by solar light flux (Section A.2.2). The unit of $I(\theta,\psi)$ is percent. Insolation can be estimated by the following expression (Shary et al., 2005):

$$I(\theta,\psi)=50\frac{\{1+\text{sign}[\sin\psi-\cos\psi(p\sin\theta+q\cos\theta)]\}[\sin\psi-\cos\psi(p\sin\theta+q\cos\theta)]}{\sqrt{1+p^2+q^2}}$$

(2.16)

where θ and ψ are azimuth and zenith solar angles, respectively (Fig. 2.8). Notice that this does not consider diffuse radiation and cloud-shadowing effects. For development of the formula, see Section A.2.2.

Maps of insolation (or hill-shading maps) are widely used in geosciences because they reflect the shape of the topographic surface. An advantage of hill-shading maps is the possibility of modeling such positions of the Sun on the celestial sphere, which are impossible in nature. This offers the potential for revealing some topographically expressed geological structures (Section 13.1). Moreover, the thermal

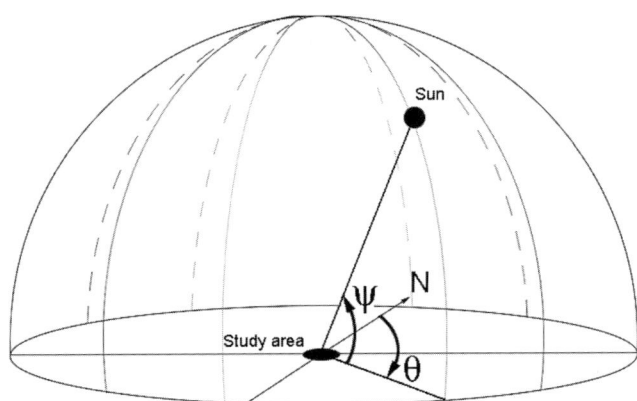

FIGURE 2.8 Illustration for solar angles used to compute insolation (hill shading). See text for explanation.

regime of slopes influencing soil properties depends in part on the incidence of solar rays to the land surface, so it depends on both G and A (Section 8.2). Insolation directly describes this incidence and therefore better considers the thermal regime. Reviews of insolation-based solar radiation models can be found elsewhere (Dubayah and Rich, 1995; Böhner and Antonić, 2009).

Examples of insolation mapping can be seen in Figs. 2.9, 5.2, 6.2a, and 6.3. Figures 7.4, 10.3, and 10.5 exemplify the combined use of hill-shading and geological/soil data. The use of hill-shading maps to reveal geological structures is shown in Fig. 13.2.

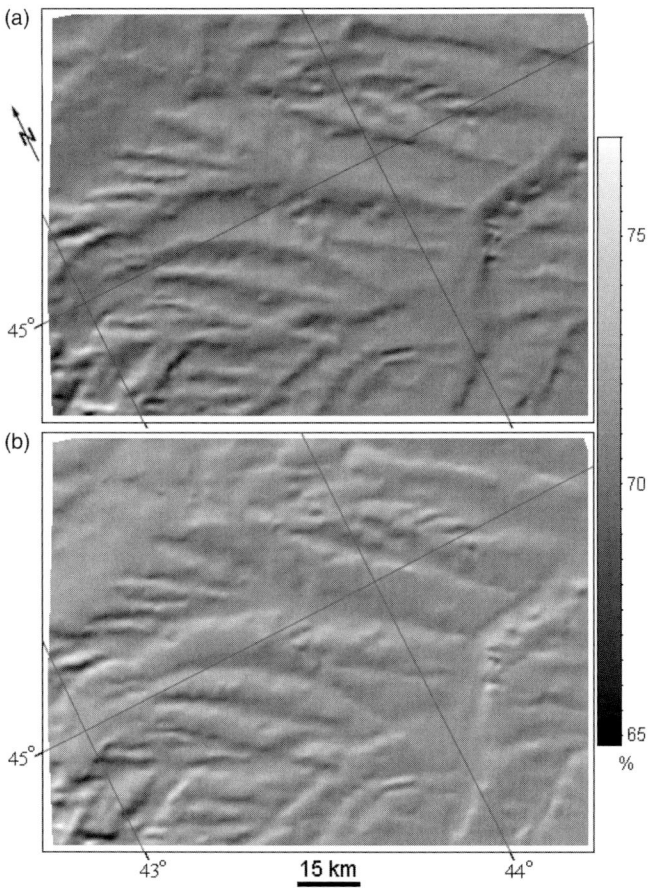

FIGURE 2.9 The Stavropol Upland, insolation for various values of zenith and azimuth solar angles: (a) 45° and 180° (south); (b) 45° and 0° (north). For the elevation map, see Fig. 2.3.

2.6 COMBINED MORPHOMETRIC VARIABLES

Morphometric variables can combine local and nonlocal ones. Combined topographic variables consider both the local geometry of the topographic surface and a relative position of a point on the surface. Among these variables are the topographic index (*TI*) and the stream power index (*SI*):

$$TI = \ln\left(\frac{CA}{\tan G}\right) \quad (2.17)$$

$$SI = \ln(CA \cdot \tan G) \quad (2.18)$$

For interpretations of *TI* and *SI*, see Table 2.1 and Section 8.3. Figure 2.10 illustrates *TI* and *SI* mapping.

2.7 LANDFORM CLASSIFICATIONS

On one hand, landform classifications based on local morphometric attributes are of fundamental importance to the theory of geomorphometry and digital terrain modeling. On the other hand, they are utilized in segmentation of the topographic surface to solve some problems of geology and soil science. There are three main quantitative approaches to classifying landforms using information on the local geometry of the surface:

1. The Gaussian classification based on the signs of the Gaussian and mean curvatures
2. The Efremov–Krcho classification using the signs of the horizontal and vertical curvatures
3. The Shary classification based on the signs of the Gaussian, mean, difference, horizontal, and vertical curvatures

2.7.1 The Gaussian Classification

The first quantitative classification of landforms was developed by Carl Friedrich Gauss (1828)[9] using the sign of the total (Gaussian) curvature. According to this classification, positive values of *K* describe an elliptic surface, negative values of *K* relate to a hyperbolic surface, while zero values of *K* define a parabolic surface. As applied to the topographic surface, its elliptic portions correspond to hills and closed

---
[9]There is an English translation of this work (Gauss, 2009).

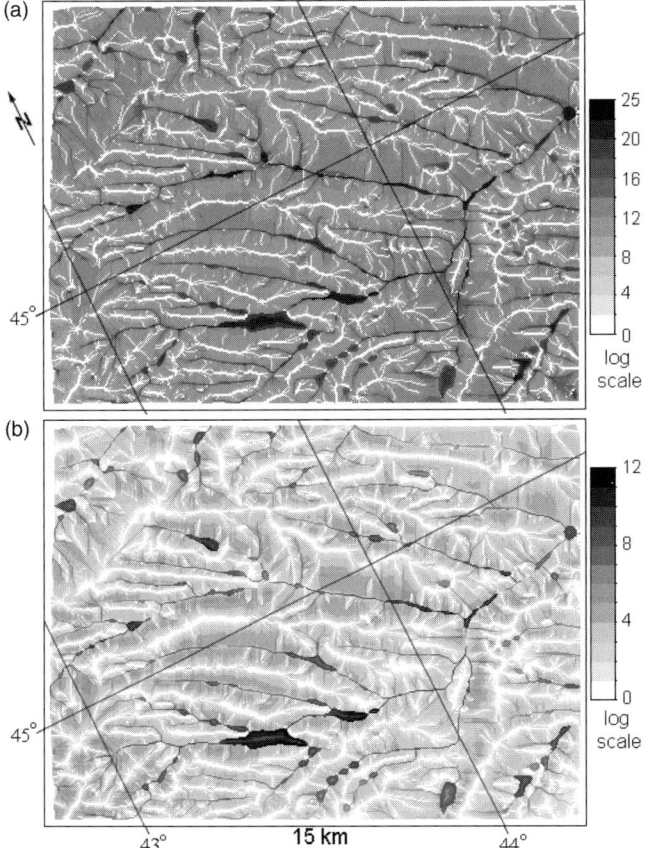

FIGURE 2.10 The Stavropol Upland: (a) topographic index, (b) stream power index. Calculation was carried out by the Martz–de Jong method (Section 4.4). For the elevation map, see Fig. 2.3. *From (Florinsky, 2010, Fig. 1.10).*

depressions, hyperbolic portions relate to saddles, and parabolic portions correlate with ridges, valleys, and plane areas.

To distinguish between hills and closed depressions, convex and concave saddles, and ridges and valleys, one should combine data on the signs of the Gaussian and mean curvatures as follows (Shary, 1991, 1995; Koenderink and van Doorn, 1992):

- $K > 0$ with $H > 0$ describe hills, while $K > 0$ with $H < 0$ define closed depressions.
- $K < 0$ with $H > 0$ relate to convex saddles, while $K < 0$ with $H < 0$ describe concave saddles.
- $K = 0$ with $H > 0$ correspond to ridges, while $K = 0$ with $H < 0$ define valleys.

Figure 2.11a illustrates the scheme of this classification using terms of structural geology, since the Gaussian classification is used in geological studies (Roberts, 2001; Bergbauer and Pollard, 2003; Bergbauer et al., 2003; Lisle and Toimil, 2007; Mynatt et al., 2007a). Note that in nature, portions of the topographic surface with zero values of the Gaussian or mean curvatures are very rarely observed (Section 2.7.3). Thus, the number of landform types in the Gaussian classification may be reduced from nine to four (i.e., antiformal saddle, synformal saddle, dome, and basin—Fig. 2.11a).

Figure 2.12a displays a map of the topographic segmentation using the Gaussian classification in its reduced form. Use of this classification in structural geology is discussed in Sections 6.2.7 and 12.2.

Koenderink and van Doorn (1992) proposed using a continual form of this discrete classification applying the shape index (IS) (Lisle and Toimil, 2007):

$$IS = \frac{2}{\pi} \arctan \frac{H}{\sqrt{H^2 - K}}. \quad (2.19)$$

IS can take values from -1 to 1. Its positive values relate to convex landforms, while negative ones correspond to concave landforms. Under this condition, absolute values from 0.5 to 1 are associated with elliptic surfaces, whereas absolute values from 0 to 0.5 relate to hyperbolic ones. Figure 2.12b illustrates a map of the shape index.

2.7.2 The Concept of Accumulation Zones. The Efremov–Krcho Classification

Local accumulation of a gravity-driven flow is controlled by two mechanisms: relative deceleration and convergence (Shary, 1991) (Table 2.1). Flow relative deceleration is determined by k_v: a flow tends to accelerate when $k_v > 0$, and to decelerate when $k_v < 0$ (Speight, 1974) (Section A.3.3). Flow convergence is controlled by k_h: a flow diverges when $k_h > 0$ and converges when $k_h < 0$ (Kirkby and Chorley, 1967) (Sections A.3.1 and A.3.2).

There is a concurrent action of convergence and relative deceleration of flows within areas characterized by both $k_h < 0$ and $k_v < 0$. These areas are said to be relative accumulation zones (Shary et al., 1991). If divergence and relative acceleration of flows act simultaneously ($k_h > 0$ and $k_v > 0$), these areas are referred to as relative dissipation zones. Areas with other combinations of the k_h and k_v signs (no concurrent action of "unidirectional" processes) are lumped together as transit zones. It should be noted that we consider zones of *relative* accumulation rather than dead-end depressions. A flow may pass through a

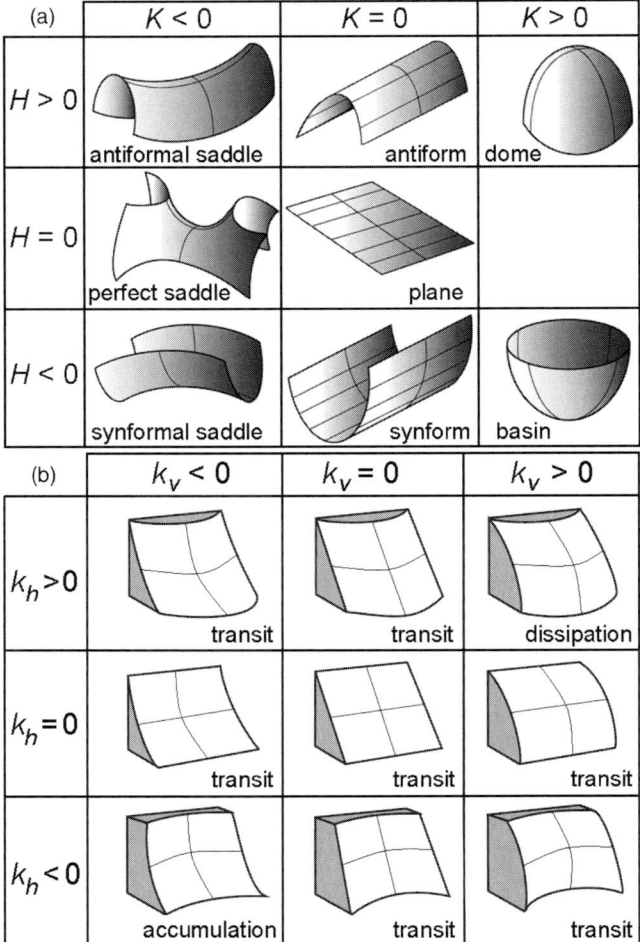

FIGURE 2.11 Landform classification schemes: (a) the Gaussian classification; shape terms used in structural geology are indicated. *From (Mynatt et al., 2007a, Fig. 2), reproduced with permission.* (b) The Efremov–Krcho classification; zones of relative accumulation, transit, and dissipation are indicated. *From (Schmidt and Hewitt, 2004, Fig. 3a), reproduced with permission.* (c) The Shary classification, 12 main types of landforms. *From (Shary et al., 2005, Fig. 23), reproduced with permission.*

great quantity of relative accumulation zones before entering into a dead-end depression.

According to the landform classifications by the signs of k_v and k_h—both qualitative (Efremov, 1949) and quantitative (Troeh, 1964; Krcho, 1983)—accumulation zones can be defined as concave–concave landforms, while dissipation zones can be described as convex–convex landforms. Another seven types of landforms can be assigned to transit

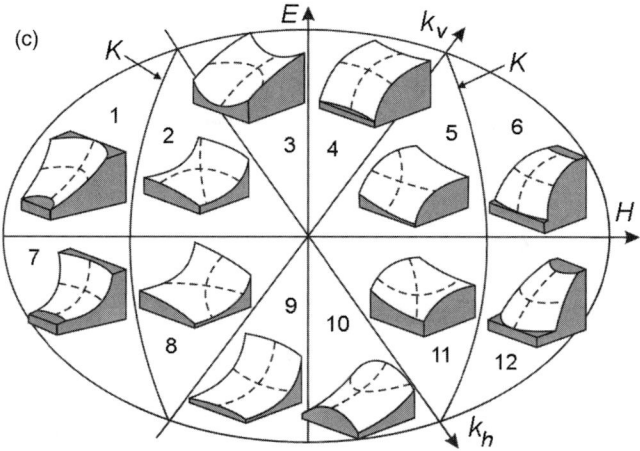

FIGURE 2.11 (Continued).

zones (Fig. 2.11b). Note that as for the Gaussian classification (Section 2.7.1), portions of the topographic surface with zero values of horizontal or vertical curvatures are rarely observed (see details in Section 2.7.3). Therefore, the number of landform types in the Efremov−Krcho classification can be also reduced from nine to four.

If mapping of accumulation, transit, and dissipation zones is carried out by a simple registration of k_h and k_v maps (Fig. 2.12c), one can visualize only spatial distribution of these zones without quantitative estimation of a probable degree of flow accumulation. To solve this problem, Shary (1995) proposed the use of data on K_a and H. Negative values of K_a correspond to transit zones, while positive values of K_a correspond to both accumulation and dissipation zones. They can be distinguished using data on H. Positive values of K_a with negative values of H correspond to accumulation zones, while positive values of K_a with positive values of H correspond to dissipation zones (Shary, 1995).

Maps of relative accumulation zones represent areas where geometrical peculiarities of the topography provide conditions for the local accumulation of gravity-driven substances, such as water (meteoric water, soil moisture), dissolved and suspended substances (salts, clay and organic particles, etc.), and other liquids (e.g., petroleum products). In this connection, maps of relative accumulation zones are used in geosciences: to predict slope instability (Lanyon and Hall, 1983; Florinsky, 2007b); to study and model soil properties at field, regional, and continental scales (Pennock et al., 1987; Florinsky et al., 1999, 2002; Florinsky and Eilers, 2002; Shary et al., 2002a; Shary, 2005); to estimate the risk for soil secondary salinization (Florinsky et al., 2000); to predict soil degradation and contamination along pipelines; to explore placer deposits

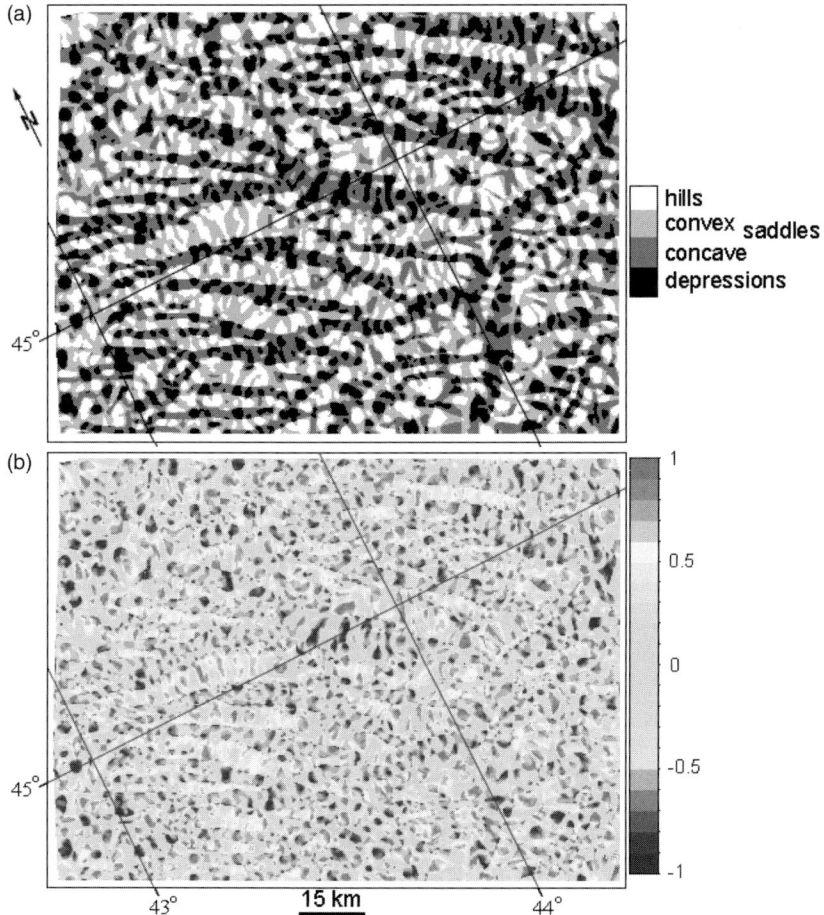

FIGURE 2.12 The Stavropol Upland: topographic segmentation performed using (a) the Gaussian classification; (b) shape index; (c) the Efremov–Krcho classification (zones of relative accumulation, transit, and dissipation); (d) the Shary classification. For the elevation map, see Fig. 2.3. (See page 4 in Color Plate Section)

(Florinsky, 2007b); to search meteoritic matter (Florinsky, 2008e); and so on. Relative accumulation zones may coincide with intersections of lineaments, faults, and fracture zones. Thus they constitute areas of contact and interaction between overland and intrasoil lateral water flows and upward flows of deep groundwaters (Florinsky, 2000) (Chapter 14).

2.7.3 The Shary Classification

It is clear that the Gaussian landform classification is based on morphometric variables belonging to the class of form attributes; that is, it

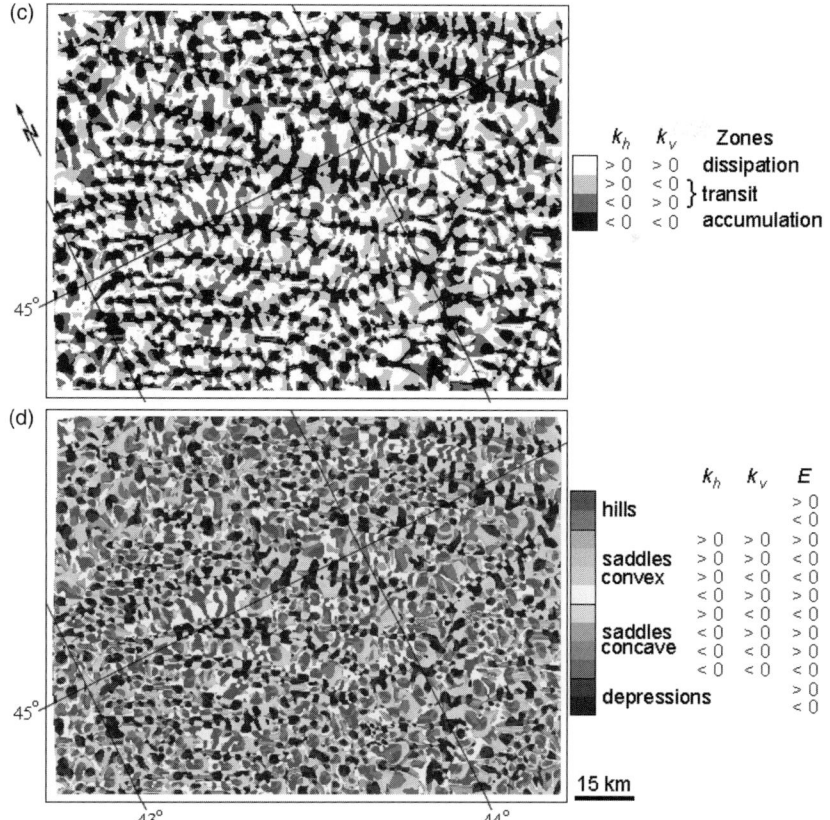

FIGURE 2.12 (Continued).

takes into account information on only principal normal sections, which directions are independent of gravity (Section 2.2). At the same time, the Efremov–Krcho classification uses topographic variables belonging to the class of flow attributes; that is, it considers information on only two normal sections controlled by gravity (Section 2.2).

Shary (1995) proved that neither classification is complete. He proposed a new classification that is free of the indicated limitations. It incorporates two previous classifications as special cases; that is, each landform type of the Gaussian and Efremov–Krcho classifications is a combination of several landform types of the Shary classification.

The Shary classification is based on the signs of five curvatures: the Gaussian, mean, difference, horizontal, and vertical curvatures. According to this classification, the topographic surface is described by 46 landform types. Twelve of them are main types (Fig. 2.11c); they can be found with a frequency of 1/12 in nature. Thirty-four landform types are rare; that is, they are observed with a frequency of less than 0.0001. These are landforms, for which one of the five curvatures has zero value. On maps, rare landforms are represented as lines and points, which are usually located between landforms belonging to the main types (Shary, 1995; Shary et al., 2005). Figure 2.12d demonstrates a topographic segmentation performed using the Shary classification.

Compared to the Gaussian and Efremov−Krcho classifications, the higher number of landform types in the Shary classification leads to its higher flexibility. For instance, Buivydaite and Mozgeris (2007) found that soil units and classes correlated stronger with the landform types derived with the Shary classification rather than the Gaussian and Efremov−Krcho ones. MacMillan and Shary (2009) have presented a detailed comparison of the Shary classification with other landform classifications including empirical ones.

CHAPTER

3

Digital Elevation Models

OUTLINE

3.1 DEM Generation	31
3.2 DEM Grid Types	36
3.3 DEM Resolution	38
3.4 DEM Interpolation	40

3.1 DEM GENERATION

A DTM is a two-dimensional discrete function of a morphometric variable, which defines the topographic surface as a set of values measured or computed at the grid nodes. Among DTMs are DEMs, digital models of slope gradient, horizontal curvatures, catchment area, and other topographic attributes (Miller and Leflamme, 1958; Doyle, 1978; Burrough, 1986). For any portion of the topographic surface, digital models of all morphometric variables are derived from a DEM of this portion.

DEMs can be generated by various field, remote, and laboratory techniques:

1. *Conventional topographic surveys* (Ghilani and Wolf, 2008; Kleim et al., 1999). These are carried out by means of optical or laser tacheometers and leveling instruments. Resulting detailed and large-scale DEMs of relatively small areas are used in large-scale

soil and geological studies (e.g., Pennock et al., 1987; King et al., 1999).
2. *Kinematic GPS surveys* (Panin and Gelman, 1997; Schmidt et al., 2003; Nico et al., 2005b; Ghilani and Wolf, 2008, ch. 15). These are employed to create large-scale and detailed DEMs with the global positioning satellite system and GPS receivers, which are mounted on an all-terrain vehicle or moved by an operator. Kinematic GPS surveys offer fast and cost-effective DTM-based solutions for many detailed and large-scale problems of soil science (e.g., Clark and Lee, 1998; Florinsky et al., 2002) and geology (Baldi et al., 2000; Pearce et al., 2006; Fiore Allwardt et al., 2007; Nazari et al., 2009).
3. *Analogue* (Wolf and Dewitt, 2000) *and digital* (Claus, 1984; Jensen, 1995; Lascelles et al., 2002; Linder, 2006) *photogrammetric approaches.* These are used to construct DEMs from stereo pairs of remotely sensed (as a rule, aerial and satellite) images in the optical range. One can utilize satellite images from various platforms, such as SPOT (Gugan and Dowman, 1988; Chen and Rau, 1993; Al-Rousan et al., 1997; Toutin, 2006), Landsat MSS, NOAA AVHRR (Akeno, 1996), ASTER (Welch et al., 1998; Hirano et al., 2003; Toutin, 2008), Ikonos (Toutin, 2004b), QuickBird (Toutin, 2004a), and the like. Photogrammetric DEMs can be used in both soil science (e.g., Odeh et al., 1991; Ziadat et al., 2003) and geology (e.g., Chorowicz et al., 1991; Babiker and Gudmundsson, 2004) in a wide range of scales. Digital photogrammetry was applied to produce two current medium-scale DEMs: the ASTER GDEM for the most part of the land surface (Hato et al., 2009; Jet Propulsion Laboratory, 2009) and the SPOT DEM for portions of Eurasia, Africa, and Central America (Spot Image, 2008). In superdetailed field and laboratory studies of soil erosion, ground-based stereo photography is implemented to create DEMs with resolution of several millimeters (Warner, 1995; Hancock and Willgoose, 2001). Superdetailed stereo photography is also applied to produce DEMs in laboratory physical modeling of tectonic processes, such as folding and faulting (Fischer and Keating, 2005; Keating and Fischer, 2008).
4. *Radar techniques*. These are carried out with synthetic aperture radars (SARs). To produce DEMs, three approaches are applied (Toutin and Gray, 2000): polarimetry (Schuler et al., 1996, 1998), stereo radargrammetry using stereo pairs of radar images (Gelautz et al., 2003; Toutin, 2010), and interferometry (Zebker et al., 1994a, 1994b; Bürgmann et al., 2000). The interferometric synthetic aperture radar (InSAR) technique is most commonly used. In particular, this method was implemented to create the global small-scale DEM of Venus (Pettengill et al., 1991; Ford, 1992) and SRTM3, the global

medium-scale DEM for a large part of the land surface (Rabus et al., 2003; NASA, 2003; Farr et al., 2007). The latter was used to describe the land topography in SRTM30_PLUS, the global DEM (Becker et al., 2009). SARs on ERS-1 and Radarsat satellites were used to produce DEMs of the ice sheets of Greenland (Joughin et al., 1996) and Antarctica (Liu et al., 2001), correspondingly. InSAR-based DEMs are utilized in medium- and small-scale soil (e.g., Hengl et al., 2007a; Mora-Vallejo et al., 2008) and geological/planetological (e.g., Solomon et al., 1991; Florinsky, 2008b) investigations. Sometimes, ground-based radar techniques are in use to generate large-scale DEMs (Nico et al., 2005a). Data from the SAR on Cassini satellite were used to produce a digital model of near-shore bathymetry for the hydrocarbon lake Ontario Lacus on Titan (Hayes et al., 2010).

5. *Laser surveys.* These use a pulse laser to determine distance between the target and the sensor. LiDAR (Light Detection And Ranging) aerial surveys are applied to create large-scale and detailed DEMs of both the land surface (Baltsavias, 1999; Wehr and Lohr, 1999; French, 2003; Brovelli et al., 2004; Lloyd and Atkinson, 2006; Liu, 2008; Bretar and Chehata, 2010) and the seafloor (down to −70 m in clear waters—Finkl et al., 2005; Gao, 2009). Although some technical problems are still unsolved (e.g., effective filtering of high-frequency noise—Aguilar and Mills, 2008; Leigh et al., 2009), LiDAR aerial surveys offer fast DTM-based solutions for large-scale problems of soil science (e.g., Ballabio and Comolli, 2010; Greve et al., 2010) and geology (e.g., Haugerud et al., 2003; Mynatt et al., 2007b). Using satellite laser altimetry, two global DEMs of the Moon were produced: the small-scale DEM (Smith et al., 1997; Zuber, 1996) and the medium-scale one (Smith D.E. et al., 2010), a series of global medium- and small-scale DEMs of Mars (Smith et al., 1999; Smith, 2003), as well as DEMs for the ice sheets of Antarctica and Greenland (Schutz et al., 2005; Dimarzio et al., 2005). Such DEMs are used in medium- and small-scale geological and planetological studies (e.g., Zuber et al., 1994, 2000; Byrne, 2007). Terrestrial LiDARs are applied to create large-scale DEMs of outcrops (Bellian et al., 2005; Buckley et al., 2008), which can be included into three-dimensional geological models (Section 7.4).

6. *Shipboard echo sounding* (Bourillet et al., 1996; Blondel and Murton, 1997). This is performed by echo sounders (e.g., multibeam ones) to produce DEMs for the floors of lakes, seas, and oceans (Hall, 1996; Sherstyankin et al., 2006; Gasperini et al., 2007; Jakobsson et al., 2008). Echo sounding data were the basis for the bathymetry in the

global SRTM30_PLUS DEM (Becker et al., 2009). Such DEMs can be utilized in large-, medium-, and small-scale geological investigations (e.g., Kukowski et al., 2001, 2008).

7. *Airborne optical sensing* (Gao, 2009). This can be used to create DEMs of shallow seafloors, which may be hazardous for navigation and echo sounding (Ji et al., 1992; Mishra et al., 2004; McIntyre et al., 2006). The depth is estimated from brightness values of the image. The approach may be applied for clear waters only. Such DEMs can be used in large- and medium-scale geological research.

8. *Satellite radar altimetry.* This technique was utilized to produce DEMs of ocean floors (Dixon et al., 1983; Smith and Sandwell, 1997; Sandwell and Smith, 2001). Data from radar altimeters of Seasat, Geosat, and ERS-1 satellites (calibrated with shipboard echo sounding data) were used to generate the bathymetry for the global DEMs ETOPO5 (World Data Center for Geophysics, 1993), ETOPO2 (National Geophysical Data Center, 2001, 2006), and ETOPO1 (Amante and Eakins, 2009). In addition, satellite radar altimetry was applied to produce DEMs of the ice sheets of Antarctica and Greenland (Zwally et al., 1983, 1987). These data were included in the DEM of the Antarctica ice sheet (Liu et al., 1999). Such DEMs are utilized in small-scale geological investigations (e.g., Florinsky, 2008a).

9. *Soil augering and geological boring* (Acker, 1974; Australian Drilling Industry Training Committee, 1997). Borehole or augerhole data are used to generate DEMs of contact surfaces of soil and stratigraphic horizons, as well as DEMs of subglacial topography (Lythe et al., 2001). Such DEMs are used in large-scale soil studies (Florinsky and Arlashina, 1998; Mendonça Santos et al., 2000; Chaplot and Walter, 2003; Zakharchenko and Zakharchenko, 2006; Scott and Needelman, 2007), as well as in medium- and small-scale geological investigations (Belonin and Zhukov, 1968; Robinson et al., 1969; Lemon and Jones, 2003; Groshong, 2006; Kaufman and Martin, 2008).

10. *Three-dimensional (3D) seismic survey* (Brown, 2004; Liner, 2004). This approach is based on the phenomenon of reflection of sound energy generated near the land surface from layer boundaries in the subsurface. The depth of the reflecting horizon as well as the sound velocity between the land surface and the horizon control the time at which the reflection returns to a recorder at the land surface. First, a user may obtain a set of digital models of "surfaces" of two-way travel time (the unit is millisecond), which relate to contact surfaces between adjacent geological horizons. Then, two-way-travel time is, as a rule, converted to the depth of a particular

geological surface.[1] 3D seismic survey is used to produce DEMs of surfaces of stratigraphic horizons and geological structures. Such DEMs are applied in medium- and small-scale geological research (Bergbauer et al., 2003; Wynn and Stewart, 2003; Ustinova and Ustinov, 2004; Roberts et al., 2009). Seismic survey data were also used to create the DEM of subglacial topography of Antarctica (Lythe et al., 2001), which was included in the ETOPO1 DEM (Amante and Eakins, 2009).

11. *Airborne ice-penetrating radar techniques* (Gogineni et al., 2001) using coherent radar systems. This approach was applied to produce the DEM of subglacial topography of Greenland (Bamber et al., 2001), which was later inserted into the ETOPO1 DEM (Amante and Eakins, 2009). Such DEMs can be utilized in medium- and small-scale geological studies.

12. *Radio-echo sounding surveys* (Robin et al., 1969) using both airborne and overland tools. This technique was applied to generate the DEM of subglacial topography of Antarctica (Lythe et al., 2001).

13. *Digitizing of contours.* Topographic and geological maps of various scales are digitized to produce DEMs of the land surface and surfaces of stratigraphic horizons, correspondingly (Mor and Lamdan, 1972; Yoeli, 1975; Eklundh and Mårtensson, 1995). In digitizing, one may use ancillary cartographic information, such as elevation values for mountain summits and depression bottoms, and structural lines. This approach was applied to generate national DEMs of the United States, Canada, and other countries (USGS, 1993; Natural Resources Canada, 1997), as well as to describe the land surface in the global DEMs GTOPO30 (USGS, 1996), GLOBE (GLOBE Task Team and others, 1999), ETOPO5 (World Data Center for Geophysics, 1993), ETOPO2 (National Geophysical Data Center, 2001, 2006), and ETOPO1 (Amante and Eakins, 2009). Isobath maps were digitized to create the most part of the DEM of the Arctic Ocean floor (Jakobsson et al., 2008), the DEM of the Baltic Sea floor (Seifert et al., 2001), and DEMs of some other seas and lakes incorporated into the ETOPO1 DEM (Amante and Eakins, 2009). Map-based DEMs are widely used to solve medium- and small-scale problems of soil science (e.g., Bell et al., 1992; Gessler et al., 1995) and geology (e.g., Morris, 1991; Onorati et al., 1992).

---

[1]Some researchers apply methods of digital terrain modeling to time-structure matrices without their conversion to DEMs (e.g., Roberts, 2001; Chopra and Marfurt, 2007b; Hart and Sagan, 2007). These studies are not discussed in this book because, from the standpoint of digital terrain analysis, such an approach is incorrect: all three coordinates of the initial matrix should have the same units (e.g., meters) to estimate correctly partial derivatives of elevation.

Many DEMs of vast territories are created by compilation of other DEMs and data, which were earlier produced with different techniques. For example, the global SRTM30_PLUS DEM is the compilation of: (1) the SRTM30 DEM for the most part of the land surface; (2) portions of the GTOPO30 DEM for the land surface of vast northern territories; (3) the ICESat data for the Antarctica ice sheet; (4) shipboard echo sounding data for the bathymetry of most oceans; and (5) the IBCAO DEM for the bathymetry of the Arctic Ocean (Becker et al., 2009).

DEMs of stratigraphic surfaces are also compiled using data obtained by various techniques, such as 3D seismic survey, geological boring, and other available geological sources (Caumon et al., 2009). In some rare cases, DEMs of surfaces of deep geological structures may be derived from indirect measurements. For instance, coordinates of earthquake hypocenters were used to generate DEMs of the oceanic lithosphere in subduction zones (Cahill and Isacks, 1992; Nothard et al., 1996).

The selection of a technique to produce a DEM for soil and geological research depends on several factors, such as the size of the study area, required accuracy and resolution of the DEM, accuracy and resolution of other maps and materials (soil, geological, geophysical, etc), as well as the cost of the DEM generation.

3.2 DEM GRID TYPES

From the mathematical point of view, two principally different approaches can be used to describe the topographic surface:

1. Two-dimensional (piecewise) continuous functions can be applied to define relatively small portions of the topographic surface (Jancaitis, 1978; Strakhov, 2007). One can also use spherical functions to define the topography of the globe, hemispheres, or other vast territories (Schröder and Sweldens, 2000; Wieczorek, 2007).
2. Two-dimensional discrete functions, that is, DEMs can be employed. In this book, we mainly discuss this approach, which gained wide acceptance because of its relative simplicity.

In the latter case, one can use various types of grids for measured or estimated elevation values. The following types of plane grids can be exploited for DEMs of relatively small portions of the topographic surface, when the curvature of the planet may be ignored (Peucker, 1980a; Davis, 1986; Carter, 1988):

1. Irregular grids based on:
    a. Random distribution of points
    b. Critical points of contour lines

3.2 DEM GRID TYPES

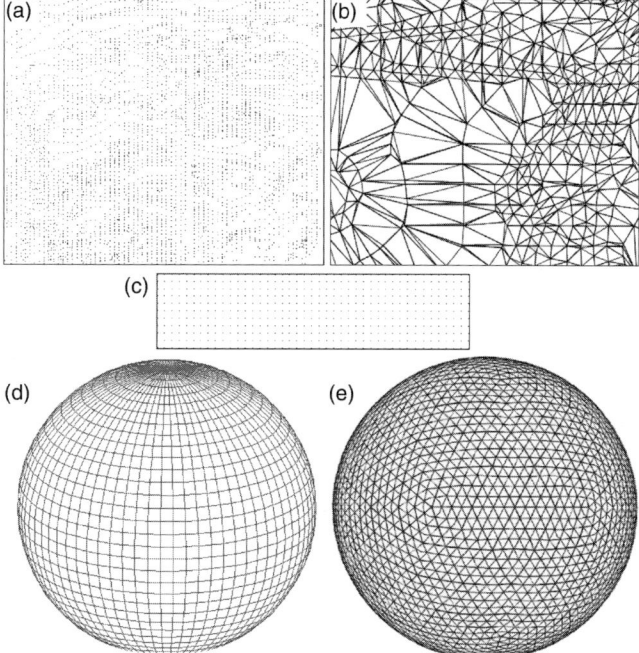

FIGURE 3.1 Main types of DEM grids: (a) Plane irregular grid generated using topographic cross sections and structural lines. *From (Florinsky, 2002, Fig. 6a).* (b) Plane triangulated irregular network. (c) Plane square grid. (d) Spheroidal equal angular grid. (e) Spheroidal grid of icosahedrons tessellated by triangles. *From (Sahr et al., 2003, Fig. 11a),* © *2003 Cartography and Geographic Information Society, reproduced with permission.*

    c. Topographic cross sections (Fig. 3.1a)
    d. Critical elements of the topographic surface, such as structural lines, local minima and maxima, and saddles (Fig. 3.1a). This grid type is commonly used to construct triangulated irregular networks (TIN)[2] (Peucker et al., 1976; van Kreveld, 1997) (Fig. 3.1b).
2. Regular grids based on triangles, squares, rectangles, hexagons, and so on. The square grid is the most popular one (Fig. 3.1c).

It has been repeatedly argued that DEMs should be generated considering critical elements of the topographic surface (e.g., Mark, 1979). Nevertheless, Kumler (1994) demonstrated that square-gridded DEMs have higher accuracy than TIN-based DEMs. The use of plane square

---

[2]The mathematical basis for TIN construction is the Delaunay triangulation (Delaunay, 1934; Hjelle and Dæhlen, 2006).

grids is also reasonable because morphometric variables are usually derived from square-gridded DEMs (Chapter 4). Conversion of irregular grids to regular ones is carried out by interpolation (Section 3.4).

Various spheroidal grids can be employed to describe the topography of the globe or vast territories, when the curvature of the planet may not be ignored. The most popular is the spheroidal equal angular grid, that is, the grid of spheroidal trapezoids with equal angular spacing along latitude and longitude (Fig. 3.1d). This grid type is the standard for global DEMs of the Earth, such as ETOPO5 (World Data Center for Geophysics, 1993), GTOPO30 (USGS, 1996), GLOBE (GLOBE Task Team and others, 1999), ETOPO2 (National Geophysical Data Center, 2001, 2006), SRTM3 (NASA, 2003), SRTM30_PLUS (Becker et al., 2009), ETOPO1 (Amante and Eakins, 2009), and ASTER GDEM (Jet Propulsion Laboratory, 2009), global DEMs of Mars (Smith, 2003), Venus (Ford, 1992; Sjogren, 1997), the Moon (Zuber, 1996), many national DEMs (USGS, 1993; Natural Resources Canada, 1997), as well as models of various geophysical characteristics (e.g., Mooney et al., 1998). Disadvantages of the spheroidal equal angular grid are: (a) the excess of information for poles, that is, each pole is described by not one but a series of points, and the number depends on the grid resolution; and (b) trapezoid areas vary depending on the latitude that may hamper further data analysis (Bjørke and Nilsen, 2004).

Some regular spheroidal grids are free of these disadvantages. Such grids are based on the initial tessellation of the sphere by spherical Platonic solids (octahedra, dodecahedra, and icosahedra) or Catalan solids (rhombic dodecahedra and triacontahedra) and subsequent subdivision of each face by spherical triangles, rhombs, or hexagons (Dutton, 1999; White, 2000; Sahr et al., 2003; Zhao et al., 2008; Yuan et al., 2010; Bernardin et al., 2011) (Fig. 3.1e). Some regular spheroidal grids used in meteorology (Umscheid and Bannon, 1977; Hortal and Simmons, 1991; Chen et al., 2008) may also be adapted to spheroidal digital terrain modeling.

3.3 DEM RESOLUTION

Selection of the DEM resolution—the average density of points of an irregular DEM, or the grid spacing (w) of a regular DEM—is one of the key problems of digital terrain modeling (Balce, 1987; Hengl, 2006). These parameters should maintain a reasonable accuracy to describe the topographic surface using the minimum number of points.

There are several empirical approaches to selecting the w value. For example, according to the Peucker criterion, w value should be 4.3 times larger than the contour interval of a digitized map (Sasowsky et al., 1992).

3.3 DEM RESOLUTION

Ivanov and Kruzhkov (1992) presented a table to choose w value depending on the type of topography and the root mean square error of a DEM.

Let us examine this issue from the theoretical standpoint. According to the sampling theorem (Kotelnikov, 1933[3]; Shannon, 1949), a one-dimensional continuous bandlimited function $y = f(x)$ with a bandwidth ν can be completely determined by a set of samples $f(k \Delta x)$ if $\Delta x \leq \frac{1}{2\nu}$, where Δx is the sampling interval, $-\infty \leq k \leq \infty$. An extension of the sampling theorem to a two-dimensional case holds that a continuous bandlimited function $z = f(x, y)$ with bandwidths ν_x and ν_y can be determined by a set of samples $f(k \Delta x, l \Delta y)$ if

$$\Delta x \leq \frac{1}{2\nu_x}, \Delta y \leq \frac{1}{2\nu_y}, \quad (3.1)$$

where Δx and Δy are sampling intervals, $-\infty \leq k,l \leq \infty$. In other words, it is possible to reconstruct a continuous function from a discrete one if at least two samples per the shortest wavelength $\lambda_{x,y}$ were collected, $\lambda_{x,y} = \frac{1}{\nu_{x,y}}$ (Rosenfeld and Kak, 1982). Practically, to avoid ambiguity in reconstruction of a function due to limitations of the sampling theorem and interpolation effects (e.g., the Gibbs phenomenon—Section 5.5), it is advisable to use a multiplicative factor $n = 2 \ldots 10$ to determine sampling intervals (Grigorenko, 1998):

$$\Delta x \leq \frac{\lambda_x}{2n}, \Delta y \leq \frac{\lambda_y}{2n}. \quad (3.2)$$

DEM generation is a discretization of the topographic surface (Makarovič, 1973, 1977; Mark, 1975a; Stefanovic et al., 1977). The spectrum of the topographic surface is unlimited, so the condition of the sampling theorem is not met. This is a tractable problem because a researcher is usually interested in studying landforms, with typical planar sizes not smaller than the threshold short wavelength $\tilde{\lambda}_{x,y}$ (Mark, 1975a). Therefore, in each given case, it is possible to consider that the topographic surface is a bandlimited function with bandwidths $\tilde{\nu}_{x,y}$.

For digital terrain analysis, the sampling theorem has three main sequences (Florinsky, 2002):

1. To keep information on short-wavelength topographic features with typical planar sizes $\tilde{\lambda}_{x,y}$ in a DEM, one should use DEM sampling intervals $\Delta x \leq \frac{\tilde{\lambda}_x}{2n}$ and $\Delta y \leq \frac{\tilde{\lambda}_y}{2n}$, or the DEM grid spacing $w \leq \frac{\tilde{\lambda}_{x,y}}{2n}$ (Mark, 1975a). These sampling intervals or the grid spacing set the resolution limit for all DTMs derived from the DEM.

---

[3]There is an English translation of this work (Kotelnikov, 2001).

2. DEM interpolation does not provide a way to increase the DEM resolution. In other words, it is impossible to reconstruct topographic features with typical planar sizes less than $\tilde{\lambda}_{x,y}$. Of course, the DEM resolution may be nominally "enhanced" by interpolation using $w << \Delta x, \Delta y$, but this procedure cannot improve the actual resolution of the DEM (Stefanovic et al., 1977; Horn, 1981; Sasowsky et al., 1992).
3. "Landforms" with typical planar sizes less than $\tilde{\lambda}_{x,y}$ generated by interpolation should be considered as a high-frequency noise caused by properties of an interpolator.

DEM errors caused by ignoring the sampling theorem are discussed in Section 5.4.

3.4 DEM INTERPOLATION

Conversion of irregular DEMs into regular ones (based on plane square grids or spheroidal equal angular girds) is carried out by interpolation. This procedure is, in particular, necessary because all effective methods for calculation of morphometric variables (Chapter 4) deal with DEMs based on regular grids. Interpolation is also used to refine regular DTMs.

The spatial interpolation problem can be formulated as follows (Mitas and Mitasova, 1999): Given the N values of elevation z_j, $j = 1, \ldots, N$, measured at discrete points $r_j = (x_j^{[1]}, x_j^{[2]}, \ldots, x_j^{[d]})$, within a certain area of a d-dimensional space, find a d-variate function $F(r)$ passing through the given points, that is,

$$F(\mathrm{r}_j) = z_j. \tag{3.3}$$

There are dozens of methods of spatial interpolation. The selection of a particular method to interpolate a particular DEM depends on the technique of DEM generation, the type of DEM grid, the type of topography, and the purpose of interpolation. Here, we describe briefly three popular approaches of DEM interpolation, which are used (or discussed) in the book:

- *Inverse distance-weighted interpolation* (Watson, 1992). An elevation value at a regular grid node is approximated by a weighted average of elevation values at points within a certain distance from the node. Weights are inversely proportional to a power of distance.
- *Triangulation-based interpolation* (Agishtein and Migdal, 1991; Watson, 1992). This includes the Delaunay triangulation (Delaunay, 1934; Hjelle and Dæhlen, 2006) of irregularly distributed points (Fig. 3.1b).

For each triangle, a bivariate function is derived and then used to estimate elevation values at regular grid nodes. For linear interpolation, flat facets are fitted to each triangle. For nonlinear interpolation, blending functions, such as polynomials, are applied to provide merging of triangles.
- *Thin-plate spline interpolation* (Mitas and Mitasova, 1999). Consider the condition to minimize the sum of the deviations from the irregular grid points and the smoothness seminorm of the spline function:

$$\sum_{j=1}^{N} |z_j - F(r_j)|^2 w_j + w_0 I(F) = \text{minimum}, \qquad (3.4)$$

where w_j, w_0 are weights, $I(F)$ is the smoothness seminorm, a measure of roughness of the function in terms of its order of derivative relative to the function. Using a bivariate smoothness seminorm with squares of second derivatives, one obtains a thin-plate spline function. It minimizes the curvature of the topographic surface and imitates a steel sheet passing through the irregular grid points. Adding the first-order derivatives into $I(F)$, one obtains a thin-plate spline with tension (Franke, 1985), which imitates an elastic membrane. A function with regular second- and higher-order derivatives can be obtained by adding higher-order derivatives into $I(F)$: the regularized spline with tension includes the sum of all derivatives with rapidly decreasing weights (Mitášová and Mitáš, 1993; Mitášová and Hofierka, 1993).

For DEMs based on plane irregular grids, classification and reviews of interpolation methods can be found elsewhere (Schut, 1976; Franke, 1982; Lam, 1983; Davis, 1986; McCullagh, 1988; Watson, 1992; Mitas and Mitasova, 1999). For DEMs based on irregular spheroidal grids, the Delaunay spherical triangulation (Gold and Mostafavi, 2000; Lukatela, 2000) can be applied to create regular spheroidal grids; interpolation of elevation values can be done with spherical interpolators (Wahba, 1981; Willmott et al., 1985; Robeson, 1997). It is obvious that all interpolators inevitably introduce new errors and artifacts in derived DEMs.

CHAPTER 4

Calculation Methods

OUTLINE

4.1 The Evans–Young Method	43
4.2 Calculation of Local Morphometric Variables on a Plane Square Grid	45
4.2.1 Motivation	*45*
4.2.2 Formulas	*46*
4.2.3 Method Validation	*49*
4.3 Calculation of Local Morphometric Variables on a Spheroidal Equal Angular Grid	54
4.3.1 Motivation	*54*
4.3.2 Formulas	*55*
4.3.3 Calculation of Linear Sizes of a Spheroidal Equal Angular Window	*57*
4.3.4 Discussion	*58*
4.4 Calculation of Nonlocal Morphometric Variables	59
4.5 Calculation of Structural Lines	61
4.5.1 Conventional Algorithms	*61*
4.5.2 Generating Function	*62*

4.1 THE EVANS–YOUNG METHOD

Local morphometric variables are functions of partial derivatives of elevation (Eqs. 2.2–2.15). There are three popular methods for computing r, t, s, p, and q (Eq. 2.1) from DEMs based on plane square grids (Young, 1978; Evans, 1979; Zevenbergen and Thorne, 1987; Shary, 1995). The methods are based on approximation of partial derivatives by finite differences

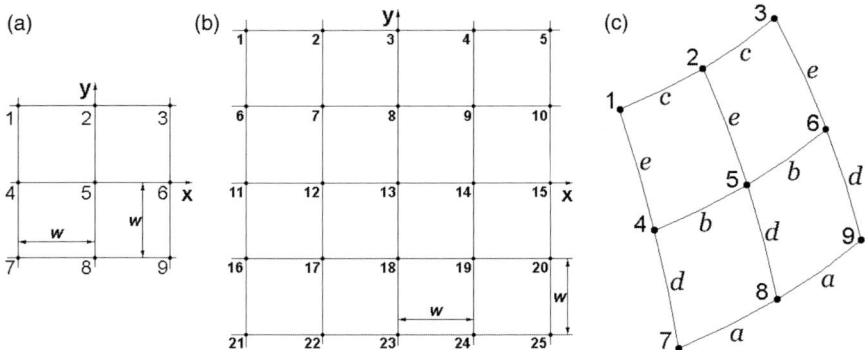

FIGURE 4.1 Main types of moving windows used to calculate local morphometric variables: (a) A 3×3 plane square-gridded window; w is a grid spacing (meters), 1, ..., 9 are numbers of the window nodes. (b) A 5×5 plane square-gridded window, w is a grid spacing (meters), 1, ..., 25 are numbers of the window nodes. (c) A 3×3 spheroidal equal angular window, a, b, c, d, and e are linear sizes (meters) of moving window elements, 1, ..., 9 are numbers of the window nodes. *From (Florinsky, 2010, Fig. 2.1).*

using the 3×3 plane square-gridded moving window (Ames, 1977). The Evans—Young method is the most accurate of the three methods (Florinsky, 1998a; Schmidt et al., 2003).

In the Evans—Young method (Young, 1978; Evans, 1979, pp. 28—29), the second-order polynomial

$$z = \frac{rx^2}{2} + \frac{ty^2}{2} + sxy + px + qy + u \tag{4.1}$$

is fitted by the least-squares approach to the nine points of the 3×3 square-spaced window with a grid spacing of w (Fig. 4.1a). The Cartesian coordinates and elevations of the topographic surface are known for the window points $(-w, w, z_1)$, $(0, w, z_2)$, (w, w, z_3), $(-w, 0, z_4)$, $(0, 0, z_5)$, $(w, 0, z_6)$, $(-w, -w, z_7)$, $(0, -w, z_8)$, and $(w, -w, z_9)$. For the point $(0, 0, z_5)$, the polynomial coefficients (which are partial derivatives of elevation) are estimated by the following formulas:

$$r = \frac{z_1 + z_3 + z_4 + z_6 + z_7 + z_9 - 2(z_2 + z_5 + z_8)}{3w^2}, \tag{4.2}$$

$$t = \frac{z_1 + z_2 + z_3 + z_7 + z_8 + z_9 - 2(z_4 + z_5 + z_6)}{3w^2}, \tag{4.3}$$

$$s = \frac{z_3 + z_7 - z_1 - z_9}{4w^2}, \tag{4.4}$$

$$p = \frac{z_3 + z_6 + z_9 - z_1 - z_4 - z_7}{6w}, \tag{4.5}$$

$$q = \frac{z_1 + z_2 + z_3 - z_7 - z_8 - z_9}{6w}. \tag{4.6}$$

Moving the 3×3 window along a DEM, one can calculate values of r, t, s, p, and q (and so values of local morphometric variables) for all points of the plane square-gridded DEM, except for boundary rows and columns.

The polynomial (Eq. 4.1) is approximated to elevation values at the window points, rather than passing exactly through them. This leads to some smoothing of the elevation function within the window, that is, to the local filtering of high-frequency noise inherent for any DEM. Such a filtering allows one to maintain a correct estimation of derivatives, which are sensitive to the high-frequency component of a signal (Section 5.4.1). A further merit of the Evans–Young method is the use of excess measurements: nine elevation values are utilized to estimate six coefficients of the polynomial (Eq. 4.1) that increases estimation accuracy.

4.2 CALCULATION OF LOCAL MORPHOMETRIC VARIABLES ON A PLANE SQUARE GRID

4.2.1 Motivation

There were proposals to use some morphometric variables, which are functions of not only the first- and second-order partial derivatives of elevation but also third-order ones (Lastochkin, 1987; Shary and Stepanov, 1991; Jenčo, 1992; Minár and Evans, 2008; Pacina, 2010). It is obvious that the Evans–Young, Zevenbergen–Thorne, and Shary methods (Section 4.1) are not suitable for estimating the third-order partial derivatives.

In this section, we describe the method that can be used to derive local topographic attributes from DEMs based on plane square grids (Florinsky, 2009b). Like previous algorithms, this method is simple and fast, but it has two advantages: First, it allows one to estimate the first- (p and q), second- (r, t, and s), and third-order (g, h, k, and m) partial derivatives of elevation (Eq. 2.1). Second, it has an increased accuracy and a better suppression of high-frequency noise.

4.2.2 Formulas

Using the well-known Taylor formula (Fichtenholz, 1966a, p. 416; Nikolsky, 1977b, § 7.13), a function $z = f(x,y)$ can be conveniently expressed as follows:

$$z = \frac{1}{6}gx^3 + \frac{1}{6}hy^3 + \frac{1}{2}kx^2y + \frac{1}{2}mxy^2 + \frac{1}{2}rx^2 + \frac{1}{2}ty^2 + sxy + px + qy + u \quad (4.7)$$

Under the least-squares approach (Bjerhammar, 1973), one should carry out n measurements of z to find N unknown coefficients of $z = f(x,y)$, with $n > (N + 1)$. Therefore, one cannot use a 3×3 window (Fig. 4.1a) to find the 10 unknown coefficients g, h, k, m, r, t, s, p, q, and u of the third-order polynomial (Eq. 4.7), because such a window includes only nine measured values of z. To find these coefficients for the central point of a window, there is a need to use, at least, a 5×5 square-gridded window (Fig. 4.1b). The Cartesian coordinates and elevations of the topographic surface are known for 25 five points of this window: $(-2w, 2w, z_1)$, $(-w, 2w, z_2)$, $(0, 2w, z_3)$, $(w, 2w, z_4)$, $(2w, 2w, z_5)$, $(-2w, w, z_6)$, $(-w, w, z_7)$, $(0, w, z_8)$, (w, w, z_9), $(2w, w, z_{10})$, $(-2w, 0, z_{11})$, $(-w, 0, z_{12})$, $(0, 0, z_{13})$, $(w, 0, z_{14})$, $(2w, 0, z_{15})$, $(-2w, -w, z_{16})$, $(-w, -w, z_{17})$, $(0, -w, z_{18})$, $(w, -w, z_{19})$, $(2w, -w, z_{20})$, $(-2w, -2w, z_{21})$, $(-w, -2w, z_{22})$, $(0, -2w, z_{23})$, $(w, -2w, z_{24})$, and $(2w, -2w, z_{25})$.

Let us fit the polynomial (Eq. 4.7) to the 25 points of the 5×5 window by the least-squares approach (Bjerhammar, 1973). Writing the polynomial (Eq. 4.7) for all points of the window, we obtain a system of 25 conditional linear equations, which can be expressed as follows:

$$\alpha = \mathbf{F}\beta, \quad (4.8)$$

where α is a 25×1 matrix of the 25 measured values of z:

$$\alpha = \begin{pmatrix} z_1 \\ \vdots \\ z_{25} \end{pmatrix}, \quad (4.9)$$

β is a 10×1 matrix of the 10 unknown coefficients of the polynomial (Eq. 4.7):

$$\beta = \begin{pmatrix} g \\ h \\ k \\ m \\ r \\ t \\ s \\ p \\ q \\ u \end{pmatrix}, \quad (4.10)$$

4.2 CALCULATION OF LOCAL MORPHOMETRIC VARIABLES ON A PLANE SQUARE GRID

and **F** is a 25×10 matrix of the known coefficients of the equation system (Eq. 4.8):

$$\mathbf{F} = \begin{pmatrix}
-\frac{4}{3}w^3 & \frac{4}{3}w^3 & 4w^3 & -4w^3 & 2w^2 & 2w^2 & -4w^2 & -2w & 2w & 1 \\
-\frac{1}{6}w^3 & \frac{4}{3}w^3 & w^3 & -2w^3 & \frac{1}{2}w^2 & 2w^2 & -2w^2 & -w & 2w & 1 \\
0 & \frac{4}{3}w^3 & 0 & 0 & 0 & 2w^2 & 0 & 0 & 2w & 1 \\
\frac{1}{6}w^3 & \frac{4}{3}w^3 & w^3 & 2w^3 & \frac{1}{2}w^2 & 2w^2 & 2w^2 & w & 2w & 1 \\
\frac{4}{3}w^3 & \frac{4}{3}w^3 & 4w^3 & 4w^3 & 2w^2 & 2w^2 & 4w^2 & 2w & 2w & 1 \\
-\frac{4}{3}w^3 & \frac{1}{6}w^3 & 2w^3 & -w^3 & 2w^2 & \frac{1}{2}w^2 & -2w^2 & -2w & w & 1 \\
-\frac{1}{6}w^3 & \frac{1}{6}w^3 & \frac{1}{2}w^3 & -\frac{1}{2}w^3 & \frac{1}{2}w^2 & \frac{1}{2}w^2 & -w^2 & -w & w & 1 \\
0 & \frac{1}{6}w^3 & 0 & 0 & 0 & \frac{1}{2}w^2 & 0 & 0 & w & 1 \\
\frac{1}{6}w^3 & \frac{1}{6}w^3 & \frac{1}{2}w^3 & \frac{1}{2}w^3 & \frac{1}{2}w^2 & \frac{1}{2}w^2 & w^2 & w & w & 1 \\
\frac{4}{3}w^3 & \frac{1}{6}w^3 & 2w^3 & w^3 & 2w^2 & \frac{1}{2}w^2 & 2w^2 & 2w & w & 1 \\
-\frac{4}{3}w^3 & 0 & 0 & 0 & 2w^2 & 0 & 0 & -2w & 0 & 1 \\
-\frac{1}{6}w^3 & 0 & 0 & 0 & \frac{1}{2}w^2 & 0 & 0 & -w & 0 & 1 \\
0 & 0 & 0 & 0 & 0 & 0 & 0 & 0 & 0 & 1 \\
\frac{1}{6}w^3 & 0 & 0 & 0 & \frac{1}{2}w^2 & 0 & 0 & w & 0 & 1 \\
\frac{4}{3}w^3 & 0 & 0 & 0 & 2w^2 & 0 & 0 & 2w & 0 & 1 \\
-\frac{4}{3}w^3 & -\frac{1}{6}w^3 & -2w^3 & -w^3 & 2w^2 & \frac{1}{2}w^2 & 2w^2 & -2w & -w & 1 \\
-\frac{1}{6}w^3 & -\frac{1}{6}w^3 & -\frac{1}{2}w^3 & -\frac{1}{2}w^3 & \frac{1}{2}w^2 & \frac{1}{2}w^2 & w^2 & -w & -w & 1 \\
0 & -\frac{1}{6}w^3 & 0 & 0 & 0 & \frac{1}{2}w^2 & 0 & 0 & -w & 1 \\
\frac{1}{6}w^3 & -\frac{1}{6}w^3 & -\frac{1}{2}w^3 & \frac{1}{2}w^3 & \frac{1}{2}w^2 & \frac{1}{2}w^2 & -w^2 & w & -w & 1 \\
\frac{4}{3}w^3 & -\frac{1}{6}w^3 & -2w^3 & w^3 & 2w^2 & \frac{1}{2}w^2 & -2w^2 & 2w & -w & 1 \\
-\frac{4}{3}w^3 & -\frac{4}{3}w^3 & -4w^3 & -4w^3 & 2w^2 & 2w^2 & 4w^2 & -2w & -2w & 1 \\
-\frac{1}{6}w^3 & -\frac{4}{3}w^3 & -w^3 & -2w^3 & \frac{1}{2}w^2 & 2w^2 & 2w^2 & -w & -2w & 1 \\
0 & -\frac{4}{3}w^3 & 0 & 0 & 0 & 2w^2 & 0 & 0 & -2w & 1 \\
\frac{1}{6}w^3 & -\frac{4}{3}w^3 & -w^3 & 2w^3 & \frac{1}{2}w^2 & 2w^2 & -2w^2 & w & -2w & 1 \\
\frac{4}{3}w^3 & -\frac{4}{3}w^3 & -4w^3 & 4w^3 & 2w^2 & 2w^2 & -4w^2 & 2w & -2w & 1
\end{pmatrix} \quad (4.11)$$

To determine the unknown coefficients of the polynomial (Eq. 4.7), we solve the equation

$$\beta = (\mathbf{F}^T\mathbf{F})^{-1}\mathbf{F}^T\alpha, \tag{4.12}$$

where $\mathbf{F}^T$ is a transposed matrix of $\mathbf{F}$, and $(\mathbf{F}^T\mathbf{F})^{-1}$ is an inverse matrix of $\mathbf{F}^T\mathbf{F}$ (see details elsewhere—Florinsky, 2009b). As a result, we obtain the formulas for the partial derivatives of elevation:

$$g = \frac{1}{10w^3}[z_5 + z_{10} + z_{15} + z_{20} + z_{25} - z_1 - z_6 - z_{11} - z_{16} - z_{21}$$
$$+ 2(z_2 + z_7 + z_{12} + z_{17} + z_{22} - z_4 - z_9 - z_{14} - z_{19} - z_{24})], \tag{4.13}$$

$$h = \frac{1}{10w^3}[z_1 + z_2 + z_3 + z_4 + z_5 - z_{21} - z_{22} - z_{23} - z_{24} - z_{25}$$
$$+ 2(z_{16} + z_{17} + z_{18} + z_{19} + z_{20} - z_6 - z_7 - z_8 - z_9 - z_{10})], \tag{4.14}$$

$$k = \frac{1}{70w^3}[z_{17} + z_{19} - z_7 - z_9 + 4(z_1 + z_5 + z_{23} - z_3 - z_{21} - z_{25})$$
$$+ 2(z_6 + z_{10} + z_{18} + z_{22} + z_{24} - z_2 - z_4 - z_8 - z_{16} - z_{20})], \tag{4.15}$$

$$m = \frac{1}{70w^3}[z_7 + z_{17} - z_9 - z_{19} + 4(z_5 + z_{11} + z_{25} - z_1 - z_{15} - z_{21})$$
$$+ 2(z_4 + z_6 + z_{12} + z_{16} + z_{24} - z_2 - z_{10} - z_{14} - z_{20} - z_{22})], \tag{4.16}$$

$$r = \frac{1}{35w^2}[2(z_1 + z_5 + z_6 + z_{10} + z_{11} + z_{15} + z_{16} + z_{20} + z_{21} + z_{25})$$
$$- 2(z_3 + z_8 + z_{13} + z_{18} + z_{23}) - z_2 - z_4 - z_7 - z_9$$
$$- z_{12} - z_{14} - z_{17} - z_{19} - z_{22} - z_{24}], \tag{4.17}$$

$$t = \frac{1}{35w^2}[2(z_1 + z_2 + z_3 + z_4 + z_5 + z_{21} + z_{22} + z_{23} + z_{24} + z_{25})$$
$$- 2(z_{11} + z_{12} + z_{13} + z_{14} + z_{15}) - z_6 - z_7 - z_8 - z_9$$
$$- z_{10} - z_{16} - z_{17} - z_{18} - z_{19} - z_{20}], \tag{4.18}$$

$$s = \frac{1}{100w^2}[z_9 + z_{17} - z_7 - z_{19} + 4(z_5 + z_{21} - z_1 - z_{25})$$
$$+ 2(z_4 + z_{10} + z_{16} + z_{22} - z_2 - z_6 - z_{20} - z_{24})], \tag{4.19}$$

$$p = \frac{1}{420w}\{44(z_4 + z_{24} - z_2 - z_{22}) + 31[z_1 + z_{21} - z_5 - z_{25}$$
$$+ 2(z_9 + z_{19} - z_7 - z_{17})] + 17[z_{15} - z_{11} + 4(z_{14} - z_{12})]$$
$$+ 5(z_{10} + z_{20} - z_6 - z_{16})\} \tag{4.20}$$

4.2 CALCULATION OF LOCAL MORPHOMETRIC VARIABLES ON A PLANE SQUARE GRID

$$q = \frac{1}{420w}\{44(z_6 + z_{10} - z_{16} - z_{20}) + 31[z_{21} + z_{25} - z_1 - z_5]$$
$$+ 2(z_7 + z_9 - z_{17} - z_{19})] + 17[z_3 - z_{23} + 4(z_8 - z_{18})]$$
$$+ 5(z_2 + z_4 - z_{22} - z_{24})\}. \qquad (4.21)$$

Moving the 5×5 window along a DEM, one can compute values of $g, h, k, m, r, t, s, p,$ and q (and so values of local morphometric variables) for all points of the DEM, except for two boundary rows and two boundary columns on each side of the plane square-gridded DEM.

We do not present a formula for the residual term u of the polynomial (Eq. 4.7), since it is not used in calculations of topographic variables. Development of Eqs. (4.13)–(4.21) was carried out with the software Maple V Release 5.0 (© Waterloo Maple Inc., 1981–1997).

4.2.3 Method Validation

4.2.3.1 Materials and Data Processing

We analyzed the accuracy of the method described using the criterion of the root mean square error (RMSE) of a function of measured variables (Florinsky, 2009b). It was proved that the method offers a higher accuracy for calculation of local morphometric attributes than the Evans–Young method, the most accurate similar technique. For details, see Section 5.3.3.1.

To exemplify application of the method described, we used a portion of a DEM of the Stavropol Upland, Russia (Fig. 2.3). An irregular DEM (Florinsky, 2002) was produced by the digitizing of a topographic map (Central Board of Geodesy and Cartography, 1968). The area selected measures about 133×100 km. The irregular DEM included 2571 points.

Using the Delaunay triangulation and a piecewise quadric polynomial interpolation with matching derivatives along triangle edges (Agishtein and Migdal, 1991), we produced a square-gridded DEM with a grid spacing of 300 m. To reduce high-frequency noise in the DEM, we applied three iterations of smoothing to the DEM using the 3×3 window with inverse distance weights. Applying the method described to the smoothed DEM, we derived digital models of $G, A, k_h, k_v, H, K, k_{min}, k_{max}, K_a, E, K_r, k_{ve}, k_{he}, M,$ and T (Figs. 2.4 and 2.5); the grid spacing was 300 m.

To compare the accuracies of the Evans–Young and author's methods, we carried out the following procedures:

- For both methods, we derived digital models of RMSE of calculation of $G, k_h, k_v, H, K, k_{min}, k_{max}, K_a, E, K_r, k_{ve}, k_{he},$ and M. RMSE formulas (Florinsky, 1998a, 2008d, 2009b) can be found in Section 5.3.2.

- Applying the Evans–Young method to the smoothed DEM, we derived digital models of G, k_h, k_v, H, K, k_{min}, k_{max}, K_a, E, K_r, k_{ve}, k_{he}, and M.
- For each morphometric variable, we estimated a DTM difference, that is, a difference between two digital models derived by the Evans–Young and author's methods.
- For each morphometric variable, we visually analyzed patterns on maps derived by both methods.
- We performed a statistical analysis of DTM differences and distributions of each morphometric attribute derived by both methods. We also carried out the Kolmogorov–Smirnov two-sample test (Daniel, 2000, ch. 8) for statistical difference between the two distributions of each topographic variable. We used samples each including 1376 points (43×32 matrices, with the grid spacing of 3000 m extracted from related digital models).

To deal with the large dynamic range of morphometric attributes and their RMSE, we logarithmically transformed these digital models by Eq. (7.1) with $n = 5$. Terrain modeling was performed by the software LandLord (Appendix B). Statistical analysis was carried out with Statgraphics Plus 3.0 (© Statistical Graphics Corp., 1994–1997).

4.2.3.2 Results and Discussion

There are essential differences in maps of RMSE of morphometric variables calculated by the different methods. In particular, RMSE patterns have distinct spatial distribution (Fig. 4.2). For both methods, RMSEs have approximately the same dynamic range widths. For example, RMSE of the horizontal curvature (m_{kh}) has the dynamic range of about 5 units (Fig. 4.2). However, the application of the author's method lowered the boundaries of the dynamic range: for m_{kh}, it changed from [1.9–6.9] to [0.0–4.9] (Figs. 4.2a, b). The statistical distribution of RMSE was also changed: for the author's method, the most part, for instance, of m_{kh} and m_{kv} values is relatively evenly distributed within a rather wide interval, from 0 to 1. For the Evans–Young method, m_{kh} and m_{kv} values are located within a narrow band ranging from 2.0 to 2.2 (Fig. 4.3).

The method described shares a common trait with the Evans–Young method: the polynomial (Eq. 4.7) is approximated to elevation values of the 5×5 window rather than passing exactly through them. This leads to a local denoising that may enhance the calculation of derivatives and local topographic variables because they are responsive to a high-frequency component of a signal (Florinsky, 2002) (Section 5.4.1). However, Wood (1996, § 4.2) supposed that calculation of derivatives fitting the third and higher-order polynomials to $n \times n$ windows

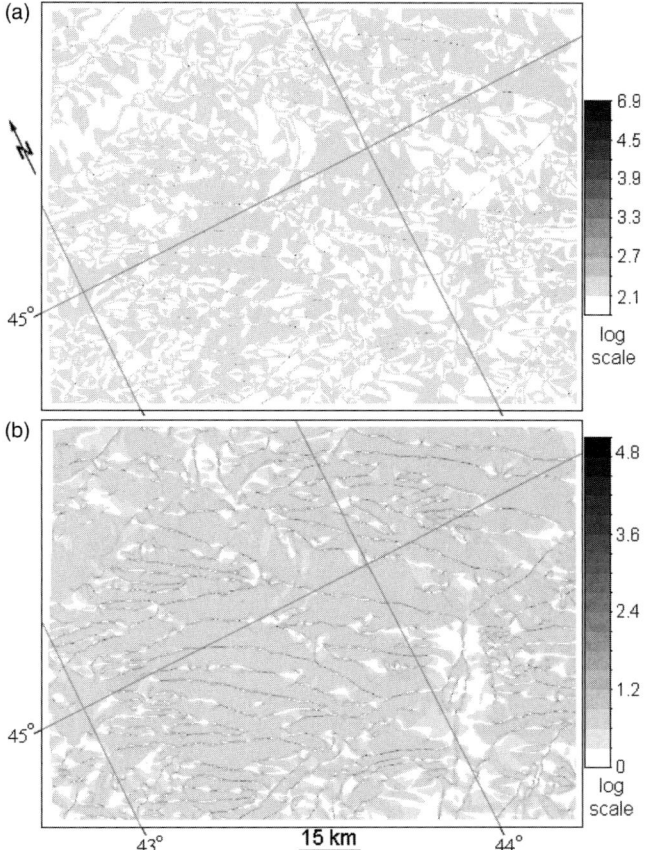

FIGURE 4.2 The Stavropol Upland: RMSE maps of k_h calculation by the Evans–Young method (a) and the author's method (b). For the elevation map, see Fig. 2.3. *From (Florinsky, 2009a, Fig. 3a).*

(for $n > 3$) can lead to an undue generalization of the topographic surface. This is not necessarily the case.

Indeed, in comparing k_h maps derived by two methods (Fig. 4.4), one can see that local changes in map patterns enclose from one to about eight pixels. This is a level of denoising rather than generalization. One can find the same level of changes in the k_h dynamic range: it varied from -2.55 to 2.44 for the Evans–Young method, and from -2.49 to 2.31 for the author's method (Fig. 4.4). Thus, the change in the k_h dynamic range was only about 4%. This is an insignificant alteration.

Histograms of the two k_h samples are very similar (Fig. 4.5a). Most changes between k_h models derived by the Evans–Young and author's

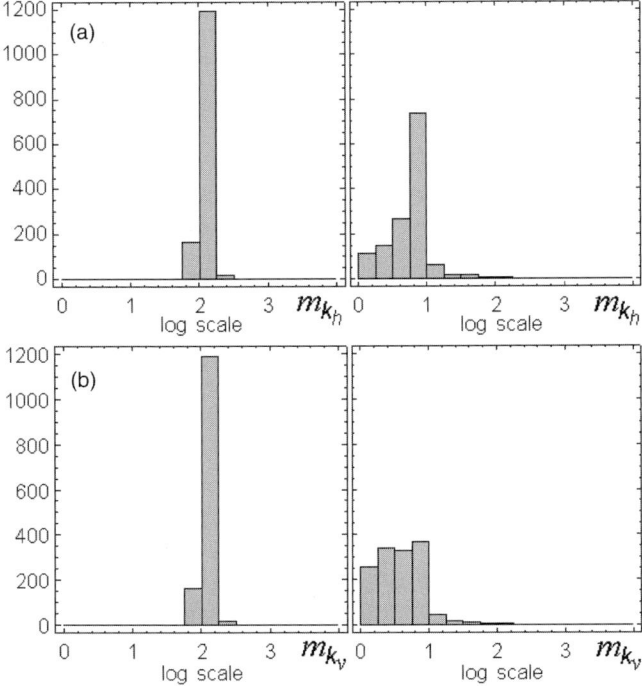

FIGURE 4.3 Statistical distribution of logarithmically transformed RMSE values of k_h and k_v calculation by the Evans–Young method (left) and the author's method (right): (a) RMSE of k_h; (b) RMSE of k_v. From (Florinsky, 2009a, Fig. 3b, c).

methods (Δk_h) were in the range from −0.1 to 0.1 (Fig. 4.5b). Again, this is 4% of the dynamic range of k_h derived by the Evans–Young method. Results of the Kolmogorov–Smirnov test for paired samples derived by different methods demonstrated that there is no statistically significant difference between the two distributions at the 95% confidence level. For k_h, for example, the estimated overall statistic DN = 0.025, the two-sided large sample K-S statistic is 0.67, and P value = 0.77. Therefore, the author's method provides additional denoising of a DEM without undue generalization of the surface.

A stronger suppression of high-frequency noise in DEMs by the author's method can be associated with two factors. First, we use the polynomial of a higher order than the Evans–Young method employs. Second, we use the 5 × 5 moving window, while the Evans–Young method utilizes the 3 × 3 one. Indeed, the increase of the moving window size may decrease the influence of high-frequency noise and interpolation errors on the computation of topographic variables (Albani et al., 2004).

4.2 CALCULATION OF LOCAL MORPHOMETRIC VARIABLES ON A PLANE SQUARE GRID 53

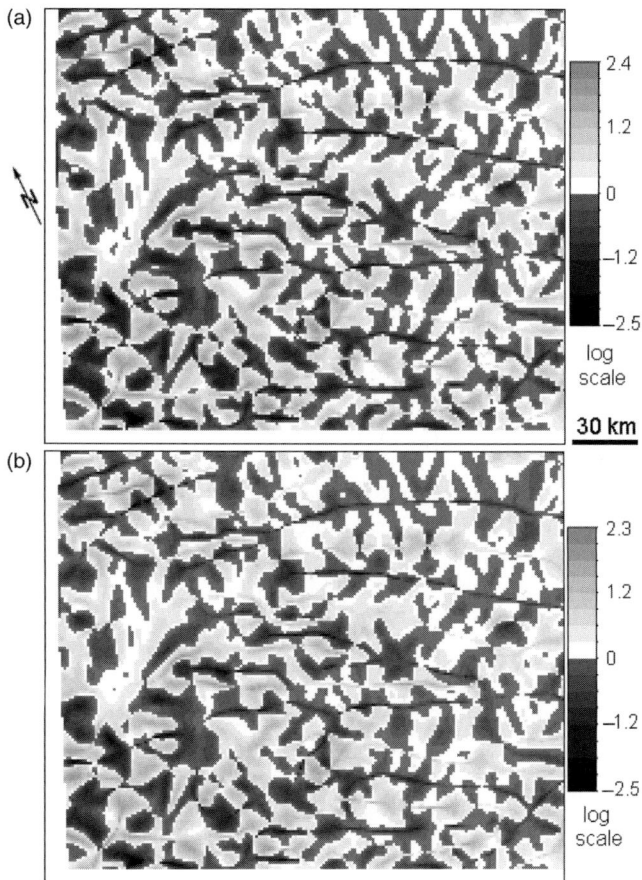

FIGURE 4.4 The Stavropol Upland: enlarged lower left portions of the k_h maps derived by the Evans–Young method (a) and the author's method (b). For the entire map, see Fig. 2.5a. *From (Florinsky, 2009b, Fig. 6b, c).*

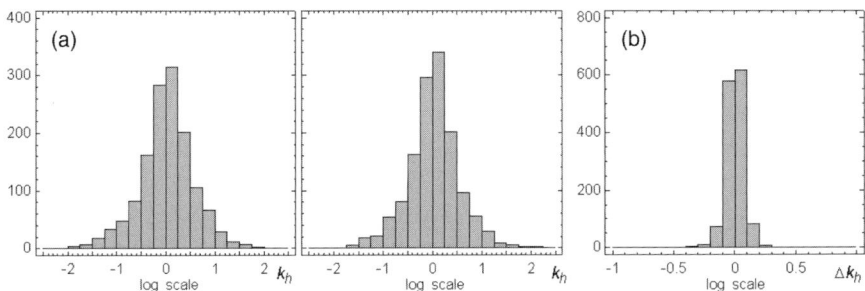

FIGURE 4.5 Statistical distribution of logarithmically transformed k_h values: (a) k_h derived by the Evans–Young method (left) and the author's method (right); (b) Δk_h. *From (Florinsky, 2009b, Fig. 7a, b, c).*

Thus, compared to the Evans–Young method, the method described has essentially higher accuracy of calculation and stronger suppression of high-frequency noise. Therefore, it is advisable to use this method as a standard tool for data processing in digital terrain modeling.

4.3 CALCULATION OF LOCAL MORPHOMETRIC VARIABLES ON A SPHEROIDAL EQUAL ANGULAR GRID

4.3.1 Motivation

Many DEMs are based on spheroidal equal angular grids (Section 3.2). Among these are national DEMs (e.g., USGS, 1993; Natural Resources Canada, 1997), global DEMs of the Earth, such as ETOPO5 (World Data Center for Geophysics, 1993), GTOPO30 (USGS, 1996), GLOBE (GLOBE Task Team and others, 1999), ETOPO2 (National Geophysical Data Center, 2001, 2006), SRTM3 (NASA, 2003), SRTM30_PLUS (Becker et al., 2009), ETOPO1 (Amante and Eakins, 2009), and ASTER GDEM (Jet Propulsion Laboratory, 2009), as well as global DEMs of other celestial bodies, such as Mars (Smith, 2003), Venus (Ford, 1992; Sjogren, 1997), and the Moon (Zuber, 1996; Smith D.E. et al., 2010). Some of these DEMs have relatively high spatial resolutions: for instance, $1''$ (Jet Propulsion Laboratory, 2009), $1''$ and $3''$ (NASA, 2003), $0.75''$ and $3''$ (Natural Resources Canada, 1997), $2''$ and $3''$ (USGS, 1993), $\sim 7''$ (Smith, 2003). Such resolutions make possible the use of these DEMs in medium-scale soil, geological, and planetological studies. Other DEMs have lower resolutions: for example, $30''$ (USGS, 1996; GLOBE Task Team and others, 1999; Becker et al., 2009), $5'$ (World Data Center for Geophysics, 1993), $4.2'$ (Smith D.E. et al., 2010), $2'$ (National Geophysical Data Center, 2001, 2006), $1'$ (Amante and Eakins, 2009), $\sim 14''$, $\sim 28''$, $\sim 1'$, $\sim 1.9'$, $3.75'$, $15'$ (Smith, 2003), $15'$ (Zuber, 1996), and $1°$ (Sjogren, 1997). Such resolutions restrict the application of these DEMs to small-scale geological and planetological studies. It is obvious that processing and analysis of spheroidal equal angular DEMs should be done considering the grid geometry.

As discussed above (Sections 4.1 and 4.2), digital models of local morphometric variables are usually derived by methods based on the approximation of partial derivatives of elevation by finite differences. Such methods have been developed and intended to calculate partial derivatives from DEMs based on plane square grids. A plane square grid and a spheroidal equal angular grid have principally different geometry (Fig. 3.1c, d). For spheroidal equal angular grids, a grid spacing, which is approximately linearly equal along meridians and parallels, exists only near the equator. For example, the $3'' \times 3''$ pixel of the

CALCULATION OF LOCAL MORPHOMETRIC VARIABLES

SRTM3 DEM (NASA, 2003) has the linear sizes of about 52 m × 93 m at the latitude of Moscow, Russia. Therefore, it is obvious that methods developed for plane square grids cannot be applied on spheroidal equal angular grids. This obstacle can be gotten around by interpolation of elevation values from a spheroidal equal angular grid to a plane square grid (e.g., Ehsani and Quiel, 2008). However, this is an additional data-processing step, and any interpolation introduces new errors into a new DEM (Willmott et al., 1985).

In this section, we describe the method for calculating local morphometric variables on spheroidal equal angular grids (Florinsky, 1998c).

4.3.2 Formulas

Let elevation be given as $z = f(x,y)$, where x and y are the orthogonal spheroidal coordinates. Let there be a 3×3 moving window formed by four adjacent spheroidal equal angular trapezoids (Fig. 4.1c). a, b, and c are linear lengths of three angularly equal arcs of parallels, while d and e are linear lengths of two angularly equal meridian arcs. The orthogonal spheroidal coordinates and elevations are known for the window points $(-c, e, z_1)$, $(0, e, z_2)$, (c, e, z_3), $(-b, 0, z_4)$, $(0, 0, z_5)$, $(b, 0, z_6)$, $(-a, -d, z_7)$, $(0, -d, z_8)$, and $(a, -d, z_9)$. Let us assume that the curvature of a planet can be ignored within the moving window; that is, the length of the window diagonal is less than 0.1 of the average radius of the planet.

To estimate the first- and second-order partial derivatives r, t, s, p, and q (Eq. 2.1) at the central point $(0, 0, z_5)$, let us fit the polynomial (Eq. 4.1) to the nine points of the 3×3 spheroidal equal angular window by the least-squares approach (Bjerhammar, 1973). Writing the polynomial (Eq. 4.1) for all points of the window, we obtain a system of nine conditional linear equations. Let us write it in the form of Eq. (4.8). In this case, α is a 9×1 matrix of the nine measured values of z:

$$\alpha = \begin{pmatrix} z_1 \\ \vdots \\ z_9 \end{pmatrix}, \qquad (4.22)$$

β is a 6×1 matrix of the six unknown coefficients of the polynomial (Eq. 4.1):

$$\beta = \begin{pmatrix} r \\ t \\ s \\ p \\ q \\ u \end{pmatrix}, \qquad (4.23)$$

and **F** is a 9×6 matrix of the known coefficients of the equation system:

$$\mathbf{F} = \begin{pmatrix} \dfrac{c^2}{2} & \dfrac{e^2}{2} & -ce & -c & e & 1 \\ 0 & \dfrac{e^2}{2} & 0 & 0 & e & 1 \\ \dfrac{c^2}{2} & \dfrac{e^2}{2} & ce & c & e & 1 \\ \dfrac{b^2}{2} & 0 & 0 & -b & 0 & 1 \\ 0 & 0 & 0 & 0 & 0 & 1 \\ \dfrac{b^2}{2} & 0 & 0 & b & 0 & 1 \\ \dfrac{a^2}{2} & \dfrac{d^2}{2} & ad & -a & -d & 1 \\ 0 & \dfrac{d^2}{2} & 0 & 0 & -d & 1 \\ \dfrac{a^2}{2} & \dfrac{d^2}{2} & -ad & a & -d & 1 \end{pmatrix}, \qquad (4.24)$$

To determine the unknown coefficients of the polynomial (Eq. 4.1), we solve Eq. (4.12); see details elsewhere (Florinsky, 1998c). As a result, we obtain the formulas for r, t, s, p, and q:

$$r = \frac{c^2(z_1 + z_3 - 2z_2) + b^2(z_4 + z_6 - 2z_5) + a^2(z_7 + z_9 - 2z_8)}{a^4 + b^4 + c^4}, \qquad (4.25)$$

$$t = \frac{2}{3de(d+e)(a^4 + b^4 + c^4)} \times \{[d(a^4 + b^4 + b^2c^2) - c^2e(a^2 - b^2)](z_1 + z_3)$$
$$- [d(a^4 + c^4 + b^2c^2) + e(a^4 + c^4 + a^2b^2)](z_4 + z_6) + [e(b^4 + c^4 + a^2b^2)$$
$$+ a^2d(b^2 - c^2)](z_7 + z_9) + d[b^4(z_2 - 3z_5) + c^4(3z_2 - z_5)$$
$$+ (a^4 - 2b^2c^2)(z_2 - z_5)] + e[a^4(3z_8 - z_5) + b^4(z_8 - 3z_5)$$
$$+ (c^4 - 2a^2b^2)(z_8 - z_5)] - 2[a^2d(b^2 - c^2)z_8 - c^2e(a^2 - b^2)z_2]\}, \qquad (4.26)$$

$$s = \frac{c[a^2(d+e) + b^2e](z_3 - z_1) - b(a^2d - c^2e)(z_4 - z_6) + a[c^2(d+e) + b^2d](z_7 - z_9)}{2[a^2c^2(d+e)^2 + b^2(a^2d^2 + c^2e^2)]}, \qquad (4.27)$$

$$p = \frac{a^2cd(d+e)(z_3-z_1) + b(a^2d^2+c^2e^2)(z_6-z_4) + ac^2e(d+e)(z_9-z_7)}{2[a^2c^2(d+e)^2 + b^2(a^2d^2+c^2e^2)]}, \quad (4.28)$$

$$q = \frac{1}{3de(d+e)(a^4+b^4+c^4)} \times \{[d^2(a^4+b^4+b^2c^2) + c^2e^2(a^2-b^2)](z_1+z_3)$$

$$- [d^2(a^4+c^4+b^2c^2) - e^2(a^4+c^4+a^2b^2)](z_4+z_6) - [e^2(b^4+c^4+a^2b^2)$$

$$- a^2d^2(b^2-c^2)](z_7+z_9) + d^2[b^4(z_2-3z_5) + c^4(3z_2-z_5)]$$

$$+ (a^4-2b^2c^2)(z_2-z_5)] + e^2[a^4(z_5-3z_8) + b^4(3z_5-z_8)$$

$$+ (c^4-2a^2b^2)(z_5-z_8)] - 2[a^2d^2(b^2-c^2)z_8 + c^2e^2(a^2-b^2)z_2]\}. \quad (4.29)$$

Moving the 3×3 window along a DEM, one can compute values of r, t, s, p, and q (and so values of local morphometric variables) for all points of the spheroidal equal angular DEM, except for boundary rows and columns. If one processes a virtually closed, global spheroidal equal angular DEM, it is possible to estimate values of topographic variables in each point of such a DEM (for examples, see Chapter 15).

We do not present a formula for the residual term u of the polynomial (Eq. 4.1), since it is not used in calculations. Development of Eqs. (4.25)–(4.29) was carried out with the software Maple V Release 5.0 (© Waterloo Maple Inc., 1981–1997).

4.3.3 Calculation of Linear Sizes of a Spheroidal Equal Angular Window

Values of a, b, c, d, and e vary depending on the latitude. Since geographic coordinates are known for every point of a spheroidal equal angular grid, so a, b, c, d, and e are easily calculated by formulas from the solution of the inverse geodetic problem for short distances (Morozov, 1979, pp. 178–179): The distance L (measured in meters) between two points (φ_1, λ_1) and (φ_2, λ_2), where φ is latitude and λ is longitude, can be found by the following equation:

$$L = \sqrt{Q^2 + P^2}, \quad (4.30)$$

where

$$Q = \Theta M_m \left[1 - (E'^2 - 2\eta_m^2)\frac{\Theta^2}{8} - (1+\eta_m^2)\frac{(\Lambda \cos \varphi_m)^2}{12} - \frac{(\Lambda \sin \varphi_m)^2}{8}\right], \quad (4.31)$$

$$P = \Lambda \cos \varphi_m N_m \left[1 + (1 - 9E'^2 + 8\eta_m^2)\frac{\Theta^2}{24} - \frac{(\Lambda \sin \varphi_m)^2}{24}\right], \quad (4.32)$$

where

$$\Theta = \varphi_2 - \varphi_1 \text{ (measured in radian),} \quad (4.33)$$

$$\Lambda = \lambda_2 - \lambda_1 \text{ (measured in radian),} \quad (4.34)$$

$$\varphi_m = \frac{1}{2}(\varphi_1 + \varphi_2), \quad (4.35)$$

$$E' = \frac{\sqrt{A^2 - B^2}}{B}, \quad (4.36)$$

where E' is the second eccentricity, and A and B are semi-major and semi-minor axes of an ellipsoid of revolution, correspondingly,

$$\eta_m^2 = E'^2 \cos^2 \varphi_m, \quad (4.37)$$

$$N_m = \frac{C}{\sqrt{1 + \eta_m^2}}, \quad (4.38)$$

$$C = \frac{A^2}{B}, \quad (4.39)$$

$$M_m = \frac{N_m}{1 + \eta_m^2}, \quad (4.40)$$

In a similar manner, one can determine the sizes of the moving window and weights of its points (Eq. 6.9) for DEM smoothing. Alternative formulas to solve the inverse geodetic problem can be found elsewhere (Sodano, 1965; Vincenty, 1975; Bowring, 1996).

4.3.4 Discussion

In the method described, as in the Evans–Young method (Section 4.1), the polynomial (Eq. 4.1) is approximated to elevation values of the 3×3 window rather than passing exactly through them. This leads to a local suppression of high-frequency noise and may enhance the calculation of partial derivatives sensitive to a high-frequency component of a signal (Florinsky, 2002) (Section 5.4.1). For the method's accuracy in terms of RMSE as a function of measured variables, see Section 5.3.3.2.

Differentiation operations are commonly applied to analyze digital models of various geophysical characteristics based on spheroidal equal angular grids. For example, Ekman (1988) calculated the Gaussian curvature of the "surface" of the isostatic uplift exploring the origin of caves in Fennoscandia. Grachev et al. (2001) studied the gradients and curvatures of a "surface" of velocity of vertical neotectonic movements for Northern Eurasia. The method described offers possibilities for the direct use of such operations on spheroidal equal angular grids, without data interpolation to plane square grids.

The method has been systematically used in medium- and small-scale soil studies (Florinsky et al., 2000; Florinsky and Eilers, 2002) (Section 10.3.3), as well as in geological research (Florinsky, 2007a, 2008a, 2008b, 2008c; Golyandina et al., 2007) (Section 6.3.2 and Chapter 15). These works demonstrated its effectiveness.

4.4 CALCULATION OF NONLOCAL MORPHOMETRIC VARIABLES

Unlike local topographic variables calculated with the approximation of elevation derivatives by finite differences, logical procedures called "flow routing algorithms" are usually applied to estimate catchment and dispersive areas. Such algorithms determine a route in which a flow is distributed from the given point of the topographic surface to downslope points. There are several flow routing algorithms and, thus, methods to derive CA and DA from DEMs based on plane square grids. There are two main groups of such techniques:

1. Eight-node single-flow direction algorithms
2. Multiple-flow direction algorithms

The first group includes methods, which use only one of the eight possible directions separated by $45°$ to model a flow from the given point (O'Callaghan and Mark, 1984; Jenson and Domingue, 1988; Martz and de Jong, 1988). The flow direction is determined by estimating a value of G to each neighbor of the point; it corresponds to the direction for which G is the greatest (Fig. 4.6a). CA at a downslope point is determined as the number of upslope points, passed by flows to reach this point, multiplied by the cell unit area (Fig. 4.6b).

The second group includes methods using the flow partitioning (Freeman, 1991; Quinn et al., 1991). In this case, a flow is directed to every adjacent downslope point of a 3×3 moving window using slope weights (Fig. 4.6c). Thus, CA at a downslope point is composed of partial contributions from different upslope points.

DA can be determined in a similar manner, but a DEM should be inverted before the processing (e.g., by multiplying the DEM by -1). As a result, crests "become" thalwegs, while thalwegs "become" crests.

Before CA or DA calculation, it is reasonable to "fill" closed pits or depressions (Martz and de Jong, 1988; Lindsay and Creed, 2005; Arnold, 2010), which can be both errors of DEM generation and actual elements of the topographic surface. In this connection, Martz and de

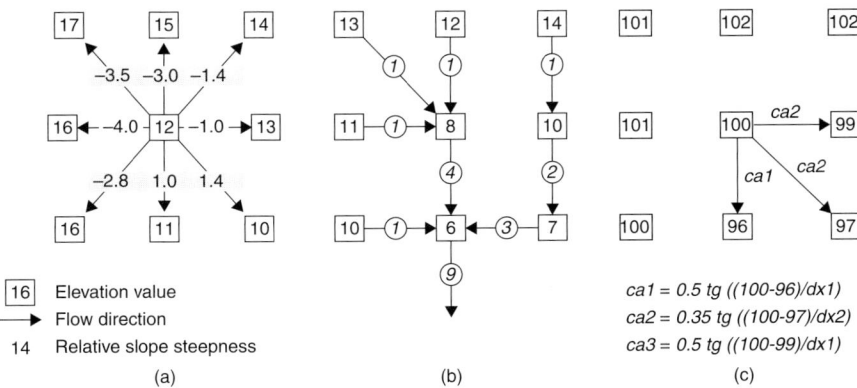

FIGURE 4.6 Illustration for principles of calculation of nonlocal morphometric variables: (a) Gradient estimation at a given point of the DEM by the Martz–de Jong method: a flow line is going from the central point to the right bottom point. (b) Estimation of the catchment area by the Martz–de Jong method: italic is the number of grid points contributing to the flow. *From (Martz and de Jong, 1988, Figs. 1 and 4), reproduced with permission.* (c) Flow partitioning (Quinn et al., 1991): *ca1*, *ca2*, and *ca3* are weights used for the flow partitioning; *dx1* and *dx2* are flow line lengths. *From (Quinn et al., 1995, Fig. 3), reproduced with permission.*

Jong (1988) distinguished two sorts of *CA* (and *DA*): minimal and maximal. Minimal *CA* is computed without pit filling. This terminates all simulated flows in closed depressions. In this case, *CA* values at points located downslope of a close pit do not consider *CA* values at points located upslope of the pit. To calculate maximal *CA*, all closed pits should be filled. In this case, *CA* values at each downslope point consider *CA* values at all connected upslope points.

Detailed description and comparison of flow routing algorithms and methods to estimate nonlocal morphometric attributes can be found elsewhere (Band, 1993; Desmet and Govers, 1996; Rieger, 1998; Wilson et al., 2008). Figure 2.6 illustrates derivation of *CA* and *DA* by the Martz–de Jong method.

Notice that all methods mentioned are intended for calculation of nonlocal topographic variables from a DEM based on a plane square grid, which has the same cell unit area (Π) in every point of the model. In the case of spheroidal equal angular grids, Π depends on the latitude. It can be estimated by the following expression (Morozov, 1979, p. 34):

$$\Pi = B^2(\lambda_2 - \lambda_1) \left| \sin \varphi + \frac{2}{3}E^2 \sin^3 \varphi + \frac{3}{5}E^4 \sin^5 \varphi + \frac{4}{7}E^6 \sin^7 \varphi + \ldots \right|_{\varphi_1}^{\varphi_2}, \quad (4.41)$$

where E is the first eccentricity:

$$E = \frac{\sqrt{A^2 - B^2}}{A}, \tag{4.42}$$

4.5 CALCULATION OF STRUCTURAL LINES

4.5.1 Conventional Algorithms

Digital terrain modeling is closely allied to image processing. Thus, delineation of ridges and thalwegs is one of the topical tasks of both digital terrain modeling and image processing. In digital terrain modeling, researchers focus on the detection of thalwegs. The delineation of ridges receives much consideration in image processing. There is a principal difference between the topographic surface, which properties are in many respects controlled by gravity, and the image intensity function. However, from the technical point of view, the algorithms of digital terrain modeling may be applied in image processing, and vice versa, algorithms of image processing are used in digital terrain analysis.

The twofold nature of structural lines (Section 2.4) is responsible for the existence of two fundamentally different groups of techniques to delineate ridges and thalwegs. Methods of the first group are based on principles of differential geometry. Originating in classical geometry (de Saint-Venant, 1852; Boussinesq, 1872; Jordan, 1872; Breton de Champ, 1877), they are predominantly used in image processing (Haralick, 1983; Koenderink and van Doorn, 1993; Eberly et al., 1994; Rieger, 1997; López et al., 1998). In these approaches, differential geometric criteria are applied to detect loci of extreme values of mean, principal, and plan curvatures. A nontrivial method by López and Serrat (1996), implementing a differential equation of Rothe (1915), may also be placed in this group.

The methods of the second group are based on logical processing of data, such as cell-to-cell flow routing and CA calculation (Section 4.4) with subsequent thresholding of its values. These approaches are commonly used in digital terrain modeling (Mark, 1984; O'Callaghan and Mark, 1984; Band, 1986; Douglas, 1986; Skidmore, 1990; Tribe, 1992; Jones, 2002).

In scale terms, methods of the first group are local since they concern a vicinity of a point within a moving window. Methods of the second group are nonlocal because determining the flow routing requires an analysis of rather large portions of the topographic surface.

The third group of algorithms combines local differential geometric and nonlocal logical approaches. In this case, differential geometrical criteria are used to detect saddle points or points of extreme contour curvature, and then logical procedures are applied to link the points for visualization of ridges and thalwegs (Gauch and Pizer, 1993; Kweon and Kanade, 1994; Steger, 1999). Detailed comparisons of the methods can be found elsewhere (Skidmore, 1990; Tribe, 1992; Gauch and Pizer, 1993; López et al., 1999).

Figure 2.7 illustrates the revealing of crests and thalwegs using a method belonging to the second group. The map was obtained with the software CatchmentSIM 1.29 (Ryan and Boyd, 2003). Flats and pits were removed from the DEM by the breaching algorithm (Jones, 2002). In flow routing, a downslope flow angle was determined by a modified multiple direction algorithm (Lea, 1992). Thalwegs were defined as pixels with CA values greater than a threshold (Jones, 2002); we used the threshold of 25 pixels. Treating an inverted DEM, the ridge network was delineated in the same way.

4.5.2 Generating Function

One method of the first group is based on the derivation of the generating function of crest and thalwegs lines (Shary and Stepanov, 1991). It is well-known that k_h is the measure of convergence and divergence of slope lines (Section A.3.2). Negative values of k_h indicate convergence areas of slope lines, while positive values of k_h correspond to their divergence areas. At the same time, k_h is the product of the curvature of contour line and slope factor (Eq. A.57) (Shary, 1991; Shary et al., 2002b). This means that k_h contains information on the behavior of both contours and slope lines. Thus, to find loci of extreme curvature of both families of spatial curves, one needs to find loci of extreme values of k_h. Clearly, they correspond to zero values of the k_h derivative, the so-called generating function (T) (Shary and Stepanov, 1991; Florinsky, 2009b):

$$T = \frac{q^3 g - 3pq^2 k + 3p^2 qm - p^3 h}{\sqrt{(p^2 + q^2)^3 (1 + p^2 + q^2)}}$$

$$+ \frac{[2 + 3(p^2 + q^2)](q^2 r - 2pqs + p^2 t)}{\sqrt{(p^2 + q^2)^5 (1 + p^2 + q^2)^3}} [(p^2 - q^2)s - pq(r - t)], \quad (4.43)$$

where g, h, k, m, r, t, s, p, and q are the third-, second-, and first-order partial derivatives of elevation (Eq. 2.1). For the development of the formula, see Section A.4.

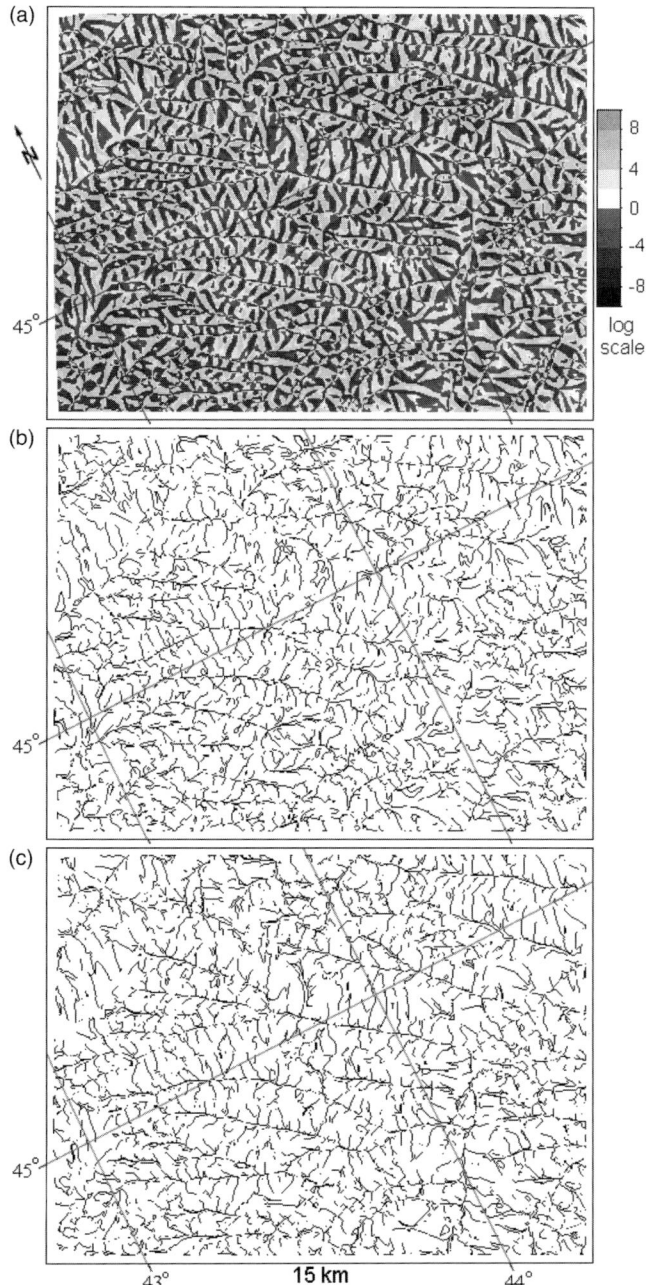

FIGURE 4.7 The Stavropol Upland, the generating function: (a) Spatial distribution of T values. (b) Zero values of T within divergence areas—the locus of the positive extreme curvature of the topographic surface. (c) Zero values of T within convergence areas—the locus of the negative extreme curvature of the topographic surface. For details of calculation, see Section 4.2.3.1; for the elevation map, see Fig. 2.3. *From (Florinsky, 2009b, Fig. 4).*

T can be interpreted as a measure for the deflection of k_h from loci of the extreme curvature of the topographic surface. The unit of T is m^{-2}. Shary and Stepanov (1991) proposed that ridges and convex break lines relate to the locus of $T = 0$ within divergence areas, while thalwegs and concave break lines correspond to the locus of $T = 0$ within convergence areas.

Figure 4.7 illustrates the revealing of crest and thalweg lines using the generating function. The map of T (Fig. 4.7a) includes all loci of the extreme curvature of the topographic surface. They appear as borders between areas of positive and negative values of T (light and dark patterns, correspondingly). The map of zero values of the generating function located within flow divergence areas ($T = 0$ with $k_h > 0$) displays the locus of the positive extreme curvature (Fig. 4.7b). The map of zero values of the generating function located within flow convergence areas ($T = 0$ with $k_h < 0$) displays the locus of the negative extreme curvature (Fig. 4.7c).

According to Shary and Stepanov (1991), the positive extreme curvature map (Fig. 4.7b) should display ridges and convex break lines, while the negative extreme curvature map (Fig. 4.7c) should represent thalwegs and concave break lines. However, even a cursory examination of the maps showed that they fail to detect these structural lines correctly. This inference is supported by comparison with the maps of ridge/thalweg networks delineated using nonlocal logical approaches (Figs. 2.7 and 4.7). The main problem is that loci of the extreme curvature are interrupted in many places (Figs. 4.7b, c). This can be caused by interpolation and smoothing errors, the remains of the high-frequency noise of the DEM, and loci of special points, namely, local maxima and minima, saddles, and minor flat horizontal areas.

Therefore, in terms of ridge/thalweg detection, T mapping has the following disadvantages. First, loci of the extreme curvature may only partially define thalweg/ridge lines. Second, the calculation of partial derivatives is sensitive to errors and noise existing copiously in DEMs (Section 5.4.1) (Florinsky, 2002). Third, T mapping cannot reveal thalweg/ridge networks as hierarchical structures of tributary orders.

CHAPTER 5

Errors and Accuracy

OUTLINE

5.1	Sources of DEM Errors	66
5.2	Estimation of DEM Accuracy	70
5.3	Calculation Accuracy of Local Morphometric Variables	71
	5.3.1 Motivation	71
	5.3.2 RMSE Formulas for Local Morphometric Variables	73
	5.3.3 RMSE Formulas for Partial Derivatives	75
	5.3.4 RMSE Mapping	78
5.4	Ignoring of the Sampling Theorem	81
	5.4.1 Motivation	81
	5.4.2 Materials and Data Processing	82
	5.4.3 Results and Discussion	86
5.5	The Gibbs Phenomenon	88
	5.5.1 Motivation	88
	5.5.2 Materials and Data Processing	90
	5.5.3 Results and Discussion	90
5.6	Grid Displacement	93
	5.6.1 Motivation	93
	5.6.2 Materials and Data Processing	94
	5.6.3 Results and Discussion	96
5.7	Linear Artifacts	98
	5.7.1 Motivation	98
	5.7.2 Isotropy of Local Morphometric Variables	100

5.1 SOURCES OF DEM ERRORS

DEM accuracy depends on the type of topography, method for DEM generation, type of DEM grid, and DEM resolution (Isaacson and Ripple, 1990; Li, 1994; Hunter and Goodchild, 1995; Aguilar et al., 2005). Rieger (1996) proposed the following classification of DEM errors:

- Low-frequency systematic errors, resulting, for example, from procedures of the photogrammetric orientation. These errors are critical for combined processing and analysis of DTMs and remotely sensed data (Florinsky, 1998b).
- Medium-frequency systematic and random errors caused, for instance, by errors of photogrammetric measurements and interpolation procedures. These errors are critical for DTM-based soil and geological studies.
- High-frequency random errors (noise), which may be, for example, connected with image defects and sensor instability.

Accuracy of DEMs, produced by conventional topographic surveys, depends essentially on systematic instrumental errors and random operator errors (Ghilani and Wolf, 2008). Accuracy of DEMs created using kinematic GPS survey is, in particular, influenced by ionospheric and tropospheric refraction, satellite ephemeris errors, satellite and receiver clock biases, and so on (Ghilani and Wolf, 2008, ch. 15).

Accuracy of DEMs generated by photogrammetric techniques is impacted by the following factors (Akovetsky, 1994; Hunter and Goodchild, 1995; Wolf and Dewitt, 2000):

- Systematic and random operator errors (Fig. 5.1)
- Systematic and random instrumental errors

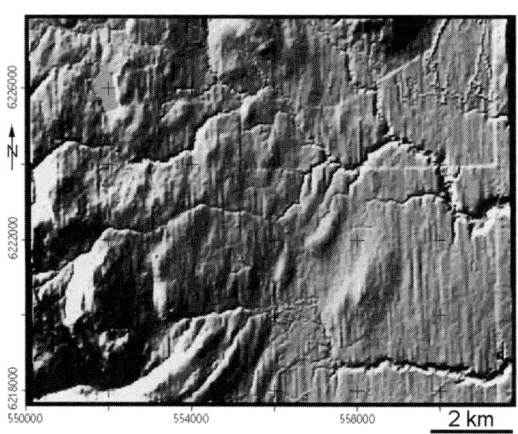

FIGURE 5.1 North–south oriented striping artifacts on the hill-shading map—the result of the DEM generation using a semiautomated profiling of aerial stereo images. Terrain Resource Information Management DEM for the Fort St. John Forest District, British Columbia, Canada; DEM grid spacing is 30 m. *From (Albani and Klinkenberg, 2003, Fig. 1), reproduced with permission.*

- Spatial resolution of aerial or satellite images depending mainly on the camera or sensor characteristics
- Vertical resolution of images dictated essentially by the base/height ratio
- Snow, vegetation, and cloud covers
- Type of topography
- Imagery accuracy depending on the curvature of the planet, atmospheric refraction, airplane or satellite stability, camera distortion or sensor stability, image processing, errors of image transformation for removing geometric distortion, and the like

Notice that in digital photogrammetry, errors of elevation measurements can account for 5 to 20% of the total measurements, depending on the image quality (Akovetsky, 1994). Aguilar et al. (2005) demonstrated that in digital photogrammetry, DEM accuracy depends essentially on three factors: type of topography, DEM grid density, and interpolation technique. Fully automated digital photogrammetric processing of stereo images (without further editing) produces, as a rule, less accurate DEMs than analogue photogrammetric techniques (Gong et al., 2000).

Accuracy of LiDAR-derived DEMs depends on sensor characteristics, vertical and horizontal error in the positioning of the laser platform, laser scan angle, flight speed and altitude, as well as physical properties of the surface, such as surface reflectivity and slope gradient (Huising and Gomes-Pereira, 1998; Wehr and Lohr, 1999; Brenner et al., 2007) (Fig. 5.2). In LiDAR survey, systematic errors range from 5 cm for flat, paved areas to 200 cm for areas with grass and shrub cover, while random errors range from 10 cm for flat areas to 200 cm for hilly areas (Huising and Gomes-Pereira, 1998). Errors also occur during the processing of the raw data, that is, data filtering, gridding, segmentation, and object reconstruction (Aguilar and Mills, 2008).

Accuracy of InSAR-derived DEMs depends on radar characteristics, type of topography, and physical properties of the surface (Jarvis et al., 2004; Falorni et al., 2005; Brenner et al., 2007). In particular, more than 15% of SRTM DEM tiles have voids for 1% of the covered area, and more than 5% of those have voids for 5% of the covered area. Large voids are typical for mountainous terrains: geometric artifacts are here caused by layover, foreshortening, and radar shadow. Large voids are also observed in sand deserts: a low dielectric constant of the surface is responsible for a low level of the signal reflectance (Elsner and Bonnici, 2007). SRTM DEMs also suffer from striping artifacts: "the stripes have a typical wavelength of about 800 m with an amplitude of about 0.2–4 m and are aligned diagonally in a pattern that suggests a close relationship to the orbital paths. . . . Striping at varying levels occurred

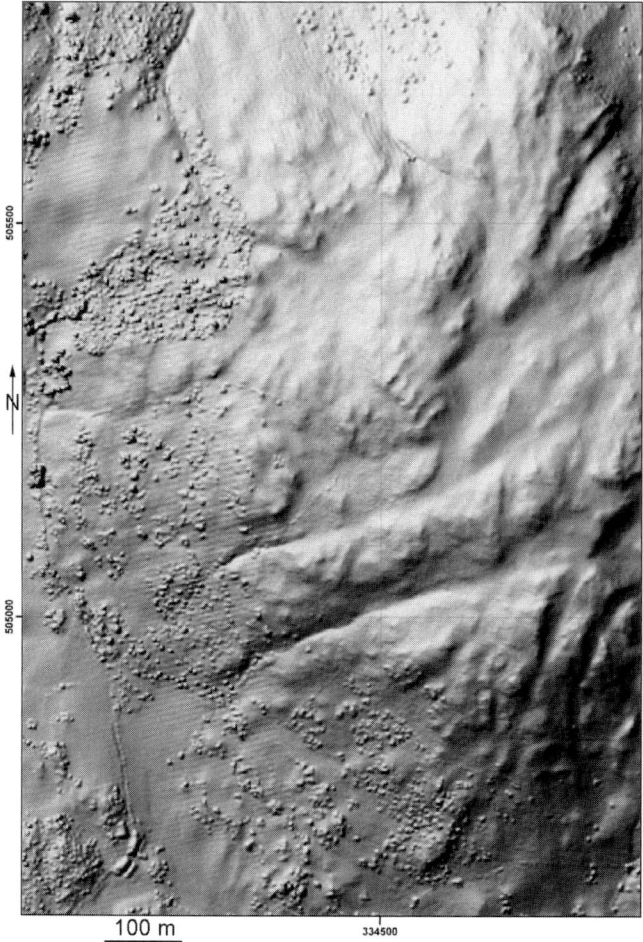

FIGURE 5.2 Hill-shading map of the LiDAR DEM for Loughrigg Fell, Lake District, UK. The artificial diagonal banding in two dominant directions ($\sim 75°$ and $143°$) is associated with aircraft pitch motion. Illumination from the northwest; DEM grid spacing is 2 m. LiDAR data © Environment Agency Science Enterprise Centre. *From (Arrell et al., 2008, Fig. 4); reproduced with permission.*

everywhere except where it was obscured by high topographic relief" (Gallant and Read, 2009, pp. 149 and 151).

It is necessary to stress that use of fully automated approaches of digital photogrammetry, LiDAR, and InSAR techniques leads, in fact, to "blind sampling" of elevations (Lemmens, 1999). It is clear that such DEMs need editing and filtering (Jarvis et al., 2004; Sithole and Vosselman, 2004; Falorni et al., 2005; Evans and Hudak, 2007; Karkee et al., 2008; Gallant and Read, 2009) because they describe not the topographic surface but a surface consisting of portions of the land surface, buildings, vegetation cover, and the like.

Errors of bathymetric DEMs produced by echo sounding may be caused by instrumental errors of an echo sounder and navigation errors. Errors are common for flat seafloor areas and steep deep-water slopes (Smith, 1993; Bourillet et al., 1996). Errors of bathymetric DEMs created with optical sensing depend on instrumental errors of sensors, the correctness of regression models for relationships between image brightness and water depth, water turbidity, and bottom reflectance relating to bottom characteristics and water depth (Gao, 2009). Accuracy of LiDAR-based bathymetric DEMs is influenced by water clarity and bottom features (Pe'eri et al., 2011).

Accuracy of bathymetric DEMs produced with satellite radar altimetry data depends on the platform and radar characteristics. Also, error sources include water surface roughness (waves), tide model errors, ocean currents, and signal delay in the ionosphere and troposphere (Sandwell and Smith, 2001). However, most errors are associated with limitations of physical mathematical models linking gravitational anomalies with seafloor topography, which are applied to convert altimetric data into seafloor DEMs. One may obtain adequate results for wavelengths of 20–200 km within areas with a sedimentary cover up to 200 m thick.

Accuracy of DEMs of soil horizon surfaces or stratigraphic surfaces, derived from augerhole or borehole data, is mostly affected by the absence of information on the shape of subsurface horizons between auger- or boreholes (Jones and Johnson, 1983; Groshong, 2006, ch. 10). There are no reasons to assume that interpolation of subsurface DEMs may provide an adequate picture of stratigraphic surfaces.

Accuracy of DEMs of stratigraphic surfaces produced with 3D seismic survey is, in particular, influenced by limitations in the survey design, coherent noise, and systematic errors in the processing (Brown, 2004; Liner, 2004). Groshong (2006, pp. 30–31) made the following observations:

> The geometry of a structure that is even moderately complex displayed in travel time is likely to be significantly different from the true geometry of the reflecting boundaries because of the distortions introduced by steep dips and laterally and vertically varying velocities. Reflections from steeply dipping units may return to the surface beyond the outer limit of the recording array and so are not represented on the seismic profile. The structural interpretation of seismic reflection data requires the conversion of the travel times to depth. This requires an accurate model for the velocity distribution, something not necessarily well known for a complex structure. The most accurate depth conversion is controlled by velocities measured in nearby wells.

Accuracy of contour-based DEMs depends on the following factors (Carter, 1988; Kumler, 1994):

- Accuracy of topographic maps
- Random operator errors (up to 90% of the total number of DEM errors)
- Systematic instrumental errors

If a DEM has been created using any interpolation technique, then the DEM obviously includes systematic and random errors of the interpolator (Carrara et al., 1997; Bater and Coops, 2009).

Two problems are typical for a DEM compiled from a set of other DEMs: (1) different portions of the DEM may have different accuracies; and (2) boundaries between the portions will be clearly seen on all morphometric maps. These problems are familiar for users of all existing global DEMs, such as GTOPO30 (USGS, 1996), GLOBE (GLOBE Task Team and others, 1999), ETOPO5 (World Data Center for Geophysics, 1993), ETOPO2 (National Geophysical Data Center, 2001, 2006), and ETOPO1 (Amante and Eakins, 2009). An example of such errors is discussed in Section 6.3.3.

5.2 ESTIMATION OF DEM ACCURACY

DEM accuracy is commonly estimated by the criterion of RMSE of elevation computed by comparing DEM points and reference points (Ivanov and Kruzhkov, 1992; Bolstad and Stowe, 1994). However, a number of reference points are limited. This can lead to improper estimation of the RMSE. This approach was elaborated by Rieger (1996), who proposed comparing a target DEM with a "reference" DEM. However, it is unreasonable to consider the "reference" DEM as the correct model.

Elevation RMSE is usually presented as the integral metrics for the accuracy of the entire DEM. However, different portions of a DEM may have distinct accuracy (Hunter and Goodchild, 1995; Carlisle, 2005; Fisher and Tate, 2006). Various approaches can be used to analyze the spatial distribution of DEM errors. In particular, Makarovič (1972) proposed estimating DEM accuracy by a mean deviation of sinusoidal amplitudes obtained after interpolation from sinusoidal amplitudes existing before interpolation. Tempfli (1980) evaluated DEM accuracy by spectral analysis. Frederiksen (1981) developed a method to predict DEM accuracy based on summing the Fourier spectrum for high-frequency components of topographic profiles. Polidori et al. (1991) attempted to estimate DEM interpolation errors by calculating the DEM fractal dimension at different scales and in different directions.

Ackermann (1978) and Li (1994) proposed formulas for elevation RMSE as a function of slope gradient. However, the accuracy of the gradient calculation also depends on elevation RMSE (Felicísimo, 1995; Florinsky, 1998a) (Section 5.3). Hannah (1981) developed a technique to detect DEM random errors. This method was based on a point comparison with adjacent points, using slope gradient and change-gradient values with specified thresholds.

Felicísimo (1994b) developed an elegant method utilizing differences between two elevation values for each DEM point: the first is recorded in the DEM, and the second is interpolated from its four neighboring points. An arithmetic mean, standard deviation, and the Student's t-criterion are then calculated. Relatively high t-values indicate possible errors. Kraus (1994) and Hunter and Goodchild (1995) described two approaches for mapping the horizontal accuracy of elevation data. In the first approach, "the epsilon band" is drawn around the position of a contour line defining the elevation z_i. The borders of the epsilon band are two contours defining elevations $z_i - \Delta z/2$ and $z_i + \Delta z/2$, where Δz is a contour interval. The epsilon band approach means that the true position of the z_i contour is somewhere between the band borders. In the second approach, the researcher should calculate and map the likelihood of the true value of elevation exceeding by a threshold. López (1997) described a method that was based on the decomposition of a regular DEM into strips and that considered it as a multivariate table. Principal components analysis was applied to select unlikely elevations.

A detailed review of methods to analyze the spatial distribution of DEM errors can be found elsewhere (Carlisle, 2005). For suppression of high-frequency noise (random errors) in DEMs, see Chapter 6.

5.3 CALCULATION ACCURACY OF LOCAL MORPHOMETRIC VARIABLES

5.3.1 Motivation

Since the inception of digital terrain modeling, considerable attention has been given to various issues of DTM accuracy (Fisher and Tate, 2006). This is connected with the fact that errors of digital models of morphometric attributes influence the accuracy and objectivity of DTM-based studies and the modeling of natural processes and phenomena (Desmet, 1997; Holmes et al., 2000; Vörösmarty et al., 2000; Thompson et al., 2001; Van Niel et al., 2004; Bishop et al., 2006; Callow et al., 2007; Erskine et al., 2007; Wise, 2007).

Usually, calculation accuracy of local topographic variables is estimated by comparing their computed and "reference" values. For reference data, researchers used hand measurements of G and A from topographic maps (Evans, 1980; Skidmore, 1989), field measurements of G, A, and k_v (Bolstad and Stowe, 1994; Giles and Franklin, 1996), values of morphometric variables derived from "reference" DEMs of actual terrains (Chang and Tsai, 1991; Sasowsky et al., 1992) and simulated terrains (Carter, 1992; Felicísimo, 1995; Hodgson, 1995; Zhou and Liu, 2004).

However, the accuracy of digital models of local topographic attributes cannot be determined by the comparison of computed and "reference" values. Indeed, it is well known that a measurement accuracy can be defined as a difference between a measured value and an actual value of a variable. However, the actual land surface is not mathematically smooth. So, it cannot have derivatives and, hence, local morphometric variables, which are functions of partial derivatives of elevation (Section 2.2). For the land surface, notions of derivatives arise only during its measurements (Shary, 1991). Therefore, because there are no actual values of local topographic variables, the accuracy of their calculation cannot be determined by a comparison of computed and "reference" values.

Moreover, such a strategy can lead to artifacts, as well as subjective and conflicting conclusions. For instance, it was reported that errors of A calculation are typical for flat areas (Chang and Tsai, 1991; Carter, 1992), while errors of G calculation are predominantly positioned on steep slopes (Chang and Tsai, 1991; Sasowsky et al., 1992; Bolstad and Stowe, 1994). However, Carter (1992) found that both G and A errors grow to high values within flat areas. At the same time, Davis and Dozier (1990) reported that G and A errors concentrate within zones of rapid change in G and A values (e.g., nearby crests or thalwegs).

Also, it is well known that the map accuracy of G, A, k_h, k_v, and other curvatures depends on DEM grid spacing (Evans, 1979, 1980; Stewart and Podolski, 1998; Shary et al., 2002b; Bergbauer and Pollard, 2003; Zhu et al., 2008). For instance, the increase of a DEM grid spacing can transform small steep zones to broad areas marked by medium G values. It was reported that G and A calculation errors increase as a grid spacing increases (Chang and Tsai, 1991). However, Carter (1992) found that G and A computed values more closely correspond to their "reference" values as grid spacing increases.

Therefore, the calculation accuracy of local morphometric variables cannot be adequately studied by a comparison of calculated and "reference" values. At the same time, it is obvious that it principally depends on the accuracy of initial data, that is, DEMs, and the accuracy of a calculation algorithm. So, attention should be focused on these two factors. Thus, Felicísimo (1995) found that G errors increase with increasing elevation RMSE (m_z). Brown and Bara (1994) and Giles and Franklin (1996) intimated that the number of errors in calculations of partial derivatives of elevation depends on the DEM noise level. Skidmore (1989) and Hodgson (1995) compared the accuracy of several methods for G and A derivation, that is, calculation of p and q (Section 2.2). Skidmore (1989) found that algorithms for p and q computation based on six points of a 3×3 moving window are more accurate than the four-point algorithms. At the same time, Hodgson (1995) argued that the four-point algorithms are more accurate than the six-point ones. This contradiction may be explained by the fact that these studies (Skidmore, 1989; Hodgson, 1995) were also carried out with a comparison of calculated and "reference" values of topographic variables.

It is clear that local morphometric variables are functions of measured variables $F = \varphi(x, y, \ldots, u)$, where $x, y, \ldots, u$ are measured variables. To estimate the accuracy of calculation of local morphometric attributes, Kuryakova (1996) proposed using the criterion of RMSE of a function of measured variables m_F (Merriman, 1899, p. 32; Bolshakov and Gaidaev, 1977, p. 117):

$$m_F = \sqrt{\left(\frac{\partial F}{\partial x}\right)^2 m_x^2 + \left(\frac{\partial F}{\partial y}\right)^2 m_y^2 + \cdots + \left(\frac{\partial F}{\partial u}\right)^2 m_u^2} \qquad (5.1)$$

where $m_x, m_y, \ldots, m_u$ are RMSE of $x, y, \ldots, u$. In this section, we describe the method for estimating the accuracy of calculation of local morphometric variables by this criterion (Florinsky, 1998a, 1998c, 2008d, 2009b).

5.3.2 RMSE Formulas for Local Morphometric Variables

Using Eq. (5.1), let us develop formulas for RMSE of calculation of local morphometric attributes (Eqs. 2.2, 2.4–2.15). Functions of measured variables are G, k_h, k_v, K, k_{min}, k_{max}, H, E, K_a, M, K_r, k_{ve}, and k_{he}; measured variables are p, q, r, s, and t. After differentiation and simple algebraic operations, we obtain the following expressions (Florinsky, 1998a, 2008d):

$$m_G = \sqrt{\left(\frac{\partial G}{\partial p}\right)_0^2 m_p^2 + \left(\frac{\partial G}{\partial q}\right)_0^2 m_q^2} = \cdots = \frac{1}{(1+p^2+q^2)}\sqrt{\frac{p^2 m_p^2 + q^2 m_q^2}{p^2+q^2}}, \qquad (5.2)$$

$$m_{k_h} = \sqrt{\left(\frac{\partial k_h}{\partial r}\right)_0^2 m_r^2 + \left(\frac{\partial k_h}{\partial t}\right)_0^2 m_t^2 + \left(\frac{\partial k_h}{\partial s}\right)_0^2 m_s^2 + \left(\frac{\partial k_h}{\partial p}\right)_0^2 m_p^2 + \left(\frac{\partial k_h}{\partial q}\right)_0^2 m_q^2}$$

$$= \frac{1}{p^2+q^2}\left(\frac{1}{1+p^2+q^2}\left\{m_p^2\left[p(q^2 r - 2pqs + p^2 t)\right]\left(\frac{2}{p^2+q^2} + \frac{1}{1+p^2+q^2}\right)\right.\right.$$

$$\left. + 2(qs - pt)\right]^2 + m_q^2 \left[q(q^2r - 2pqs + p^2t)\left(\frac{2}{p^2+q^2} + \frac{1}{1+p^2+q^2}\right)\right.$$

$$\left.\left. + 2(ps - qr)\right]^2 + m_r^2 q^4 + 4m_s^2 p^2 q^2 + m_t^2 p^4 \right\}\right)^{1/2}, \tag{5.3}$$

$$m_{k_v} = \sqrt{\left(\frac{\partial k_v}{\partial r}\right)_0^2 m_r^2 + \left(\frac{\partial k_v}{\partial t}\right)_0^2 m_t^2 + \left(\frac{\partial k_v}{\partial s}\right)_0^2 m_s^2 + \left(\frac{\partial k_v}{\partial p}\right)_0^2 m_p^2 + \left(\frac{\partial k_v}{\partial q}\right)_0^2 m_q^2}$$

$$= \frac{1}{(p^2+q^2)(1+p^2+q^2)}$$

$$\times \left(\frac{1}{1+p^2+q^2}\left\{m_p^2\left[p(p^2r + 2pqs + q^2t)\left(\frac{2}{p^2+q^2} + \frac{3}{1+p^2+q^2}\right)\right.\right.$$

$$\left. -2(pr+qs)\right]^2 + m_q^2\left[q(p^2r + 2pqs + q^2t)\left(\frac{2}{p^2+q^2} + \frac{3}{1+p^2+q^2}\right)\right.$$

$$\left.\left.\left. -2(ps+qt)\right]^2 + m_r^2 p^4 + 4m_s^2 p^2 q^2 + m_t^2 q^4\right\}\right)^{1/2}, \tag{5.4}$$

where m_p, m_q, m_r, m_s, and m_t are RMSE of calculation of p, q, r, s, and t, respectively; m_G, m_{k_h}, and m_{k_v} are RMSE of calculation of G, k_h, and k_v, respectively. Similarly, we obtain equations for other local topographic attributes:

$$m_K = \frac{\sqrt{16(p^2 m_p^2 + q^2 m_q^2)(rt - s^2)^2 + (r^2 m_t^2 + 4s^2 m_s^2 + t^2 m_r^2)(1 + p^2 + q^2)^2}}{(1 + p^2 + q^2)^3}, \tag{5.5}$$

$$m_{k_{max}} = m_{k_{min}} = \sqrt{m_H^2 + m_M^2}, \tag{5.6}$$

$$m_H = m_E = \frac{1}{2}\sqrt{m_{k_h}^2 + m_{k_v}^2}, \tag{5.7}$$

$$m_{K_a} = \sqrt{k_v^2 m_{k_h}^2 + k_h^2 m_{k_v}^2}, \tag{5.8}$$

$$m_M = \frac{1}{2}\sqrt{\frac{4H^2 m_H^2 + m_K^2}{H^2 - K}}, \tag{5.9}$$

$$m_{K_r} = \sqrt{k_{ve}^2 m_{k_{he}}^2 + k_{he}^2 m_{k_{ve}}^2}, \tag{5.10}$$

$$m_{k_{he}} = \sqrt{m_{k_h}^2 + m_{k_{min}}^2}, \tag{5.11}$$

$$m_{k_{ve}} = \sqrt{m_{k_v}^2 + m_{k_{min}}^2}, \tag{5.12}$$

where m_K, m_{kmin}, m_{kmax}, m_H, m_E, m_{Ka}, m_M, m_{Kr}, m_{kve}, and m_{khe} are RMSE of calculation of K, k_{min}, k_{max}, H, E, Ka, M, Kr, k_{ve}, and k_{he}, respectively.

The RMSE formula for the generating function can be found elsewhere (Florinsky, 2009b).

5.3.3 RMSE Formulas for Partial Derivatives

According to Eqs. (5.2–5.12), one should calculate p, q, r, s, and t, as well as m_p, m_q, m_r, m_s, and m_t to derive a digital model of m_F. Functions of measured variables are p, q, r, s, and t; measured variables are elevation values at nodes of a moving window z_i, where $i = 1, 2 \ldots 9$ for the Evans–Young method (Section 4.1) and the author's method for spheroidal equal angular grids (Section 4.3); $i = 1, 2 \ldots 25$ for the author's method for plane square grids (Section 4.2). So, formulas for m_p, m_q, m_r, m_s, and m_t can also be determined by the general expression for RMSE of a function of measured variables (Eq. 5.1).

5.3.3.1 Calculation on a Plane Square Grid

Using Eq. (5.1), let us develop formulas for RMSE of calculation of p, q, r, s, and t for the Evans–Young method (Eqs. 4.2–4.6), and g, h, k, m, r, t, s, p, and q for the author's method (Eqs. 4.13–4.21).

For the Evans–Young method, we obtain, in particular:

$$m_r = \sqrt{\left(\frac{\partial r}{\partial z_1}\right)_0^2 m_{z_1}^2 + \left(\frac{\partial r}{\partial z_2}\right)_0^2 m_{z_2}^2 + \cdots + \left(\frac{\partial r}{\partial z_9}\right)_0^2 m_{z_9}^2}, \quad (5.13)$$

where $m_{z_1}, m_{z_2}, \cdots, m_{z_9}$ are RMSE of $z_1, z_2, \ldots, z_9$, correspondingly. In the strict sense, $m_{z_i} = \psi(x, y)$ depends on the type of topography as well as methods of DEM generation and interpolation (Hunter and Goodchild, 1995). There are expressions to estimate m_z as a function of G (Li, 1994). However, this approach is incorrect because the accuracy of G calculation also depends on m_z (Eq. 5.2) (Florinsky, 1998a). At the same time, Li (1994) demonstrated that for contour-based DEMs, one can assume:

$$m_{z_i} = const = B\Delta z, \quad (5.14)$$

where Δz is the contour interval, $B = (0.16–0.33)$ depending on the type of topography. So, let us consider $m_{z_1} = m_{z_2} = \cdots = m_{z_9} = m_z$. Substituting m_z into Eq. (5.13), we obtain:

$$m_r = m_z \sqrt{\left(\frac{\partial r}{\partial z_1}\right)_0^2 + \left(\frac{\partial r}{\partial z_2}\right)_0^2 + \left(\frac{\partial r}{\partial z_3}\right)_0^2 + \cdots + \left(\frac{\partial r}{\partial z_9}\right)_0^2}. \quad (5.15)$$

TABLE 5.1 RMSE of Partial Derivatives Calculated by the Evans–Young Method (Section 4.1) and the Author's Method (Section 4.2)

RMSE	The Evans–Young Method	The Author's Method
m_p and m_q	$\frac{m_z}{\sqrt{6}w}$	$\sqrt{\frac{527}{70}}\frac{m_z}{6w}$
m_r and m_t	$\frac{\sqrt{2}m_z}{w^2}$	$\sqrt{\frac{2}{35}}\frac{m_z}{w^2}$
m_s	$\frac{m_z}{2w^2}$	$\frac{m_z}{10w^2}$
m_g and m_h	—	$\frac{m_z}{\sqrt{2}w^3}$
m_k and m_m	—	$\frac{m_z}{\sqrt{35}w^3}$

From (Florinsky, 2009b, Table 1).

After differentiation and simple algebraic operations, we obtain expressions of m_r, m_t, m_s, m_p, and m_q for the Evans–Young method (Table 5.1). Similarly for the author's method, we obtain formulas of m_r, m_t, m_s, m_p, and m_q, as well as m_g, m_h, m_k, and m_m (viz., RMSE of calculation of g, h, k, and m, correspondingly) (Table 5.1).

The formulas (Table 5.1) show that RMSEs of partial derivatives of elevation are in direct proportion to m_z and in inverse proportions to w (for m_p and m_q), w^2 (for m_r, m_t, and m_s), and w^3 (for m_g, m_h, m_k, and m_m). Thus, the third-order partial derivatives are the most sensitive, while the first-order ones are the least sensitive to the grid spacing value.

Earlier, we proved that the Evans–Young method has the highest accuracy among its analogues, which are based on the approximation of partial derivatives by finite differences on the 3×3 moving window (Florinsky, 1998a). Let us compare the formulas of m_r, m_t, m_s, m_p, and m_q related to the Evans–Young and the author's methods.

The formulas of m_r, m_t, m_s, m_p, and m_q (Table 5.1) show that under the same m_z and w values, application of the author's method results in significantly lower values of m_r, m_t, and m_s, than those related to the Evans–Young method. Indeed, m_r and m_t for the author's method are almost six times less than those for the Evans–Young method. m_s for the author's method is five times less than that for the Evans–Young method. m_p and m_q for the author's method are a mere 10% higher than those for the Evans–Young method. This means that the author's method can provide a higher accuracy of derivation of topographic variables than the Evans–Young method.

However, G is calculated using two first-order partial derivatives, p and q. Table 5.1 shows that m_p and m_q for Eqs. (4.20 and 4.21) are 10% higher than those for Eqs. (4.5 and 4.6). Therefore, for G calculation, the Evans–Young method is more accurate than the author's method.

5.3.3.2 Calculation on a Spheroidal Equal Angular Grid

Similarly, we obtain formulas of m_p, m_q, m_r, m_s, and m_t for calculation of partial derivatives of elevation on spheroidal equal angular grids (Eqs. 4.25–4.29) (Florinsky, 1998c, 2008d):

$$m_p = m_z \sqrt{\frac{a^2 d^2 + c^2 e^2}{2[a^2 c^2 (d+e)^2 + b^2(a^2 d^2 + c^2 e^2)]}}, \tag{5.16}$$

$$\begin{aligned}
m_q = &\frac{m_z}{3de(d+e)(a^4+b^4+c^4)} \\
&\times (2\{[d^2(a^4+b^4+b^2c^2)+c^2e^2(a^2-b^2)]^2 + [e^2(a^4+c^4+a^2b^2) \\
&- d^2(a^4+c^4+b^2c^2)]^2 + [e^2(b^4+c^4+a^2b^2)+a^2d^2(c^2-b^2)]^2\} \\
&+ [d^2(a^4+b^4+3c^4-2b^2c^2)+2c^2e^2(b^2-a^2)]^2 \\
&+ [e^2(a^4+3b^4+c^4-2a^2b^2)-d^2(a^4+3b^4+c^4-2b^2c^2)]^2 \\
&+ [2a^2d^2(c^2-b^2)-e^2(3a^4+b^4+c^4-2a^2b^2)]^2)^{1/2},
\end{aligned} \tag{5.17}$$

$$m_r = m_z \sqrt{\frac{6}{a^4+b^4+c^4}}, \tag{5.18}$$

$$m_s = m_z \frac{\sqrt{c^2[a^2(d+e)+b^2e]^2 + b^2(a^2d-c^2e)^2 + a^2[c^2(d+e)+b^2d]^2}}{\sqrt{2}[a^2c^2(d+e)^2 + b^2(a^2d^2+c^2e^2)]}, \tag{5.19}$$

$$\begin{aligned}
m_t = &\frac{2m_z}{3de(d+e)(a^4+b^4+c^4)} \\
&\times (2\{[d(a^4+b^4+b^2c^2)+c^2e(b^2-a^2)]^2 + [d(a^4+c^4+b^2c^2) \\
&+ e(a^4+c^4+a^2b^2)]^2 + [e(b^4+c^4+a^2b^2)+a^2d(b^2-c^2)]^2\} \\
&+ [d(a^4+b^4+3c^4-2b^2c^2)+2c^2e(a^2-b^2)]^2 \\
&+ [d(a^4+3b^4+c^4-2b^2c^2)+e(a^4+3b^4+c^4-2a^2b^2)]^2 \\
&+ [e(3a^4+b^4+c^4-2a^2b^2)+2a^2d(c^2-b^2)]^2)^{1/2},
\end{aligned} \tag{5.20}$$

where a, b, c, d, and e are linear lengths of elements of the 3 × 3 spheroidal equal angular moving window (Fig. 4.1c).

The development of Eqs. (5.2–5.5, 5.16–5.20) and formulas presented in Table 5.1 was carried out with the software Maple V Release 5.0 (© Waterloo Maple Inc., 1981–1997). A generalization of the method described was proposed by Zhou and Liu (2004).

Notice that the RMSE formula used (Eq. 5.1) is not complete. In this form, it is usually applied to functions of independent measured variables. For a function of correlated or dependent variables, the RMSE formula includes residual terms of the Taylor series (Bolshakov and Gaidaev, 1977, p. 116):

$$m_F = \sqrt{\left(\frac{\partial F}{\partial x}\right)^2 m_x^2 + \left(\frac{\partial F}{\partial y}\right)^2 m_y^2 + \cdots + \left(\frac{\partial F}{\partial u}\right)^2 m_u^2 + 2\left(\frac{\partial F}{\partial x}\right)\left(\frac{\partial F}{\partial y}\right) m_x m_y + 2\left(\frac{\partial F}{\partial x}\right)\left(\frac{\partial F}{\partial u}\right) m_x m_u + \cdots}.$$

(5.21)

It is obvious that residual terms can influence RMSE values. In some cases, elevation values at the nodes of a moving window may be correlated. In such cases, it is reasonable to use Eq. (5.21) to estimate RMSE of calculation of partial derivatives and morphometric variables. An interested reader may want to develop related RMSE formulas, as well as to estimate an influence of residual terms.

5.3.4. RMSE Mapping

Equations (5.2)–(5.12) can be easily applied to produce digital models of m_F. Indeed, values of r, t, s, p, q, m_r, m_t, m_s, m_p, and m_q (and, hence, m_F) are estimated for the central point of the 3 × 3 (or 5 × 5) moving window. Moving the window along a DEM, it is possible to calculate m_F values for each point of the DEM except for boundary rows and columns.

Mapping is a convenient and pictorial strategy for visualizing the propagation of errors in spatial modeling (Heuvelink et al., 1989; Kraus, 1994; Hunter and Goodchild, 1995). Let us illustrate m_F mapping with two examples:

1. RMSE estimation on plane square grids is exemplified by a portion of the Kursk Region in Russia near the Kursk nuclear power plant.
2. RMSE estimation on spheroidal equal angular grids is exemplified by the Moscow Region in Russia.

A DEM of the Kursk Region (Fig. 5.3a) is described in Section 13.4.2.2. A DEM of the Moscow Region (Fig. 5.4a) was extracted from the GLOBE DEM (GLOBE Task Team and others, 1999). This DEM includes 59,400 points (the matrix 330 × 180); the grid spacing is 1′.

To suppress high-frequency noise, both DEMs were smoothed three times with the 3 × 3 moving window. Using the author's method (Section 4.2), we calculated some morphometric attributes including H (Fig. 5.3b) for the Kursk Region; the grid spacing of 150 m was utilized. Using the other author's method (Section 4.3), we calculated k_h (Fig. 5.4b) and some other topographic variables for the Moscow

5.3 CALCULATION ACCURACY OF LOCAL MORPHOMETRIC VARIABLES 79

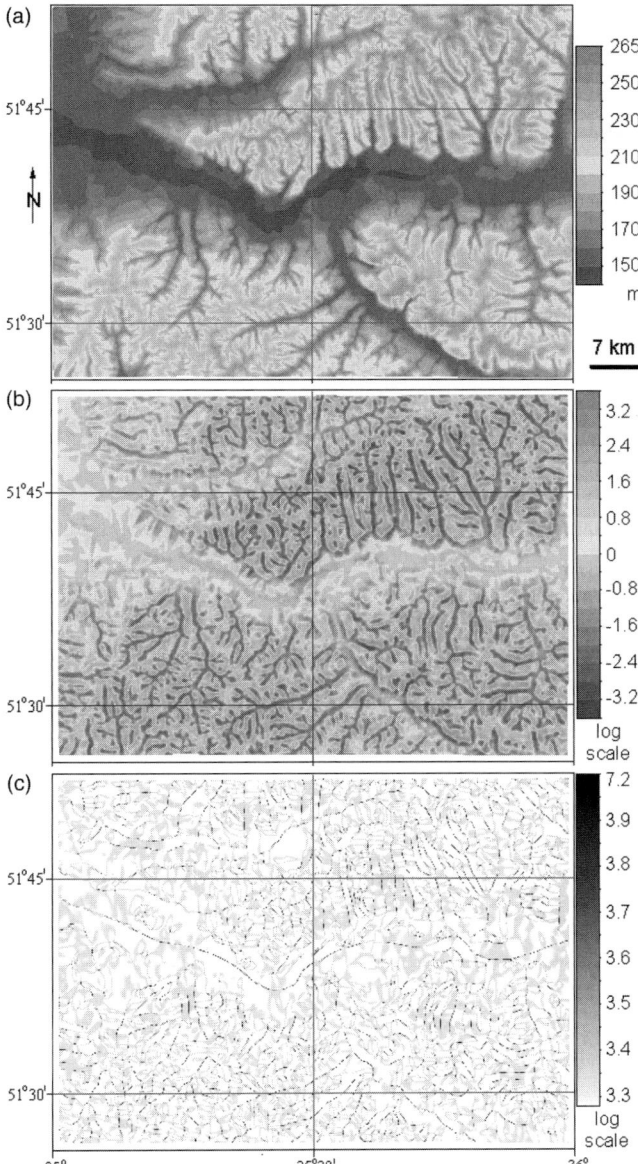

FIGURE 5.3 The Kursk Region near the Kursk nuclear power plant: (a) elevation, (b) mean curvature, (c) RMSE of the mean curvature calculation. *From (Florinsky, 2010, Fig. 2.6).* (See page 6 in Color Plate Section)

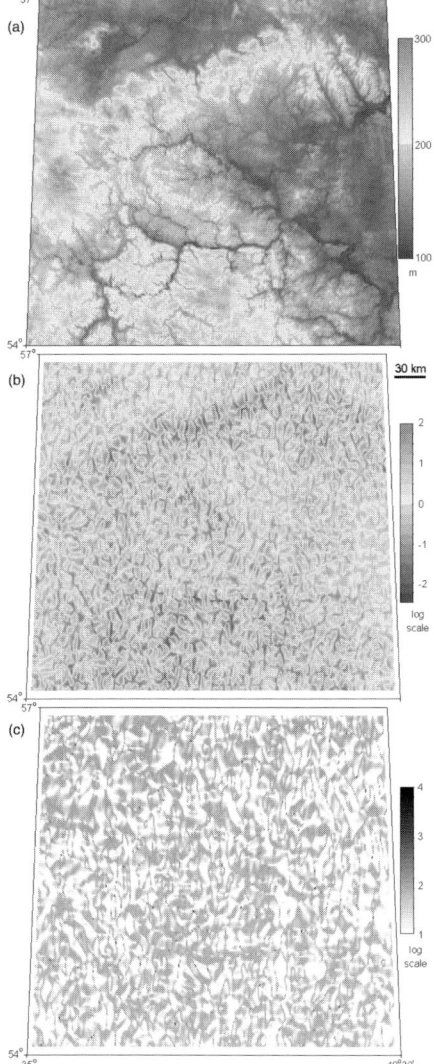

FIGURE 5.4 The Moscow Region and adjacent territories: (a) elevation, (b) horizontal curvature, (c) RMSE of the horizontal curvature calculation. *From (Florinsky, 2008d, Figs. 1 and 2a, c).* (See page 7 in Color Plate Section)

Region; the grid spacing of 1′ was applied. Applying the formulas (Eqs. 5.3, 5.7, 5.16–5.20, and Table 5.1) with $m_z = 5$ m, we derived the m_H digital model for the Kursk Region (Fig. 5.3c) and the m_{kh} digital model for the Moscow Region (Fig. 5.4c). To deal with the large dynamic range of H, k_h, m_H, and m_{kh}, we logarithmically transformed their digital models by Eq. (7.1) with $n = 5$ for H and m_H, and $n = 7$ for k_h and m_{kh}. Data processing was performed by the software LandLord (Appendix B).

Analysis of m_F maps allowed us to determine some regularity for spatial distribution of m_F values. Generally, the flatter topography, the higher the m_F value (Florinsky, 1998a, 1998c). m_F values can be in excess of maximum absolute values of F within flat areas. However, this does not mean that there are errors in DTMs within such areas: m_F is a statistical property of a function F; that is, m_F values represent a possibility for errors.

To improve the impartiality of DTM-based soil and geological studies, researchers have to treat data on local topographic variables with criticism, as well as consider the mentioned effects in data interpretation. Models and maps of m_F can be used, for instance, as follows:

- To account for a spatial distribution of m_F in analysis and interpretation of F map;
- To refine a DEM within areas having high m_F values, and then to recalculate F within these areas.

5.4 IGNORING OF THE SAMPLING THEOREM

5.4.1 Motivation

Let us recall the third sequence of the sampling theorem (Section 3.3): "landforms" with typical planar sizes less than $\tilde{\lambda}_{x,y}$, occurring in a DEM after its interpolation, should be considered as interpolator-induced high-frequency noise. Every so often, such a noise arises in a DEM with "enhanced," or overdetailed resolution—that is, a DEM produced by interpolation using a grid spacing essentially less than (1) an average density of points in an initial irregular DEM; or (2) a grid spacing in an initial regular DEM.

As this takes place, high-frequency noise may not be clearly seen in elevation maps derived from DEMs with "enhanced" resolution. This is because a vertical magnitude of noise may be too low compared with a contour interval. However, these minor false "landforms" may be dramatically increased in the subsequent derivation of topographic variables using differentiation procedures.

It is common knowledge of signal and image processing that differentiation of a signal increases noise manifestation in a derivative (Baker, 1982; Rosenfeld and Kak, 1982). In other words, differentiation can impair the signal-to-noise ratio. In a general case, the noise can have a higher derivative than the signal because the noise is less "smooth" than the signal and can fluctuate more randomly than the signal.

The higher the order of the derivative, the higher a noise manifestation. The following example can clarify this point for the

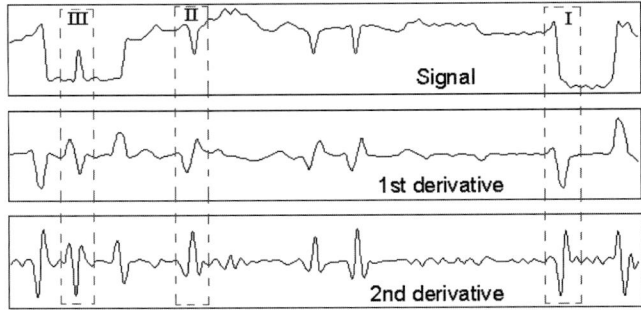

FIGURE 5.5 Differentiation of a one-dimensional signal; an edge (I), pit (II), and peak (III) of the signal. *From (Florinsky, 2002, Fig. 3).*

one-dimensional case (Fig. 5.5). For an edge of the signal, the first-order derivative has one extremum (a peak or a pit), while the second-order derivative has two extrema (a peak and a pit). Also, for a peak or a pit of the signal, the first-order derivative has two extrema (a peak and a pit), while the second-order derivative has three extrema (a peak surrounded by two pits, or a pit surrounded by two peaks).

Derivation of local topographic variables from DEMs is carried out using the first- and second-order partial derivatives of the elevation function (Section 2.2). As any DEM includes noise (i.e., random and systematic errors), its propagation with a differentiation-induced magnification is typical for digital terrain modeling (Brown and Bara, 1994; Giles and Franklin, 1996; Desmet, 1997; Florinsky, 2002; Oksanen and Sarjakoski, 2005) (see, for example, illustrations in Sections 6.2.5 and 6.3.3).

In particular, scholars noticed errors typical of DEMs with "enhanced" resolution, such as terraces, "traces" of contours and triangular patterns (see below) (Batson et al., 1975; Wood and Fisher, 1993; Eklundh and Mårtensson, 1995; Desmet, 1997). However, these authors explained artifacts observed by the lack of interpolation methods and did not associate them with overdetailed resolution of DEMs. In fact, such errors are the result of ignoring the sampling theorem. In this section, we discuss this type of DTM error.

5.4.2 Materials and Data Processing

To illustrate errors caused by overdetailed resolution, we used DEMs of two areas: (1) the part of the Crimean Peninsula and adjacent sea bottom; and (2) the Severny Gully (Pushchino, Russia). The generation of DEMs is described in Sections 13.4.1.2 and 9.3.2, correspondingly.

5.4 IGNORING OF THE SAMPLING THEOREM

Using the inverse distance weighting interpolation (Watson, 1992) of the irregular DEM of the Crimean Peninsula (Fig. 5.6a), we produced two regular DEMs with $w = 500$ m (Fig. 5.6b) and $w = 3000$ m. Application of $w = 500$ m gives an "enhanced" resolution, since such w value is less than average distance between points within all parts of the irregular DEM (Fig. 5.6a). The grid spacing of 3000 m corresponds approximately to average distances between points describing the seafloors of the Black Sea and the Sea of Azov as well as some areas of the Crimean Plain. The Crimean Mountains can be treated with a smaller grid spacing, such as 1000 m, but we applied the single grid spacing value fitting all parts of the DEM. Digital models of G (Fig. 5.6c, d) and k_v (Fig. 5.6e, f) were derived from the two regular DEMs of the Crimean Peninsula by the Evans–Young method (Section 4.1).

Applying the Delaunay triangulation and a piecewise quadric polynomial interpolation with matching derivatives along triangle edges (Agishtein and Migdal, 1991) to the irregular DEM of the Severny Gully

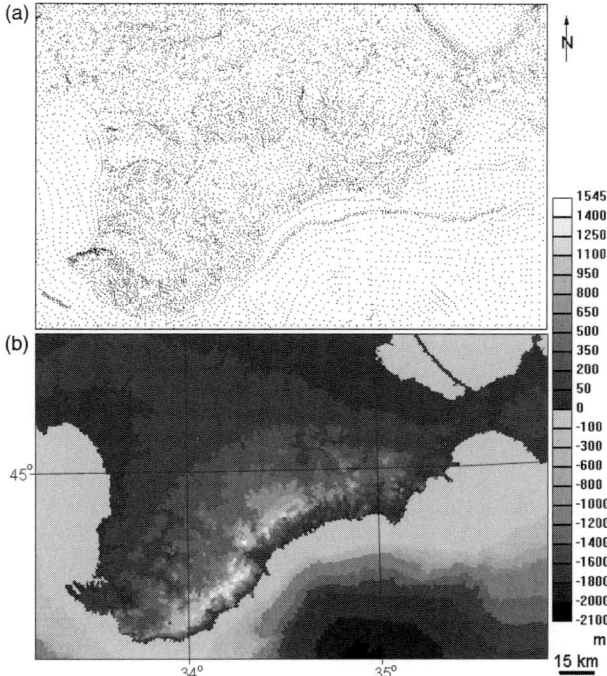

FIGURE 5.6 The Crimea and the adjacent sea bottom: (a) distribution of DEM points; (b) elevation, $w = 500$ m; (c) slope gradient, $w = 500$ m; (d) slope gradient, $w = 3000$ m; (e) vertical curvature, $w = 500$ m; (f) vertical curvature, $w = 3000$ m. *From (Florinsky, 2002, Fig. 4).*

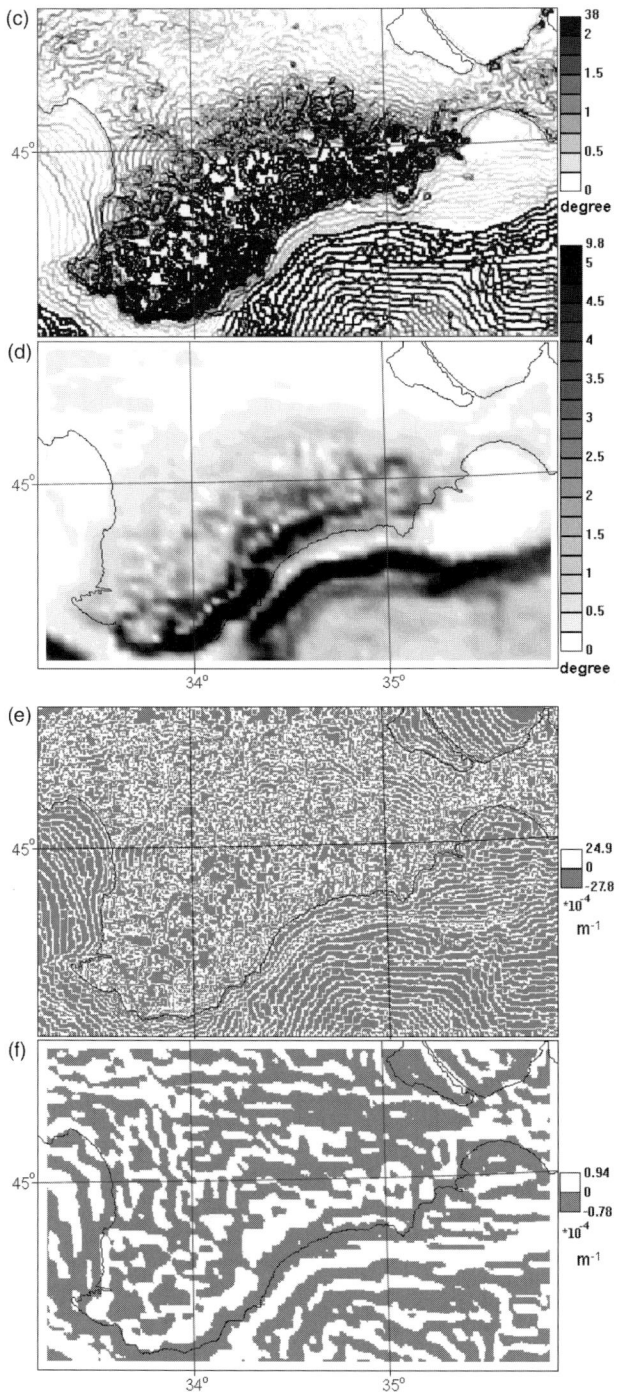

FIGURE 5.6 Continued

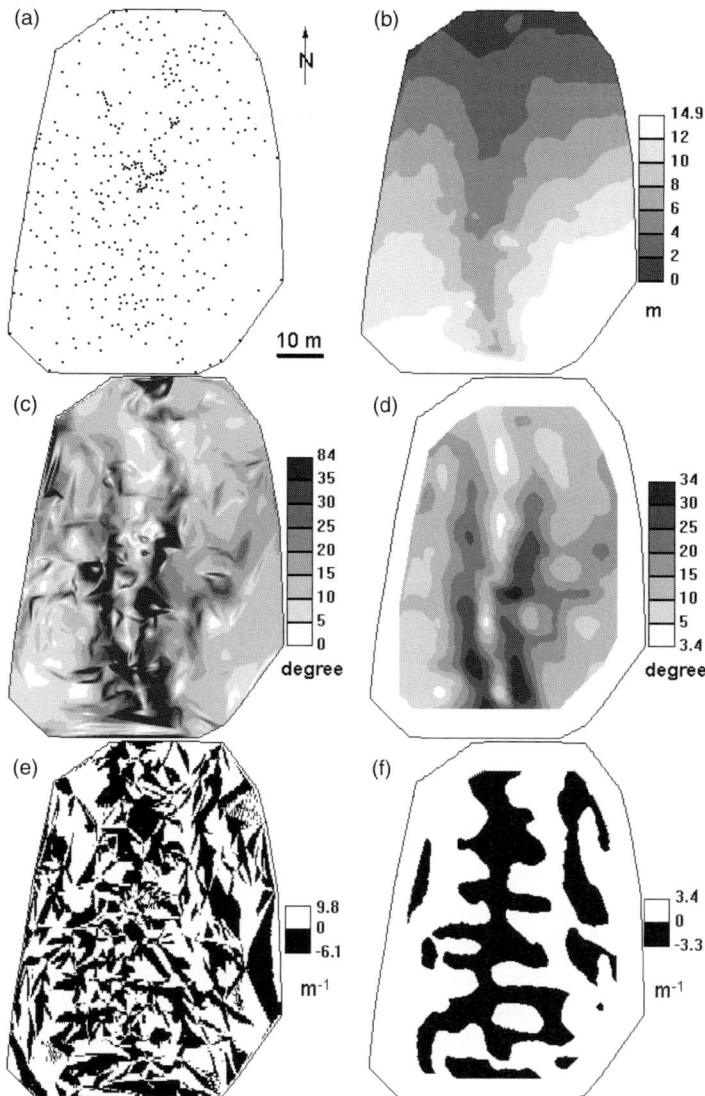

FIGURE 5.7 The Severny Gully: (a) distribution of DEM points; (b) elevation, $w = 0.25$ m; (c) slope gradient, $w = 0.25$ m; (d) slope gradient, $w = 3$ m; (e) horizontal curvature, $w = 0.25$ m; (f) horizontal curvature, $w = 3$ m. *From (Florinsky, 2002, Fig. 5).*

(Fig. 5.7a), we produced two regular DEMs with $w = 0.25$ m (Fig. 5.7b) and $w = 3$ m. Application of $w = 0.25$ m certainly gives "enhanced" resolutions since this w value is well less than average distances between points of the irregular DEM. The grid spacing of 3 m approximately relates to average distances between points of the irregular DEM (Fig. 5.7a). Digital models of G (Fig. 5.7c, d) and k_h (Fig. 5.7e, f) were derived from the two regular DEMs of the Severny Gully by the Evans–Young method (Section 4.1).

To highlight artifacts on the k_v and k_h maps, we subdivided k_v and k_h values into two intervals relative to the zero value (Figs. 5.6e, f, and 5.7e, f). Data processing was performed by the software LandLord (Appendix B).

5.4.3 Results and Discussion

On the G and k_v maps of the Crimea derived from the DEM with $w = 500$ m, one can see "traces" of contours (viz., curvilinear artifacts in form and location related to the contour lines) within the seafloor and relatively flat areas of the Crimean Plain (Fig. 5.6c, e). These maps are obviously unfit for use in any purpose, since they display a system of false curvilinear "landforms." However, G and k_v maps derived from the DEM with $w = 3000$ m (Fig. 5.6d, f) can be useful in regional geomorphic and geological studies. For example, the k_v map (Fig. 5.6f) was successfully employed to recognize and classify lineaments of the Crimea (Florinsky, 1996) (Section 13.4.1).

On the G and k_h maps of the Severny Gully derived from the DEM with $w = 0.25$ m, one can see "marks" of triangulation, such as triangles and other relatively regular features (Fig. 5.7c, e). These maps are also unfit for use in any DTM-based application, as they reveal a set of false triangular "landforms." At the same time, G and k_h data derived from the DEM with $w = 3$ m (Fig. 5.7d, f) were successfully applied to study the dependence of soil moisture on topography (Florinsky and Kuryakova, 2000) (Section 9.3).

"Traces" of contours (Fig. 5.6c, e) are portraits of slender upright artificial "escarps" of broad flat artificial "terraces". "Escarps" may arise along contour lines, while "terraces" may occur between contours in DEM interpolation by the inverse distance weighting method. This method is sensitive to clustering of points: the search of nearest neighbors finds a lot of points along contours while none across ones. As a result, an elevation of a contour is assigned to regular DEM points along the contour, while circa average elevation of two neighbor contours is assigned to regular DEM points between these contours. It has been suggested that the method is not appropriate to interpolate contour-based DEMs (Wood and Fisher, 1993; Eklundh and Mårtensson, 1995). However, this algorithm imperfection makes itself evident in the case of over detailed interpolation only. So, this is, in fact, connected with a lack of information between contours.

5.4 IGNORING OF THE SAMPLING THEOREM

Triangular artifacts (Fig. 5.7c, e) are the results of some inaccuracy of the matching derivatives along edges of triangles in the piecewise interpolation (Agishtein and Migdal, 1991). However, this imperfection of the interpolation algorithm makes itself evident in the case of overdetailed interpolation only. Therefore, an actual root of the artifacts is a lack of information about elevations between triangulation nodes.

Such inaccuracies lead to relatively minor errors in the regular DEMs with "enhanced" resolution. At least, it is impossible to see them on the elevation maps (Figs. 5.6b and 5.7b). Therefore, one can ignore these artifacts in some cases: for example, an elevation map derived from a DEM with "enhanced" resolution can be used for illustration goals. However, these errors are dramatically increased after derivation of local morphometric variables as they are calculated by differentiation (Section 2.2).

There are several ways to avoid these artifacts. First, it has been suggested that more sophisticated and smoother interpolation techniques should be used to prevent the formation of "terraces" in DEMs (Eklundh and Mårtensson, 1995), contour "traces" on G, A, k_v, and hillshading maps (Batson et al., 1975; Wood and Fisher, 1993; Desmet, 1997), and triangle patterns on G and A maps (Desmet, 1997). Second, Eklundh and Mårtensson (1995) proposed generalizing contours in order to prevent terraces in DEMs. Third, high-frequency filtering and smoothing is applied before differentiation to reduce the noise level in signal and image processing (Baker, 1982; Rosenfeld and Kak, 1982). Similar filtering procedures are used in digital terrain modeling (Chapter 6) to suppress high-frequency noise in DTMs derived from DEMs (Horn, 1981), or in DEMs before DTM derivation (Brown and Bara, 1994; Giles and Franklin, 1996; Desmet, 1997; Wise, 2000).

All these approaches may improve the results of DEM interpolation and hide a structure of an irregular DEM grid. However, this cannot clearly increase an actual resolution of a DEM (conversely, the second option can decrease it). It is incorrect to use DEMs with "enhanced" resolution for DTM derivation because an overdetailed resolution of a DTM does not relate to information on actual landforms kept in a DEM before interpolation. Treating DTMs beyond a resolution limit can lead just to some abstract investigation of the geometry of a matrix of interpolated values $z = f(x, y)$ rather than the geometry of the topographic surface.

Common sense leads us to assume that there is only one appropriate solution of the problem discussed. Let a spatial resolution of an irregular DEM correspond to $\tilde{\lambda}_{x,y}$ (Section 3.3). In this case, one should use w value relating, at least, to $\tilde{\lambda}_{x,y}$ in DEM interpolation, if an interpolated DEM will then be utilized to derive other DTMs. Practically, this w value should relate to an average distance between points in the irregular DEM. Other w values chosen from considerations of the study tasks (Chapter 9) (Florinsky and Kuryakova, 2000) should be higher than this minimal w value.

5.5 THE GIBBS PHENOMENON

5.5.1 Motivation

The Gibbs phenomenon is a specific behavior of some functions manifested as over- and undershoots around a jump discontinuity (Nikolsky, 1977b, § 15.9; Hewitt and Hewitt, 1980; Jerri, 1998). The Gibbs phenomenon is typical for the Fourier series, orthogonal polynomials, splines, wavelets, and some other approximation functions. It appears in many scientific problems and applications involving signal and image processing (Rosenfeld and Kak, 1982, p. 158).

For the one-dimensional case, the simplest mathematical illustration of the Gibbs phenomenon is an approximation of a square wave function (Fig. 5.8a)

$$F(x) = \begin{cases} 1, & 0 < x < \pi \\ 0, & x = 0, \pm\pi \\ -1, & -\pi < x < 0 \end{cases} \quad (5.22)$$

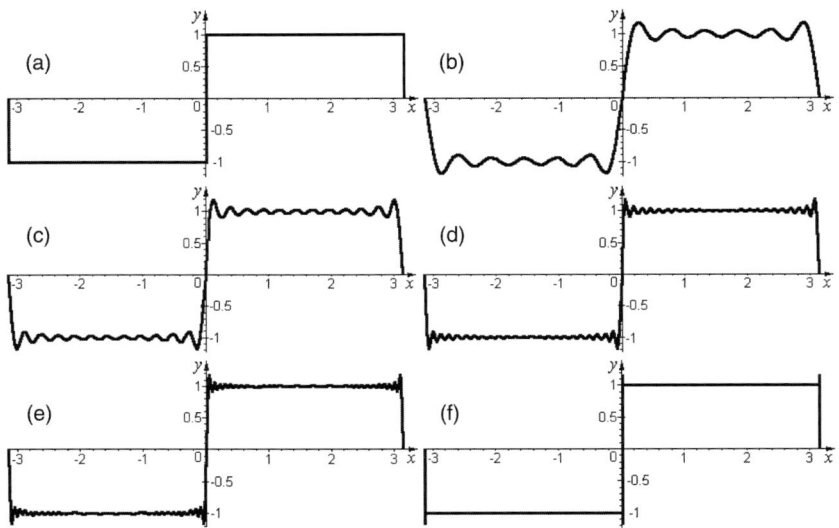

FIGURE 5.8 Approximation of a square wave function by a trigonometric polynomial: (a) square wave function; partial sums: (b) $k = 6$, (c) $k = 12$, (d) $k = 24$, (e) $k = 36$, (f) $k \to \infty$. Approximation was carried out by Maple V Release 5.0 (© Waterloo Maple Inc., 1981–1997). From (Florinsky, 2002, Fig. 1).

by a trigonometric polynomial of the form

$$f_{2k-1}(x) = \frac{4}{\pi} \sum_{k=1}^{\infty} \frac{\sin(2k-1)x}{2k-1} = \frac{4}{\pi}\left(\sin x + \frac{\sin 3x}{3} + \cdots + \frac{\sin(2k-1)x}{2k-1}\right).$$

(5.23)

The polynomial (Eq. 5.23) converges uniformly to the square wave function (Eq. 5.22), except for points of discontinuity of $F(x)$, $x = 0$, $\pm\pi$. This means that as k increases, the graphs of the partial sums $f_{2k-1}(x)$ approach arbitrarily closely to the graphs of $F(x) = \pm 1$, except for the vicinities of the points $x = 0$, $\pm\pi$ (Fig. 5.8b—e). This convergence defect manifests itself as over- and undershoots around the points $x = 0$, $\pm\pi$. It is important that as k increases, a vertical size of the over- and undershoots does not change (Fig. 5.8b—e). It is equal to 17.9% of half the jump size, or 8.95% of the jump size. With $k \to \infty$, the ultimate graph of $f_{2k-1}(x)$ is a jagged line including extended vertical legs as big as 8.95% of the jump size (Fig. 5.8f). By increasing k, it is possible to reduce the horizontal extension of the Gibbs phenomenon only (Fig. 5.8) (Fichtenholz, 1966b, pp. 490—497).

In the Gibbs phenomenon, an over- and undershoot size depends on the size of a jump discontinuity and a sort of function. In the classical case of the Fourier series, the over- and undershoot size is equal to 8.95% of the jump size (Jerri, 1998). Fichtenholz (1966b) established that the Gibbs phenomenon takes place for any piecewise smooth function (e.g., splines); the over- and undershoot size is 8.95% of the jump size for a spline of the order $k \to \infty$. Richards (1991) demonstrated that the over- and undershoot size increases if a spline order decreases. For example, it is 9.49% and 13.39% of the jump size for $k = 8$ and $k = 2$, respectively. The higher the jump, the higher the over- and undershoots.

Since splines and other approximation functions are widely used for DEM interpolation (Watson, 1992), understanding of the Gibbs phenomenon becomes substantial to provide correct digital terrain modeling—for instance, to choose an appropriate interpolation method to reduce or eliminate over- and undershoots. Practically, any DEM includes many jump discontinuities of the elevation function wherein the Gibbs phenomenon can arise after interpolation. These are areas with sharp changes of elevation (or steep gradient), such as terraces, escarpments, abrupt slopes, peaks, and pits. Jump discontinuities can also occur near pronounced systematic and random DEM errors—for example, near an abrupt linear step in elevations resulting from processing an orthophoto as separate patches and subsequent joining them to assemble a DEM (Hunter and Goodchild, 1995), or near a point with false elevation of 100 m within an area with an average altitude of 10 m.

90   5. ERRORS AND ACCURACY

For instance, the following two artificial landforms may appear near a steep escarpment with a relative altitude of 50 m: a "bank" (an overshoot) along the escarpment edge, and a "ditch" (an undershoot) along the escarpment foot. In this case, a relative height of the "bank" as well as a relative depth of the "ditch" can be about 4.5 m (8.95% of 50 m) if the interpolation is carried out with high-order splines. Artificial closed "banks" (overshoots) and "ditches"''' (undershoots) may arise around isolated pits and peaks, correspondingly.

DEM interpolation has been commonly used in the past five decades, so researchers have noted over- and undershoots near steep gradients in interpolated DEMs, and have also tried to solve this problem (Akima, 1974; McCullagh, 1981; Nielson and Franke, 1984; Mitášová and Mitáš, 1993); see detailed discussion in Section 5.5.3. However, they did not recognize over- and undershoots as manifestations of the Gibbs phenomenon. In this section, we discuss properties of the Gibbs phenomenon as applied to digital terrain modeling.

5.5.2 Materials and Data Processing

A DEM of an imaginary site was produced. The site measures 7×7 m. The square-gridded DEM consists of 64 points. The site includes two jump discontinuities: (1) a scarp with an elevation difference of about 26 m and a horizontal distance between edge and foot of about 1 m; and (2) a false elevation of -100 m within an area with an average elevation of -4 m (Fig. 5.9a).

Two DEMs with the grid spacing of 0.03 m (Fig. 5.9b) were derived from the initial DEM by (1) the Delaunay triangulation and a linear interpolation; and (2) the Delaunay triangulation and a piecewise quadric polynomial interpolation with matching derivatives along triangle edges (Agishtein and Migdal, 1991). Cross sections $A - A'$ and $B - B'$ were constructed through jump discontinuities (Fig. 5.9c, d). The linear interpolation was performed by the software Surfer 6.04 (© 1993–1996, Golden Software Inc.). The smooth interpolation and data visualization were carried out by the software LandLord (Appendix B).

5.5.3 Results and Discussion

On the elevation map produced with the smooth interpolation, one can see three marks ("knolls") of the Gibbs phenomenon around the false elevation (Fig. 5.9b). With the exception of a small "pit," the elevation map does not demonstrate the Gibbs phenomenon marks along the scarp due to the too high contour interval value. At the same time, there

5.5 THE GIBBS PHENOMENON

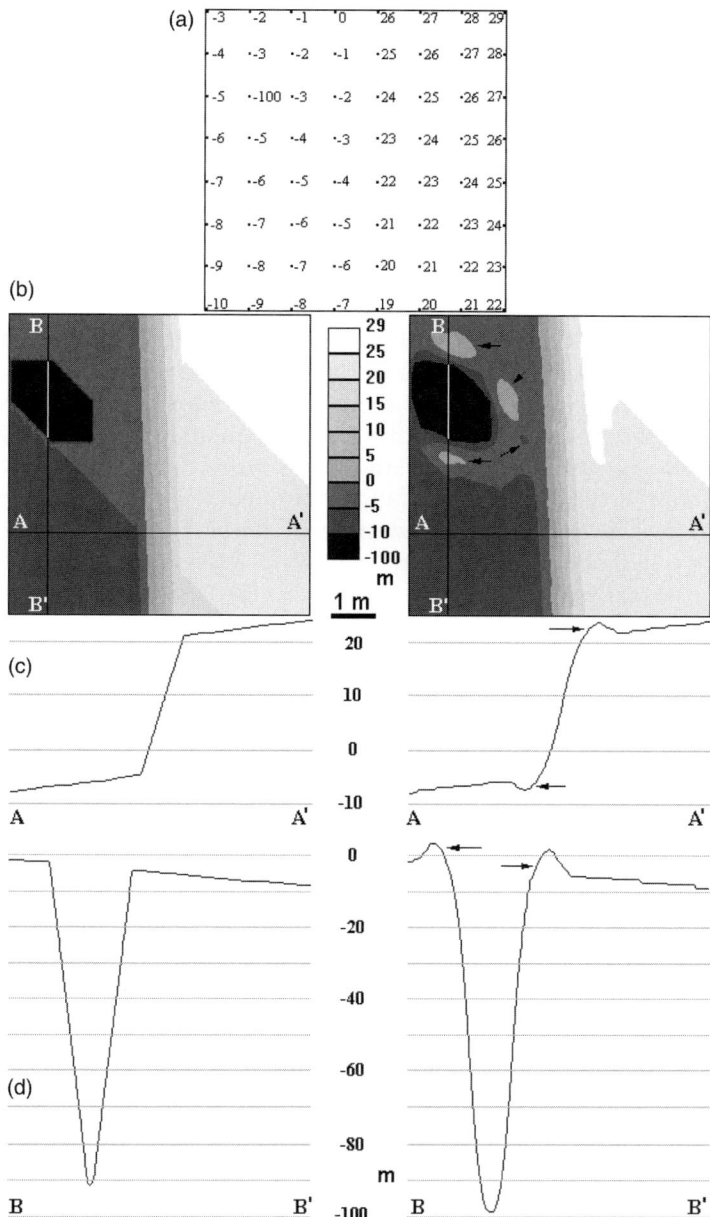

FIGURE 5.9 The Gibbs phenomenon in DEM interpolation: (a) distribution of DEM points; (b) elevation, A − A′ and B − B′ are cross sections through areas with jump discontinuities; (c) cross sections A − A′; (d) cross sections B − B′. Left—linear interpolation, right—smooth interpolation. Arrows mark the manifestations of the Gibbs phenomenon. *From (Florinsky, 2002, Fig. 2).*

are no Gibbs phenomenon marks on the elevation map produced with the linear interpolation (Fig. 5.9b).

Figure 5.9c presents the cross section A − A′ without the Gibbs phenomenon marks. It relates to the DEM produced with the linear interpolation. At the same time, there are the Gibbs phenomenon marks on the cross section A − A′ corresponding to the DEM produced with the smooth interpolation. These are over- and undershoots to the left and to the right of the jump discontinuity (Fig. 5.9c). So, two "landforms" arose due to the Gibbs phenomenon after the smooth interpolation: the "bank" is along the edge, and the "ditch" is along the foot.

The cross section B − B′ without the Gibbs phenomenon marks is presented in Fig. 5.9d. It relates to the DEM produced with the linear interpolation. At the same time, there are two marks of the Gibbs phenomenon on the cross section B − B′ corresponding to the DEM produced with the smooth interpolation (Fig. 5.9d). These overshoots are part of the "bank" around the hole associated with the false elevation of −100 m. The results demonstrate that the Gibbs phenomenon can arise after a smooth interpolation rather than a linear one.

It is obvious that DEM errors caused by the Gibbs phenomenon can propagate through the processing and can produce new errors in "secondary" DTMs derived from a DEM. Indeed, differentiation increases the manifestation of noise in derivation of local morphometric variables (Section 5.4.1). For catchment area and combined topographic attributes (Sections 2.3 and 2.6), errors can arise in each point located downslope from the Gibbs phenomenon rather than near jump discontinuities only. As vertical errors adversely affect flow-path determination (Veregin, 1997), the Gibbs phenomenon can disturb a design of crest and thalweg network maps (Section 2.4).

There are four main ways to prevent or reduce DEM errors caused by the Gibbs phenomenon:

1. Decreasing the jump discontinuity before DEM interpolation
2. Using interpolators that do not generate the Gibbs phenomenon
3. Omitting over- and undershoots after DEM interpolation
4. Filtering the Gibbs phenomenon

The first way can be carried out by DEM refining, that is, inserting additional points into an irregular DEM within the area with a jump discontinuity. This should be done before DEM interpolation. For example, the irregular DEM includes an escarp described by two sets of points along its edge and foot. To reduce the jump discontinuity, it would be rational to insert at least one additional set of points along the midslope of the escarp. This sort of DEM refining can decrease over- and undershoots or may hold an interpolation function in position without over- and undershoots, because the function passes through (or approximated to) additional points. Since the refining is carried out before interpolation, it neither violates a regular grid nor leads to a refining of the entire regular DEM. A "drawback" of this way is the necessity of careful control of the generation of the irregular DEM, including a choice of a proper distribution of irregular grid nodes. However, this is a common problem of DEM generation.

The second option may be performed using algorithms of linear interpolation as we demonstrated above. However, they may generate discretization errors in the regular DEM (i.e., a raster grid) if irregular grid nodes are sparsely distributed. To avoid such artifacts, methods of smooth interpolation are commonly used, but they can generate errors caused by the Gibbs phenomenon. To break this problem, one may use some smooth interpolation schemes based on splines with tension (Nielson and Franke, 1984; Franke, 1985; Mitášová and Mitáš, 1993; Mitas and Mitasova, 1999) (Section 3.4). These schemes were specially developed to protect DEM interpolation from over- and undershoots near jump discontinuities.

The third way can be carried out by interactive editing of DEMs (Weibel and Brändli, 1995). Its drawback is the necessity of a very careful control of the regular DEM generated by interpolation. An operator should search over- and undershoots near steep terraces, escarpments, abrupt slopes, peaks, pits, and so on. To do this correctly, the operator should be well aware of the actual topography of a studied area. Notice that the Gibbs phenomenon can be useful for the DEM editing because, as a rule, it introduces additional "banks" and "ditches" around pronounced random and systematic errors of DEM production. These over- and undershoots may assist the operator to find and eliminate such errors.

The fourth option may be conducted with some filtering procedures, such as the Fejer averaging and the Lanczos local averaging (Jerri, 1998). In this case, the Gibbs phenomenon may be eliminated or reduced since these filters smooth the edges of jump discontinuities. To do this, however, the exact location of jump discontinuities must be known.

The choice of a particular way depends on the user qualification and available software.

5.6 GRID DISPLACEMENT

5.6.1 Motivation

It is obvious that somewhat different point sets can be sampled using different positions of a discretization grid about the original signal or image. Subsequently, one can produce slightly different reconstructions of the signal or image for different positions of the discretization grid. As a rule, these minor discretization errors can be considered as a high-frequency noise, and one may in most cases ignore them. However, they may be increased in secondary products derived from these slightly different signals or images using differentiation (Section 5.4.1) (Zlatopolsky, 1992).

Discretization of the two-dimensional function of elevation is a key procedure of digital terrain modeling (Makarovič, 1973, 1977; Stefanovic et al., 1977) (Sections 3.2 and 3.3). The geometry of a DEM grid may affect patterns on maps of topographic variables (Mark, 1975a; Carter, 1988; Wilson et al., 1998). It is also apparent that slightly different DEMs of the same area can be produced by displacement (in a general case, by rotation) of a grid, in which nodes elevation values were measured or interpolated in nodes of the grid. We mean that the geometry of an irregular grid, or the spacing of a regular grid, remains constant; it is only the grid position and/or orientation that changes. Such DEMs can have minor differences, but all of them may be used to describe the area. Moreover, grid displacement may sometimes improve the DEM accuracy: Endreny et al. (2000) found that rotation of a SPOT-derived DEM about the axis may improve DEM RMSE.

The following question arises: what differences could be found in digital models of a topographic attribute derived from a DEM after displacement of a DEM grid? In other words, how does displacement of a DEM grid, wherein elevations are interpolated or measured, influence derivation of topographic variables? Of particular interest is the effect of the different position or orientation of a DEM grid on the revealing of topographically expressed geological features, such as lineaments (Chapter 13). In this section, we discuss the influence of DEM grid displacement on the map patterns of two topographic variables used in geological studies.

5.6.2 Materials and Data Processing

The study area—the Stavropol Upland in Russia—measures about 172×143 km (Fig. 5.10). An irregular DEM including 4459 points was produced by digitizing a topographic map (Central Board of Geodesy and Cartography, 1968). Using the Delaunay triangulation and a piecewise quadric polynomial interpolation (Agishtein and Migdal, 1991), we produced three regular DEMs with $w = 2500$ m. Regular grids of the DEMs have distinct orientations relative to the irregular DEM: rotation angles of the reference axes of the regular grid are $0°$, $25°$, and $90°$ in

5.6 GRID DISPLACEMENT

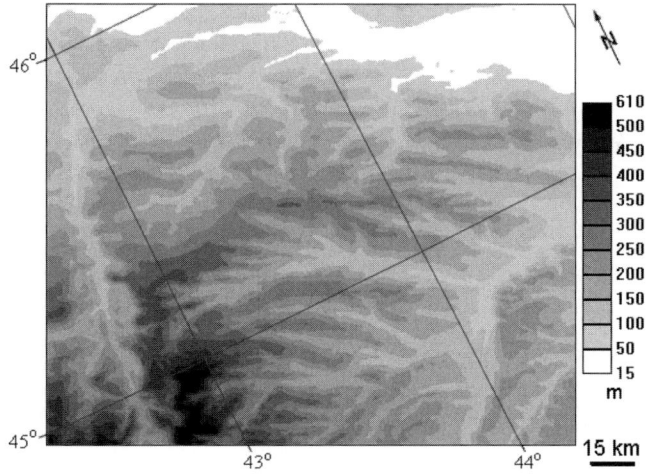

FIGURE 5.10 The Stavropol Upland: elevation. *From (Florinsky, 2002, Fig. 6).*

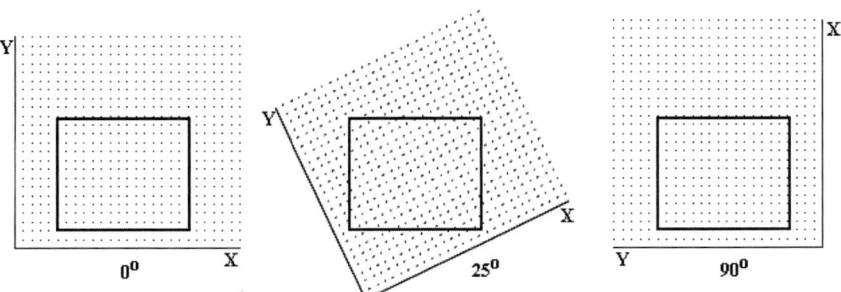

FIGURE 5.11 Rotation of the discretization grid around the irregular DEM (black frame). *From (Florinsky, 2002, Fig. 7).*

the counterclockwise direction in relation to the reference axes of the irregular grid (Fig. 5.11).

Digital models and maps of k_h and k_v (Fig. 5.12) were derived from the three regular DEMs by the Evans–Young method (Section 4.1). We subdivided k_h and k_v values into two intervals relative to the zero value, because this type of presentation of curvature data is used to reveal lineaments (Florinsky, 1992, 1996) (Chapter 13). Data processing was performed by the software LandLord (Appendix B).

Using samples extracted from k_h and k_v models, we conducted pairwise comparison of the DTMs obtained. The sample size was

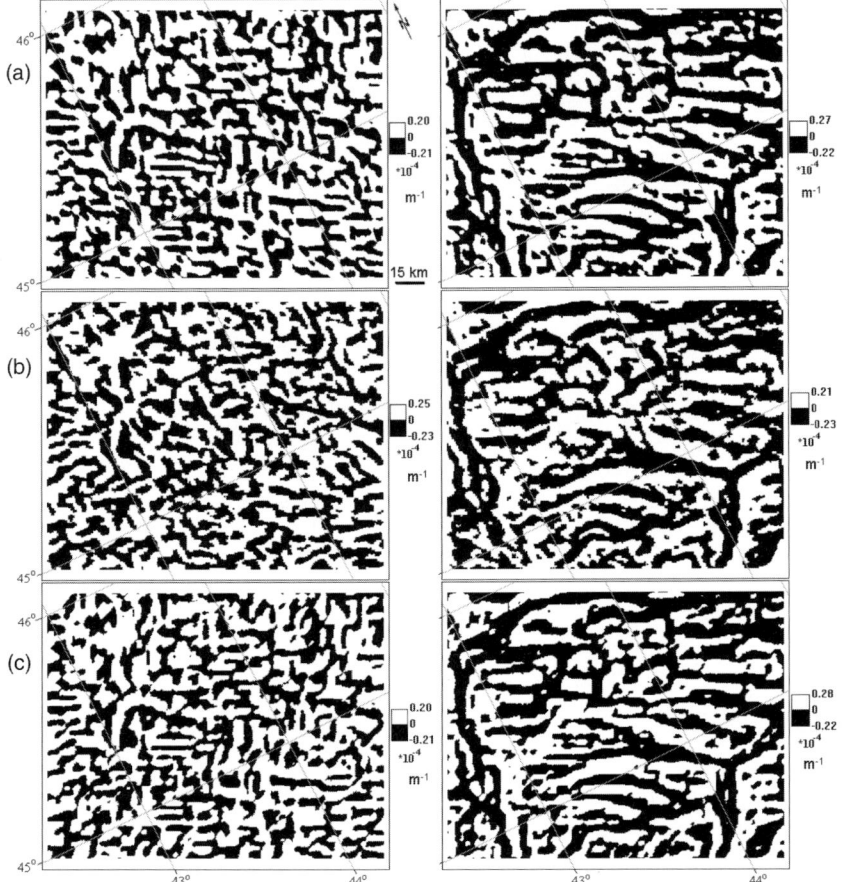

FIGURE 5.12 The Stavropol Upland: horizontal curvature (left) and vertical curvature (right), derived from the DEMs with (a) 0°-, (b) 25°-, and (c) 90°-rotation of the discretization grid. *From (Florinsky, 2002, Figs. 8 and 9).*

3600 points. Statistical analysis was carried out by Statgraphics Plus 3.0 (© Statistical Graphics Corp., 1994–1997).

5.6.3 Results and Discussion

DEM rotation leads to the revealing of slightly different topographic patterns. Although k_h and k_v maps derived from the variously oriented DEMs have many similarities, they also include a number of distinctions (Fig. 5.12). In particular, some patterns break or merge, and other patterns change their width and length. Most of patterns remain in their

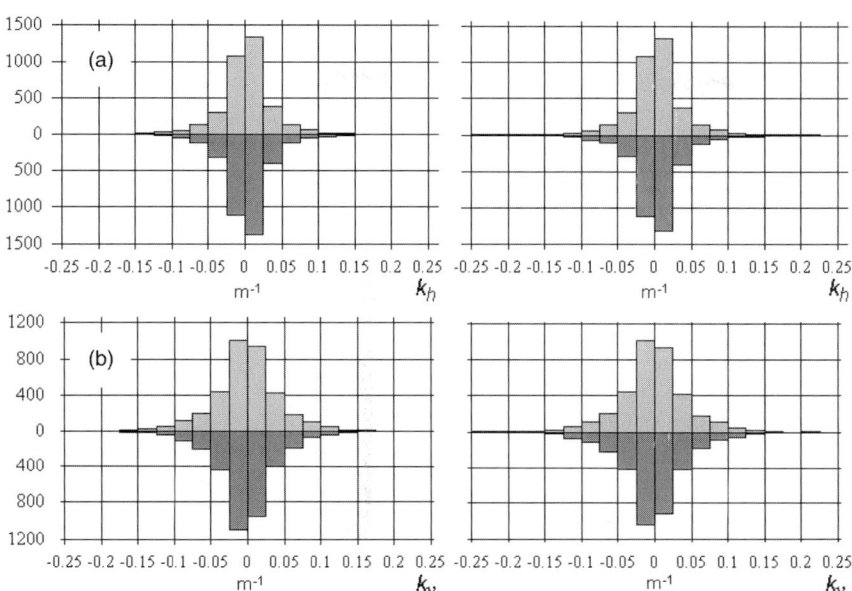

FIGURE 5.13 Statistical distribution of (a) k_h and (b) k_v values derived from the regular DEMs interpolated with various rotation angles of the discretization grid: all upper histograms are for 0°-rotation, lower histograms are for 25°- (left) and 90°-rotation (right). *From (Florinsky, 2010, Fig. 1.21).*

position. However, small dots, narrow lines, and small particles of big patterns appear and disappear on maps. Notice that the typical sizes of these patterns are less than the grid spacing. Therefore, they fit into a high-frequency noise. These artifacts are, within certain limits, manifestations of DEM discretization errors increased by differentiation (Section 5.4.1).

Similar effects arise in lineament revealing by digital processing of satellite images: some linear structures can appear and disappear owing to the image rotation in relation to the discretization grid (Zlatopolsky, 1992). There is no way to prevent such effects completely when dealing with discrete functions, which describe the surface continuum.

Results of the DTM pairwise comparison are presented in Fig. 5.13 and Table 5.2. Histograms of both k_h and k_v samples are very similar (Fig. 5.13). The Kolmogorov–Smirnov test applied to k_h and k_v pair samples demonstrated that no statistically significant difference exists between the distributions of k_h (or k_v) calculated with different position of the discretization grid at the 95% confidence level (Table 5.2). In other words, visually observable discretization errors are statistically insignificant.

TABLE 5.2 Pairwise Kolmogorov–Smirnov Test of the k_h and k_v Samples Derived from the Regular DEMs Interpolated with Various Rotation Angles of the Discretization Grid (0°-, 25°-, and 90°-rotation)

Sample Pairs		Statistics	
		K-S	P
k_h	0° and 25°	0.71	0.70
	0° and 90°	0.52	0.95
k_v	0° and 25°	1.20	0.11
	0° and 90°	0.60	0.86

From (Florinsky, 2010, Table 1.3).

5.7 LINEAR ARTIFACTS

5.7.1 Motivation

In DTM-based revealing of geological lineaments, indicators of these structures are linear patterns on maps of local morphometric variables (Chapter 13). Approximately north-, west-, northeast-, and northwest-striking linear patterns are commonly observed on such maps. These orthogonal and diagonal lineaments can reflect topographically expressed geological structures (Chebanenko, 1963; Moody, 1966; Shults, 1971; Katterfeld and Charushin, 1973; Trifonov et al., 1983; Bryukhanov and Mezhelovsky, 1987; Anokhin and Odesskii, 2001; Maslov and Anokhin, 2006) (Fig. 5.14). However, they may also be artifacts. There are several possible causes for orthogonal and diagonal lineaments: (1) DTM grid geometry; (2) DEM production errors; and (3) DEM interpolation errors. Let us briefly discuss these sources of artifacts.

1. It is obvious that DTM mapping can, more or less, visualize the DTM grid geometry, that is, orthogonal and diagonal directions. However, it is possible to ignore this effect using DTMs with a sufficiently high resolution, when the human visual system does not perceive the grid geometry. It is also obvious that the grid geometry may affect calculations (e.g., interpolation—see below).
2. There are at least five sources of linear artifacts in DEM generation:
    a. Photogrammetric generation of DEMs can lead to linear artifacts if one uses semiautomated profiling of aerial stereo images (Brown and Bara, 1994; Albani and Klinkenberg, 2003) (Fig. 5.1).
    b. There are abrupt linear altitude steps resulting from the processing of aerial photos as separate patches and the joining of them to assemble a DEM (Hunter and Goodchild, 1995).

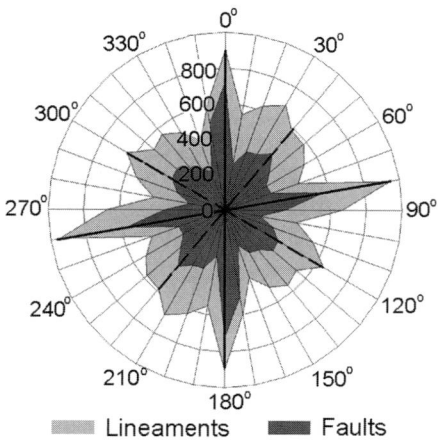

FIGURE 5.14 Rose diagram for direction azimuths of the global distribution of lineaments and faults. Radial coordinate is the number of lineaments and faults; solid lines denote the orthogonal system of lineaments and faults, dashed lines denote the diagonal system. *From (Maslov and Anokhin, 2006, Fig. 1), reproduced with permission.*

  c. Artificial orthogonal linear scarps can be found in small-scale and global DEMs, the result of assembling several DEMs varying in origin, scale, and accuracy (see illustrations in Section 6.3).
  d. Artificial lineaments may occur in detailed DEMs produced by laser and radar surveys. Such artifacts relate to man-made linear elevation differences located along some roads, dams, and so on. Sometimes such lineaments may be observed even in SRTM DEM: they are straight highways crossing large forest territories. An example of such a road is a portion of the Moscow Perimeter Highway crossing Russia's Losiny Ostrov National Park (Fig. 6.5): a northwest-striking lineament is detected in northeastern parts of k_v maps; the corresponding portion of the highway can be seen on the elevation map.
  e. Artificial lineaments may relate to striping artifacts of LiDAR and InSAR DEMs associated with air- or spacecraft motion, for example, aircraft pitch motion or orbital paths (Fig. 5.2 and Section 5.1) (Arrell et al., 2008; Gallant and Read, 2009). Linear artifacts (roll errors) can be found at deep-water areas in bathymetric DEMs produced by echo sounding (Bourillet et al., 1996).
3. DEM interpolation along a series of profiles of the four cardinal directions, orthogonal and diagonal, can lead to linear artifacts striking in these directions (Wood and Fisher, 1993). For sparsely sampled areas, interpolation based on polynomials, splines, and triangulation with linear interpolation may also produce artificial linear patterns of the cardinal directions (Declercq, 1996).

It is common knowledge of the theory of image processing that derivative operators for transformation of two-dimensional signals may be isotropic and anisotropic (Rosenfeld and Kak, 1982). By isotropic operators are meant rotation invariants: rotating a function $z = f(x,y)$ by an angle φ about the z-axis and then applying an operator Ω to $z = F(x', y')$ gives the same result as applying Ω to $z = f(x,y)$ and then rotating a result by the angle φ about the z-axis; that is, $F(x',y') = f(x,y)$, where x, y and x', y' are the unrotated and rotated Cartesian coordinates, respectively. An operator is anisotropic if this condition is not met. An example of isotropic operators is the Laplacian, whereas anisotropic operators are exemplified by compass operators measuring gradients in several directions.

Derivation of a digital model of a local topographic variable Θ from a DEM by related formulas (Eqs. 2.2–2.15) can be considered as applying an operator Ω of the variable Θ to an elevation function z. The operator Ω transforms the elevation function $z = f(x,y)$ into the function of the local topographic variable $\Theta = \psi(x,y)$, or, what is the same, Ω transforms a DEM into the digital model of Θ.

Local morphometric variables are functions of the partial derivatives of elevation. It would therefore be reasonable to suggest that operators of these variables may also be isotropic or anisotropic. We should emphasize that from mathematical definitions of local morphometric attributes (Table 2.1) it is clear that all of them, except for A, are independent of the orientation of the x- and y-axes. These variables are associated with directions related to intristic properties of the surface (Section 2.2 and Appendix A) rather than to those of the coordinate axes. However, some researchers prefer to ascribe an occurrence of linear patterns of cardinal directions in maps of topographic variables to probable anisotropic properties of their operators, ignoring mathematical definitions of these variables. This suggests that it is reasonable to present proof of the isotropy for operators of local topographic variables (Florinsky, 2005).

5.7.2 Isotropy of Local Morphometric Variables

To prove the isotropy of operators of local topographic variables (Eqs. 2.2–2.15), we apply a principle for testing derivative operators in image processing (Rosenfeld and Kak, 1982, p. 238). First, let us write the first and second partial derivatives of $z = F(x',y')$, that is, p', q', r', s', and t', in terms of the first and second partial derivatives of $z = f(x,y)$, that is, p, q, r, s, and t (Eq. 2.1). On differentiation, simple algebraic operations, and substitutions of the well-known expressions of coordinate rotation

$$x = x' \cos\varphi - y' \sin\varphi, \qquad y = x' \sin\varphi + y' \cos\varphi, \qquad (5.24)$$

5.7 LINEAR ARTIFACTS

we obtained the following formulas (Florinsky, 2005):

$$p' = \frac{\partial z}{\partial x'} = \frac{\partial z}{\partial x}\frac{\partial x}{\partial x'} + \frac{\partial z}{\partial y}\frac{\partial y}{\partial x'} = p\cos\varphi + q\sin\varphi, \quad (5.25)$$

$$q' = \frac{\partial z}{\partial y'} = \frac{\partial z}{\partial x}\frac{\partial x}{\partial y'} + \frac{\partial z}{\partial y}\frac{\partial y}{\partial y'} = q\cos\varphi - p\sin\varphi, \quad (5.26)$$

$$r' = \frac{\partial^2 z}{\partial x'^2} = r\cos^2\varphi + 2s\cos\varphi\,\sin\varphi + t\sin^2\varphi, \quad (5.27)$$

$$s' = \frac{\partial^2 z}{\partial x'\partial y'} = (t-r)\cos\varphi\,\sin\varphi + s(\cos^2\varphi - \sin^2\varphi), \quad (5.28)$$

$$t' = \frac{\partial^2 z}{\partial y'^2} = r\sin^2\varphi - 2s\cos\varphi\,\sin\varphi + t\cos^2\varphi. \quad (5.29)$$

As $p \neq p'$, $q \neq q'$, $r \neq r'$, $s \neq s'$, and $t \neq t'$, these partial differential operators are not rotation invariants.

Second, let us write the k_h formula (Eq. 2.4) for the rotated Cartesian coordinates, that is, in terms of p', q', r', s', and t' (Eqs. 5.25–5.29) (Florinsky, 2005):

$$\begin{aligned}
k'_h &= -\frac{q'^2 r' - 2p'q's' + p'^2 t'}{(p'^2 + q'^2)\sqrt{1 + p'^2 + q'^2}} \\
&= -\frac{(q\cos\varphi - p\sin\varphi)^2 (r\cos^2\varphi + 2s\cos\varphi\,\sin\varphi + t\sin^2\varphi)}{[(p\cos\varphi + q\sin\varphi)^2 + (q\cos\varphi - p\sin\varphi)^2]\sqrt{1 + (p\cos\varphi + q\sin\varphi)^2 + (q\cos\varphi - p\sin\varphi)^2}} \\
&\quad \frac{-2(p\cos\varphi + q\sin\varphi)(q\cos\varphi - p\sin\varphi)\{(t-r)\cos\varphi\,\sin\varphi + s(\cos^2\varphi - \sin^2\varphi)\}}{} \\
&\quad + (p\cos\varphi + q\sin\varphi)^2(r\sin^2\varphi - 2s\cos\varphi\,\sin\varphi + t\cos^2\varphi) = -\frac{q^2 r - 2pqs + p^2 t}{(p^2 + q^2)\sqrt{1 + p^2 + q^2}}
\end{aligned} \quad (5.30)$$

Comparing the k_h formulas for the rotated and unrotated coordinates (Eqs. 5.30 and 2.4), it is possible to see that $k'_h = k_h$. These mean that the k_h operator is isotropic. Similarly, it was proved that operators of G, k_v, H, K, K_a, K_r, M, E, k_{min}, k_{max}, k_{he}, and k_{ve} are isotropic, while the operator of A is anisotropic (Florinsky, 2005). This means that:

- Rotating an elevation function about the z-axis and then applying operators of G, k_h, k_v, H, K, K_a, K_r, M, E, k_{min}, k_{max}, k_{he}, and k_{ve}, as well as any of their linear transformations, cannot lead to variations in both values of the variables and patterns on their maps, compared with results of applying these operators to an unrotated elevation function.
- Application of these operators to DEMs cannot be responsible for the occurrence of artificial lineaments of cardinal directions in morphometric maps.

CHAPTER 6

Filtering

OUTLINE

6.1 Tasks of DTM filtering	103
6.1.1 Decomposition of the Topographic Surface	104
6.1.2 Denoising	104
6.1.3 Generalization	105
6.2 Methods of DTM filtering	109
6.2.1 Trend-Surface Analysis	109
6.2.2 The Filosofov Method	109
6.2.3 Two-Dimensional Discrete Fourier Transform	112
6.2.4 Two-Dimensional Discrete Wavelet Transform	113
6.2.5 Smoothing	115
6.2.6 Row and Column Elimination	117
6.2.7 The Cutting Method	119
6.3 Two-Dimensional Singular Spectrum Analysis	122
6.3.1 Algorithm	122
6.3.2 Materials and Data Processing	125
6.3.3 Results and Discussion	130

6.1 TASKS OF DTM FILTERING

A surface can be viewed as a sum of surfaces. This triviality forms the basis for solving a variety of problems in digital terrain analysis and in DTM-based soil and geological studies. The most popular tasks are as follows:

1. Separation of low- and high-frequency components of the topographic surface to study regularities in its structure and its elements of different scales

2. DTM denoising
3. DTM generalization, that is, removal of non-noise high-frequency components of the topographic surface from DTMs

6.1.1 Decomposition of the Topographic Surface

It is well known that topography is a result of the interaction of both endogenous and exogenous processes of different scales. One of the problems involved in geomorphology is reconstructing the history and characteristics of these processes from the current properties of topography. Another important problem of geomorphology is determining the hierarchical structure of topography (Phillips, 1988; De Boer, 1992), which can be used as a basis for describing the hierarchical structure of ecosystems (Puzachenko et al., 2002).

Researchers usually try to solve these two problems by decomposition of the topographic surface into its components of different scales. Mathematically, such a task is not complex and can be performed by various methods. In particular, these are trend-surface analysis (Section 6.2.1), the Filosofov method (Section 6.2.2), two-dimensional (2D) discrete Fourier transform (Section 6.2.3), 2D discrete wavelet transform (Section 6.2.4), smoothing (Section 6.2.5), row and column elimination from DTMs (Section 6.2.6), 2D singular spectrum analysis (Section 6.3.1), and so on.

It is obvious, however, that some low-frequency harmonics of the elevation function do not necessarily relate, for instance, to some tectonic structure, while some high-frequency harmonics do not relate to some exogenic (e.g., eolian) landforms. Nevertheless, decomposition of the topographic surface may be useful for initial estimation of a hierarchical structure of the terrain and the origin of its component.

6.1.2 Denoising

Any DEM contains many random errors and artifacts (Bourillet et al., 1996; Hastings and Dunbar, 1998; Arabelos, 2000; Holmes et al., 2000; Bergbauer et al., 2003; Smith and Sandwell, 2003; Rodríguez et al., 2006). The occurrence of high-frequency noise in a DEM leads to production of much noisier models of morphometric attributes derived from the DEM (Brown and Bara, 1994; Giles and Franklin, 1996; Steen et al., 1998; Stewart and Podolski, 1998; Roberts, 2001; Florinsky, 2002; Bergbauer and Pollard, 2003; Bergbauer et al., 2003; Oksanen and Sarjakoski, 2005). For example, curvature maps derived from undenoised DEMs are almost unreadable and cannot be used in research. This is associated with the increase of high-frequency noise by computation of partial derivatives of elevation (Section 5.4.1).

Suppression of high-frequency noise in DEMs can be performed by low-frequency spatial filtering based on the 2D discrete Fourier transform (Section 6.2.3), 2D discrete wavelet transform (Section 6.2.4), smoothing (Section 6.2.5), cutting method (Section 6.2.7), 2D singular spectrum analysis (Section 6.3.1), depression filling (Section 4.4), and so on.

6.1.3 Generalization

Any research usually includes special stages of data processing dedicated to reduce an excess of information and to reveal the most important features of objects, phenomena, and processes under study. In order to do this, each science has qualitative and quantitative approaches to analyze and transform initial data. In sciences dealing with spatially distributed data as well as using cartographic representation of data and cartographic method of investigation (Salishchev, 1955; Dury, 1962; Berlyant, 1986), a set of such approaches is known as generalization (Eckert, 1908; Filippov, 1955; Steward, 1974; Brassel and Weibel, 1988; Florinsky, 1991; McMaster and Shea, 1992; Weibel, 1997; Mackaness et al., 2007). Since a map is one of the key outputs of digital terrain modeling, next we will briefly discuss the main notions of cartographic generalization.

6.1.3.1 Generalization in Cartography

The term *cartographic generalization* usually refers to generalization of map content according to the map purpose and scale. There are two types of generalization: the scale-oriented generalization and the aim-oriented one (Shiryaev, 1977).

The scale-oriented generalization is a necessary condition to producing any map. This is the cartographic manifestation of reducing the excess of information. "Generalization is immanent for a map; a cartographer, in fact, performs generalization during an initial survey and production of the very first map (which can be as detailed as one likes) without thinking of any generalization" (Baransky, 1946, p. 188). There are two reasons for this: (1) the complexity of the form, configuration, or structure of mapped features does not allow generating their "perfect" images; and (2) any graphical tool for representing information should have clear and laconic language.

The scale-oriented generalization deals with the following three tasks:

1. Simplification of the map structure, preserving a general similarity to the initial map
2. Preservation of a maximum precision for location of mapped features

3. A compromise between a maximum informativeness of a map and its readability

The aim-oriented generalization is a necessary condition for revealing the most important or particular features of objects, phenomena, and processes concerned. Perception and understanding of such information using a nongeneralized map may be difficult owing to the integrative nature of data representation.

The aim-oriented generalization deals with the following two tasks:

1. Revealing and representation of particular attributes, properties, and relations of objects, phenomena, and processes under study
2. Transition from individual characteristics to group ones

Resolution of these tasks can improve the results of the cartographic analysis (Lawrence, 1971), as well as allow one to obtain qualitatively new information (Salishchev, 1955; Berlyant, 1986) that can be of great fundamental and practical importance. A prominent example of the role of cartographic generalization in fundamental research is the production of a small-scale hypsometric map of the East European Plain: its analysis allowed Tillo (1890) to reveal regularities in the topographic structure at the subcontinental scale. The generalized nature of small-scale maps and globes made it possible to develop geological theories of the continental drift (Wegener, 1915) and the expanding Earth (Carey, 1988).

To carry out the aim-oriented generalization, one uses concepts and algorithms of the scale-oriented generalization.

6.1.3.2 DTM Generalization

The topographic surface is an extremely complicated natural object, so its investigation must incorporate data generalization. Otherwise, a researcher risks obtaining informationally overloaded, poorly readable maps. Recasting Baransky (1946) (Section 6.1.3.1), generalization is immanent for a DTM. Indeed, discretization of the continuous function of elevation is data generalization. Models of any topographic variables derived from a DEM consider the generalized nature of any DEM.

Weibel (1992) recognized three classes of DTM generalization techniques: global filtering, selective filtering, and heuristic generalization:

1. *Global filtering*. This class includes trend-surface analysis (Section 6.2.1), a low-frequency spatial filtering based on the 2D discrete Fourier transform (Section 6.2.3) and wavelet transform (Section 6.2.4), smoothing (Section 6.2.5), row and column elimination from DTMs (Section 6.2.6), cutting method (Section 6.2.7), 2D singular spectrum analysis (Section 6.3.1), fractal interpolation schemes (Bindlish and Barros, 1996), and so on. From the standpoint

of traditional cartography, these approaches are too formal because they do not consider important features of the topography (e.g., structural lines) in data processing.
2. *Selective filtering.* Methods of this class remove "insignificant" elevation points from a DEM using some thresholds. Among these techniques is adaptive triangular mesh filtering (Heller, 1990), which allows preserving structural lines if their points are assigned higher weights than other points (Weibel, 1997).
3. *Heuristic generalization.* Techniques of this class are based on an emulation of principles used in manual cartography. In particular, they consider important topographic features (e.g., structural lines, peaks, and pits), which are generalized and preserved in output DEMs (Weibel, 1997). Among these are the methods proposed by

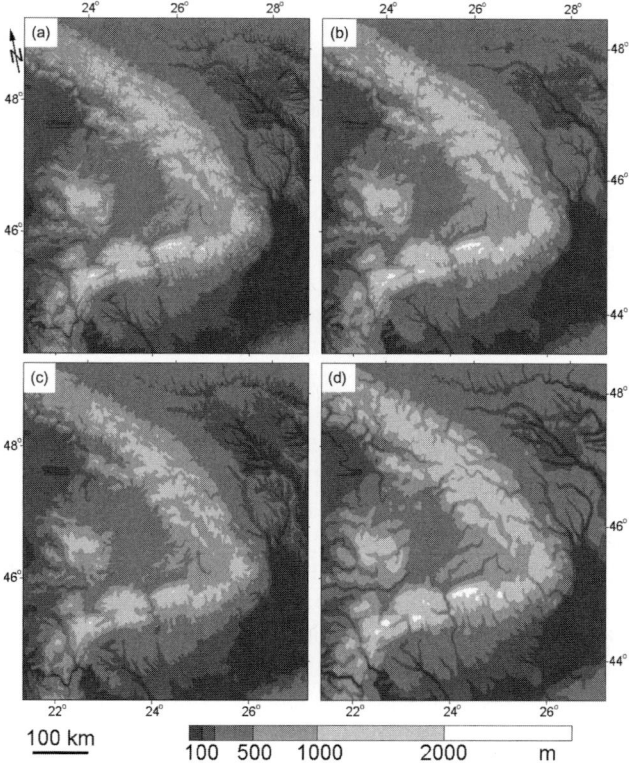

FIGURE 6.1 DEM generalization by the method of Leonowicz et al. (2009). Elevation maps of the Carpathian Mountains: (a) before generalization; (b) after application of an upper quartile filter; (c) after application of a lower quartile filter; (d) after combination of filtered maps with simplified drainage network and removing pits. A DEM was extracted from the global DEM GTOPO30 (USGS, 1996). The grid spacing is 30". *From (Leonowicz et al., 2009, Figs. 4, 6, and 12), reproduced with permission.*

Leonowicz et al. (2009), Ai and Li (2010), Samsonov (2010), and Zhou and Chen (2011). For example, the method of Leonowicz et al. (2009) uses upper and lower quartile filters that tend to preserve original elevations of crests and thalwegs (Fig. 6.1). First, an initial DEM is filtered with an upper quartile filter to produce a model appropriate for ridge areas. Second, the initial DEM is filtered with a lower quartile filter to produce a model appropriate for valley bottoms. Third, the two filtered DEMs are combined. Finally, a generalization level is adjusted by simplifying the drainage network (the shortest streams are removed).

There are several peculiarities of the scale-oriented generalization in the context of digital terrain analysis, notably:

- The grid density of an irregular DEM or the grid spacing of a regular DTM formally determines a generalization level or resolution of a DTM. If different portions of an irregular DEM have different grid densities, such distinctions are transmitted to all DTMs derived from the DEM: their generalization level will also be a function of planar coordinates.
- Interpolation of an irregular DEM may play a role of generalization, if the grid spacing of a regular grid is much larger than the density of the irregular grid (Rhind, 1971).
- There are minor errors (high-frequency noise) in any DEM. This dictates a need for low-frequency spatial filtering or smoothing of a DEM (Sections 6.2.3–6.2.5) that can be considered a form of generalization (Florinsky, 1991). In the derivation of local morphometric variables, some denoising takes place if one uses the Evans–Young method (Section 4.1) or the author's methods (Sections 4.2 and 4.3), since these approaches *approximate* polynomials (Eqs. 4.1 and 4.7) to elevation values of the moving windows.
- Numerical results of morphometric calculations depend on the DEM grid spacing, that is, on the DEM generalization level (Chapter 4) (Evans, 1972, 1979, 1980; Zhang and Montgomery, 1994; Stewart and Podolski, 1998; Shary et al., 2002b; Bergbauer and Pollard, 2003; Wynn and Stewart, 2003; Hancock, 2005; Smith et al., 2006; Sørensen and Seibert, 2007; Zhu et al., 2008). The accuracy of such calculations also depends on the DEM grid spacing (Section 5.3).
- To generalize maps of morphometric variables, one can aggregate (decrease the number of) quantitative classes of mapped surfaces (Jenks, 1963). For example, classifying values of horizontal curvature into two levels—more or less than zero—it is possible to represent clearly flow divergence and convergence areas (e.g., Fig. 5.7f).
- Generalization of topographic segmentation maps (Section 2.7) is carried out by the aggregation of landform classes. For example,

mapping of accumulation, transit, and dissipation zones is the generalization in reference to segmentation using the Efremov−Krcho classification. Topographic segmentation by the Gaussian classification can be considered as the generalization with respect to segmentation with the Shary classification. Similar approaches can be used for other segmentation models (MacMillan and Pettapiece, 2000; MacMillan et al., 2000).
- DTMs are commonly used in statistical analyses of relations between soil and topographic characteristics and DTM-based prediction of soil properties (Chapters 9−11). It is known that data generalization may increase the absolute values of correlation coefficients between related processes or phenomena owing to removal of minor variations (Berlyant, 1986). For example, the correlation coefficient describing the association between soil moisture content and mean curvature changed from −0.58 to −0.88 after the DEM smoothing (Kuryakova et al., 1992). Thus, it is natural that DTM smoothing may improve the prediction of soil taxonomic classes in medium-scale soil mapping (Moran and Bui, 2002; Grinand et al., 2008; Behrens et al., 2010).

6.2 METHODS OF DTM FILTERING

6.2.1 Trend-Surface Analysis

Trend-surface analysis, a form of regression analysis, is in regular use in the geosciences (Chorley and Huggett, 1965; Davis, 1986). Using the least-squares approach, a digital model of some property is approximated by a global function of the form

$$z = f_1(x,y) + f_2(x,y) + \cdots + f_k(x,y) + R(x,y), \qquad (6.1)$$

where $f_k(x, y)$ are trend components and $R(x,y)$ is a residual. Trend-surface analysis can be based on various functions, such as algebraic, orthogonal, and trigonometric polynomials (Tobler, 1969).

Trend components are mapped separately and are then visually analyzed and interpreted (Fig. 6.2). Trend-surface analysis of the land surface DEMs was applied to study geological ring structures (Gosteva et al., 1983; Zverev and Strykov, 1985) and lineaments (Fedorov, 1991).

6.2.2 The Filosofov Method

The method of Filosofov (1960) was initially designed for manual analysis of topographic maps. The method allows revealing some tectonic structures, as well as modeling the interaction of erosion and vertical tectonic movements. Over many years, structural and petroleum

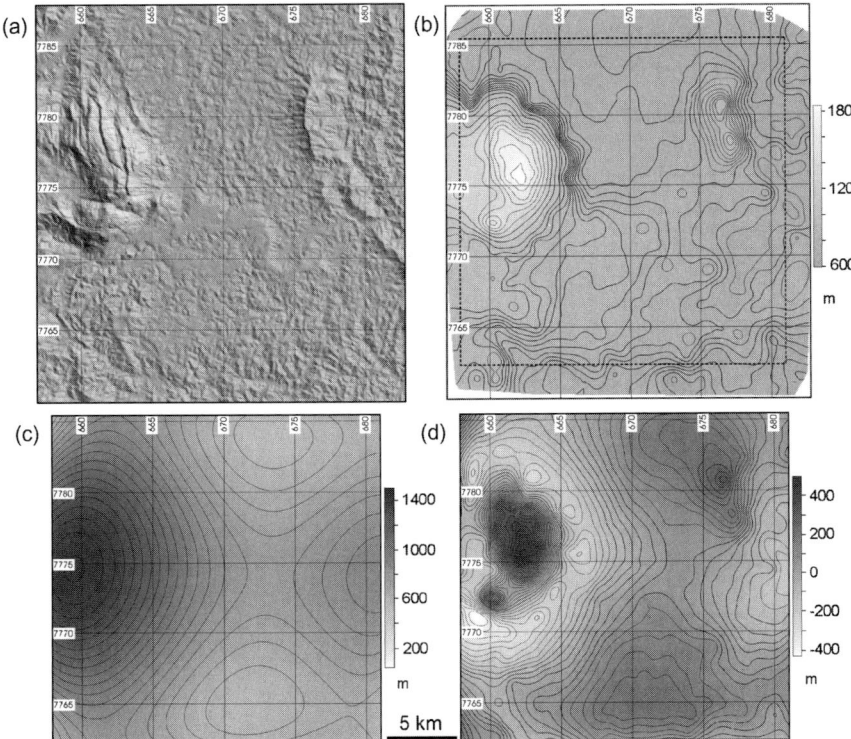

FIGURE 6.2 Use of the Filosofov method and trend-surface analysis in geological studies of the Quadrilátero Ferrífero region, the São Francisco Craton, Brasil: (a) Hill shading (illumination: the solar azimuth angle is 45°, the solar zenith angle is 30°). A DEM was generated by contour interpolation from a topographic map, scale 1 : 50,000. The DEM includes 292,357 points (the matrix 559 × 523); the grid spacing is 50 m. *From (Grohmann, 2004b, Fig. 2), reproduced with permission.* (b) Original isobase surface; dotted line indicates the area used for the trend-surface analysis. (c) The sixth-order trend surface. (d) The residual of the sixth-order trend surface. *From (Grohmann, 2005, Fig. 3a, c, d), reproduced with permission.*

geologists have used this manual method in a wide range of scales, from 1 : 10,000 to 1 : 2,500,000 (Filosofov, 1960, 1975; Gvin, 1963; Zuchiewicz, 1989; Golts and Rosenthal, 1993; Verkhovtsev, 2008). It makes sense to provide a reader with a brief description of the original technique since its digital versions were recently developed (Grohmann, 2004a, 2004b; Jordan, 2007; Grohmann et al., 2007, 2011) (Fig. 6.2).

The Filosofov method uses the hypothesis that valleys of different orders were formed in different times, and they are associated with tectonic motions and structures of different spatial scales. The method is based on the graphical decomposition of the topographic surface via manual transformations of the topographic contours. It may be considered to be a sort of manual trend-surface analysis. The method includes the following basic steps (Filosofov, 1960):

1. Map preparation of thalweg orders using the Strahler numbering system (Strahler, 1952, 1957). The first-order thalweg is a thalweg without tributaries; the second-order thalweg is a thalweg formed by the joining of two first-order thalwegs; and so on. In a similar manner, a map of crest orders can be produced.
2. Map generation of a base-level surface of some order by interpolating elevation values related to thalwegs of this and higher orders. For example, to produce a map of the third-order base-level surface, one should interpolate elevations associated with the third- and higher-order thalwegs. The higher the order of the base-level surface, the higher the generalization level of topography. It is recommended that the second-order base-level maps be used to reveal local anticlinal structures, the third- and fourth-order base-level maps to recognize platform structures (e.g., flexures and depressions), and the fifth- or sixth-order base-level maps to analyze anteclises and syneclises.
3. Map preparation of a residual topography of some order by subtracting the base-level surface of this order from the topographic surface. These maps can be used to reveal local tectonic structures.
4. Map production of a summit-level surface of some order by interpolating elevation values related to crests of this and higher orders. Summit-level maps are generated similarly to base-level maps (see above). They can also be used to detect areas of neotectonic uplift and/or subsidence of different scales.
5. Map generation of a difference of two base-level surfaces of different orders. Such maps may characterize a change of elevations in a period between the formation of valleys of different orders. The greater the difference in elevation between neighboring base-level surfaces, the more intensive the vertical tectonic movements.

Simplified computerized approaches are available to calculate digital models of the base- and summit-level surfaces as well as relative relief. In this case, relative relief is a difference between summit level—the highest altitude for a given area—and base level—the lowest altitude for a given area (Dury, 1962, p. 174). Calculations are performed with moving windows. Relative relief may be used as an index of the relative velocity of vertical tectonic movements. Digital models of such attributes have been applied to reveal active tectonic structures (Ioffe and Kozhurin, 1997; Szynkaruk et al., 2004), to recognize palaeosurfaces (Johansson, 1999), to estimate seismic activity (Zamani and Hashemi, 2000), and to study the interaction between endogenous and exogenous processes of orogenesis (Kühni and Pfiffner, 2001a, 2001b; Zhang et al., 2006).

6.2.3 Two-Dimensional Discrete Fourier Transform

The 2D discrete Fourier transform (Sundararajan, 2001) is widely used in the geosciences. Within the framework of the 2D Fourier transform, the topographic surface is represented as a linear combination of sinusoidal surfaces of various frequencies, phase shifts, and amplitudes. The problem of the 2D Fourier transform is determining the coefficients of these sinusoidal surfaces. The problem of the 2D inverse Fourier transform is reconstructing the topographic surface using the coefficients of sinusoidal surfaces.

Let a 2D discrete function of elevation $z = f(x, y)$ that is, a DEM is a matrix $N \times M$ of $z(k, l)$; $k = 0, 1, 2, \ldots, N - 1$; $l = 0, 1, 2, \ldots, M - 1$. For this function, the 2D discrete Fourier transform is

$$Z(n, m) = \sum_{k=0}^{N-1} \sum_{l=0}^{M-1} z(k, l) e^{-i2\pi(nk/N + ml/M)}, \qquad (6.2)$$

where $Z(n, m)$ are the coefficients of sinusoidal surfaces, i is the imaginary unit, n and m are frequencies: $n = 0, 1, 2, \ldots, N - 1$; $m = 0, 1, 2, \ldots, M - 1$.

The $Z(n, m)$ coefficients represent the frequency domain of the terrain. A matrix $N \times M$ of the modules of the $Z(n, m)$ coefficients is the power spectrum of the terrain.

The 2D inverse discrete Fourier transform is

$$z(k, l) = \frac{1}{N}\frac{1}{M} \sum_{n=0}^{N-1} \sum_{m=0}^{M-1} Z(n, m) e^{i2\pi(nk/N + ml/M)}. \qquad (6.3)$$

Wavelengths of nth and mth harmonics ω_n and ω_m are

$$\omega_n = \frac{Nw}{n}, \ 0 < n \leq N/2; \quad \omega_n = \frac{Nw}{N - n}, \ N/2 \leq n < N; \qquad (6.4)$$

$$\omega_m = \frac{Mw}{m}, \ 0 < m \leq M/2; \quad \omega_m = \frac{Mw}{M - m}, \ M/2 \leq m < M, \qquad (6.5)$$

where w is a DEM grid spacing (Papo and Gelbman, 1984).

To filter a DEM, $Z(n, m)$ coefficients of the harmonics concerned should be considered in Eq. (6.3) during the reconstruction of the topographic surface. For low-pass filtering, for example, $Z(n, m)$ coefficients corresponding to wavelengths shorter than some threshold should be set to zero. Calculations are usually performed by the well-known algorithm of the Fast Fourier Transform (Cooley and Tukey, 1965).

The 2D discrete Fourier transform has been periodically used to generalize DTMs (Papo and Gelbman, 1984; Rudy, 1999), to describe hierarchical structure of the topography (Harrison and Lo, 1996; Gallant and

Hutchinson, 1997; Puzachenko et al., 2002), and to denoise DEMs (Arrell et al., 2008; Gallant and Read, 2009). In structural geological analysis, the 2D discrete Fourier transform of DEMs has been applied to both stratigraphic surfaces (Bergbauer et al., 2003) and the land surface (Chigirev, 1976; Dhont and Chorowicz, 2006).

For instance, the 2D discrete Fourier transform was utilized to reconstruct the topography of the ancient Ethiopian dome comprising the Ethiopian and Yemeni Plateaus (Fig. 6.3) (Collet et al., 2000a, 2000b). The DEM of the current topographic surface of the region (Fig. 6.3a) was used as the initial data. First, the authors "closed" three rifts (the Red Sea, Gulf of Aden, and Ethiopian rifts), which have been initiated by mantle-plume activity during Oligocene times. To do this, portions of the initial DEM, corresponding to the Arabian and Somalian plates, were shifted relative to the stationary Nubian plate (Fig. 6.3b). Second, elevation values were masked and recalculated for the areas of the recent volcanic edifices and gaps between the rift shoulders remaining after the first step. Then, the topography was smoothed using the Fast Fourier Transform. As a result, a model of the pre-rift topographic surface of the Ethiopian dome was obtained (Fig. 6.3c).

6.2.4 Two-Dimensional Discrete Wavelet Transform

The wavelet transform is signal decomposition using a system of wavelets, that is, functions each of which is a shifted and scaled copy of a function, the mother wavelet. The term *wavelet transform* defines a class of decomposition methods. Existing types of the wavelet transform differ widely in definitions and properties (Chui, 1992; Daubechies, 1992; Pereberin, 2001).

The wavelet transform of a one-dimensional discrete signal $\mathbf{s} = \{s_j\}$, $j \in Z$ is carried out by the following recursive formulas (Pereberin, 2001):

$$\mathbf{v}_i = \downarrow_2[\mathbf{v}_{i+1} * \tilde{\mathbf{h}}], \quad \mathbf{w}_i = \downarrow_2[\mathbf{v}_{i+1} * \tilde{\mathbf{g}}], \quad i = i_1 - 1, i_1 - 2, \ldots, i_0, \tag{6.6}$$

where $\mathbf{v}_i$ is the approximation of the signal with a lower resolution (a low-frequency component), $\mathbf{w}_i$ is the detailing information (a high-frequency component), $\tilde{\mathbf{h}}$ and $\tilde{\mathbf{g}}$ are a low-frequency and high-frequency analysis filters, respectively, $\downarrow_2[\mathbf{x}]$ is the operator removing every second element from the signal $\mathbf{x}$, * denotes convolution, and i is the scale.

Some scale value i_1 is related to the initial signal $\mathbf{s}$: $\mathbf{v}_{i_1} = \mathbf{s}$. On each step of the wavelet transform, the signal is separated into two components, $\mathbf{v}_i$ and $\mathbf{w}_i$. Elements of high-frequency components are wavelet coefficients. The lowest scale i_0 corresponds to the case when a low-frequency component consists of the only element.

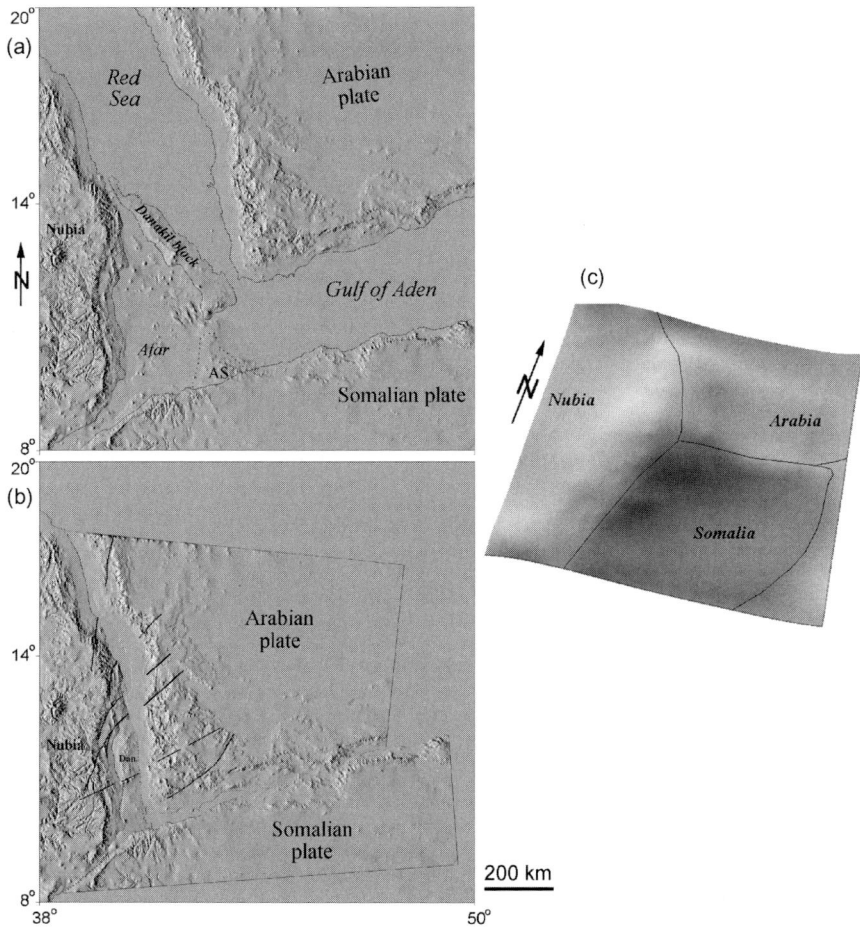

FIGURE 6.3 Application of the 2D discrete Fourier transform to model the ancient Ethiopian dome: (a) Hill shading of the current location of tectonic blocks. Black lines correspond to the block limits; the dotted lines are the supposed limit of the Ali-Sabieh (AS) block. A DEM was derived from the Digital Chart of the World (US DMA, 1991) by contour interpolation. The grid spacing is 2000 m. (b) Reconstitution of the territory for the Oligocene time (pre-rifting period). Dan.—Danakil block, solid lines are lineaments; dotted lines are the Ethiopian rift axis and the Arabia/Danakil limit. *From (Collet et al., 2000b, Figs. 1 and 3), reproduced with permission.* (c) 3D model of pre-rifting topography of the Ethiopian dome after applying the Fast Fourier Transform to the DEM. *From (Collet et al., 2000a, Fig. 2), reproduced with permission.*

The inverse wavelet transform, or reconstruction of the signal **s** is carried out by the following formula (Pereberin, 2001):

$$\mathbf{v}_{i+1} = \uparrow_2[\mathbf{v}_i] * \mathbf{h} + \uparrow_2[\mathbf{w}_i] * g, \quad i = i_0, i_0 + 1, \ldots, i_1 - 1, \qquad (6.7)$$

where **h** and **g** are a low- and high-frequency synthesis filters, correspondingly, $\uparrow_2[\mathbf{x}]$ is the operator inserting the zero element between elements of the signal **x**.

Equations (6.6) and (6.7) are general expressions for the direct and inverse wavelet transforms of discrete signals. A particular type of the transform is defined by the four filters: $\tilde{\mathbf{h}}$, $\tilde{\mathbf{g}}$, **h**, and **g**. Mathematical description of particular filters (e.g., Haar, D_4, biorthogonal, B-spline) can be found elsewhere (Chui, 1992; Daubechies, 1992; Pereberin, 2001).

The 2D wavelet transform can be performed by combining 1D wavelet transforms. Let there be a finite-size matrix. Let us apply one step of the 1D wavelet transform to each row of the matrix. We obtain two matrices, which rows include low- and high-frequency components of rows of the initial matrix. Also, let us apply one step of the 1D wavelet transform to each column of the both matrices. As a result, we obtain four matrices. The first of them is the low-frequency component of the initial 2D signal, while the other three include detailing information.

In digital terrain analysis, the 2D wavelet transform has been used for the denoising of DEMs (Fig. 6.4) (Bergbauer et al., 2003; Falorni et al., 2005; Lark, 2007), decomposition of the topographic surface into components of different scales and description of the hierarchical structure of the terrain (Gallant and Hutchinson, 1996, 1997; McArthur et al., 2000; Sulebak and Hjelle, 2003), as well as DTM generalization (Bjørke and Nilsen, 2003; Wu, 2003). Audet (2011) applied spherical wavelets to global topographic and gravitational data to analyze the lithospheric isotropy for the Earth, Venus, Mars, and the Moon.

6.2.5 Smoothing

DEM smoothing with moving windows is one of the most popular approaches for removal of artifacts from DTMs, DEM denoising, and DEM generalization (Tobler, 1966; Evans, 1972; Florinsky, 1991; Weibel, 1992; Schmid-McGibbon, 1995; MacMillan and Pettapiece, 2000; Albani and Klinkenberg, 2003; Leonowicz et al., 2009). DEM smoothing is widely used in DTM-based geological (Stewart and Podolski, 1998; Korsakova, 2002; Bergbauer and Pollard, 2003; Florinsky, 2007a) and soil (Moran and Bui, 2002; Grinand et al., 2008; Behrens et al., 2010) studies. There are various types of smoothing filters, such as moving average smoothing, median smoothing, and so on.

For the moving average smoothing, the general formula is as follows:

$$z'_{(n+1)/2} = \frac{\sum_{i=1}^{n} W_i z_i}{\sum_{i=1}^{n} W_i}, \qquad (6.8)$$

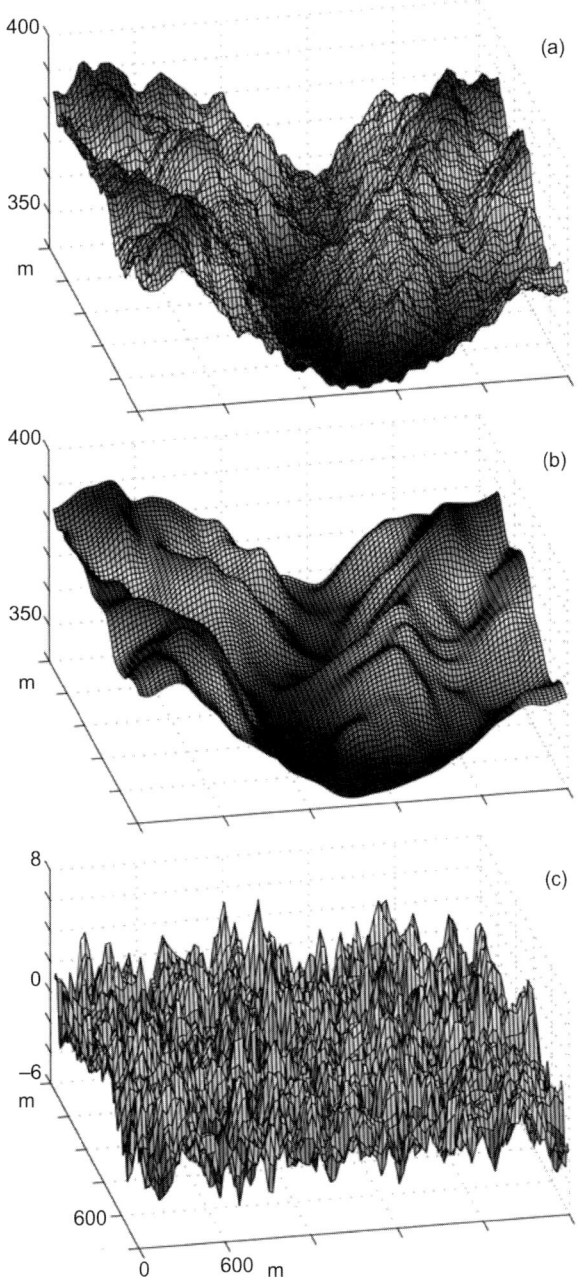

FIGURE 6.4 Application of a wavelet filter to denoise a SRTM DEM of a portion of the Little Washita Basin, Oklahoma, USA: 3D models of (a) the raw SRTM DEM; (b) the denoised DEM; (c) the noise removed from the SRTM DEM. A biorthogonal wavelet with five levels of decomposition was used. *From (Falorni et al., 2005, Fig. 11a, c, d), reproduced with permission.*

where $z'_{(n+1)/2}$ is the smoothed elevation value in the central point of the moving window, $n = 9, 25, 49, \ldots$ for 3×3, 5×5, $7 \times 7, \ldots$ windows, respectively; z_i is the elevation value in the ith point of the window, $i = 1, \ldots n$, and W_i is the weight of the ith point (Wood, 1996, § 4.4.5):

$$W_i = \frac{1}{(1 + d_i)^m}, \qquad (6.9)$$

where d_i is the distance from the ith point to the central point, $m = 0, 1, 2$.

The size of a moving window and the number of smoothing iterations are selected empirically. However, this approach has to be used with caution as it may produce high-frequency artifacts, the phase inversion (Chigirev, 1976). For instance, a pit may replace a hill, or a blockage may arise in a valley (Armstrong and Martz, 2003).

The moving average smoothing is illustrated Fig. 6.5. A DEM of the city of Moscow and adjacent areas was extracted from the SRTM3 DEM (NASA, 2003). The DEM includes 284,544 points (the matrix 624 × 456); the DEM grid spacing is 3″. Horizontal и vertical curvatures were derived from the DEM by the author's method (Section 4.3). To deal with the large dynamic range of k_h and k_v, we logarithmically transformed their digital models by Eq. (7.1) with $n = 4$.

One can see that maps of horizontal и vertical curvatures derived from the unsmoothed initial DEM (Fig. 6.5a) are almost unreadable. It is highly unlikely that they may be used in DTM-based research. This is because the DEM includes high-frequency noise, which is increased by calculation of derivatives (Section 5.4.1). Note, however, that the elevation map derived from the initial DEM is quite readable and can be used as an illustration (Fig. 6.5a). Maps of horizontal и vertical curvatures became readable after several iterations of the DEM smoothing. The greater the number of smoothing iterations, the more generalized maps can be obtained (Fig. 6.5b, c).

6.2.6 Row and Column Elimination

Point, row, and column elimination from a DEM is one of the most popular approaches of DTM generalization (Rhind, 1971; Florinsky, 1991; Stewart and Podolski, 1998; Bergbauer and Pollard, 2003). The simplest technique of the scale-oriented DEM generalization is the elimination of every nth row and every nth column from a regular DEM; n is integer (Rhind, 1971). This approach can be utilized as the aim-oriented generalization since the selection of a DTM grid spacing depends on a particular problem in soil or landscape research (Chapter 9).

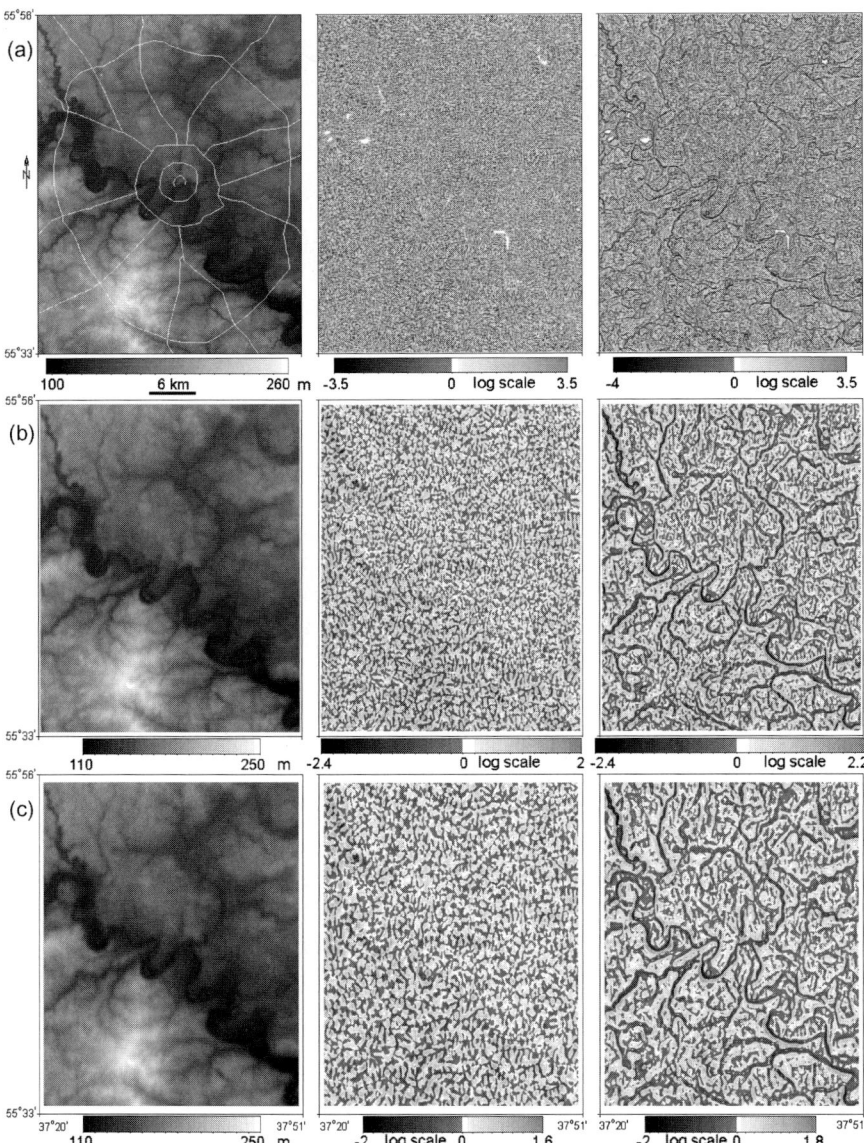

FIGURE 6.5 Moscow's city and adjacent areas, Russia: (a) the nonsmoothed DEM; (b) the 5-times smoothed DEM; (c) the 10-times smoothed DEM. Left—elevation (white lines are main highways), center—horizontal curvature, right—vertical curvature. A 3×3 smoothing window was used. *From (Florinsky, 2010, Fig. 1.22).*

Although elevation values in the DEM grid nodes are unchanged by row and column elimination, such an approach generates artifacts in a generalized DEM: for example, neighboring valleys belonging to different basins may merge into one valley (Armstrong and Martz, 2003).

Obviously, row and column elimination is the equivalent to the change of the grid spacing value. This approach has been used to decompose the topographic surface for description of the hierarchical structure of the terrain (Schmidt and Andrew, 2005), as well as to determine the adequate resolution for DTMs in soil studies (Florinsky and Kuryakova, 2000; Zhu et al., 2008; Roecker and Thompson, 2010); see details in Chapter 9. The change of the grid spacing value was applied to suppress high-frequency noise and to reveal geological structures of different scales using digital models of principal curvatures of stratigraphic surfaces (Stewart and Podolski, 1998; Bergbauer and Pollard, 2003; Wynn and Stewart, 2003). Notice that the grid spacing value influences the calculation accuracy of morphometric variables (Section 5.3).

Hierarchical DEMs have the built-in possibility of generalizing a model by removing particular hierarchical layers. Among such models are DEMs based on hierarchical TINs (de Floriani et al., 1993, 1997, 2000; Abdelguerfi et al., 1998) and spheroidal DEMs based on the hierarchical tessellation of a sphere by spherical polygons (Dutton, 1999; Sahr et al., 2003; Bernardin et al., 2010).

DTM generalization by elimination of every nth row and every nth column is illustrated in Fig. 6.6. Two DEMs of the East European Plain and adjacent territories (Fig. 6.6a) with different resolutions, 4' and 10', were compiled using several sources. Land elevations were extracted from the global DEM GLOBE (GLOBE Task Team and others, 1999); the bathymetry of all oceans and most seas was taken from the global DEM ETOPO2 (National Geophysical Data Center, 2001); and the bathymetry of the Caspian Sea and some large lakes was digitized from topographic maps (Florinsky, 2007a).

Models of vertical curvature (Fig. 6.6b) were derived from three times smoothed DEMs by the author's method (Section 4.3). To deal with the large dynamic range of k_v, we logarithmically transformed its digital models by Eq. (7.1) with $n = 7$ and 8 for the DEM grid spacing of 4' and 10', respectively. The generalization effect is especially seen on maps of vertical curvature.

6.2.7 The Cutting Method

Generalization of curvature maps and visualization of landforms or geological structures of a higher hierarchical level can be achieved

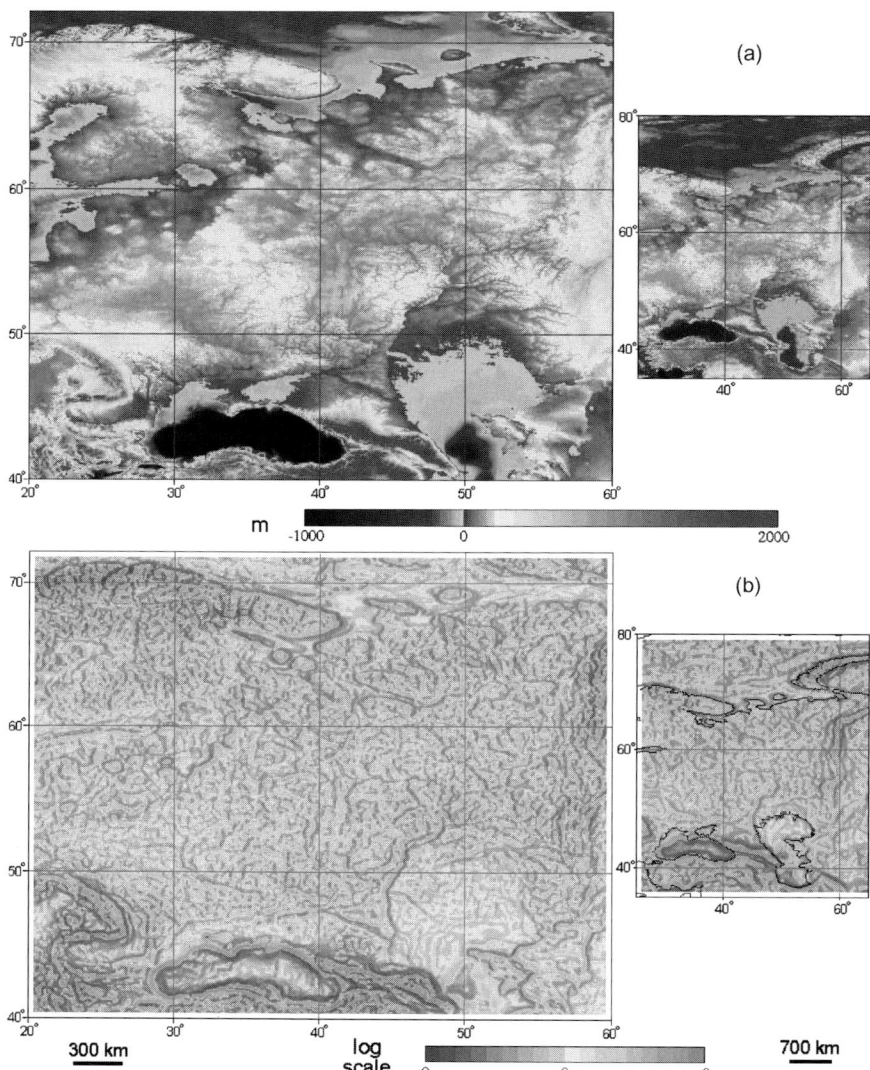

FIGURE 6.6 The East European Plain and adjacent territories: (a) elevation, (b) vertical curvature. Grid spacings are 4′ (left) and 10′ (right). DEMs include 289,081 points (the matrix 601 × 481) and 64,530 points (the matrix 239 × 270), respectively. *From (Florinsky, 2010, Fig. 1.23).* (See page 8 in Color Plate Section)

using the curvature threshold (k_t). It specifies an absolute value of a curvature, below which the curvature is set to zero.

Koshkarev (1982) introduced a method of such cutting or lopping off negligible or minor curvature values. Mynatt et al. (2007a) applied this approach to classify forms of a geological structure (Fig. 6.7) by signs of

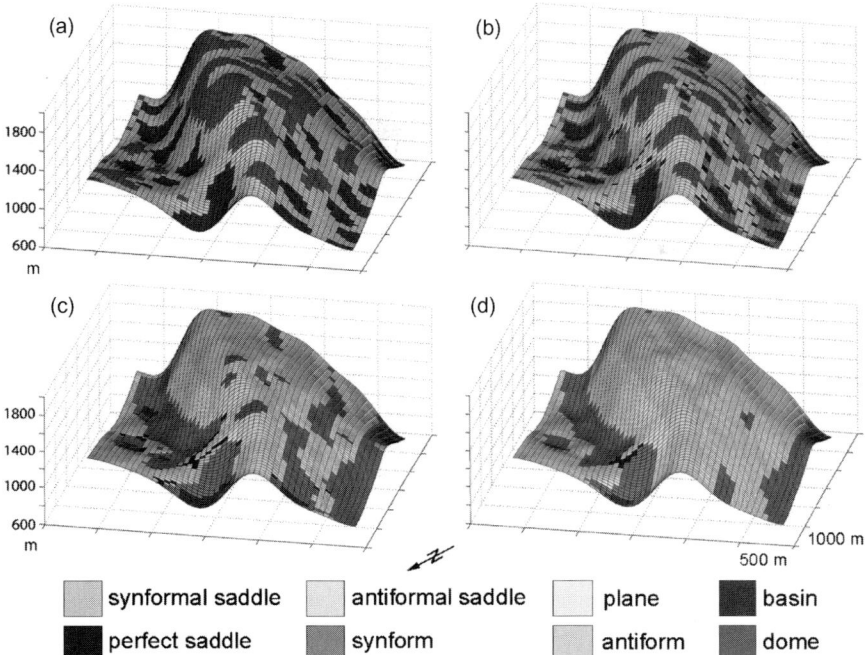

FIGURE 6.7 DTM generalization by the cutting method. Classification of the Sheep Mountain Anticline, Wyoming, USA, by signs of the Gaussian and mean curvatures with varying values of the curvature threshold: (a) $k_t = 0$ m^{-1}, (b) $k_t = 10^{-4}$ m^{-1}, (c) $k_t = 5 \times 10^{-4}$ m^{-1}, (d) $k_t = 10^{-3}$ m^{-1}. A DEM was created by digitizing a structure contour map. The regular DEM includes 4641 points; the grid spacing is 50 m. *From (Mynatt et al., 2007a, Fig. 5), reproduced with permission.* (See page 9 in Color Plate Section)

the Gaussian and mean curvatures (Section 2.7.1). The cutting was applied to the principal curvatures, which then were used to estimate K and H (Eqs. 2.6 and 2.7). These authors noted a generalization effect, which should be considered in interpretation of such maps:

For $k_t = 0$ m^{-1} (k_{min} and k_{max} were not set to zero), forms of the geological surface were classified as domes, basins, as well as antiformal and synformal saddles (Fig. 6.7a). There were no cylindrical and planar forms.[1] "The anticlinal hinge of the fold and the subparallel undulations on the backlimb are identified as convex up domes and antiformal saddles, while the synclinal hinge is composed of basins and synformal saddles" (Mynatt et al., 2007a, p. 1261). For $k_t = 10^{-4}$ m^{-1}, minor values of k_{min} and k_{max} were set to zero, and as a result, small cylindrical areas and local perfect saddles appeared on the map (Fig. 6.7b). For $k_t = 5 \times 10^{-4}$ m^{-1} and 1×10^{-3} m^{-1},

---

[1]This is the expected result: cylindrical and planar landforms are "rare forms"; they do not practically occur in nature (Shary, 1995). For details, see Section 2.7.3.

larger cylindrical and planar areas appeared (Fig. 6.7c, d). "The anticlinal hinge and adjacent backlimb undulations are composed almost entirely of antiformal points and the synclinal hinge is composed primarily of synclinal points. Both limbs are approximated as composed predominately of planar points" (Mynatt et al., 2007a, p. 1262).

6.3 TWO-DIMENSIONAL SINGULAR SPECTRUM ANALYSIS

Singular spectrum analysis (SSA) was originated as a model-free technique to analyze one-dimensional time series (Elsner and Tsonis, 1996; Danilov and Zhigljavsky, 1997; Golyandina et al., 2001). The SSA can be used to decompose a time series into a sum of a trend, oscillations, and a noise, to detect periodicities, to smooth and denoise signals, to forecast time series, to impute missing data, and so on.

There are several multidimensional extensions of the SSA. For example, a multichannel SSA is intended to analyze simultaneously a set of time series with common features (Elsner and Tsonis, 1996; Danilov and Zhigljavsky, 1997; Golyandina and Stepanov, 2005). The multichannel SSA can be formally applied to 2D scalar fields if one dimension is considered to be time. The two-dimensional singular spectrum analysis (2D-SSA) was specially designed to process 2D scalar fields (Danilov and Zhigljavsky, 1997; Golyandina and Usevich, 2010). Unlike the multichannel SSA, the 2D-SSA is invariant regarding field rotation.

6.3.1 Algorithm[2]

Let us consider a 2D discrete field $f:\{1, \ldots, N_r\} \times \{1, \ldots, N_c\} \mapsto \mathbf{R}$, given by a matrix

$$\mathbf{F} = \begin{pmatrix} f(1,1) & f(1,2) & \cdots & f(1,N_c) \\ f(2,1) & f(2,2) & \cdots & f(2,N_c) \\ \vdots & \vdots & \ddots & \vdots \\ f(N_r,1) & f(N_r,2) & \cdots & f(N_r,N_c) \end{pmatrix}. \quad (6.10)$$

Algorithm parameters are window sizes (L_r, L_c), where $1 \leq L_r \leq N_r$, $1 \leq L_c \leq N_c$, $1 < L_r L_c < N_r N_c$. Set $K_r = N_r - L_r + 1$ and $K_c = N_c - L_c + 1$. The algorithm includes two stages—decomposition and reconstruction—each of which consists of two steps (see details on matrix calculus elsewhere—Magnus and Neudecker, 1999, ch. 1).

[2]The 2D-SSA algorithm developed by N.E. Golyandina is described after our co-authored papers (Golyandina et al., 2007, 2008).

Stage 1: Decomposition. The first step—embedding—consists of constructing the trajectory matrix of the field F by moving $L_r \times L_c$ windows. In the 1D case, one transforms a 1D object into a 2D matrix (Elsner and Tsonis, 1996; Golyandina et al., 2001). Here, we embed a 2D object into a four-dimensional (4D) space. To flatten the 4D object, we transform moving windows

$$F_{i,j} = \begin{pmatrix} f(i,j) & f(i,j+1) & \cdots & f(i,j+L_c-1) \\ f(i+1,j) & f(i+1,j+1) & \cdots & f(i+1,j+L_c-1) \\ \vdots & \vdots & \ddots & \vdots \\ f(i+L_r-1,j) & f(i+L_r-1,j+1) & \cdots & f(i+L_r-1,j+L_c-1) \end{pmatrix}, \quad (6.11)$$

where $1 \leq i \leq K_r$, $1 \leq j \leq K_c$, to columns of the flattened trajectory matrix **W** ($F_{i,j}$ transfers to the $(i+(j-1)K_r)$th column). For example, if $L_r = L_c = 2$, then the window

$$\begin{pmatrix} f(1,1) & f(1,2) \\ f(2,1) & f(2,2) \end{pmatrix} \quad (6.12)$$

is transformed into the first column $(f(1,1), f(2,1), f(1,2), f(2,2))^T$.

It is appropriate to use vectorizing and matricizing operations: for a $M \times N$ matrix **B**, vec**B** $\in \mathbf{R}^{MN}$ is the vector constructed from stacked columns of **B**. If we fix matrix sizes M and N, then (M, N)-matricizing will be opposite to vectorizing: matr vec**B** = **B**. Thus, the trajectory matrix **W** of the field F consists of $K_r K_c$ columns vec$F_{i,j}$, where $1 \leq i \leq K_r$, $1 \leq j \leq K_c$. Furthermore, the matrix **W** of the size $L_r L_c \times K_r K_c$ can be presented in a more structured form:

$$\mathbf{W} = \begin{pmatrix} \mathbf{H}_1 & \mathbf{H}_2 & \cdots & \mathbf{H}_{K_c} \\ \mathbf{H}_2 & \mathbf{H}_3 & \cdots & \mathbf{H}_{K_c+1} \\ \vdots & \vdots & \ddots & \vdots \\ \mathbf{H}_{L_c} & \mathbf{H}_{L_c+1} & \cdots & \mathbf{H}_{N_c} \end{pmatrix}, \quad (6.13)$$

where

$$\mathbf{H}_i = \begin{pmatrix} f(1,i) & f(2,i) & \cdots & f(K_r,i) \\ f(2,i) & f(3,i) & \cdots & f(K_r+1,i) \\ \vdots & \vdots & \ddots & \vdots \\ f(L_r,i) & f(L_r+1,i) & \cdots & f(N_r,i) \end{pmatrix} \quad (6.14)$$

The matrix **W** has the block-Hankel structure with the same blocks along secondary diagonals. Each block $\mathbf{H}_i$ is Hankel itself: it is the trajectory matrix of the 1D series $f(\ ,i)$ (the ith column of the initial field F). The matrix **W** will be further called the block-Hankel trajectory matrix of the field F. Note that there is a one-to-one correspondence between $N_r \times N_c$ fields and block-Hankel matrices with $L_r \times K_r$ Hankel blocks.

The next step is a singular value decomposition of the block-Hankel trajectory matrix:

$$\mathbf{W} = \sum_{i=1}^{d} \mathbf{W}_i = \sum_{i=1}^{d} \sqrt{\lambda_i} U_i V_i^{\mathrm{T}}, \tag{6.15}$$

where $\lambda_1, \ldots, \lambda_d$ are nonzero eigenvalues of the matrix $\mathbf{WW}^{\mathrm{T}}$ arranged in decreasing order of magnitudes ($\lambda_1 \geq \lambda_2 \geq \cdots \geq \lambda_d > 0$); $\{U_1, \ldots, U_d\}$, $U_i \in \mathbf{R}^{L_r L_c}$ is the corresponding orthonormal system of the eigenvectors; $\{V_1, \ldots, V_d\}$, $V_i \in \mathbf{R}^{K_r K_c}$ is the orthonormal system of the corresponding factor vectors

$$V_i = \mathbf{W}^{\mathrm{T}} U_i / \sqrt{\lambda_i}. \tag{6.16}$$

By analogy with principal component analysis, the vectors $\sqrt{\lambda_i} V_i$ are called principal component vectors. They are conveniently considered as matrices: the (L_r, L_c)-matricizing of an eigenvector is called an eigenfield, the (K_r, K_c)-matricizing of a factor vector is called a factor field, and the (K_r, K_c)-matricizing of a vector of principal components is called a principal component field. A set of square root of ith eigenvalue, ith eigenfield, and ith factor field is called ith eigentriple (ET).

Stage 2: Reconstruction. The next step consists in the grouping of addends in the decomposition (Eq. 6.15), that is, the corresponding eigentriples. Let us divide the set $\{1, \ldots, d\}$ into m disjoint subsets $I_1, \ldots, I_m$. Summing $\mathbf{W}_i$, $i \in I_k$, we come to the expansion

$$\mathbf{W} = \sum_{k=1}^{m} \mathbf{W}_{I_k}. \tag{6.17}$$

Grouped matrices $\mathbf{W}_{I_k}$ do not necessarily have the block-Hankel form. Therefore, one needs an additional step to transfer the decomposition (Eq. 6.17) of the block-Hankel trajectory matrix $\mathbf{W}$ into a decomposition of the initial field F. This can be done by the orthogonal projection (in the Frobenious norm) of the matrices $\mathbf{W}_{I_k}$ on the set of the block-Hankel matrix with Hankel blocks, like Eq. (6.13). After projection, we obtain

$$\mathbf{W} = \sum_{k=1}^{m} \tilde{\mathbf{W}}_k, \tag{6.18}$$

where $\tilde{\mathbf{W}}_k$, $k = 1, \ldots, m$, have a form of Eq. (6.13). Using the one-to-one correspondence between block-Hankel trajectory matrices and 2D fields, we come to the final decomposition of the initial field:

$$F = \sum_{k=1}^{m} \tilde{F}_k. \tag{6.19}$$

The algorithm decomposes the initial field into a sum of components. It is expected that if the field F is a sum of a smooth surface, oscillations, and a noise, there exists such a grouping that the resultant decomposition (Eq. 6.19) is close to the initial field decomposition. This gives an opportunity for smoothing, denoising, removing periodic noise, and the like.

Rules to select 2D-SSA parameters are mostly similar to the 1D-SSA (Golyandina et al., 2001). In particular, small window sizes produce adaptive smoothing. The most detailed decomposition and the better separation of the field components can be obtained if window sizes are close to ($N_r/2$, $N_c/2$). However, use of large window sizes can lead to a mixing problem caused by too detailed decomposition of components.

6.3.2 Materials and Data Processing

To demonstrate the abilities of the 2D-SSA as a tool of DTM filtering, we selected a portion of the Northern Andes measuring $4° \times 4°$, located between 2°S and 2°N, and 78°30'W and 74°30'W (Fig. 6.8). The area covers some regions of Ecuador, Colombia, and Peru, including parts of the Coastal plain, the Andean Range, and the Upper Amazon Basin.

A DEM of the study area was extracted from the global DEM GTOPO30 (USGS, 1996). The DEM includes 230, 880 points (the matrix 480 × 481); the grid spacing is 30" (Fig. 6.9a).

We selected this area and GTOPO30 for two reasons. First, it is well known that this DEM incorporates a high-frequency noise caused by interpolation errors and inaccurate merging of topographic charts having different accuracy. Spatial distribution of the noise in GTOPO30 is

FIGURE 6.8 Geographical location of the portion of the Northern Andes. Some deep-seated faults are shown after the global fault scheme compiled by Chebanenko (1963, Fig. 28). *From (Florinsky, 2010, Fig. 1.24).*

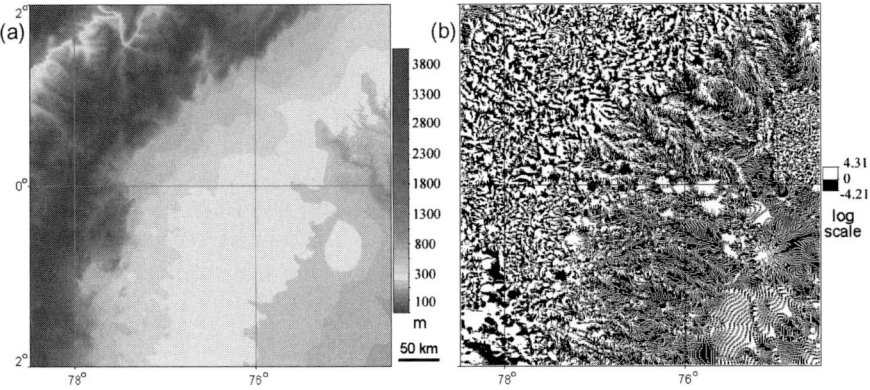

FIGURE 6.9 The Northern Andes: (a) elevation, (b) horizontal curvature derived from the initial DEM. *From (Golyandina et al., 2008, Fig. 2a, d).* (See page 10 in Color Plate Section)

uneven and depends on the accuracy of cartographic sources. In particular, the potent noise is typical for forested regions of Africa and South America because reasonably detailed and accurate topographic data were unavailable for such areas. Thus, interpolation of sparse contours has been used to compile these portions of GTOPO30. Although DEM noise is no obstacle to producing realistic maps of elevation, it leads to derivation of noisy and unreadable maps of local morphometric variables (computation of the first and second partial derivatives of elevation dramatically increases the noise—Section 5.4.1). The study area, consisting of two main zones—high mountains and forested foothills—which have a different signal-to-noise ratio, is ideally suited for validating the 2D-SSA as a tool to denoise DEMs. Second, use of this DEM allows us to demonstrate the ability of the 2D-SSA to decompose the topographic surface into components of different scales under complex geomorphic conditions.

To reduce the huge range of elevations (about 6080 m; see Fig. 6.9a), the initial DEM was logarithmically transformed. Logarithmic DEMs were used in the further processing and mapping.

Using the window size of 30 × 30, the initial DEM was decomposed into 900 eigentriples (Fig. 6.10). We evaluated various combinations of eigentriples to reconstruct DEMs. Finally, we selected ET combinations producing interpretable reconstructed DEMs. To denoise the initial DEM (Fig. 6.9a), a DEM was reconstructed from the ET 1−100 (Fig. 6.11a). To exemplify DTM generalization, a DEM was reconstructed from the ET 1−50 (Fig. 6.11b) and ET 1−25. To separate continental, regional, and local components of topography, six DEMs were reconstructed from the ET 1 (Fig. 6.12a), ET 2, ET 3, ET 2−3 (Fig. 6.12c),

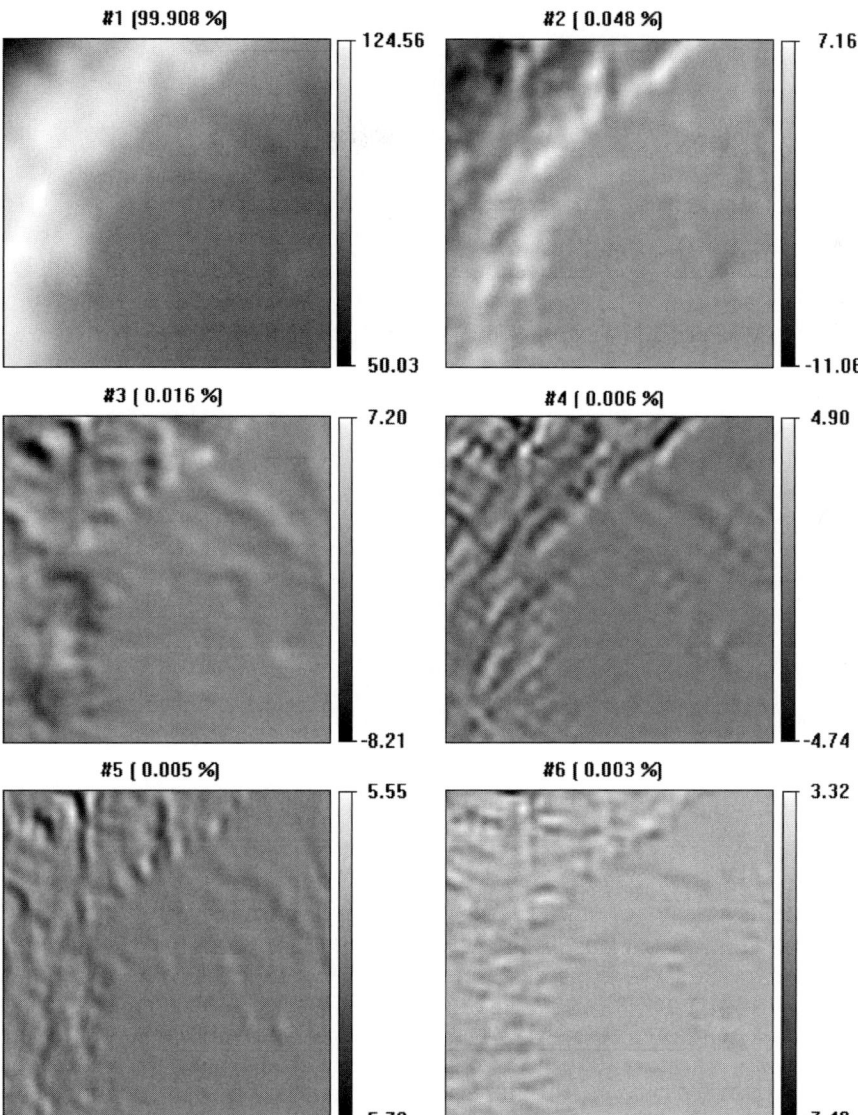

FIGURE 6.10 The Northern Andes, principal component fields 1−6 of the DEM decomposition. The percentage reflects shares of the corresponding eigentriples in the singular value decomposition. Logarithmic scale is used. *From (Golyandina et al., 2008, Fig. 1).*

ET 4−25 (Fig. 6.12d), and ET 51−100 (Fig. 6.13a). To visualize a noise component, a DEM was reconstructed from the ET 101−900 (Fig. 6.13b). DEM processing was done by the software 2D-SSA 1.2 (© K. Usevich and N. Golyandina, 2005−2007).

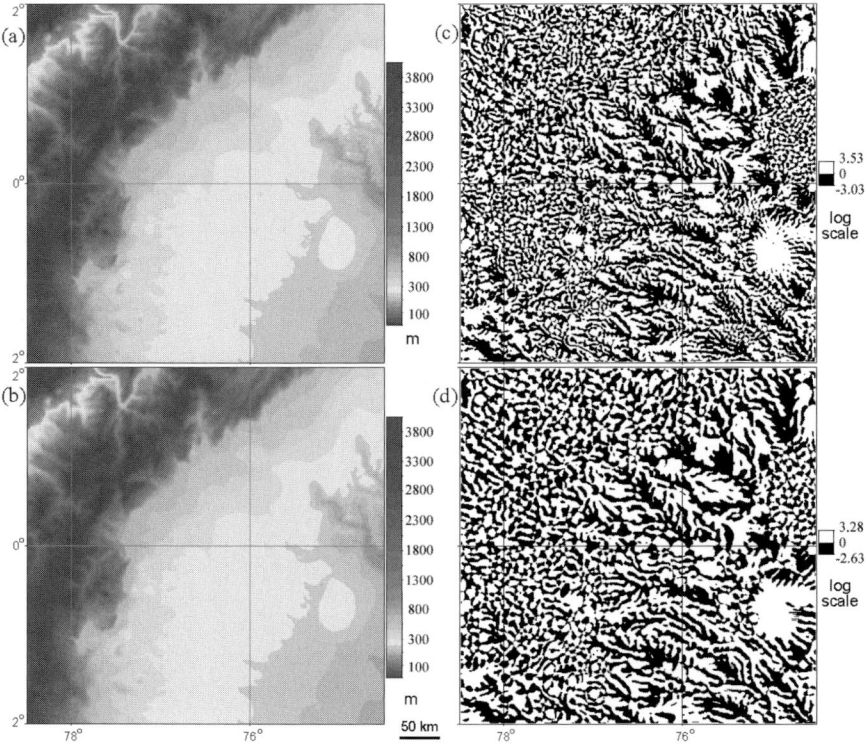

FIGURE 6.11 The Northern Andes, DTM denoising. Elevation reconstructed from: (a) the ET 1–100, (b) the ET 1–50. Horizontal curvature derived from: (c) the ET 1–100 DEM, (d) the ET 1–50 DEM. *From (Golyandina et al., 2008, Fig. 2b, c, e, f)*. (See page 10 in Color Plate Section)

To return to elevation values, reconstructed logarithmic DEMs were exponentiated. Using the author's method (Section 4.3), we derived k_h models from the initial DEM and some reconstructed DEMs. To deal with the large dynamic range of k_h, we logarithmically transformed its digital models by Eq. (7.1) with $n = 8$. To clarify the effects of denoising and generalization, we mapped k_h, subdividing its values into two levels: $k_h > 0$ and $k_h < 0$ (areas of flow divergence and convergence, correspondingly).

The DTMs produced had a grid spacing of 30". Exponentiation, k_h calculation, and mapping were done by the software LandLord (Appendix B).

Using samples extracted from each DEM and k_h models, we performed a pairwise comparison of DTMs obtained. The sample size was 2209 points (the matrix 47×47); the grid spacing was 5'. Statistical analysis was carried out with the software Statgraphics Plus 3.0 (© Statistical Graphics Corp., 1994–1997).

6.3 TWO-DIMENSIONAL SINGULAR SPECTRUM ANALYSIS

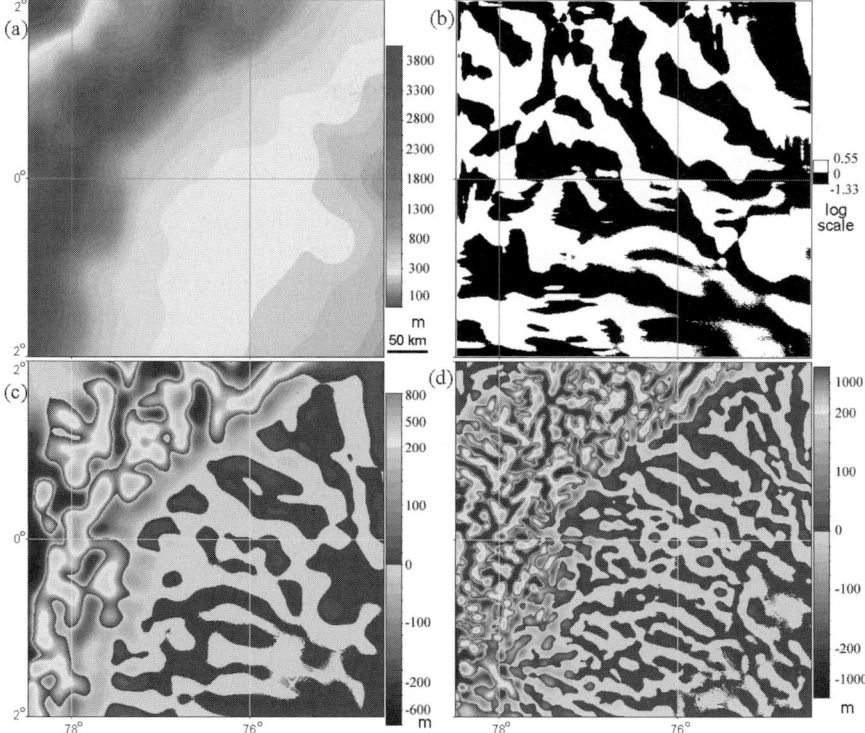

FIGURE 6.12 The Northern Andes, low-frequency components. (a) Elevation reconstructed from the ET 1. (b) Horizontal curvature derived from the ET 1 DEM. Elevation reconstructed from: (c) the ET 2–3, (d) the ET 4–25. *From (Golyandina et al., 2008, Fig. 3a–d).* (See page 11 in Color Plate Section)

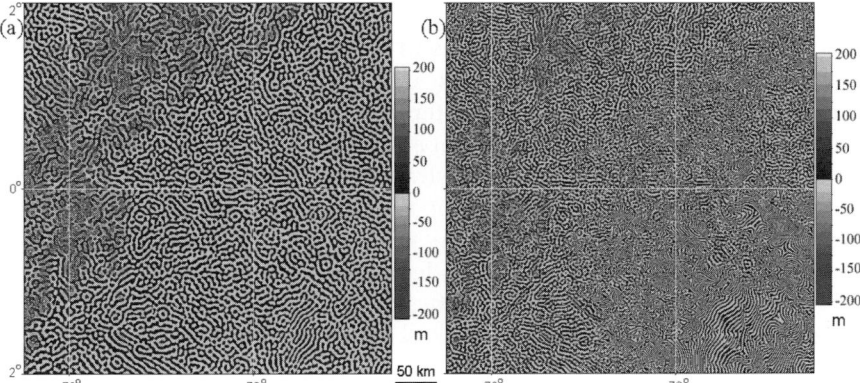

FIGURE 6.13 The Northern Andes, high-frequency components. Elevation reconstructed from: (a) the ET 51–100, (b) the ET 101–900. *From (Golyandina et al., 2008, Fig. 3e, f).* (See page 12 in Color Plate Section)

6.3.3 Results and Discussion

A visual comparison of elevation maps derived from the initial DEM (Fig. 6.9a) and two DEMs reconstructed from the ET 1–100 and ET 1–50 (Fig. 6.11a, b) permits one to see nothing but marginal changes in image patterns. A cursory examination may lead to underestimation of the DEM denoising and generalization results.

k_h maps provide better insight into these outcomes. Indeed, typical manifestations of interpolation errors—contour "tracks" (Florinsky, 2002) (Section 5.4)—can be found on the k_h map derived from the initial DEM (Fig. 6.9b). These tracks are typical for the Andean foothills covered by dense rain forests. It is not likely that this map may be used for any application. However, there are no error tracks on the k_h maps derived from the DEMs, which were reconstructed from the ET 1–100 and ET 1–50 (Fig. 6.11c, d). One can see so-called flow structures formed by convergence and divergence areas (black and white image patterns, correspondingly). These maps may be used for geomorphic and geological interpretation.

A comparison between the k_h maps derived from the DEMs reconstructed from the different eigentriple combinations (Figs. 6.9b and 6.11c, d) shows a pronounced effect of the map generalization. The smaller the number of eigentriples used to reconstruct a DEM, the smoother and simpler k_h maps obtained. Reducing the number of eigentriples used for DEM reconstruction leads to a marked reduction in the range of k_h values (cf. Figs. 6.9b and 6.11c, d), but it only slightly influences the range of elevation values (cf. Figs. 6.9a and 6.11a, b). This is clearly demonstrated by quantile-quantile plots (Fig. 6.14).

The DEM reconstructed from the ET 1 has the highest level of generalization (Fig. 6.12a). This DEM represents a generalized morphostructure of the continental scale, the Andean Range with foothills. k_h derivation from this DEM allows us to reveal a system of approximately northwest-striking lineaments (Fig. 6.12b), which may indicate strike-slip faults (Florinsky, 1996) (Chapter 13). Although there are no structures of this sort in the recent database of the Quaternary faults (Eguez et al., 2003), this does not demonstrate that the lineaments are of erosional origin. First, the geology of the Upper Amazon Basin is little known. Second, the lineaments may indicate pre-Quaternary structures. Indeed, Chebanenko (1963) has described a system of deep-seated northwest-striking faults southeastward from the study area (Fig. 6.8). The lineaments detected may be associated with northwestern extensions of these faults.

The DEM reconstructed from the ET 2–3 (Fig. 6.12c) represents landforms that are probably connected with regional tectonic structures. For the DEM reconstructed from the ET 2 only, the elevation map shows

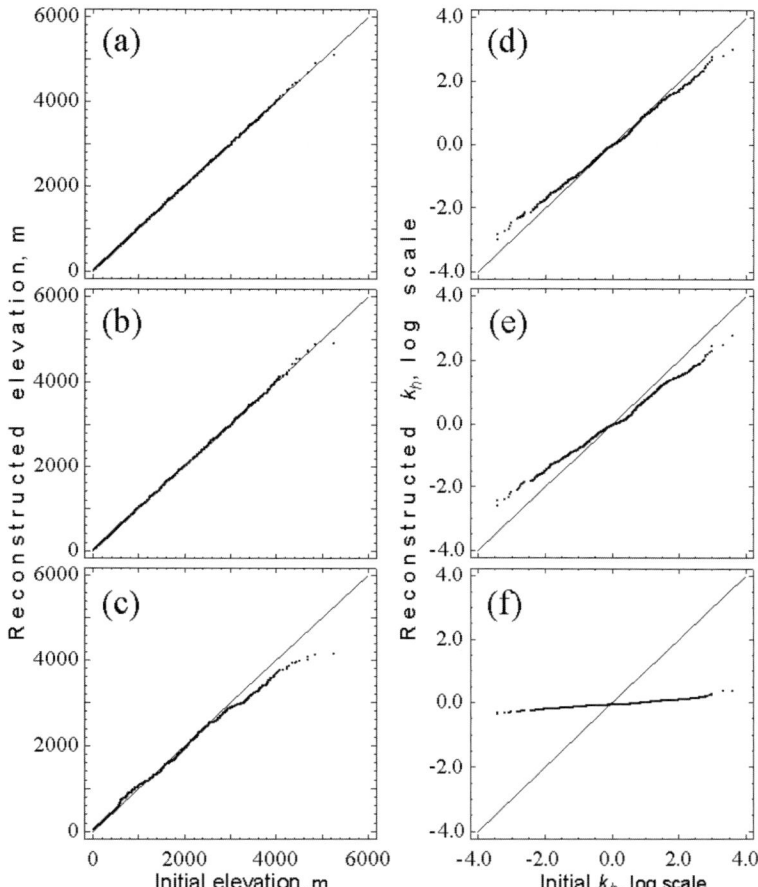

FIGURE 6.14 The Northern Andes, quantile–quantile plots for initial (x-axis) and reconstructed (y-axis) values: Elevation reconstructed from: (a) the ET 1–100, (b) the ET 1–50, (c) the ET 1. Horizontal curvature derived from: (d) the ET 1–100 DEM, (e) the ET 1–50 DEM, (f) the ET 1 DEM. The greater the distance between points and the diagonal line, the greater the difference between reconstructed and initial models. *From (Florinsky, 2010, Fig. 1.29).*

approximately northeast-striking patterns, which agree closely with the direction of the Andean Range within the study area. For the DEM reconstructed from the ET 3 only, the elevation map represents approximately northwest-striking structures, probably related to the k_h lineaments (Fig. 6.12b). It is conceivable that these structures control the spatial distribution of river valleys of the Upper Amazon Basin. The DEM reconstructed from the ET 4–25 represents landforms that are probably associated with geomorphic processes of a regional scale. For example, one can see generalized side valleys of the Amazon River (Fig. 6.12d).

The DEM reconstructed from the ET 51−100 (Fig. 6.13a) represents high-frequency components, which cannot be considered as noise. However, it is impossible to find familiar patterns of the drainage network on this map. These components might represent topographic manifestations of local (exogenous) geomorphic processes. This DEM is the difference between two DEMs reconstructed from the ET 1−100 and ET 1−50 (Fig. 6.11a, b).

The DEM reconstructed from the ET 101−900 (Fig. 6.13b) is a residual of the initial DEM (Fig. 6.9a) after its reconstruction from the ET 1−100 (Fig. 6.11a). Among other high-frequency components, this model includes the noise inherent in the initial DEM. This is precisely the noise that rendered the k_h map derived from the initial DEM unsuitable for applications. Among contour "tracks," one can see a rectangular feature measuring $1° \times 30'$, which is located eastward from 75°30′W and northward from the equator (Fig. 6.13b). The "rectangle" can also be seen on three maps of k_h (Figs. 6.9b and 6.11c, d). This is a trace of the inaccurate merging of adjacent topographic charts with different accuracies during the GTOPO30 compilation.

The DEM reconstructions from the ET 1, ET 2, ET 3, and ET 2−3 may be considered an application of low-pass filters to the initial DEM, while the DEM reconstruction from the ET 101−900 and ET 51−100 can be regarded as an application of high-pass filters.

The results demonstrate that the 2D-SSA is a powerful method to denoise DTMs. Indeed, 2D-SSA-based denoising leads to very fine changes in elevations. These changes cannot be captured except through computation of partial derivatives. This suggests that the 2D-SSA is an exceptionally important method for preliminary treatment of noisy DEMs. This opens the way to utilizing them for derivation of local and nonlocal morphometric variables.

The 2D-SSA can be used to decompose the topographic surface into components of continental, regional, and local scales. At the same time, eigentriple selection to reconstruct topographic components of different scales is nonunique and ambiguous. Notice that a similar problem arises if trend-surface analysis or the 2D Fourier transform is used to extract regional or local topographic components. However, the problem is largely associated with the qualitative nature of the concepts of "scale" and "hierarchical level" in geomorphology rather than with mathematical features of the methods. Finally, the 2D-SSA is a model-free approach; that is, it does not use *a priori* assumptions or statistical hypotheses about DEM structures.

CHAPTER

7

Mapping and Visualization

OUTLINE

7.1 Peculiarities of Morphometric Mapping	133
7.2 Combined Visualization of Morphometric Variables	135
7.3 Cross Sections	135
7.4 Three-Dimensional Topographic Modeling	136
7.5 Combining Hill-Shading Maps with Soil and Geological Data	141

7.1 PECULIARITIES OF MORPHOMETRIC MAPPING

Several general rules should be considered to produce readable and interpretable morphometric maps:

1. Even in the late nineteenth century, Tillo (1890, p. 10) noted that "contours by themselves do not form the integral or even partial picture of topography," although they describe mathematically spatial distribution of elevations. It is also undesirable to use isoline (or, isopleth) maps to represent other morphometric attributes; see, for example, numerous maps in the work by Krcho (1973). This is connected with very complex shapes of morphometric surfaces. As a result, human eyes poorly perceive isopleth morphometric maps. For the correct perception of topographic peculiarities by the human visual system, there is a need to use layer tinting (or, layer coloring). In this well-known cartographic method, a surface $z = f(x,y)$ is subdivided into a set of z-value intervals; each interval is colored with a graded set of colors. This enables surface peculiarities to be clearly displayed. In this book, all maps of morphometric attributes are produced with layer tinting.

2. For all morphometric variables, whose values can be both positive and negative, it is necessary to subdivide their values relative to the zero value. This is connected, in part, with the fact that values of different signs can measure different directions of a physical process associated with a particular variable (Table 2.1). For example, positive values of k_h describe flow divergence, while negative ones define flow convergence.
3. Correct selection of color scale is essential for morphometric mapping (Roberts, 2001). Our experience shows that there is a need to use contrast color schemes for all morphometric variables, whose values can be both positive and negative. This offers a means to distinguish between areas of positive and negative values (e.g., areas of flow divergence and flow convergence, respectively). We systematically use the blue and yellow scale (blue and yellow halftones relate to negative and positive values, respectively) (e.g., Fig. 2.5). For gray scales, it is reasonable to apply darker halftones to negative values, and lighter ones to positive values (e.g., Fig. 6.5).
4. Wide dynamic ranges usually characterize topographic variables. To avoid loss of information on the spatial distribution of values of morphometric attributes in mapping, it makes sense to apply a logarithmic transform using the following expression (Shary et al., 2002b):

$$\Theta' = \text{sign}(\Theta)\ln(1 + 10^{nm}|\Theta|), \tag{7.1}$$

where Θ is a value of a morphometric variable; $n = 0$ for nonlocal variables and $n = 2, \ldots, 9$ for local variables; $m = 2$ for K, K_a, and K_r, $m = 1$ for other variables. Selection of the n value depends on the DTM grid spacing; recommended values are presented in Table 7.1. Such a form

TABLE 7.1 Values of n for Eq. (7.1) Depending on the DEM Grid Spacing

DEM Grid Spacing, m	n
<1	2
1–10	3
10–100	4
100–1000	5
1000–5000	6
5000–10,000	7
10,000–75,000	8
>75,000	9

From (Florinsky, 2010, Table 1.2).

of logarithmic transformation considers that dynamic ranges of some topographic attributes include both positive and negative values.

7.2 COMBINED VISUALIZATION OF MORPHOMETRIC VARIABLES

Sometimes, researchers try to represent several morphometric variables on a single map. This is a difficult task from a technical point of view (to avoid the overlap of layers), as well as in the context of the human visual system's perception of multilayer graphical information.

In the precomputer era, Kvietkauskas (1963–1964) proposed a method to produce four-color morphometric maps with the following color scheme: a yellow scale was used for elevation values, a gray scale for slope gradient, a red scale for positive values of vertical curvature, and a blue scale for its negative values. Elevation, gradient, and vertical curvatures were mapped together using four nonoverlapping color raster grids.

Currently, two morphometric layers—elevation and hill shading— are most commonly combined (e.g., Sterner, 1995). Input elevation maps are displayed by color layer tinting, while input hill-shading data are presented as monochrome gray-scale maps. To combine color information on elevation with intensity information on insolation, one can use an intensity-hue-saturation transformation (Edwards and Davis, 1994).

Samsonov (2010) developed an approach for combined mapping of slope gradient and aspect. According to this method, both G and A are represented by hachures. The hachure thickness depends on the slope gradient value (as in classical hachure techniques; see Lehmann, 1816); this creates a hill-shading effect. Unlike monochromatic classical techniques, the hachure color depends on the slope aspect in the proposed method (Fig. 7.1). The resulting map is highly informative: Quantitatively, it integrates the properties of several morphometric maps, such as maps of elevation, gradient, aspect, and flow lines. Qualitatively, such a topographic image has a high degree of plasticity that allows easy perception of the shape of the topographic surface.

7.3 CROSS SECTIONS

Cross-section construction using DEMs of both the land and stratigraphic surfaces is a standard option of many softwares (Ichoku et al., 1994; Groshong, 2006, ch. 6). Cross sections allow one to display the peculiarities of relations between the land surface and the surfaces of

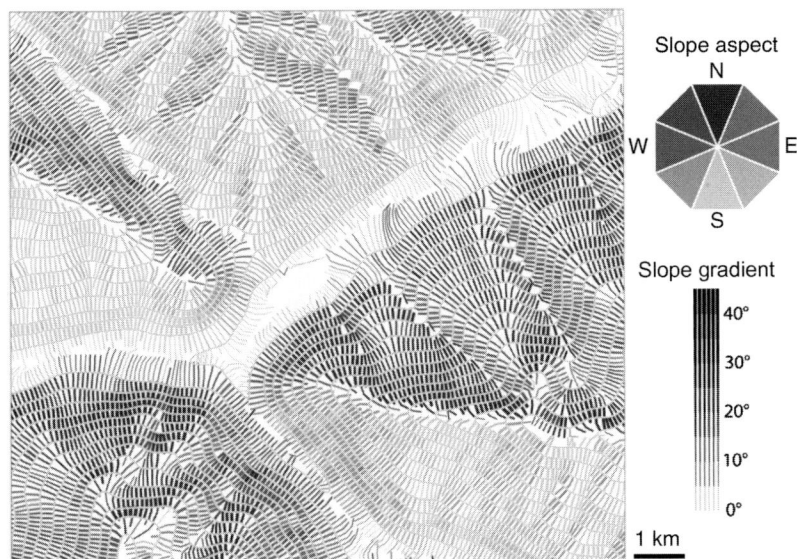

FIGURE 7.1 Combined visualization of slope gradient and aspect by colored hachures. *From (Samsonov, 2010, Fig. 9), reproduced with permission.* (See page 13 in Color Plate Section)

soil or stratigraphic horizons. Groshong (2006, p. 133) distinguishes two types of cross sections:

> illustrative or predictive. The purpose of an illustrative cross section is to illustrate the cross section view of an already-completed map or 3D interpretation. A slice through a 3D interpretation is a perfect example. The purpose of a predictive cross section is to assemble scattered information and, utilizing appropriate rules, predict the geometry between control points. A predictive cross section can be used to predict the geometry of a horizon for which little or no information is available.... Cross sections constructed for the purpose of structural interpretation are usually oriented perpendicular to the fold axis, perpendicular to a major fault, or parallel to these trends.

To construct interpretable cross sections, one should use an exaggeration of the vertical scale depending on the type of topography and the size of territory concerned. Figure 7.2 illustrates the use of DEM-derived cross sections in geological studies.

7.4 THREE-DIMENSIONAL TOPOGRAPHIC MODELING

Three-dimensional modeling is a useful data-processing step concerning the human perception of spatially distributed phenomena and

7.4 THREE-DIMENSIONAL TOPOGRAPHIC MODELING

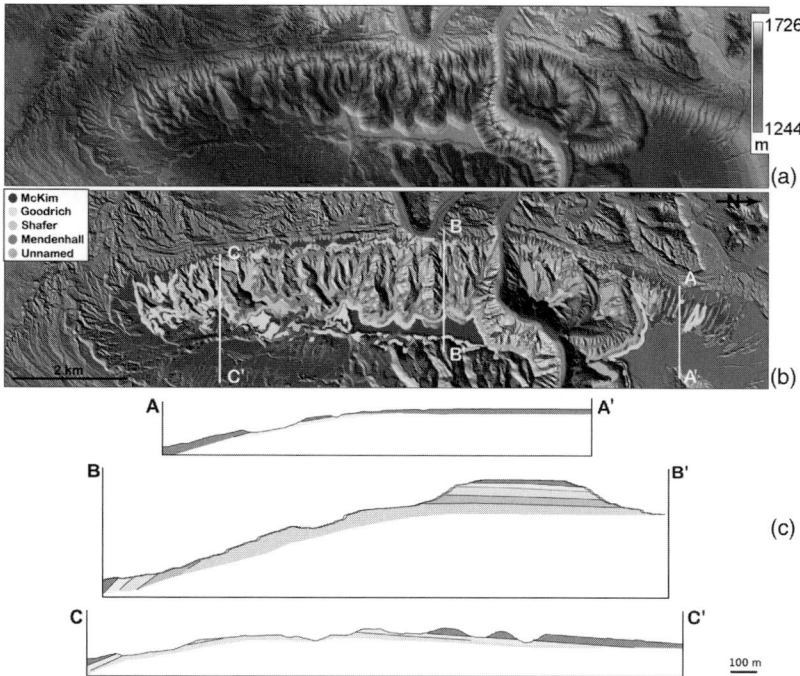

FIGURE 7.2 The use of DEM-derived cross sections in geological studies. The Raplee Ridge monocline, Utah, USA: (a) Elevation. (b) Hill shading with the spatial distribution of the five bedding-plane surfaces exhumed within the fold. (c) Cross sections (no vertical exaggeration). A DEM was derived from the Airborne Laser Swath Mapping data. The grid spacing is 1 m. *From (Hilley et al., 2010, Figs. 2a and 3), reproduced with permission.* (See page 14 in Color Plate Section)

information. Perspective viewing of a 3D model from different points in space can enhance the visual interpretation of remotely sensed images and maps as well as understanding of the relationships between elements of a geosystem.

There are three main types of DEM-based 3D modeling:

1. 3D topographic modeling, that is, 3D modeling of the land surface, a geological surface, or a soil horizon surface (Evans, 1972; Peucker, 1980b; Eyton, 1986; Patterson, 2001) (Fig. 7.3a). Currently, 3D modeling using simple orthographic and perspective projections is a standard option of many commercial and research softwares (Wood, 2009). Also, there are more complex, perspective panoramic 3D models (Patterson, 2001; Jenny et al., 2010). As for cross sections, to construct interpretable 3D models, one should use an exaggeration of the vertical scale depending on the type of topography and the size of territory concerned.

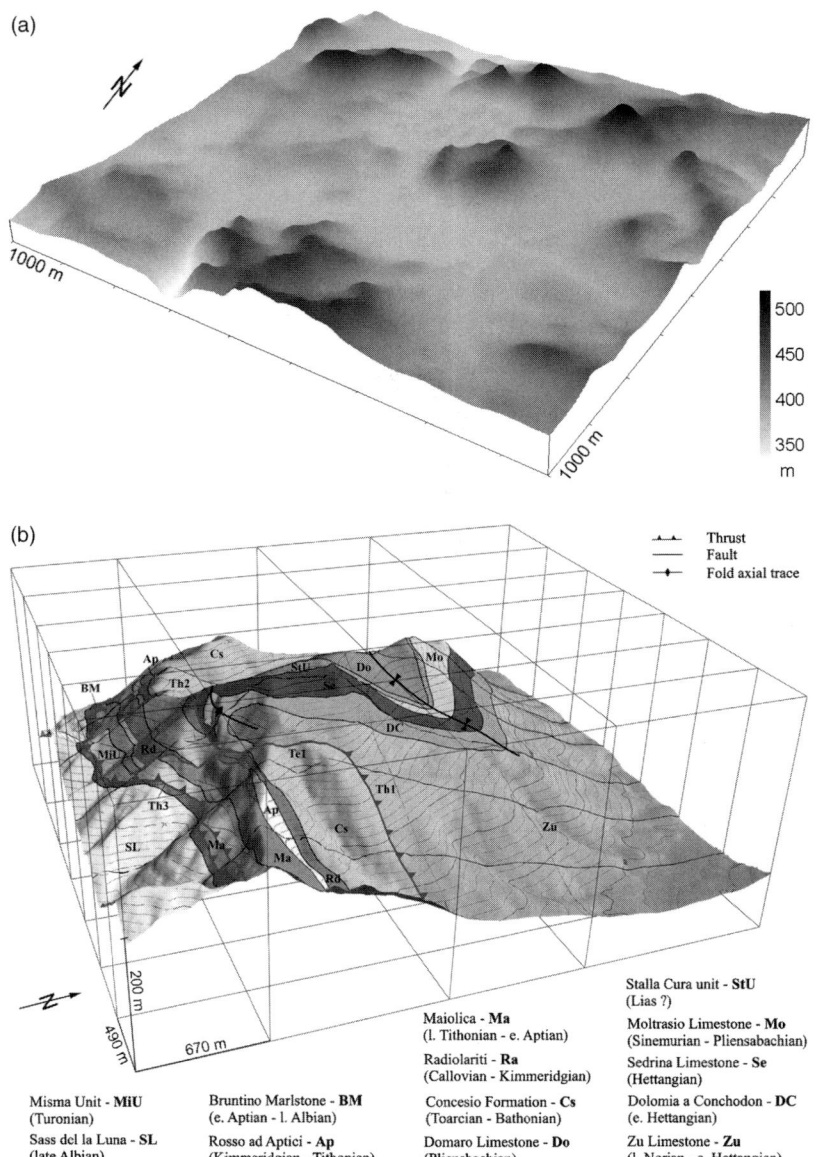

FIGURE 7.3 Examples of 3D topographic models: (a) 3D model of the Tunguska event area (60°53′09″N, 101°53′40″E). A DEM was derived from a topographic map, scale 1:100,000. The DEM includes 182,213 points (the matrix 431 × 423); the grid spacing is 20 m. The lower Triassic Kulikovsky paleovolcanic complex is visible due to the 5^x vertical exaggeration. *From (Florinsky, 2008e, Fig. 1).* (b) 3D geological map of Mount Misma, Italian Alps. A DEM was derived from 10-m contour interval topographic maps. The grid spacing is 2.5 m. Geological information was derived from a 1:5000 scale geological map. *From (Zanchi et al., 2009, Fig. 10), reproduced with permission. (See pages 15 and 16 in Color Plate Section)*

FIGURE 7.3 (Continued). (c) 3D geological model of the Beirut watershed, Lebanon. The model was cut along two cross sections (see right bottom corner), and each side of the cross section was pushed away from the other. The layers younger than the Aptian horizon are shown floating. The topographic surface was derived from a 30-m DEM obtained from the Terra Aster Product. Geological surfaces were derived from a 1:50,000 scale geological map. *From (Dhont et al., 2005, Fig. 4), reproduced with permission.* (See page 16 in Color Plate Section)

2. Overlaying thematic data (e.g., remotely sensed scenes, soil and geological maps) on 3D topographic models (e.g., Eyton, 1986; Ziadat et al., 2003; Palyvos et al., 2006; Qi and Zhu, 2006; Kukowski et al., 2008) (Figs. 7.3b and 10.4). This approach is also a standard option of many softwares.
3. 3D geological modeling (Morris, 1991; Pflug and Harbaugh, 1992; McMahon and North, 1993; Breunig, 1996; Maerten et al., 2001; Lemon and Jones, 2003; Masumoto et al., 2004; Dhont et al., 2005; Guillaume et al., 2008; Kaufmann and Martin, 2008; Caumon et al., 2009; De Donatis et al., 2009; Fernandez et al., 2009; Jones et al., 2009; Zanchi et al., 2009) and 3D soil modeling (Pereira and FitzPatrick, 1998; Grunwald et al., 2000, 2007; Mendonça Santos et al., 2000; Chaplot and Walter, 2003; Ramasundaram et al., 2005; Smith et al., 2008; Delarue et al., 2009) that uses various combinations of DEM-derived 3D models of the land surface, geological surfaces, or surfaces of soil horizons, as well as other data on spatially distributed geological and

soil features (Fig. 7.3c). Caumon et al. (2009, p. 928) explained the necessity and importance of 3D geological modeling as follows:

> Understanding the spatial organization of subsurface structures is essential for quantitative modeling of geological processes. It is also vital to a wide spectrum of human activities, ranging from hydrocarbon exploration and production to environmental engineering. Because it is not possible to directly access the subsurface except through digging holes and tunnels, most of this understanding has to come from various indirect acquisition processes. 3D subsurface modeling is generally not an end, but a means of improving data interpretation through visualization and confrontation of data with each other and with the model being created, as well as a way to generate support for numerical simulations of complex phenomena (i.e., earthquakes, fluid transport) in which structures play an important role. As the interpretation goes, the 3D framework forces us to make interpretive decisions that would be left on the side in map or cross section interpretations. Skilled geologists know how to translate 3D into 2D and vice versa, but, no matter how experienced one can be, this mental translation is bound to be qualitative, hence inaccurate and sometimes incorrect. 3D model building calls for a complex feedback between the interpretation of the data and the model.

The general scheme for construction of 3D geological models consists of the following steps (Fernandez et al., 2009):

1. Georeferencing and digitizing data in three-dimensional space, including:
   a. Deriving a DEM of the land surface (Section 3.1)
   b. Digitizing geological borders from geological maps and field data
   c. Digitizing faults and other structural data from geological maps and field data
   d. Deriving DEMs of geological surfaces from geological maps, cross sections, and borehole data
2. For each geological surface and fault, estimating local strikes and dips by geometrical analysis of areal extents of geological contacts (Section 13.3.2).
3. Generating a set of digital models for all geological surfaces and faults known in the study area. To do this, one should create partial surfaces honoring local strikes and dips for each geological surface and fault. Assembling partial surfaces associated with each geological surface or fault should also be done.
4. Analyzing relative positions of obtained models of geological surfaces and faults, in particular, to detect their intersections; and correcting positional errors.
5. Assembling a complete 3D model using constructed geological surfaces and faults.

Three-dimensional geological models are widely used, for example, in oil and gas exploration (Durham, 1999; Clarke and Phillips, 2003).

Unlike 3D geological models, 3D soil models are not in common use. However, Smith et al. (2008, p. 190) noted that "spatial soil horizon modelling can use methods and software developed originally for shallow geological modelling, when soil horizons follow a unique super positional order. The modeling of horizons that do not have one single super positional order or are even overturned and convoluted—for example, in periglacially disturbed soils—can be solved using techniques developed for modeling faulted and overturned geology."

7.5 COMBINING HILL-SHADING MAPS WITH SOIL AND GEOLOGICAL DATA

Technically, combining some thematic and morphometric layers on a single map is a rather difficult task. The readability of such a map governs the effectiveness of its analysis studying interrelationships between topography and other components of a geosystem.

Currently, the most popular approach in geological and soil visualization is to overlay geological or soil maps on hill-shading maps (Section 2.5) (Vigil et al., 2000a, 2000b; Barton et al., 2003; Giasson et al., 2008; Marchetti et al., 2010). This technique can improve the visual

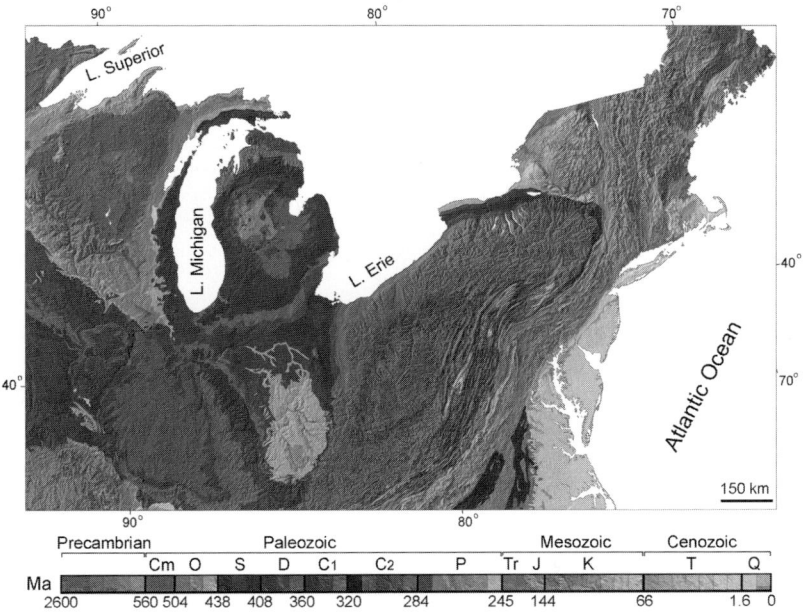

FIGURE 7.4 Combining geological information and hill-shading maps: a northeastern portion of the geological map of the United States. *From (Vigil et al., 2000a), courtesy of the U.S. Geological Survey.* (See page 17 in Color Plate Section)

perception of relationships between geological/soil and geomorphic features. Figure 7.4 illustrates this technique: the small-scale geological map of the United States (King and Beikman, 1974) is superimposed on the hill-shading map of this territory (Thelin and Pike, 1991a, 1991b). The overlaying of soil data on hill-shading maps is shown in Figs. 10.3 and 10.5.

PART II

DIGITAL TERRAIN MODELING IN SOIL SCIENCE

CHAPTER

8

Influence of Topography on Soil Properties

OUTLINE

8.1 Introduction	145
8.2 Local Morphometric Variables and Soil	146
8.3 Nonlocal Morphometric Variables and Soil	148
8.4 Discussion	149

8.1 INTRODUCTION

It is well known that topography is one of the soil-forming factors (Dokuchaev, 1883, 1891; Sibirtsev, 1899; Vysotsky, 1906; Zakharov, 1911, 1913; Neustruev, 1915, 1927, 1930; Jenny, 1941; Huggett, 1975; Fridland, 1976; Gerrard, 1981; Schaetzl and Anderson, 2005). Topography influences (micro)climatic and meteorological characteristics, which affect the hydrological and temperature regimes of soils (Neustruev, 1927, 1930; Geiger, 1927; Romanova, 1977; Kondratyev et al., 1978; Raupach and Finnigan, 1997; Böhner and Antonić, 2009; Emeis and Knoche, 2009), the prerequisites of the gravity-driven lateral overland and intra-soil transport of water and other substances (Kirkby and Chorley, 1967; Young, 1972; Speight, 1980), as well as the spatial distribution of the vegetation cover (Yaroshenko, 1961; Franklin, 1995; Florinsky and Kuryakova, 1996). Thus, it is natural that topography directly or indirectly controls the spatial distribution of physical, chemical, and biological soil properties (Moore et al., 1991; Florinsky et al., 2002; Shary et al., 2002b;

Schaetzl and Anderson, 2005, ch. 13). An in-depth understanding of this control is required for further modeling and mapping of soil properties based on topographic data (Chapter 10).

Topography influences soil properties through two main "tools":

1. The gravity-driven lateral migration and accumulation of water; and
2. Spatial differentiation of the temperature regime of slopes.

Even in the early twentieth century, Vysotsky (1906, pp. 3–4) noted:

> Water for soil is like blood for a living organism. It dissolves and transports various more or less soluble and organic substances, which are created by the weathering of parent rocks or decomposition of organic residuals, or moved through the atmosphere as dust and precipitation. But apart from such an internal action, water is an important external actor conducting (in known cases, together with wind action) a lot of work, such as erosion, denudation, and accumulation, as well as separation of soil matter.... The direction of such actions depends, first and foremost, on topography ... because its forms influence not only overland water circulation but also intrasoil one.

Recently, Legates et al. (2011, p. 65) stressed the importance of soil moisture:

> Soil moisture is a critical component of the earth system and plays an integrative role among the various subfields of physical geography.... Soil moisture affects atmospheric, geomorphic, hydrologic, and biologic processes ... it lies at the intersection of these areas of scientific inquiry. Soil moisture impacts earth surface processes in such a way that it creates an obvious synergistic relationship among the various subfields of physical geography. The dispersive and cohesive properties of soil moisture also make it an important variable in regional and microclimatic analyses, landscape denudation and change through weathering, runoff generation and partitioning, mass wasting, and sediment transport.

Thus, it is reasonable to use the example of soil moisture to discuss the principal aspects of the topographic influence on soil properties.

The following quantitative topographic characteristics are responsible for the spatial distribution and redistribution of water in the landscape: slope gradient, slope aspect, horizontal, vertical, and mean curvature (local morphometric variables), as well as catchment area (the nonlocal topographic attribute) (Table 2.1). Let us discuss their role in details.

8.2 LOCAL MORPHOMETRIC VARIABLES AND SOIL

Slope gradient controls soil moisture content as follows: as G increases, the slope area and velocity of water flow increase, so the precipitation received per unit area and its infiltration decrease, the runoff and evaporation area increase, and hence soil moisture content decreases (Zakharov, 1913, 1940). This leads to the usual negative correlations between the soil moisture content and G (see results of correlation analyses in Chapters 9 and 11).

Slope aspect influences the soil water balance since A, in association with G, affects insolation (Kondratyev et al., 1978) and evapotranspiration (Romanova, 1977). It is well known that in the northern hemisphere, soil moisture content tends to be the highest on north slopes, intermediate on west and east slopes, and least on south slopes (Sibirtsev, 1899; Zakharov, 1913; Ponagaibo, 1915; Neustruev, 1927, 1930). Also, A affects soil moisture content controlling the impact of neighboring geographical objects (i.e., mountains, seas, and deserts), which determine the character and direction of atmospheric flows (Neustruev, 1915, 1930; Zakharov, 1940).

Slope gradient and aspect affect redistribution of snow over the land surface. Thus, these morphometric variables influence the spatial differentiation and dynamics of freezing and melting of soils and, in turn, the spatial differentiation of soil water storage (Taychinov and Fayzullin, 1958).

Horizontal and vertical curvatures are the key topographic factors determining overland and intrasoil water dynamics (Kirkby and Chorley, 1967), being the measures of flow convergence/divergence and flow deceleration/acceleration, respectively (Table 2.1). Lateral intrasoil flow of the saturation zone and soil moisture content increase when flows converge (k_h takes negative values) and decrease when flows diverge (k_h takes positive values) (Kirkby and Chorley, 1967; Carson and Kirkby, 1972). Moreover, horizontal curvature influences hydrological processes in unsaturated soils: infiltration flux diverges when $k_h > 0$ and converges when $k_h < 0$ (Zaslavsky and Rogowski, 1969). It was found experimentally that the dynamics of lateral flows of the saturation zone and soil moisture content depend essentially on horizontal curvature (Anderson and Burt, 1978). This topographic attribute can play the key role in the formation of saturation zones: they are the most stable in areas of flow convergence (O'Loughlin, 1981). Soil moisture content also increases when flows decelerate (k_v takes negative values) and decreases when flows accelerate (k_v takes positive values) (Kirkby and Chorley, 1967; Kuryakova et al., 1992). These facts explain the usual negative correlations of the soil moisture content with k_h and k_v (see results of correlation analyses in Chapters 9 and 11).

For the arid climatic conditions and relatively flat topography of Israel, Sinai et al. (1981) observed a high correlation (-0.9) of soil moisture content of the root zone with mean curvature approximated by the Laplacian (Section A.3.4). Such a relation results from microtopographic control of the lateral intrasoil water transport rather than from redistribution of overland water flows (they are not typical for that landscape). A strong dependence of soil moisture on mean curvature was also found in Russia's southern Moscow Region, which has a contrast topography and is located in a moderate continental climate zone (Kuryakova et al., 1992) (Section 9.3).

Saturation zones are often correlated with landforms, for which both horizontal and vertical curvatures are negative (Feranec et al., 1991). These are relative accumulation zones, where both flow convergence and relative deceleration of flows act together (for details, see Section 2.7.2). There are landsliding (Lanyon and Hall, 1983), soil gleying, maximum thickness of the A horizon, and maximum depth to calcium carbonate (Pennock et al., 1987) in these zones due to an increased water content in soils and grounds.

8.3 NONLOCAL MORPHOMETRIC VARIABLES AND SOIL

Relationships between soil moisture content and a relative slope position (upslope, midslope, and downslope) were qualitatively understandable even in the early twentieth century (Zakharov, 1913). Quantitatively, the dependence of soil moisture content on catchment area (which, in fact, describes the relative position of a point on the topographic surface—Section 2.3) was probably first described by Zakharov (1940, p. 384) as follows: "water amount per unit area increases from upslope to downslope due to additional water supply." Thus, as CA increases, soil moisture content also increases. This explains the usual positive correlations of the soil moisture content with CA (see results of correlation analyses in Chapters 9 and 11). Speight (1980) argued that catchment area rather than horizontal curvature is of first importance for the soil moisture dynamics, because catchment area considers a relative position of a given point in the landscape.

The topographic index (Section 2.6), combining the metrics of slope gradient with catchment area, can further improve the description of morphometric prerequisites for the spatial distribution of soil moisture (Moore et al., 1986). This is because TI takes into account both the local geometry of a slope and the relative location of the given point in the landscape. As CA increases and G decreases, TI and soil moisture content increase. This results in higher absolute values of correlation coefficients of soil moisture content with TI than with CA and G (Thompson and Moore, 1996).

However, topographic index and horizontal curvature cannot separately offer the prospect of predicting soil moisture dynamics. Saturation zone depth may have higher correlations with some other empirically determined variables, such as a product of horizontal curvature and catchment area (Burt and Butcher, 1985). This generates a need for the use of a representative set of morphometric attributes in soil studies (Section 10.4).

8.4 DISCUSSION

It is obvious that soil moisture content depends not only on topography but also on some physical and hydraulic characteristics of soils, such as soil texture and soil water retention. However, spatial distribution of these parameters also depends on morphometric variables (Moore et al., 1993; Pachepsky et al., 2001) because they are, in one way or another, controlled by the intensity and direction of gravity-driven overland and intrasoil transport of substances.

Notice that the spatial distribution of moisture in a soil layer may sometimes depend on characteristics of the top surface of parent rocks. Among these are dense, water poor- or impermeable rocks (e.g., clays, granites). In such cases, topographic variables of the top surface of the C horizon may play similar roles as those of the land surface (Florinsky and Arlashina, 1998; Chaplot and Walter, 2003).

Results of the author's studies of topographic influence on soil moisture can be found in Section 9.3 and Chapter 11.

Finally, we should note that the influence of topography on soil properties depends on the management or tillage practice, for instance, zero tillage versus conventional tillage (Farenhorst et al., 2003; Senthilkumar et al., 2009). Most of the works dealing with relationships between topography and soil properties in agricultural landscapes have been conducted in Canada, the United States, and Australia on fields tilled over a 50- to 150-year period without dramatic modifications of the land surface and soil cover. This may be one reason why high correlations have been systematically observed for the system "topography—soil" in agrolandscapes. A strong, long-term agricultural load can seriously reduce the topographic control of soil properties (Venteris et al., 2004; Samsonova et al., 2007).

CHAPTER 9

Adequate Resolution of Models

OUTLINE

9.1 Motivation	151
9.2 Theory	153
9.3 Field Study	157
9.3.1 Study Site	157
9.3.2 Materials and Methods	158
9.3.3 Results and Discussion	162

9.1 MOTIVATION

In landscape studies the following question usually arises: What density of sampling points should be used to depict adequately the spatial distribution of a landscape property concerned for a given scale, measurement accuracy, and minimum of samples? (Lidov, 1949; Mueller-Dombois and Ellenberg, 1974; Campbell, 1979; Kershaw and Looney, 1985; Burrough, 1993). If one carries out a study with a regular grid of sampling points, the problem reduces to determination of a grid spacing. The success of an investigation depends on the correct solution of this problem.

At first sight, the problem would seem to be a technical task. In fact, this is the fundamental problem since it is connected with determining a spatial scale of an object, phenomenon, and process under study. This is a critical question because different physical laws and landscape processes dominate at different spatial scales. Extension of one or other concept or model to all scales can result in invalid description of actual relationships (Haggett et al., 1965; Schumm and Lichty, 1965; Klemeš, 1983; Phillips, 1988; de Boer, 1992; Band and Moore, 1995; Bierkens et al., 2000).

An adequate description of a landscape property with a minimum number of samples implies that grid spacing corresponds to an area wherein property values vary smoothly, or are assumed to have a constant value for a given scale and measurement accuracy. Different sciences use different terms for such an area. In this book, these area and grid spacing are denoted as *the adequate area* of a landscape property and *the adequate grid spacing*, respectively (see details in Section 9.2).

In DTM-based geomorphic and geological studies, one can determine an adequate grid spacing (w) of a DTM from a typical size S of landforms or geological structures concerned. According to the sampling theorem (Section 3.3), the adequate w cannot be more than $S/2$. In studies of other landscape components (e.g., soils, plants) and landscape processes (e.g., lateral intrasoil transport of substances), determination of the adequate w is less trivial. This is due to the high spatial variability of landscape properties (Campbell, 1979; Kershaw and Looney, 1985; Oliver and Webster, 1986; Trangmar et al., 1987). In addition, a landscape property can have several adequate areas associated with different natural processes (Sitnikov, 1978, 1980; Kershaw and Looney, 1985) (Section 9.2). There are some closely related methods for determining adequate areas of landscape properties. In these methods, an indicator of an adequate area is a smooth portion of a graph describing the dependence of a property or its statistical parameter on the area or grid spacing used to measure the property (Mead, 1974; Mueller-Dombois and Ellenberg, 1974; Kershaw and Looney, 1985; Oliver and Webster, 1986; Trangmar et al., 1987; Wood et al., 1988; Famiglietti and Wood, 1995).

The problem can be further complicated if one analyzes data on two or more landscape properties in combination, or predicts an attribute through an analysis of other attributes. This is because different landscape properties may have different adequate areas (Trangmar et al., 1987; Phillips, 1988, 1995). Moreover, existing *a priori* relationships between two landscape properties (justified theoretically or observed at other places) can be manifested at only certain adequate areas of these properties (Phillips, 1988). For instance, it is well known that a water table may look like a generalized land surface. Such a regularity may be observed by an analysis of water table depths and a DEM with w omitting minor landforms. However, one can, at best, find a low correlation of the water table with the topography using a too detailed or an overgeneralized DEM (Thompson and Moore, 1996).

Thus, one may establish invalid statistical regularities (e.g., low correlation coefficients), as well as incorrect predictions and conclusions for *a priori* related landscape properties using high-quality data. This problem is typical in the analysis of data sets that have different resolutions and incongruous grids (Lidov, 1949; Band and Moore, 1995). However, the problem cannot be solved merely by using data sets with the same resolutions and grids. The situation is complicated by the fact that adequate areas are not correlated with "basic" landscape units, such as stow and facies (Phillips, 1988).

Determining the grid spacing is always required in combined analysis of DTMs and data on other landscape components, as well as in DTM-based modeling of landscape properties. Expert opinion usually estimates w (Anderson and Burt, 1980; Burt and Butcher, 1985; Moore et al., 1993; Quinn et al., 1995; Florinsky and Kuryakova, 1996); hence a solution can be subjective. Arbitrary choice of w can result in incorrect results and artifacts. For example, Speight (1980) did not find a relationship between soil moisture content and k_h because w was too small (Anderson and Burt, 1980). Sinai et al. (1981) did not observe a correlation between soil salinization and H because w was too large. In TI-based hydrological modeling, the average depth of the water table decreases and the overland runoff increases when w increases (Wolock and Price, 1994). For forest ecosystems of the British Columbia, use of TI calculated with $w = 4$ m resulted in the correct prediction of the water table depth, while totally invalid prediction was obtained using $w = 16$ m (Thompson and Moore, 1996). Slopes of DTM-based hydrographs essentially depend on w (Zhang and Montgomery, 1994; Da Ros and Borga, 1997). DTMs with $w < 30$ m can provide the correct prediction of soil hydromorphy, while coarser DEM resolutions deteriorate the prediction (Chaplot et al., 2000).

Therefore, a correct choice of w is one of the main problems of DTM-based soil studies. In this chapter, we describe the author's method for determining the adequate grid spacing of DTMs used to analyze and model soil properties (Florinsky and Kuryakova, 2000). Use of the method is exemplified by studying relationships between topography and soil moisture.

9.2 THEORY

The method is largely based on principles of the concept of (representative) elementary volume used in mass transfer description, in particular, in hydrogeology (Sitnikov, 1978, 1980). Some principles of this concept can be applied to solve a wide range of two-dimensional problems of the geosciences if one switches from elementary volume to (representative) elementary area (Wood et al., 1988; Famiglietti and Wood, 1995).

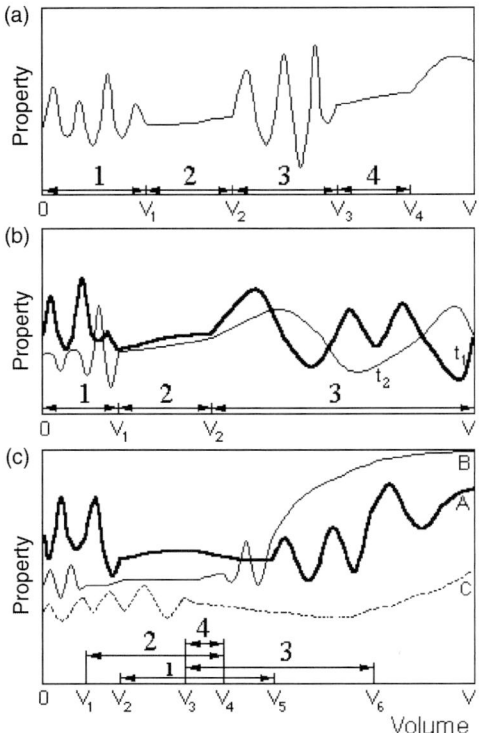

FIGURE 9.1 Possible dependence of a landscape property on volume or area of observation (Sitnikov, 1978, 1980): (a) dependence on volume or area; (b) dependence on volume or area at points in time t_1 and t_2; (c) dependence on volume or area for heterogeneous patches A, B, and C. *From (Florinsky and Kuryakova, 2000, Fig. 1).*

Assume that a response presented in Fig. 9.1a describes the variation in the value of some landscape property (e.g., a rock property) with the volume or area for which the property is observed (Sitnikov, 1978, 1980). For the interval 1, the large extent of variability of property values can be a result of some pronounced spatial heterogeneity, for example, pore effects. Property values vary smoothly from the volume or area V_1 to V_2. Then, another distinct spatial heterogeneity (e.g., macro-cracks) leads to abrupt variations of property values within the interval 3. Again, property values vary smoothly from the volume or area V_3 to V_4.

By the elementary volume or area, we mean a minimum volume or area in which a property is independent of heterogeneities; that is, the property values vary smoothly, or have a low variability (Sitnikov, 1978, 1980). For instance, V_1 and V_3 are elementary volumes or areas for the given property (Fig. 9.1a). Therefore, for a landscape property one can distinguish a set of elementary volumes or areas associated with several intervals of smooth relationships between property values and volume or area. Sometimes, intervals of smooth variation of a landscape property cannot be found due to peculiarities of natural processes or measurement errors.

An elementary volume or area can also depend on time due to temporal variability of landscape properties (Fig. 9.1b) (Sitnikov, 1978, 1980). In addition, an elementary volume or area of a property can depend on specific natural conditions. If a terrain includes heterogeneous patches, a landscape property can have different elementary volumes or areas at adjacent patches. For example, assume that a terrain consists of three heterogeneous patches A, B, and C. Assume also that dependencies of property values on volume or area are smooth within intervals 1, 2, and 3 in patches A, B, and C, respectively (Fig. 9.1c). Therefore, a property has elementary volumes or areas V_2, V_1, and V_3 in patches A, B, and C, respectively. If one can find borders between the patches, it is desirable to study this property separately at each patch. However, one can observe the interval 4 wherein property values have low variability at all three patches (Fig. 9.1c). The elementary volume or area of this interval is V_3. It can be used as a common elementary volume or area for the entire terrain (Sitnikov, 1978, 1980; Kershaw and Looney, 1985).

By the adequate interval of volumes or areas, we mean an interval of volumes or areas wherein property values vary smoothly or have a constant value. Value variability may be ignored if it is within the limits of the measurement accuracy (Sitnikov, 1980). For example, intervals 2 and 4 are adequate intervals of volumes or areas for the property discussed above (Fig. 9.1a). It is mandatory to use adequate intervals of volumes or areas. Otherwise, one can obtain nonreproducible and uninterpretable results due to the high and unpredictable variability of property values within nonadequate intervals of volumes or areas.

By adequate volume or area, we mean a volume or an area belonging to an adequate interval of volumes or areas (Sitnikov, 1978, 1980). By adequate grid spacing, we mean a grid spacing corresponding to an adequate area. According to the sampling theorem (Section 3.3), if s^2 is the adequate area of a landscape property, then $s/2$ is the adequate grid spacing, which relates to s^2 and provides adequate description of the property.

An adequate grid spacing can be defined by (1) measurement of a landscape property at grid nodes using different values of w; and (2) plotting of values measured against w (similarly to Fig. 9.1a). Smooth portions of this graph will indicate intervals of adequate w (Sitnikov, 1980).

In combined analysis of two landscape properties and in prediction of one property through analysis of another property, one should work with an adequate interval of areas common to two properties. Values of both properties are constant or vary smoothly within this interval, by definition of the adequate interval (see above). Therefore, correlation coefficients between values of two properties also can vary smoothly within the common adequate interval. At the same time, one can observe a high variability of correlation between two properties within adjacent nonadequate intervals marked by a high variability of property values. So, to determine a common adequate interval of areas and grid spacing providing the correct combined study of two properties, one should (1) analyze the correlation between these properties observed with different grid spacings; and (2) plot correlation coefficients versus w (similarly to Fig. 9.1a). A smooth portion of this graph can indicate an interval of adequate grid spacings and hence an adequate interval of areas common to both properties.

By adequate w, we mean the adequate grid spacing if a morphometric variable (described as a DTM) is one of two landscape properties under study. The adequate area corresponding to the adequate w determines a typical planar size of landforms providing a topographic control of a landscape property concerned. To define adequate w, one should carry out the following procedures:

- Derive a set of DTMs using a series of w.
- Perform a correlation analysis of data on a landscape property and a morphometric variable estimated with various values of w.
- Plot correlation coefficients between the landscape property and the morphometric variable versus w.
- Determine smooth portions of the graph, which indicate intervals of adequate w.

There are three main variants for implementing the method described, depending on types of initial data:

1. DEM and data on a landscape property are obtained using the same square grids with a grid spacing w. One can then derive a set of digital models of a morphometric variable concerned from the DEM using grid spacings $w, 2w, 3w, \ldots, nw$; n is an integer. Then, one should analyze the correlations between the landscape property and the morphometric variable using samples with these grid spacings.
2. DEM and data on a landscape property are obtained using distinct grids and are then interpolated to a common square grid with a grid spacing w. Further steps are the same as those in the first variant.
3. DEM and data on a landscape property are obtained using distinct grids, and spatial interpolation of data on the landscape property is undesirable or impossible (e.g., landscape data are collected along a transect or a contour—see Section 9.3.2). In this case, one should:
   a. Produce several DEMs by interpolation with different w.
   b. Derive a set of digital models of a morphometric variable concerned from these DEMs.

c. Interpolate values of the topographic attribute calculated with different w for points in which the landscape property was observed.
d. Analyze correlations between the landscape property and the morphometric variable computed with different w.

From the viewpoint of minimization of interpolation errors, the first variant is the best since interpolation is not used. Next is the third variant in which one should interpolate only DTMs. The second variant may result in the greatest number of interpolation errors. In this study, we used the third variant of the implementation method.

Different morphometric variables can be connected with landscape processes of different scales (Anderson and Burt, 1980; Florinsky and Kuryakova, 1996). Thus, a landscape property can be "controlled" by different topographic attributes at different adequate areas. Therefore, different morphometric variables can offer different adequate w suited for each landscape property. In landscape studies, particularly in determining the regression dependence of a landscape property on morphometric variables, it is desirable to find the "main" interval of an adequate w common to all morphometric variables under consideration. Obviously, it makes sense to perform a regression analysis only with an adequate w of this interval.

It is also obvious that one should consider statistically significant correlations only. Therefore, it is advisable to use a sample size of 40, as well as to ignore smooth portions of correlation coefficient graphs if these portions include statistically insignificant coefficients.[1]

9.3 FIELD STUDY

9.3.1 Study Site

The study site is located in the center of the East European Plain, south of the Moscow Region, near the city of Pushchino (Fig. 9.2). This zone has a moderate continental climate with warm summers and prolonged cold winters. Average temperatures in January and July are $-10°C$ and $18.6°C$, respectively. Precipitation is about 640 mm per year, 350–450 mm of which are rainfall.

---

[1]It is commonly recommended to use a minimum sample size of 50–60 (in geostatistical studies, it should be >100 – Webster and Oliver, 1992). The sample size of 40 is chosen as a compromise: collection of 50–60 soil samples may be difficult due to laboriousness and cost of field and laboratory works. At the same time, correlation coefficients of 0.3 and higher are statistically significant for the sample size of 40.

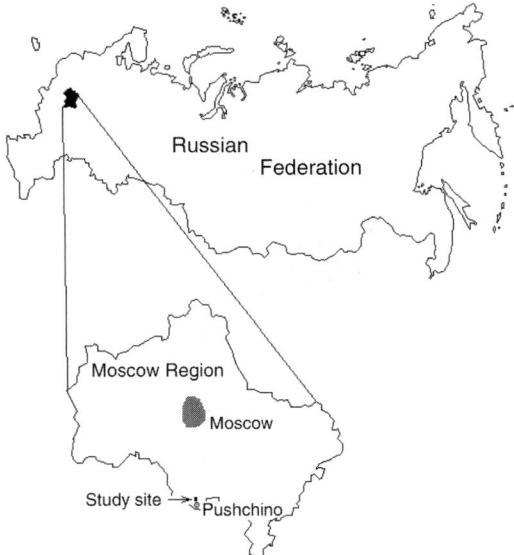

FIGURE 9.2 Geographical location of the Severny Gully (37°34′30″N, 54°41′10″E). *From (Florinsky and Kuryakova, 2000, Fig. 2).*

The site is situated on the soddy, partially forested landslide slope of the valley of the Oka River. Elevations are about 130 m above sea level. The slope has a gradient of ~10° and a northerly aspect. Middle Carboniferous fractured and karstified limestones covered by Quaternary loams lie at a depth of about 6 m. Groundwater lies at a depth of about 6 m (Lyubashin and Lisitsin, 1981).

The study site includes a portion of the north-striking Severny Gully and adjacent crests. It measures about 58 × 77 m; the variation in altitude is about 15 m (Fig. 9.3). The soil complex includes gray forest soils at crests and slopes, and meadow hydromorphic soils in the gully bottom. Vegetation cover consists of birches and herbs on crests, hazels and common horsetails on slopes, and nettles in the gully bottom.

9.3.2 Materials and Methods

We obtained an irregular DEM of the study site through a tacheometric survey with an optical tacheometer TaN. The survey was performed by G.A. Kuryakova and the author in June 1990 (Kuryakova et al., 1992).

The irregular DEM consists of 374 points. It is constructed in a relative Cartesian coordinate system. As the local datum, we used the minimum elevation value within the study site (Fig. 9.3).

To estimate the surficial soil moisture content (*Moist*), we carried out a soil sampling on June 20, 1990. Precipitation was about 60 mm from

9.3 FIELD STUDY

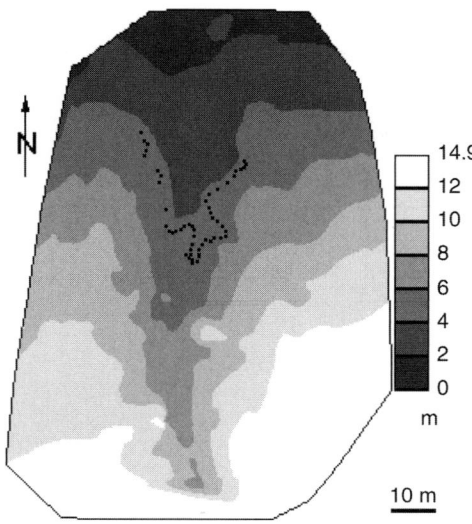

FIGURE 9.3 The Severny Gully, elevations. Dots are locations of soil-sampling sites along the 4.25-m contour. *From (Florinsky and Kuryakova, 2000, Fig. 3b).*

June 1 to 20, 1990 in Pushchino and suburb. There was a drizzling rain (about 2 mm) on the eve of the soil sampling. To prevent a strong influence of evaporation on M, we performed the soil sampling in the morning, within one hour, at the air temperature of 20–22°C.

The soil sampling was performed at 62 points (included in the irregular DEM), which are located along the 4.25-m contour (Fig. 9.3). These points are rather evenly distributed along the contour, with distance ranging from 0.5 m to 1.5 m. On the west side of the gully, three larger distances correspond to two tree falls and a track. We chose the 4.25-m contour because (1) it passes over the main landforms within the site: two crests, slopes, and the gully bottom (Fig. 9.3); and (2) it is suitable for the soil sampling because higher elevation slopes are too steep.

We took three soil samples at each of the 62 points at a depth of about 0.1 m. All samples had the similar soil texture (loam). The soil sampling was conducted by G.A. Kuryakova, P.A. Shary, and the author (Kuryakova et al., 1992).

We evaluated *Moist* for each of these samples by weighing of pre- and postdrying samples on an analytical damper balance ADV-200. Drying of samples was carried out by a drying box 2 V-151 for 24 hours at 105°C (Carter, 1993). To reduce the influence of random deviations of *Moist* values, we used the average of *Moist* for the three samples collected at each of the 62 points as net values of *Moist* (Fig. 9.4). Laboratory analysis was carried out by the author.

We performed the soil sampling along one contour in order to eliminate from consideration an *apparent* influence of elevation on soil

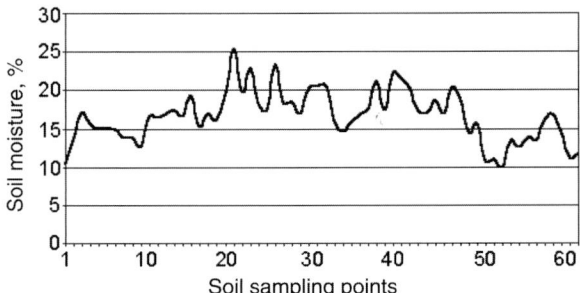

FIGURE 9.4 Distribution of the soil moisture content along the 4.25-m contour. *From (Florinsky and Kuryakova, 2000, Fig. 4).*

properties. Obviously, the actual influence of elevation on soil moisture observed in mountainous regions (due to the altitudinal zonality) cannot occur within the study site. G, A, k_h, k_v, H, and CA are responsible for physical mechanisms of distribution and redistribution of moisture in landscapes (Chapter 8). Elevation is not responsible by itself for physical mechanisms of gravity-driven moisture movement. Dependence of soil moisture on elevation observed in plain landscapes is in fact a result of the influence of CA on the spatial distribution of water in soils.

Using the Delaunay triangulation and a piecewise quadric polynomial interpolation (Agishtein and Migdal, 1991) of the irregular DEM, we produced 13 square-gridded DEMs with the following w (the numbers in parentheses are points in the regular DEMs): 1 m (3312), 1.5 m (1421), 2 m (742), 2.5 m (461), 3 m (301), 3.5 m (202), 4 m (152), 4.5 m (110), 5 m (85), 5.5 m (63), 6 m (48), 6.5 m (35), and 7 m (31).

We decided on this range of w for the following reasons. To estimate local morphometric variables, one has to calculate the first and second derivatives of elevation (Section 2.2). These calculations increase errors of DEM generation and interpolation (Section 5.4.1). Preliminary testing of the irregular DEM suggests that these errors can be ignored for $w \geq 1$ m. Considering the size of the site, $w = 7$ m is the maximum w, which can be used to derive local morphometric variables. $\Delta w = 0.5$ m was of interest, and its use allowed us to determine the adequate w for the study site (Section 9.3.3).

Digital models of G, k_h, k_v, and H (Fig. 9.5) were derived from all regular DEMs by the Evans–Young method (Section 4.1). Altogether, we produced 52 regular DTMs. Then, we used the Delaunay triangulation and a piecewise quadric polynomial interpolation (Agishtein and Migdal, 1991) of these regular DTMs to determine values of G, k_h, k_v, and H at the 62 sampling points.

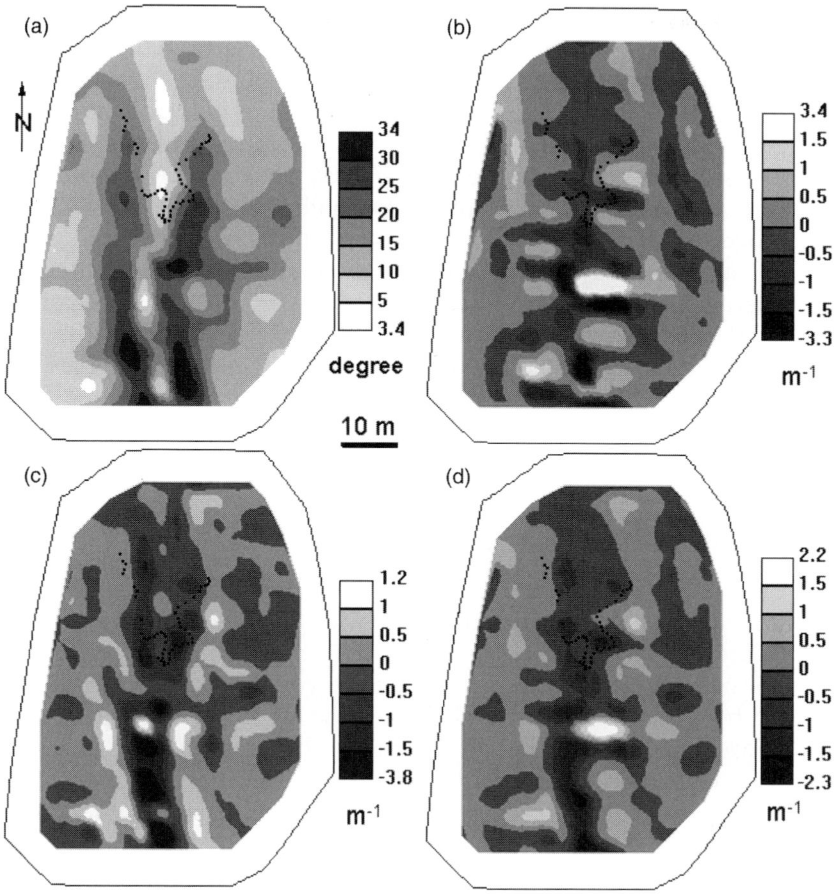

FIGURE 9.5 The Severny Gully, morphometric variables derived with $w = 3$ m: (a) slope gradient, (b) horizontal curvature, (c) vertical curvature, (d) mean curvature. Dots are locations of soil-sampling sites along the 4.25-m contour. *From (Florinsky and Kuryakova, 2000, Fig. 5).*

To determine an adequate w and to estimate the dependence of *Moist* on morphometric variables, we carried out a linear multiple correlation analysis of *Moist* with G, k_h, k_v, and H calculated for the 13 values of w. 62-point samples were used for DTMs with $w = 1,...,5$ m, 59-point samples—for DTMs with $w = 5.5$ and 6.5 m, 56-point samples—for DTMs with $w = 6$ m, and 53-point samples—for DTMs with $w = 7$ m (boundary effects were omitted). Then, we presented the dependence of correlation coefficients on w in the form of graphs for G, k_h, k_v, and H. To describe an effect of topography on the distribution of *Moist*, the "best" combination of topographic variables was chosen by the stepwise linear regression (Kleinbaum et al., 2008). We used the 62-point samples corresponding to two adequate w determined for regression analysis.

We did not study the influence of A on *Moist* because the study site is located (1) on the slope with a uniform northerly aspect and (2) in the forest where sunlight is diffused by trees. Therefore, we can consider A as a background value in the study site, although the dependence of *Moist* on A is apparent (Ponagaibo, 1915; Zakharov, 1940; Romanova, 1977). We did not study the influence of CA and TI on *Moist* because we did not perform the tacheometric survey for the gully head; derivation of CA and TI from DEMs without data on catchment heads can lead to invalid results.

Statistical analysis was carried out by the software Statgraphics Plus 3.0 (© Statistical Graphics Corp., 1994–1997). Digital terrain modeling was performed by the software LandLord (Appendix B).

9.3.3 Results and Discussion

Table 9.1 presents the results of the correlation analysis, and Figure 9.6 provides the graphs of correlation coefficients between *Moist* and G, k_h, k_v, and H versus w. As expected, these graphs include some portions of smooth variation of correlation coefficients with w, as well as some portions of high variability for these relationships (Fig. 9.6). All four graphs present two main portions: the left portion (between $w = 1$ m and $w \cong 4$ m) with relatively smooth variations, and the right portion (between $w \cong 4$ m and $w = 7$ m) with pronounced fluctuations.

Using smooth portions of the graphs as indicators of adequate intervals of w, we can distinguish:

- The adequate interval W_1 for H, k_v, and k_h ranging between $w \cong 2.25$ and $w \cong 4$ m
- The adequate interval W_2 for G, H, k_v, and k_h ranging between $w \cong 2.5$ and $w \cong 3$ m

W_2 is the "main" adequate interval of w because all correlations peak there except spurious correlations of *Moist* with G for $w = 4.5$ and 6 m (see explanation below). For example, for $w = 3$ m correlation coefficients of *Moist* with G, k_h, k_v, and H are -0.28, -0.52, -0.50 and -0.60, respectively (Table 9.1). These results generally conform to previous findings on the topographic control of soil moisture (see details in Chapter 8).

Intervals of nonadequate w differ greatly in appearance from adequate intervals of w: there are drastic variations of the dependence of correlation coefficients on w (Fig. 9.6). Correlations of *Moist* with k_h and H are much lower within the nonadequate interval of w (between

TABLE 9.1 The Severny Gully: Correlation Coefficients between the Soil Moisture Content and Some Morphometric Variables Depending on the DEM Grid Spacing.

Morphometric Variable	DEM Grid Spacing, m												
	1	1.5	2	2.5	3	3.5	4	4.5	5	5.5	6	6.5	7
G	−0.29	−0.24	—	−0.32	−0.28	—	—	0.34	0.24	—	0.58	—	—
k_h	—	—	−0.34	−0.38	−0.52	−0.36	−0.36	—	−0.35	—	−0.29	—	−0.42
k_v	−0.32	−0.40	−0.37	−0.53	−0.50	−0.44	−0.47	—	−0.52	—	−0.53	—	−0.33
H	−0.27	−0.29	−0.46	−0.58	−0.60	−0.53	−0.49	—	−0.46	—	−0.41	—	−0.39

$P \leq 0.05$ for statistically significant correlations; dashes are statistically insignificant correlations.
From (Florinsky and Kuryakova, 2000, Table 2).

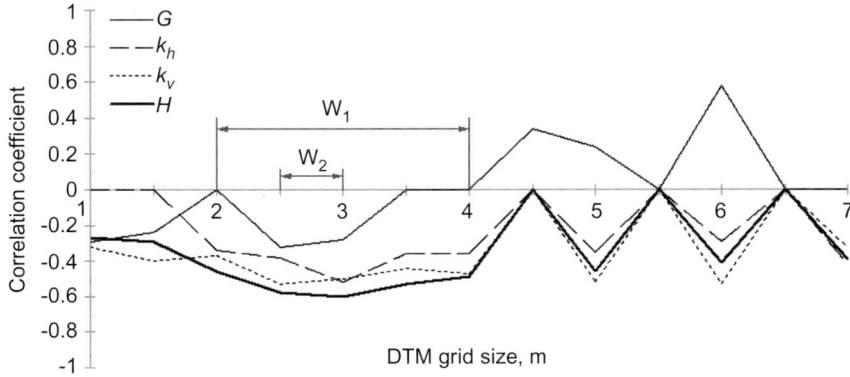

FIGURE 9.6 Correlation between the soil moisture content and morphometric variables versus the DEM grid spacing. *From (Florinsky and Kuryakova, 2000, Fig. 6).*

$w \cong 4$ m and $w = 7$ m) than within the main adequate interval (Table 9.1 and Fig. 9.6). Moreover, within this nonadequate interval, we observe positive correlations between *Moist* and G (Table 9.1 and Fig. 9.6). This is the evident artifact caused by use of the nonadequate w. Indeed, the soil moisture content decreases when G increases (Section 8.2). So, if there is a dependence of *Moist* on G, correlation coefficients between *Moist* and G can take negative values, as we found for $w = 1-3.5$ m (Table 9.1 and Fig. 9.6). This demonstrates the importance of the correct determination of w. This is also a pictorial example of how one may obtain invalid conclusions using statistics only, without an understanding of the physical senses of morphometric variables and relationships between topography and landscape processes.

Results of the regression analysis are presented in Table 9.2. We obtained regression equations for two adequate w, 2.5 and 3 m. In both cases, G and H entered the equations as independent variables. The regression equations describe 45% and 39% of *Moist* variability for $w = 2.5$ m and $w = 3$ m, respectively. These R^2 values are not high. Possibly, this is because we did not analyze the dependence of *Moist* on *CA* essentially affecting the spatial distribution of soil moisture (Zakharov, 1940; Speight, 1980) (Section 8.3).

The method allows one to estimate not only adequate w, but also an adequate area of landforms "supporting" topographic control of landscape properties. From the relation between an adequate area and an adequate grid spacing (Section 9.2), the adequate area of landforms controlling *Moist* lies in the range from 20 m^2 to 40 m^2 within the study site. Notice that the adequate $w \cong 3$ m is close to $w = 4$ m, providing the *TI*-based prediction of the water table depth in forests of the British Columbia (Thompson and Moore, 1996). This can indicate that the

TABLE 9.2 The Severny Gully: Parameters of the Regression Equations for the Dependence of the Soil Moisture Content on Morphometric Variables under Adequate w.

w, m	Independent Variables	Coefficient	P	R^2
2.5	G	−0.20	0.00	0.45
	H	−4.77	0.00	
	Constant	18.05	0.00	
3.0	G	−0.14	0.02	0.39
	H	−5.16	0.00	
	Constant	17.13	0.00	

From (Florinsky and Kuryakova, 2000, Table 3).

topographic control of some soil-hydrological processes occurs on landforms with the adequate area ranging between 36 m² and 64 m² in subboreal forested regions. Therefore, 3–4 m are suitable values of the adequate w for DTM-based soil and hydrological studies in these climatic conditions.

Using the method described, some difficulties can arise in choosing the minimum w and Δw. It is easy to see (Fig. 9.6) that for $\Delta w = 1$ m, it would be impossible to find fluctuations of the dependence of correlation coefficients on w within the right nonadequate interval (between $w \cong 4$ m and $w = 7$ m). Using $\Delta w = 1$ m, these fluctuations cannot be found with the minimum w of 1 m or 1.5 m. We suppose that as a minimum w one should use such a w value that DTM errors caused by calculation of derivatives can be ignored (Section 5.4). Δw should be less than half of the minimum w.

CHAPTER

10

Predictive Soil Mapping

OUTLINE

10.1 The Dokuchaev Hypothesis as a Central Idea of Soil Predictions	167
10.2 Early Models	170
10.3 Current Predictive Methods	172
10.3.1 Classification of Methods	*173*
10.3.2 Mathematical Approaches	*174*
10.3.3 Small-Scale Predictive Models and Upscaling	*182*
10.3.4 Prediction Accuracy	*185*
10.4 Topographic Multivariable Approach	187

10.1 THE DOKUCHAEV HYPOTHESIS AS A CENTRAL IDEA OF SOIL PREDICTIONS

In the mid-1880s, Vasily Dokuchaev (1886, pp. 352–353) formulated the following hypothesis:

> Any ... soil is always and everywhere a mere *function*[1] of the following factors of soil formation[2] : 1) the nature (content and structure) of the parent rock, 2) the

[1] The italic text was highlighted by Dokuchaev.
[2] The fundamental idea of soil science that soil is a result of interaction of the five principal forming factors was first postulated by Dokuchaev in 1883: "Soils being a result of a very complicated interaction between local climate, plant and animal organisms, content and structure of parent rocks, topography, and, finally, age of the terrain" (Dokuchaev, 1883, p. III). Zakharov (1927, p. 8) called this idea "the first basic law of soil science".

climate of the given terrain, 3) the mass and character of vegetation, 4) the age of the terrain, and finally, 5) the terrain topography. It immediately follows that (a) if the mentioned *factors* are the same in two different localities (however far apart they might be), the soils in the two localities should also be similar, and vice versa; consequently, (b) if we have thoroughly studied these factors, *we may predict in advance* what the soil itself should be like. Next, (c) it is well known that *momentum* should not change if one force component increases or decreases by some value, while another force component changes by the opposite value; thus, there should be a similar, *to some extent*, relationship between *character of soil* and *character* of its forming factors. Hence, it is clear that it is theoretically possible to state and solve, for example, the following problem: How would a given soil change if there is an increase of the terrain temperature by say 1–2°, with a synchronous increase of meteoric water by 1–2 inches? How would soil change if there is an annual increase of the vegetative mass by 20 poods per desyatina[3] with a decrease of the temperature by 1–2°?

Again, all these are so clear, so logically legitimate and necessarily, that *nobody* will probably argue with that.

For now unfortunately, it is difficult to prove all these propositions *using facts* and *with desirable completeness*, particularly to answer the last question (c) *in detail*. The reasons are quite obvious. Firstly, there is a great complexity of conditions affecting soil; secondly, these conditions have no absolute values, and, therefore, it is difficult to express them numerically; finally, we possess very few data for some factors and none whatever for others. Nevertheless, we may hope that all these difficulties will be overcome with time, and then soil science will truly become an exact science.

Later, Dokuchaev (1899, p. 3) carried out the first step towards the formalization of the problem. He proposed the first soil formation equation:

$$\Pi = f(K, O, \Gamma)B, \quad (10.1)$$

where Π is soil, K is climate, O is organisms, Γ is parent material, and B is age of the soil (topography was not included into the expression due to a stenographer's mistake). Zakharov (1927, p. 8) presented a general soil formation equation ideally describing the first law of soil science:

$$\pi = f(\text{М.Г.П.}, \text{Р.Ж.Орг.}, \text{Кл.}, \text{Возр.стр.}, \text{Р–ф}), \quad (10.2)$$

where π is soil, М.Г.П. is parent rock material, Р.Ж.Орг. is plant and animal organisms, Кл. is climate, Возр.стр. is the age of the terrain, and Р–ф is topography.

In 1927, the Dokuchaev hypothesis and Eqs. (10.1 and 10.2) became known in the West due to two circumstances. First, Afanasiev (1927, p. 10) presented an English translation of the hypothesis on the First International Congress of Soil Science hold in Washington, D.C. in June 1927.

---

[3]Pood is the unit of weight of the Russian Imperial measurement system, 1 pood = 16.38 kg; desyatina is the unit of area of the Russian Imperial measurement system, 1 desyatina = 1.0925 ha.

10.1 THE DOKUCHAEV HYPOTHESIS AS A CENTRAL IDEA OF SOIL PREDICTIONS

Second, Sergey Zakharov took part in the Congress. Western soil scientists including Hans Jenny (Waksman and Deemer, 1928) had opportunity to learn about the hypothesis and the equations during the Congress and the transcontinental soil excursion. Jenny (1941, p. 16) adopted the Zakharov equation (10.2) replacing Russian abbreviations with English ones:

$$S = f(cl, o, r, p, t, \ldots), \tag{10.3}$$

where S is soil; cl, o, r, p, and t are soil-forming factors: climate, organisms, topography, parent material, and time, respectively. The ellipsis indicate that additional soil formers may be included in Eq. (10.3).

McBratney et al. (2003) proposed a Zakharov-like formulation for empirical quantitative descriptions of relationships between soil and other spatially distributed factors, using them as soil predictors:

$$S_c, S_a = f(s, c, o, r, p, a, n), \tag{10.4}$$

where S_c is soil classes, S_a is soil attributes; s is soil, other properties of the soil at a point; c is climate, local climatic properties; o is organisms, vegetation, fauna, and human activity; r is topography, morphometric variables; p is parent material, lithology; a is age, time; and n is space, spatial position.

One of the main obstacles to implementation of such models is the complexity involved in producing reasonably accurate, detailed quantitative descriptions of most soil-forming factors (i.e., climate, parent material, and biota) and properties of the soil. However, detailed and relatively accurate quantitative description of topography—the most temporally stable component of geosystems—can be quickly and easy performed. Indeed, for many territories, there exist DEMs with resolutions of 25–30 m (e.g., USGS, 1993; Natural Resources Canada, 1997) and higher (up to 3 m; see Kuhn and Fedorko, 2006). Moreover, advances in GPS kinematic survey and LiDAR aerial survey led to the situation in which DEMs with a resolution of 1 m and higher could be generated for virtually any terrain. However, it is unlikely that similar (in terms of resolution and coverage) quantitative descriptions of soil properties will be available even in the long term. This is associated with spatial and temporal variability of soil properties (Beckett and Webster, 1971; Campbell, 1979; Burrough, 1993; Heuvelink and Webster, 2001), as well as the labor-intensive and high-cost soil surveys or samplings and soil laboratory analyses.

Thus in recent approaches to predictive digital soil mapping (DSM) (McBratney et al., 2000, 2003; Scull et al., 2003; Lagacherie and McBratney, 2007; Walter et al., 2007; Lagacherie, 2008; MacMillan, 2008; Grunwald, 2009), morphometric variables are commonly used as key predictors of soil properties[4] (for details, see Section 10.3). Such approaches are based on the following general assumption: Let us assume that (1) dependence of a soil property on morphometric variables is found using a relatively small sample set of soil and topographic variables; and (2) statistical relationships between two subsets of variables—a soil property and morphometric factors-prerequisites—are reasonably strong. In this case, it is possible to predict the spatial distribution of soil property within a field, landscape, and, sometimes, small region using quantitative topographic data.

The objects of predictive soil mapping are morphological, physical, chemical, and biological quantitative properties of soil, as well as soil taxonomic units. As predictors, researchers use different sets of morphometric variables: from the only elevation (e.g., Leenaers et al., 1990; Baxter and Oliver, 2005) up to 69 (!) morphometric indices[5] (Behrens et al., 2005). The most popular set includes elevation, G, k_h, k_v, and TI.

As predictors, one may use morphometric variables of not only the land surface, but also the top surface of parent material. This can be useful if rock properties (e.g., density) influence the intrasoil lateral migration and accumulation of water and, hence, the spatial distribution of soil properties (Florinsky and Arlashina, 1998; Chaplot and Walter, 2003).

10.2 EARLY MODELS

In early predictive soil models, topographic maps were used as a source of initial topographic data.

Fedoseev (1959) used (1) a coefficient describing the water storage in the root zone for different landscape positions relative to a reference hillslope; and (2) data on the seasonal dynamics of soil moisture content depending on G, A, and the slope shape (convex, concave, and flat) to predict the spatial distribution of soil moisture content. Kirkby and Chorley (1967) proposed semi-empirical mathematical models of intrasoil lateral migration of water using data on slope gradients and horizontal curvature. Romanova (1970, 1971) developed a method for predictive mapping of seasonal moisture distribution using maps of G, A, and the

---

[4]Besides DTMs, the list of soil predictors includes remotely sensed data (Mulder et al., 2011), data on soil electrical conductivity, and other thematic layers (e.g., bedrock lithology, climate, and vegetation).

[5]As independent predictors, Behrens et al. (2005) used k_h, k_v, k_{max}, and k_{min}; each of them was estimated by several calculation methods. It is unlikely that such an approach is reasonable. Selection of (topographic) predictors was, for example, discussed by Lark et al. (2007).

slope shape, and derived empirical graphs describing the dependence of soil moisture content on these attributes. However, these approaches took into account only local peculiarities of topography. They did not consider the contribution of upslope areas to the formation and dynamics of soil moisture at downslope areas (i.e., the role of catchment area).

Romanova (1963) also developed a method to predict and map the redistribution of meteoric waters considering the catchment area and slope gradient. On a topographic map, a slope was separated into sections having similar soil properties and G values. Water input to a slope section was a sum of water precipitated to this section and runoff from upslope sections. The runoff was calculated as a difference between the water precipitated there and the water infiltrated into the soil, depending on the upslope area. Each slope section was assigned a value of the empirical "runoff coefficient," which considered the dependence of runoff and meteoric water infiltration on G, initial soil moisture, and vertical transmissivity of the soil. The main obstacle of this method was its empirical basis, which hindered its formalization and further development. Eventually, Svetlitchnyi et al. (2003) computerized the methods of Romanova (1963, 1970, 1971).

Ideologically, Romanova's method (1963) is closed to TOPMODEL, a concept of distributed hydrological modeling at a scale of small watersheds. The first version of TOPMODEL was based on the manual analysis of topographic maps (Beven and Kirkby, 1979). Later, several computerized versions were developed (Quinn and Beven, 1993; Kirkby, 1997; Beven and Freer, 2001). TOPMODEL uses three main approximations:

1. Saturation zone dynamics can be approximated by a sequence of stable states.
2. At the given point of the land surface, the hydraulic gradient of saturation zone can be approximated by the local slope gradient.
3. The dependence of lateral soil transmissivity on the depth can be described by an exponent function of water table depth (or soil storage deficit).

The first two approximations allow use of simple relations between the moisture capacity (or soil storage deficit) of a watershed and local water table depths (or local soil storage deficits). The main variable of these relations is the topographic index (Section 2.6). Various versions of TOPMODEL have been used in digital modeling of overland flow dynamics, evapotranspiration, soil moisture, and water table depth in subboreal zones (Beven and Kirkby, 1979; Beven et al., 1984; Quinn and Beven, 1993; Thompson and Moore, 1996; Blazkova et al., 2002), the Mediterranean (Durand et al., 1992; Piñol et al., 1997), the humid tropics (Moličová et al., 1997), and so on.

Like the method of Romanova (1963), the basic TOPMODEL does not consider the spatial variability of soil moisture caused by the influence of slope aspect. To solve this problem, (1) catchment slopes are classified into several groups depending on A; (2) TOPMODEL is calibrated for each slope group; and (3) the soil hydrological characteristics of each slope groups are separately modeled (Band et al., 1993). Besides, TOPMODEL is a pretty simple model developed for the natural conditions particular to central England and Wales. Therefore, utilization of TOPMODEL in other soil and hydrological conditions may produce incorrect results (Quinn et al., 1995; Beven, 1997).

Information on plan curvature (k_p) (Section A.3.1) has been widely used to produce soil maps by the Relief Plasticity method, a manual geomorphometric technique (Anisimov et al., 1977; Stepanov et al., 1984, 1987; Stepanov and Loshakova, 1998). According to this method, soil contours are determined using isolines of $k_p = 0$ called morphoisographs.[6] These isolines separate divergence zones of overland and intrasoil flows ($k_p > 0$) from their convergence zones ($k_p < 0$). Then, soils of crests and upslopes are set to divergence zones; soils of valleys, downslopes, and depressions are set to convergence zones; midslope soils are set to intermediate positions. Soil data are usually derived from existing soil maps. The Relief Plasticity method has been applied to produce a lot of medium- and small-scale soil maps for vast territories of Central Asia, the East European Plain, and West Siberia[7] (Stepanov, 1984, 1989; Satalkin, 1988).

10.3 CURRENT PREDICTIVE METHODS

MacMillan (2008, p. 114) made the following observations about DSM methods:

> Almost all efforts to develop and apply DSM techniques can be seen to follow approximately the same basic steps.... Differences in protocols arise from differences

---

[6]Morphoisograph is the approximation for isoline of $k_h = 0$ (Sections A.3.1 and A.3.2). A graphical technique to derive isolines of $k_p = 0$ from topographic maps was first described by Sobolevsky (1932).

[7]Stepanov et al. (1998, p. 30) presented a list of the 17 published and the 13 unpublished soil maps of Russian regions (mostly, at 1 : 200,000 and 1 : 300,000 scales), which were produced by the Relief Plasticity method. The published maps covered the following territories: Bashkiria, Bryansk, Chelyabinsk, Ivanovo, Kaluga, Kostroma, Kurgan, Moscow, Novosibirsk, Omsk, Ryazan, Sverdlovsk, Smolensk, Tver, Tula, Vladimir, and Yaroslavl Regions. The unpublished maps cover the following territories: Altai, Archangel, Kaliningrad, Kemerovo, Kirov, Leningrad, Mordovia, Nizhny Novgorod, Novgorod, Perm, Pskov, Udmurtia, and Vologda Regions.

in the kinds of outputs that are to be predicted, the kinds of input data layers selected to support predictions and the kinds of equations or rules developed to make predictions. These steps are inter-related such that decisions on what to predict (individual soil properties or soil classes) influence both the selection of input variables and the development of predictive equations and vice versa.

The six basic steps of DSM are as follows (MacMillan, 2008):

1. Conceptualization of (a) a discrete (categorical or quantitative) area-class entity of a soil map, or (b) a quantitative continuous individual soil property to be predicted.
2. Identification and obtaining or derivation of relevant input data sets including DTMs, remotely sensed data, thematic maps, and data of field surveys.
3. Development of classification rules or predictive equations using related mathematical approaches (Section 10.3.2).
4. Application of classification rules or predictive equations to input data.
5. Evaluation of accuracy or efficiency of prediction in terms of root mean square error, positional error, and so on.
6. Production of final maps.

10.3.1 Classification of Methods

DTM-based methods of predictive digital soil mapping are naturally classified into two main groups:

1. Methods applied to predict quantitative soil properties. There are two subgroups (Florinsky et al., 1999, 2002; Pachepsky et al., 2001):
   a. Methods for prediction of a quantitative soil property in each point of the land surface. They are used to produce quantitative *continuous* maps of quantitative continuous soil properties. Such methods are usually based on multiple regression analysis, hybrid geostatistical techniques, and regression kriging (for details, see Section 10.3.2).
   b. Methods for prediction of a quantitative soil property for distinct landforms of the land surface (e.g., upper and lower slopes, shoulders, footslopes). Such methods commonly use topographic segmentation approaches (Section 2.7). In modeling, a transition occurs from a quantitative continuous scale of a soil property to a quantitative discrete (categorical) one. Such methods are utilized to derive quantitative *discrete unit* maps of quantitative continuous soil properties, such as soil horizon thickness, depth to calcium carbonate, soil organic carbon content, and soil moisture content (Pennock et al., 1987; Florinsky et al., 1999, 2002; Park et al., 2001; Pennock and Corre, 2001; Florinsky and Eilers, 2002; Park and van de Giesen, 2004).

2. Methods applied to predict categorical variables (e.g., soil map classes or units). There are three subgroups (McBratney et al., 2003; MacMillan, 2008):
   a. Unsupervised classification approaches. These approaches are used to derive new soil classes from the observed soil properties. The type and nature of predicted classes are greatly dependent on the type of predictors and sample locations rather than on local expert knowledge. Such methods can be, for example, based on decision tree models (Section 10.3.2).
   b. Supervised classification approaches. These approaches are used to develop classification rules for recognizing soil classes. The rules describe relationships between each output class and a set of environmental predictors from representative sites (so-called training data). Such approaches can be, for instance, based on decision tree models, fuzzy logic, discriminant analysis, and Bayesian analysis (Section 10.3.2). Also, supervised classification approaches include deterministic methods (Walter et al., 2007). They are used if a simplified physical law may explain the spatial distribution of a soil property. For example, the genesis and spatial distribution of hydric-like soils are associated with (periodical) waterlogging, areas that may be derived from DTMs. Thus, digital models of *TI* and *DA* (Sections 2.6 and 2.3) have been applied to predict spatial distribution of such soils (Merot et al., 1995; Bedard-Haughn and Pennock, 2002; Mourier et al., 2008; Penizek and Boruvka, 2008).
   c. Heuristic classification approaches. These approaches can be used if one is able to identify (i) all soil classes to be predicted; and (ii) conditions under which all classes may occur. Such expert knowledge may arise from (i) local field experience; or (ii) theoretical considerations about relationships between morphometric variables (or quantitatively segmented landforms) and soil units (MacMillan et al., 2000; Zhu et al., 2001; Buivydaite and Mozgeris, 2007; Barringer et al., 2008; Smith S. et al., 2010).

10.3.2 Mathematical Approaches

DTM-based methods of predictive digital soil mapping can use a variety of mathematical approaches. Below we briefly describe the most common of them.[8]

---

[8]Detailed description of these approaches can be found elsewhere (McBratney et al., 2000, 2003; Bishop and McBratney, 2001; Scull et al., 2003; Walter et al., 2007; MacMillan, 2008).

1. Multiple regression analysis, the classical least-squares regression technique (Kleinbaum et al., 2008); a dependent variable is a soil property, and independent predictors are morphometric variables.[9] The approach has been used to predict and map the following quantitative soil properties:
   a. At field scales: soil horizon thickness (Moore et al., 1993; Bell et al., 1994b; Odeh et al., 1994, 1995; Florinsky et al., 1999, 2002; Bourennane et al., 2000; Gessler et al., 2000; Park et al., 2001; Herbst et al., 2006; Sumfleth and Duttmann, 2008), depth to calcium carbonate (Bell et al., 1994b; Florinsky et al., 1999, 2002), depth to gleyed horizon (Park et al., 2001), soil organic matter, soil pH (Moore et al., 1993), soil particle size distribution (Moore et al., 1993; Odeh et al., 1994, 1995; McBratney et al., 2000; Lark et al., 2007; Sumfleth and Duttmann, 2008), soil moisture (Florinsky et al., 1999, 2002; Sulebak et al., 2000; Chaplot and Walter, 2003; Lark et al., 2007) (Chapter 11), soil organic carbon (Florinsky et al., 1999, 2002; Chaplot et al., 2001; Terra et al., 2004; Lark et al., 2007), profile mass of soil carbon (Gessler et al., 2000), soil total carbon (Sumfleth and Duttmann, 2008), soil total nitrogen (Florinsky et al., 1999, 2002; Sumfleth and Duttmann, 2008), extractable phosphorus (Moore et al., 1993), residual phosphorus (Florinsky et al., 1999, 2002), exchangeable sodium (Lark et al., 2007), soil microbial biomass carbon, denitrifier enzyme activity, denitrification rate (Florinsky et al., 2004) (Chapter 11), soil hydraulic properties (Herbst et al., 2006), and cation exchange capacity (McBratney et al., 2000; Bishop and McBratney, 2001; Lark et al., 2007).
   b. At scales of several watersheds: soil horizon thickness (Gessler et al., 1995), total carbon content (McKenzie and Ryan, 1999), and cation exchange capacity (McBratney et al., 2000; Bishop and McBratney, 2001).
   c. At regional scales: soil organic matter content (Dobos et al., 2007), soil clay content, electrical conductivity as a measure of soil salinity (Odeh et al., 2007), and cation exchange capacity (McBratney et al., 2000; Bishop and McBratney, 2001).

   Hybrid methods have also been applied. For instance, Thompson and Kolka (2005) predicted and mapped the organic carbon content at a watershed scale using (i) stepwise linear regression and (ii) regression trees to identify predictors, and (iii) robust linear regression to develop models. Huang et al.

---

[9]Section 10.4 includes a detailed description of an algorithm to predict soil properties using multiple regression.

(2007) mapped the total soil carbon at a field scale using stepwise linear regression and principal component regression.
2. Hybrid geostatistical methods (cokriging, kriging with external drift, etc.), which use morphometric attributes as ancillary variables in kriging of a predictand (Goovaerts, 1997; Webster and Oliver, 2007). These methods have been utilized to predict and map the following quantitative properties of soils:
   a. At field scales: soil horizon thickness (Odeh et al., 1994, 1995; Bourennane et al., 2000; Herbst et al., 2006), soil organic carbon (Terra et al., 2004), mineral nitrogen content (Baxter and Oliver, 2005), clay and gravel content (Odeh et al., 1994, 1995; McBratney et al., 2000; Bishop et al., 2006), cation exchange capacity (McBratney et al., 2000; Bishop and McBratney, 2001), and soil hydraulic properties (Herbst et al., 2006);
   b. At scales of several watersheds: soil organic matter (Pei et al., 2010), cation exchange capacity (McBratney et al., 2000; Bishop and McBratney, 2001), and soil zinc content (Leenaers et al., 1990).
3. Regression kriging, the combination of a multiple linear regression model with kriging of the regression residuals (Odeh et al., 1995). The method has been used to predict and map the following quantitative properties and class entities of soils:
   a. At field scales: soil horizon thickness (Odeh et al., 1994, 1995; Herbst et al., 2006; Sumfleth and Duttmann, 2008; Kuriakose et al., 2009; Zhu and Lin, 2010), soil organic matter (Zhu and Lin, 2010), soil organic carbon (Terra et al., 2004), soil mineral nitrogen (Baxter and Oliver, 2005), total carbon and nitrogen content (Sumfleth and Duttmann, 2008), soil manganese content (Zhu and Lin, 2010), soil particle size distribution (Odeh et al., 1994, 1995; McBratney et al., 2000; Sumfleth and Duttmann, 2008; Zhu and Lin, 2010), soil pH (Zhu and Lin, 2010), cation exchange capacity (McBratney et al., 2000; Bishop and McBratney, 2001), and soil hydraulic properties (Herbst et al., 2006);
   b. At scales of several watersheds: soil horizon thickness (Hengl et al., 2004), soil organic matter (Hengl et al., 2004; Chai et al., 2008; Marchetti et al., 2010), soil pH (Hengl et al., 2004), and cation exchange capacity (McBratney et al., 2000; Bishop and McBratney, 2001);
   c. At regional scales: soil organic matter (Hengl et al., 2007a) (Fig. 10.1a), soil organic carbon (Mora-Vallejo et al., 2008; Mendonça-Santos et al., 2010), soil clay content (Odeh et al., 2007; Minasny and McBratney 2007; Mora-Vallejo et al., 2008), soil texture classes, soil taxonomic classes (Hengl et al., 2007b) (Fig. 10.2c), soil electrical conductivity as a measure of soil salinity (Odeh et al., 2007), and soil salinity risk classes (Taylor and Odeh, 2007).

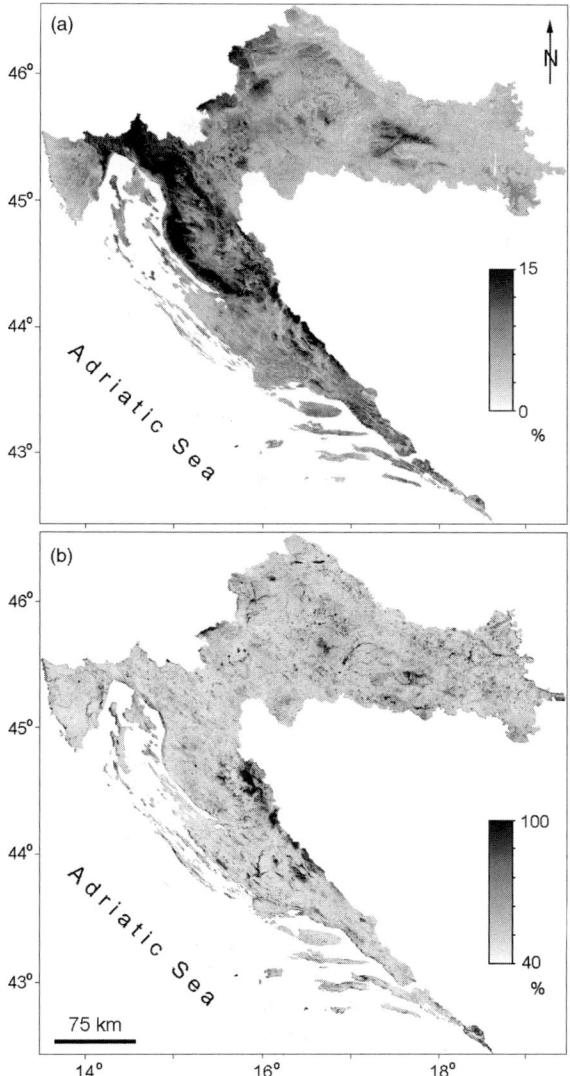

FIGURE 10.1 Small-scale predictive mapping of topsoil organic matter content using regression kriging for the territory of Croatia: (a) predicted values, (b) prediction error in terms of regression kriging variance. Original sampled values (2087 measurements) were obtained from the Croatian Soil Database. Sixteen environmental predictors were derived from a DEM, generated using the Shuttle Radar Topography Mission (SRTM) data (NASA, 2003) and 1 : 25,000 topographic maps, and multitemporal satellite radiometric images of the Moderate Resolution Imaging Spectroradiometer (MODIS). Among six predictors selected by the stepwise regression, the best topographic predictors were *TI* and elevation. Predictors explained 40% of the variation in soil organic matter. *From (Hengl et al., 2007a, Fig. 3c, d), reproduced with permission.*

178　10. PREDICTIVE SOIL MAPPING

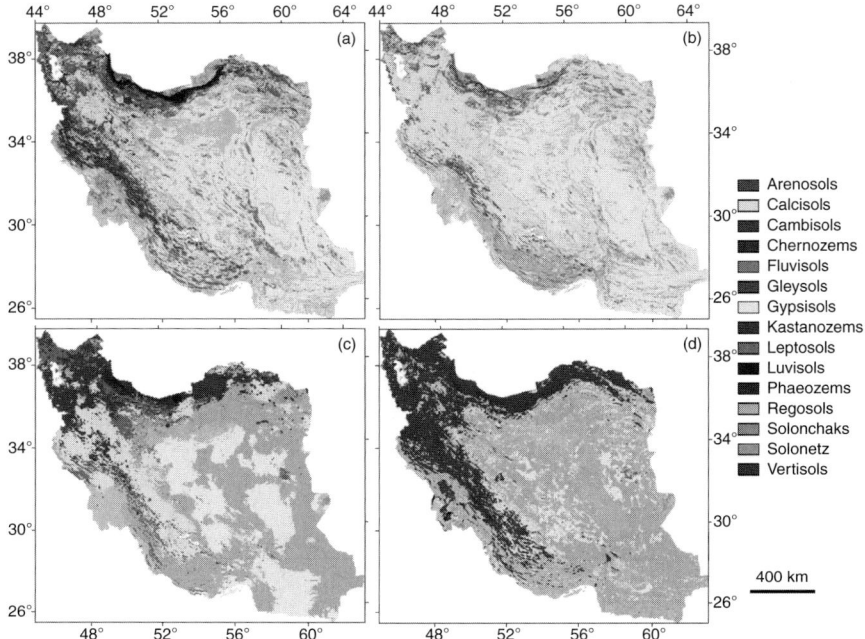

FIGURE 10.2　Small-scale predictive mapping of World Reference Base soil groups for the territory of Iran. Soil groups were interpolated by (a) supervised classification using maximum likelihood; (b) multinomial logistic regression; (c) regression kriging on memberships; and (d) per-pixel classification of interpolated taxonomic distances. A DEM was produced combining SRTM data (NASA, 2003) and GTOPO30 (USGS, 1996). The DEM grid spacing is 1 km. G, H, TI, and some other topographic attributes were used as predictors. *From (Hengl et al, 2007b, Fig. 5), reproduced with permission.* (See page 18 in Color Plate Section)

4. Fuzzy logic (Zadeh, 1965; McBratney and Odeh, 1997). The central idea of fuzzy logic is that a function, defining if an element belongs to a set, can take any value between 0 and 1, not only 0 or 1. As soil landscapes are characterized by continuous nature, fuzzy logic is useful in predictive soil mapping. The approach has been applied to predict soil taxonomic classes in large-scale soil mapping (Odeh et al., 1992; Zhu et al., 1996, 2001; Lark, 1999; MacMillan et al., 2000; Barringer et al., 2008) (Fig. 10.3), to delineate colluvial soils (Zádorová et al., 2011), to predict and map the topsoil clay content (de Bruin and Stein, 1998), soil moisture, soil particle size distribution, soil mineral nitrogen, soil organic matter, soil pH, and available phosphorus (Lark, 1999) at a field scale, as well as the topsoil particle size distribution and soil horizon thickness at a watershed scale (Zhu et al., 2001; Shi et al., 2004; Qi and Zhu, 2006) (Fig. 10.4).

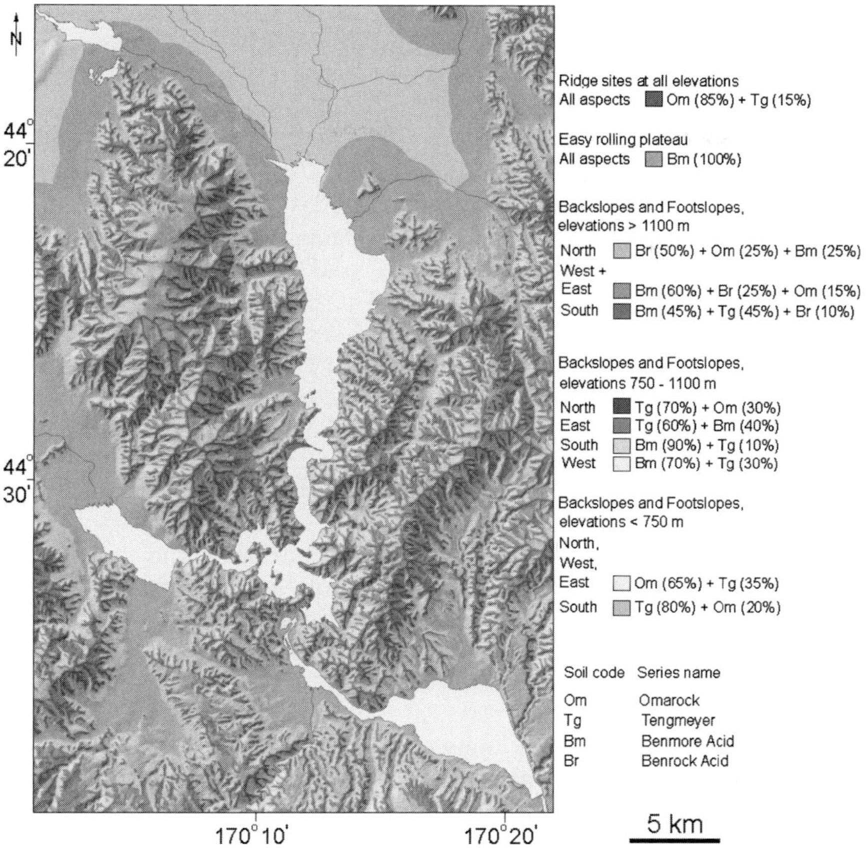

10.3 CURRENT PREDICTIVE METHODS

FIGURE 10.3 Large-scale predictive soil mapping by means of landscape segmentation and fuzzy logic (exemplified by the Benmore Range, South Island, New Zealand). The national DEM of New Zealand was used; the DEM grid spacing is 25 m. *From (Barringer et al., 2008, Fig. 3), © 2008 Springer-Verlag Berlin Heidelberg, reproduced with kind permission from Springer Science + Business Media.* (See page 19 in Color Plate Section)

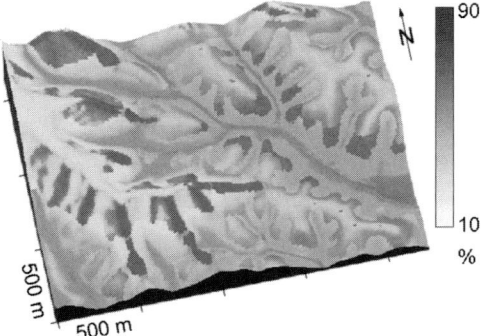

FIGURE 10.4 Large-scale predictive mapping of the distribution of the A horizon silt percentage using fuzzy logic (exemplified by the Raffelson watershed, La Crosse, Wisconsin, USA). A DEM was produced by digitizing a 1 : 24,000 topographic map; the grid spacing is 30 m. Elevation, G, k_h, k_v, and TI were used for prediction. *From (Qi and Zhu, 2006, Fig. 5-right), reproduced with permission.*

To predict soil taxonomic classes in large- and medium-scale soil mapping, hybrid approaches are sometimes utilized. They may combine fuzzy logic with regression kriging (Carré and Girard, 2002; Carré and McBratney, 2005) and discriminant analysis (Sorokina and Kozlov, 2009).

5. Decision trees (Clark and Pregibon, 1992). These methods work by splitting data into homogeneous subsets. Two main types of the decision tree analysis are used in DSM: classification tree analysis (a predictand is a categorical variable) and regression tree analysis (a predictand is a numeric variable). The approach has been employed to predict soil taxonomic classes in large-, medium-, and small-scale soil mapping (Lagacherie and Holmes, 1997; McBratney et al., 2000; Bui and Moran, 2001, 2003; Moran and Bui, 2002; Zhou et al., 2004; Scull et al., 2005; Behrens and Scholten, 2007; Cole and Boettinger, 2007; Hollingsworth et al., 2007; Saunders and Boettinger, 2007; Grinand et al., 2008; Mendonça-Santos et al., 2008; Behrens et al., 2010; Moonjun et al., 2010), to predict and map the soil cation exchange capacity at a field scale (Bishop and McBratney, 2001), soil profile thickness, total phosphorus (McKenzie and Ryan, 1999), and soil drainage classes (Cialella et al., 1997) at a scale of several watersheds, and the extent of the organic soils at a regional scale (Greve et al., 2010).

6. Discriminant analysis, the classical statistical technique to predict group membership using knowledge on interval variables (Webster and Burrough, 1974). This approach has been used to predict soil taxonomic classes in large-, medium-, and small-scale soil mapping (Thomas et al., 1999; Dobos et al., 2000; Hengl et al., 2007b) (Fig. 10.2a), to map soil drainage classes at a field scale (Kravchenko et al., 2002) and a scale of several watersheds (Bell et al., 1992, 1994a), soil horizon thickness at a field scale (Puzachenko et al., 2006), as well as classes of soil texture (Hengl et al., 2007b) and soil salinity risk (Taylor and Odeh, 2007) at a regional scale.

7. Multinomial logistic regression, which is used to predict a categorically distributed predictand from a set of real-, binary-, or categorical-valued predictors (Kleinbaum et al., 2008). The method has been utilized to predict soil taxonomic classes in large-, medium-, and small-scale soil mapping (Hengl et al., 2007b; Giasson et al., 2008; Debella-Gilo and Etzelmüller, 2009) (Figs. 10.2b and 10.5), to map the presence or absence of argillic horizon, secondary carbonates, calcic horizon, durinodes, duripan (Howell et al., 2007), and the E horizon (Gessler et al., 1995) at a scale of several watersheds, as well as classes of soil texture (Hengl et al., 2007b) and soil salinity risk (Taylor and Odeh, 2007) at a regional scale.

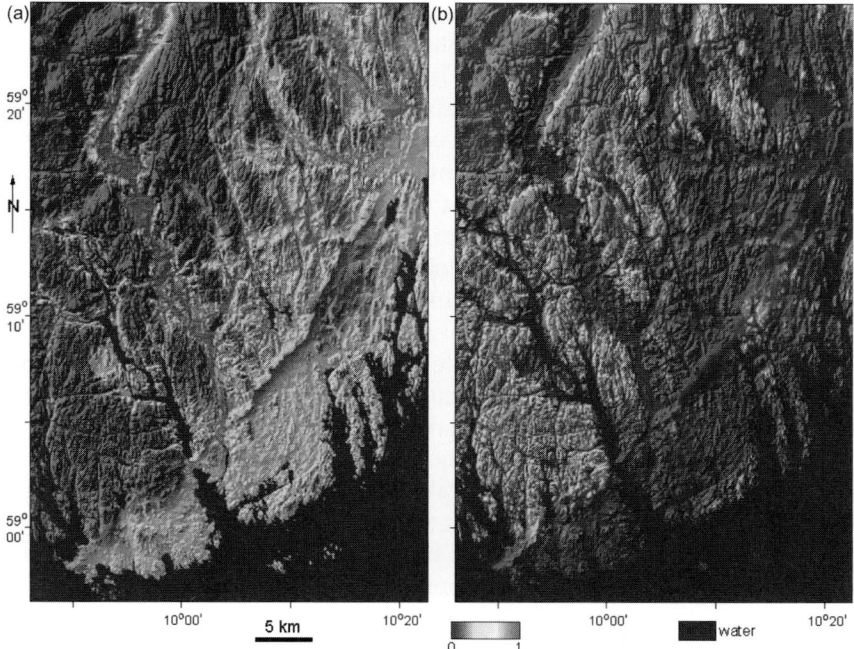

FIGURE 10.5 Large-scale mapping of the probability distribution, p-value, of (a) Cambisols and (b) Leptosols in the Vestfold County, southeastern Norway. Prediction was carried out by multinomial logistic regression. Leptosols occupy hills and rocky areas, and Cambisols occupy valleys. A DEM was created using 20-m contour topographic maps; the DEM grid spacing is 25 m. *From (Debella-Gilo and Etzelmüller, 2009, Fig. 5), reproduced with permission.* (See page 20 in Color Plate Section)

8. Bayesian analysis of evidence (Ghosh et al., 2006), an approach quantifying a predictive strength of an input subset of data relative to all other subsets. The approach has been employed to extract rules to recognize and map soil classes (Skidmore et al., 1991; Zhou et al., 2004) and some individual soil properties (Cook et al., 1996).
9. Artificial neural networks (Graupe, 2007), a mathematical model consisting of interconnected "neurons." Such models can change their structures using information flowing through the network in training. This approach has been used to predict soil taxonomic classes in large- and medium-scale soil mapping (Zhu, 2000; Behrens et al., 2005; Behrens and Scholten, 2007; Moonjun et al., 2010), to map the soil clay content at a field scale (McBratney et al., 2000), as well as the whole-soil profile carbon storage and available water capacity at a regional scale (Malone et al., 2009).
10. Multiscale support vector regression, a group of nonparametric machine learning techniques (Vapnik, 2000). The method has been

applied to map the topsoil organic carbon content, extractable aluminum in the B horizon, and soil horizon thickness (Ballabio, 2009), as well as soil heavy metal content (Ballabio and Comolli, 2010) at a watershed scale.

11. Empirical best linear unbiased predictor with residual maximum likelihood (Lark et al., 2006). This approach has been employed to map the organic matter content at a scale of several watersheds (Chai et al., 2008) and to predict the topsoil clay content at a regional scale (Minasny and McBratney, 2007).
12. Generalized linear models, an extension of linear regression allowing models to fit data that follow a non-normal probability distribution (McCullagh and Nelder, 1999). The approach has been used to predict the soil clay content at a field scale (McBratney et al., 2000), as well as to estimate the depth to the occurrence of argillic horizon, secondary carbonates, calcic horizon, durinodes, and duripan at a scale of several watersheds (Howell et al., 2007).
13. Generalized additive models, a regression approach in which nonlinearity can be introduced by replacing linear combinations of the predictors with combinations of nonparametric smoothing functions (Hastie and Tibshirani, 1990). The method has been utilized to predict the soil clay content (McBratney et al., 2000) and soil cation exchange capacity (Bishop and McBratney, 2001) at a field scale, as well as the soil thickness at a watershed scale (Tesfa et al., 2010).

10.3.3 Small-Scale Predictive Models and Upscaling

The papers cited above demonstrate that DTM-based predictive mapping of quantitative soil properties is typical for a field and small watershed scales. However, it is rarely used at large regional or country scales (e.g., predictions of the soil organic matter for territories of Hungary [Dobos et al., 2007] and Croatia [Hengl et al., 2007a]; see Fig. 10.1a). DTM-based predictive mapping of categorical entities—soil taxonomic classes—is common for production of large-, medium-, and small-scale soil maps (viz., at a field, watershed, and regional or country scales). However, Iran was the largest predictively mapped country (Hengl et al., 2007b) (Fig. 10.2).

In this connection, the Homosoil method (Mallavan et al., 2010) is a promising technique for deriving predictive soil maps of vast territories, whose soils are poorly known. The Homosoil is intended for small-scale extrapolation of soil taxonomic classes across the globe. Using available data on parent materials, topography (elevation, slope gradient, and topographic index), and climate (solar radiation, rainfall, temperature, and evapotranspiration), the Homosoil identifies a so-called homosoil area with similar soil-forming factors for an area with no soil data. If soils of the homosoil area are known, models describing relationships between soils and selected lithologic, topographic, and climatic predictors can be applied to the area with no soil data. In fact, the Homosoil method directly implements Dokuchaev's idea (1886) on predictive soil mapping (Section 10.1).

Upscaling procedures (Bierkens et al., 2000) may be applied to produce predictive soil maps for vast territories. However, it is obvious that such procedures are complex and ambiguous, at least in the context of DTM-based soil prediction (Thomas et al., 1999; Lagacherie et al., 2001). For instance, Florinsky and Eilers (2002) used a concept of accumulation, transit, and dissipation zones (Section 2.7.2) to predict the spatial distribution of the topsoil organic carbon content at a field scale, and to upscale the prediction to regional and subcontinental scales within the Black Soils Zone of the Canadian prairies. First, relationships between the organic carbon content of the A horizon and accumulation, transit, and dissipation zones of microtopography were found at the field scale. Second, the field-scale relations were applied to meso- and macrotopography of ecological districts with similar soils, parent materials, and landforms. Finally, DEM-derived maps of accumulation, transit, and dissipation zones were used to produce predictive maps at three different scales (Fig. 10.6). Generalization of organic carbon content contours came directly from generalization of accumulation, transit, and dissipation zones due to the use of DEMs of various resolutions and DEM smoothing.

Carré et al. (2008) compared two principally different approaches to generalize soil contours on DTM-based predictive soil maps in transition from 1 : 25,000 to 1 : 250,000 scale. The first, "bottom−up" approach is based on (1) a taxonomic aggregation of soil properties (carbonate rate, hydromorphic rate, texture, parent material) and topographic attributes (elevation and *TI*) to define taxonomic units; and (2) generalization of their contours to produce pedolandscape mapping units. The second, "top−down" approach includes (1) a taxonomic aggregation of the topographic attributes to delineate taxonomic units; and (2) generalization of their contours to produce pedolandscape mapping units. These approaches produce principally distinct compositions of soil contours. The first approach produces patchwork-like soil contours (Fig. 10.7a), while soil contours constructed by the second approach reflect the main peculiarities of the regional topography and landscape structure (Fig. 10.7b). Although there are no objective criteria to choose a particular approach to generalize soil contours, the second approach is intuitively preferred. Carré et al. (2008, p. 210) also accepted the idea that the second "approach can be recommended for

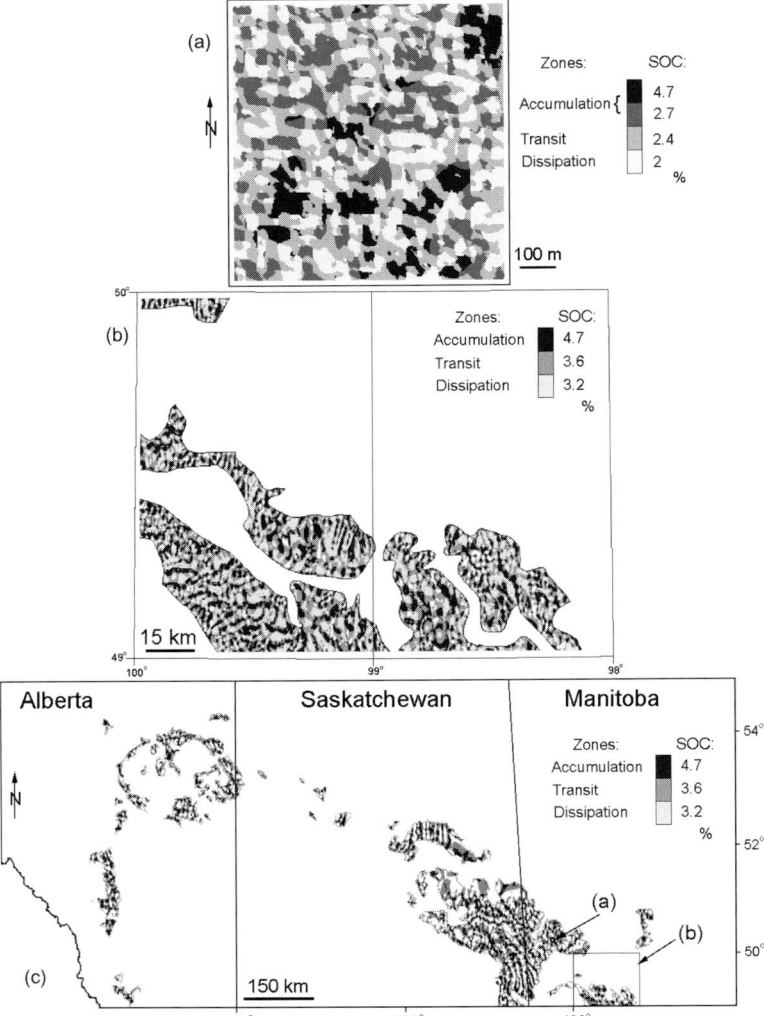

FIGURE 10.6 Upscaling prediction of the topsoil organic carbon content within the Black Soils Zone of the Canadian prairies: (a) A field scale—the Miniota Precision Agriculture Research Site. (b) A regional scale—a part of southern Manitoba. (c) A subcontinental scale—southern portions of Alberta, Saskatchewan, and Manitoba provinces of Canada. The dot in the circle shows the location of the field-scale study site; the frame shows the location of the regional-scale study area. For the field scale, a DEM was produced by the kinematic GPS survey (see details in Section 11.3.1); the DEM grid spacing is 15 m. For the regional scale, we used a DEM derived from the Canadian Digital Elevation Data files based on the 62 G topographic chart, 1 : 250,000 scale (Natural Resources Canada, 1997). The DEM includes 321,201 points (the matrix 801 × 401); the grid spacing is 9". For the subcontinental scale, we used a DEM derived from the global DEM GLOBO (GLOBE Task Team and others, 1999). The DEM includes 1,080,000 points (the matrix 1800 × 600); the grid spacing is 1'. To suppress high-frequency noise, the regional and subcontinental DEMs were smoothed. *Modified from (Florinsky and Eilers, 2002, Figs. 2c, 3c, and 4c).*

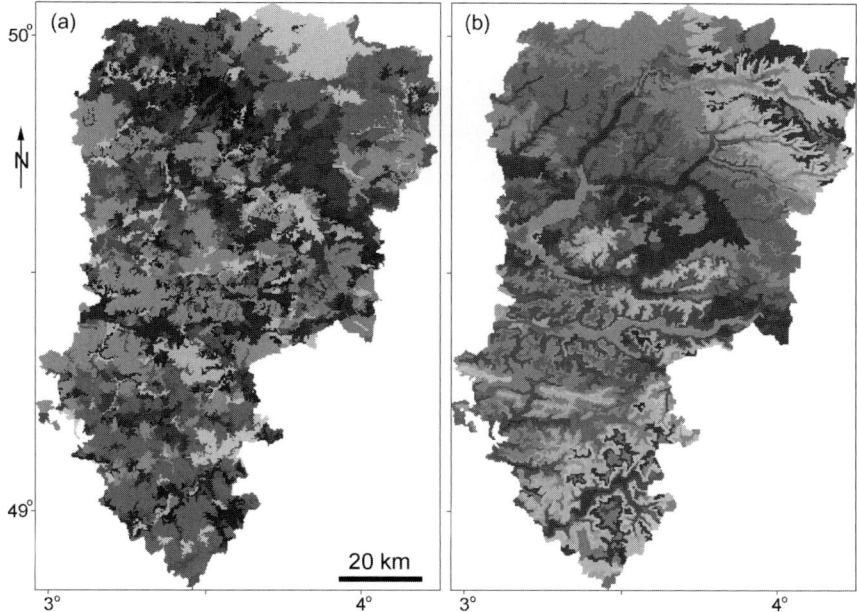

FIGURE 10.7 Upscaling soil contours for medium-scale soil mapping (exemplified by the Aisne Department, France): pedolandscape units resulting from the "bottom–up" (a) and "top-down" (b) approaches. The DEM grid spacing is 50 m. Map legends are not displayed as we discuss the geometry of soil contours only. *Modified after (Carré et al., 2008, Fig. 17.2), © 2008 Springer Science + Business Media B.V., reproduced with kind permission from Springer Science + Business Media.* (See page 21 in Color Plate Section)

digital soil mapping with limited data since" topography "is the most relevant factor of the pedogenesis."

This problem is not a new one for soil mapping. There was a discussion on the manual delineation of soil contours using the Relief Plasticity method[10] versus conventional approaches (Stepanov et al., 1984, 1987; Dmitriev, 1998; Stepanov and Loshakova, 1998; Volkova and Zhuchkova, 2000) (Fig. 10.8). Soil maps compiled by this method are positively different from conventional patchwork-like soil maps because of the clear representation of relationships "topography–soil cover" (compare soil contours in Figs. 10.8b and 10.8a).

10.3.4 Prediction Accuracy

The problem of accuracy or effectiveness of predictive soil mapping is rather ambiguous. On the one hand, all mathematical approaches

---

[10]For the description of the Relief Plasticity method, see Section 10.2.

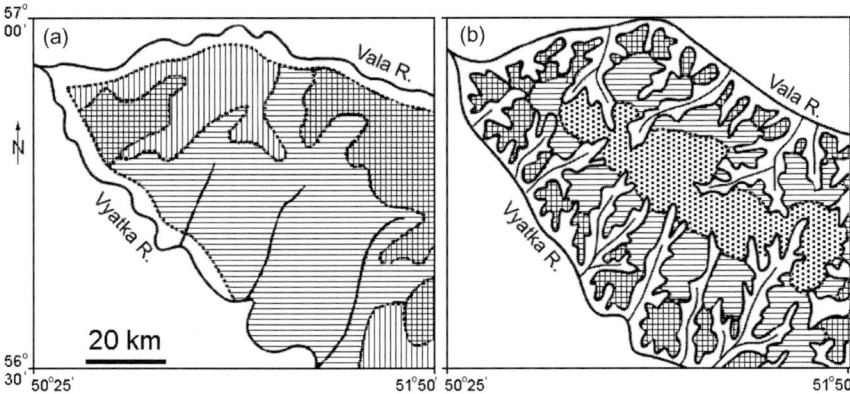

FIGURE 10.8 Different geometry of soil contours on two small-scale soil maps for the Kirov Region, Russia, produced by (a) a conventional manual technique, when a soil scientist implicitly considers topographic peculiarities (Fridland, 1988); (b) the Relief Plasticity method, when a soil scientist considers topographic peculiarities using a map of zero isolines of plan curvature (Stepanov et al., 1987). Map legends are not displayed as we discuss the geometry of soil contours only. *From (Stepanov and Loshakova, 1998, Fig. 1e, f), reproduced with permission.*

discussed in Section 10.3.2 have some criteria of prediction quality. Various validation indices can be used to measure prediction accuracy (e.g., RMSE and mean error; Fig. 10.1b). On the other hand, the situation can be described in the terms McBratney et al. (2000) used concerning the accuracy of quantitative methods in soil survey:

> Application of each of the pedometric techniques depends on the purpose, resolution and setup of the survey as the ultimate use of soil survey information determines the accuracy required. Different techniques produce different error of interpolation ... In all the cases, since the purposes are different, the risk of taking wrong decisions due to survey error is also different. Therefore, the pedometric techniques described above cannot just be applied to any situation without consideration of the specific needs and appropriateness of the inherent assumptions of the techniques.

Particular attention should be given to pronounced differences in predictive soil patterns produced by different DSM techniques. For example, see Fig. 10.2 for categorical soil entities; for quantitative soil properties, see Fig. 3 in (Terra et al. 2004) and Fig. 4 in (Hengl et al. 2007b). However, when different soil scientists use conventional soil-mapping techniques, they also usually produce significantly different maps in terms of soil contours (patterns) and soil taxonomic classes (Fig. 10.8). This effect is associated with (1) a high spatial variability of the soil cover; (2) spatiotemporal variability of the most soil properties;

and (3) application of different concepts of soil formation and classification.

Comparative analyses of prediction accuracy or effectiveness of various methods of digital predictive soil mapping can be found elsewhere (Odeh et al., 1994, 2007; McBratney et al., 2000; Bishop and McBratney, 2001; Terra et al., 2004; Behrens and Scholten, 2007; Hengl et al., 2007b; Minasny and McBratney, 2007; Taylor and Odeh, 2007).

In closing this section, we should stress that although "a range of data analysis methods can be applied to develop models for spatial prediction using environmental correlation," "the success of environmental correlation depends on the strength of relationships between soil and environmental variables" (McKenzie and Ryan, 1999, pp. 81 and 92).

10.4 TOPOGRAPHIC MULTIVARIABLE APPROACH

The author's approach described in this section relates to predictive soil-mapping methods based on multiple regression analysis (Section 10.3.2). The approach includes the following steps:

1. Generation of a DEM of a site (field). One can use any appropriate technique to derive a DEM, such as the GPS kinematic survey, LiDAR aerial survey, or conventional topographic survey. It is obvious that a particular type of survey affects the accuracy and resolution of DTMs and soil maps to be derived, but it does not influence the sequence of the further data processing.
2. Derivation of a representative set of digital models of morphometric variables from the DEM. The recommended set includes elevation, 14 local attributes (k_h, k_v, H, K, k_{min}, k_{max}, K_a, E, K_r, k_{he}, k_{ve}, M, G, and A), two nonlocal attributes (CA and DA, or SCA and SDA), as well as two combined variables (TI and SI). Definitions, interpretation, and formulas can be found in Chapter 2. It is *a priori* impossible to know which particular morphometric variables control a soil property under given natural conditions. Thus, it is reasonable to utilize this representative set of attributes.
3. Selection of a plot within the site, which is representative relative to the entire site in terms of the spatial distribution of morphometric variables. The location and size of a representative plot are defined by expert opinion. The author's experience shows that the plot area should be no more than 10% of the entire site area. Validation of the morphometric representativeness of the plot (that is, estimation if there is a significant difference between two distributions of topographic variables—within the plot and the entire site) can be, for

example, performed by the Kolmogorov–Smirnov two-sample test (Daniel, 2000, ch. 8).
4. Soil sampling within the plot and soil laboratory analyses.
5. Correlation and multiple regression analyses of soil properties versus topographic variables for samples collected at the plot (the sample size should be no less than 40; see footnote 1 in Chapter 9). Regression equations for soil properties are derived using morphometric variables as independent predictors. Notice that one may want to normalize topographic attributes in the 0 to 1 range in correlation and regression analyses. This allows estimation of the relative contribution of topographic variables directly from values of regression coefficients (Tomer and Anderson, 1995).
6. Derivation of predictive maps of soil properties for the entire site using regression equations obtained for the plot and DTMs of the entire site.

Notice that validation of the prediction using an independent data set is important, but it may be problematic because of the labor and cost involved in field and laboratory work. Validation can be done through the following steps:

- Collect soil test samples in several points of the site located outside the plot.
- Measure soil properties under study for the test samples.
- Predict values of soil properties for the test points by DTMs and regression equations obtained previously.
- Perform statistical analysis to compare measured and predicted values of soil properties.

Alternatively, jackknifing (Efron, 1982; Bishop and McBratney, 2001) can be applied to validate regression models. To provide an independent assessment of the prediction accuracy, the data set is randomly split into prediction and validation subsets. The first subset is used to create a model, which is validated with the second subset.

Several peculiarities are associated with the use of morphometric variables in predictive soil modeling:

- Slope aspect is a circular variable: its values range from $0°$ to $360°$, and both of these values correspond to the north direction. Thus, A cannot be used in linear statistical analysis. To estimate statistical relations between A and other variables, one can use either the approach of circular statistics (Mardia, 1972; Batschelet, 1981), or $\sin A$ and $\cos A$ instead A (King et al., 1999).
- Many soil and morphometric characteristics have non-normal distributions. Therefore, it is correct to use the Spearman rank

correlation coefficients (Daniel, 2000, ch. 9) to estimate statistical relationships between topography and soil properties.
- Some topographic attributes are linear combinations of others. For example, mean curvature is a combination of horizontal and vertical curvatures (Section 2.2), while topographic and stream power indices are combinations of catchment area and slope gradient (Section 2.6). In this connection, it is incorrect to use together horizontal, vertical, and mean curvatures (or catchment area, slope gradient, and topographic index) in multiple linear regression analysis.
- Every so often, a dynamic range of a topographic variable for a plot can be less than a dynamic range of the variable for the entire site. This is typical for catchment and dispersive areas. However, it is technically difficult to select a plot satisfying dynamic ranges of all morphometric attributes within the site. In the case of an imperfect representativeness of the plot, regression equations cannot be applied to zones, where values of a topographic predictor differ significantly from its dynamic ranges within the plot.

Application examples of the method described are presented in Chapter 11.

CHAPTER

11

Analyzing Relationships in the "Topography—Soil" System

OUTLINE

11.1 Motivation	191
11.2 Study Sites	192
11.3 Materials and Methods	195
11.3.1 Field work	*196*
11.3.2 Laboratory Analyses	*197*
11.3.3 Data Processing	*198*
11.4 Results and Discussion	214
11.4.1 Variability in Relationships between Soil and Morphometric Variables	*214*
11.4.2 Topography and Denitrification	*216*

11.1 MOTIVATION

The correctness of DTM-based modeling and predictive mapping of soil properties depends not only on spatial and temporal variability of soil properties by themselves (Campbell, 1979; Burrough, 1993), but also on two related factors.

First, there exists temporal variability in "soil—topography" relationships. For example, Burt and Butcher (1985) described temporal variability in the dependence of saturation depth and slope discharge on k_h and TI, but there were no explanations of this phenomenon. Heddadj and Gascuel-Odoux (1999) found seasonal variations in the dependence of unsaturated hydraulic conductivity on slope position.

Second, there exist variations in the topographic control of soil properties depending on the depth of a soil layer. It is essential to recognize an effective soil layer, wherein relationships between soil and topography are observable and statistically significant. For example, TOPMODEL (Section 10.2) uses an assumption that soil moisture content decreases with the soil depth due to a decline in hydraulic conductivity (Beven and Kirkby, 1979).

Study of denitrification—a process of biological conversion of nitrate (NO_3^-) into nitrous oxide (N_2O) and nitrogen (N_2) gases (Payne, 1981)—is important for understanding two fundamental interdisciplinary problems: (1) the nitrogen dynamics at regional and global scales (Mishustin and Shilnikova, 1971; Khalil and Rasmussen, 1992); and (2) the contribution of N_2O emission to global warming, stratospheric ozone depletion, and photochemical air pollution (Conrad, 1996; Meixner and Eugster, 1999). Denitrification is also of concern in agricultural practice because it is responsible for most of the nitrogen loss from fall-applied fertilizers: about 30% of nitrogen fertilizers applied to agricultural soils are lost to the atmosphere as a result of the activity of denitrifying bacteria (Murray et al., 1989). It is well known that denitrification is influenced by soil water content; the hydrological differences in the landscape control patterns of the spatial distribution of denitrifying activity (Groffman and Hanson, 1997). Thus, it is not surprising that denitrification processes depend on topography (Pennock et al., 1992; Van Kessel et al., 1993; Corre et al., 1996; Whelan and Gandolfi, 2002; Florinsky et al., 2004).

In this chapter, we demonstrate the possibilities of DTM-based analysis, modeling, and predictive mapping of the spatial distribution of soil properties using the approach described in Section 10.4. We studied:

- Temporal variability in the influence of topography on dynamic properties of soil (Florinsky et al., 1999, 2002)
- Changes in the influence of topography on dynamic properties of soil depending on the soil layer depth (Florinsky et al., 1999, 2002)
- The effect of topography on the activity of denitrifiers under different soil moisture conditions (Florinsky et al., 2004, 2009c)

11.2 STUDY SITES

The studies were carried out in two sites, Miniota and Minnedosa, situated in southern Manitoba, Canada (Fig. 11.1). The sites are located about 260–280 km west of the city of Winnipeg. The Miniota site measures 809 × 820 m; the maximum elevation difference is about 6 m

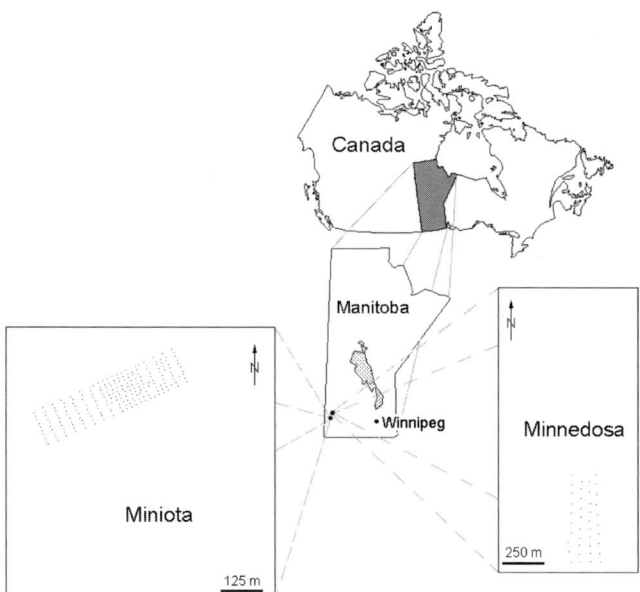

FIG. 11.1 Geographical location of the study sites: the Miniota site—50°13′40″N 100°51′20″W, the Minnedosa site—50°14′43″N 99°50′34″W. Dots indicate soil sampling sites. *From (Florinsky, 2010, Fig. 3.7)*

(Fig. 11.2a). The Minnedosa site measures 1680 × 820 m; the maximum elevation difference is about 13 m (Fig. 11.3a).

The study sites are located in the continental climate zone with warm summers and prolonged, cold winters. Mean annual temperature is 2.5°C, mean summer temperature is 16°C, mean winter temperature is −11°C. Mean annual precipitation is 460 mm including 310 mm of rainfall and 150 mm of snowfall (Fitzmaurice et al., 1999).

The study sites are representative of a broad region of undulating-to-hummocky glacial till landscapes in Western Canada. They are situated on the Newdale Plain, at an elevation of 500–580 m above sea level. The parent material consists of loamy textured glacial till deposits (Clayton et al., 1977). For most of the sites, soils are Black Chernozems and Gleysols (Soil Classification Working Group, 1998). The Newdale Orthic Black Chernozems series is predominant on well-drained crests and midslopes. The Varcoe Gleyed Carbonated Rego series is indicative of imperfectly drained downslopes, often in association with the Angusville Gleyed Eluviated series. The Penrith, Hamiota, and Drokan Gleysols series predominate in poorly drained depressions (Fitzmaurice et al., 1999; Bergstrom et al., 2001b).

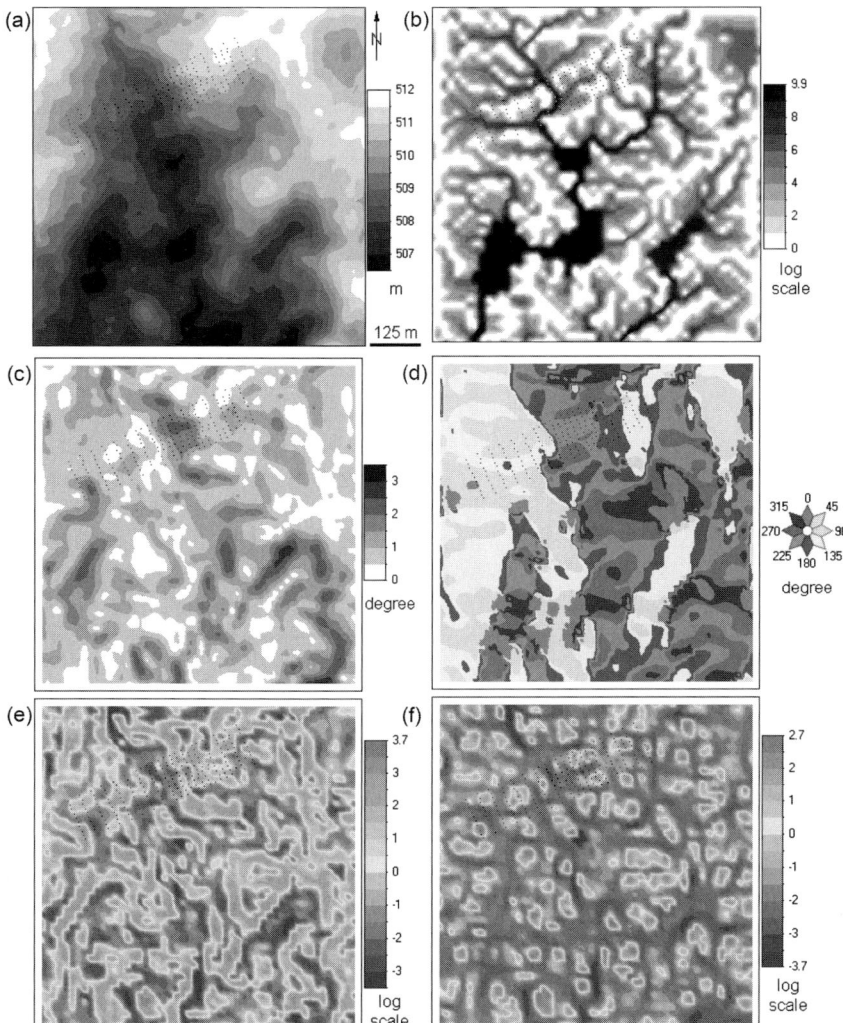

FIG. 11.2 The Miniota site, some morphometric variables: (a) elevation, (b) specific catchment area, (c) slope gradient, (d) slope aspect, (e) vertical curvature, (f) minimal curvature. Dots indicate soil-sampling sites. *From (Florinsky et al., 2009a, Fig. 1).* (See page 22 in Color Plate Section)

The sites are located in the aspen parkland of the Canadian prairies, the northern extension of open grasslands in the Great Plains of North America. Native vegetation of willows (*Salix sp.*), aspen (*Populus tremuloides*), and sedges (*Carex sp.*) surrounds water-saturated depressions. Most of the sites have been cropped for over 50 years. Before 1976, the Miniota site was farmed in a wheat–fallow rotation. In 1976,

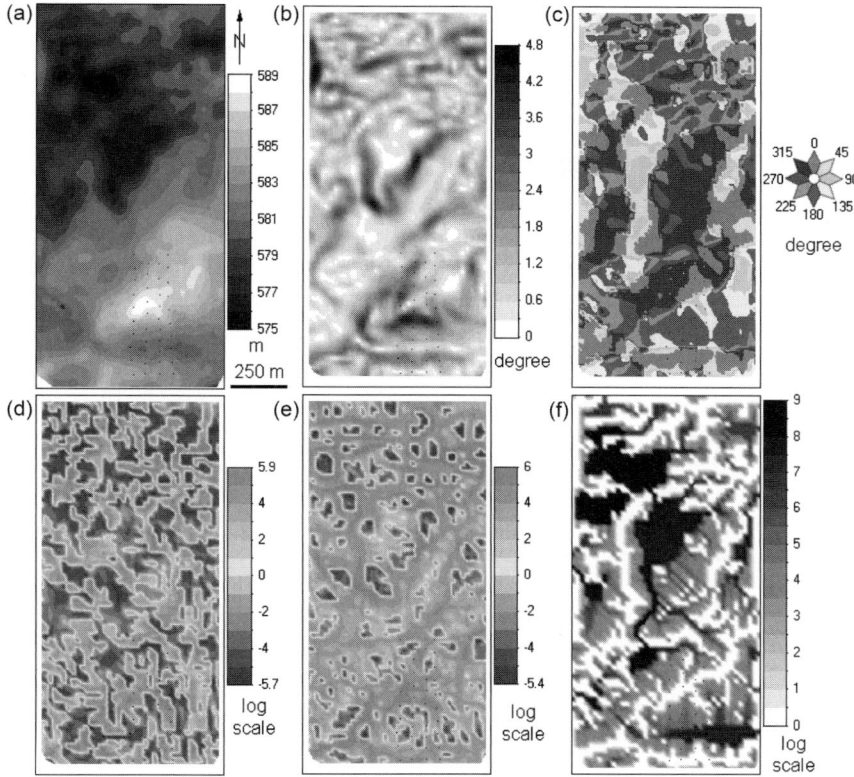

FIG. 11.3 The Minnedosa site, some morphometric variables: (a) elevation, (b) slope gradient, (c) slope aspect, (d) horizontal curvature, (e) maximal curvature, (f) specific catchment area. Dots indicate soil-sampling sites. *From (Florinsky et al., 2009a, Fig. 2)*. (See page 23 in Color Plate Section)

continuous cropping was initiated, with a cereal–broadleaf rotation. Since 1988, a zero-tillage management system has been employed. The Minnedosa site is a conventional tilled field cultivated by a deep-tiller, with one pass in autumn and one or two passes in spring. Historically, crops have included wheat, barley, oats, rape, and flax (Bergstrom et al., 2001a). The site was cropped to rape in 2000 and wheat in 2001.

11.3 MATERIALS AND METHODS

Field work and laboratory analyses were performed by employees and students of the Manitoba Land Resource Unit, Agriculture and Agri-Food Canada, as well as the Department of Soil Science, University of Manitoba, Winnipeg, Canada (Manning, 1999; Bergstrom et al., 2001b; McMahon, 2001). Standard field and laboratory techniques were utilized (see below).

11.3.1 Field work

Irregular DEMs of the study sites were constructed using kinematic GPS surveys with single-frequency Trimble 4600LS Surveyors mounted on all-terrain vehicles. The irregular DEMs of the Miniota and Minnedosa sites include 4211 and 7193 points, respectively.

A plot was selected within each study site to include a typical soil catena (Fig. 11.1).

On the Miniota site, the plot measures about 450×150 m with a difference in elevation of about 4.2 m (Fig. 11.2a). Soil samples were collected along 10 parallel transects about 450 m long, each about 11 m apart. There were 21 sampling points along each transect: points spaced at intervals of approximately 30 m on the crest and northeastern slope, and at intervals of approximately 15 m on the southwestern steeper slope. In total, there were 210 sampling points in the plot.

On the Minnedosa site, the plot measures about 500×200 m with a difference in elevation of about 8 m (Fig. 11.3a). Soil samples were collected along four parallel transects about 500 m long, each about 50 m apart. There were 10 sampling points along transects spaced at intervals of approximately 50 m, for a total of 40 sampling points in the plot.

All sampling points were georeferenced by GPS and flagged.

On the Miniota site, soil samples were collected at 210 points from four depth increments—0–0.3 m, 0.3–0.6 m, 0.6–0.9 m, and 0.9–1.2 m—using a soil auger (Carter, 1993). Among other soil properties, these samples were used to determine gravimetric soil moisture content (%). For each depth increment, the soil moisture content was determined six times: in early May, early July, and late August in both 1997 and 1998 (in August 1997—for depth increments of 0–0.3 m and 0.3–0.6 m).

On the Minnedosa site, two sets of soil attributes were sampled:

1. Two environmental properties influencing soil microbial activity: gravimetric soil moisture content (%) and soil bulk density (g cm^{-3})
2. Six indices of soil microbial activity (Table 11.1): most probable number of denitrifiers, microbial biomass carbon content, denitrifier enzyme activity, denitrification rate, microbial respiration rate, and N$_2$O flux

Soil samples were collected at 40 points using aluminum soil cores 5 cm in diameter and 5 cm in height. The sampling depth was approximately 10 cm because each core pressed into the ground passed surface litter and discontinuities (about 2 cm).

TABLE 11.1 Interpretations of Soil Microbial Variables

Variable, Unit	Interpretation
Most probable number of denitrifiers, $\#_{organisms}\,g_{soil}^{-1}$	A measure of the number of denitrifiers in the soil.
Microbial biomass carbon, $\mu g_C g_{soil}^{-1}$	A measure of the microbial biomass expressed as carbon.
Denitrifier enzyme activity, $\mu g_N g_{soil}^{-1} h^{-1}$	A measure of the amount of denitrifying enzymes in the soil.
Denitrification rate, $\mu g_N g_{soil}^{-1} h^{-1}$	A measure of the total gas N production from the soil.
Microbial respiration rate, $\mu g_{CO_2} g_{soil}^{-1} h^{-1}$	A measure of the rate of the total microbial respiration in the soil.
N_2O flux, $n g_{N_2O} m^{-2} s^{-1}$	A measure of the rate of N_2O emission from the soil.

From (Florinsky et al., 2004, Table 2)

To minimize the impact of temporal variability of denitrification and storage of samples, collection of soil core and N_2O flux measurements occurred simultaneously. N_2O flux was estimated using vented static chambers (Hutchinson and Mosier, 1981). Chambers were inserted within 1 m of each sample point. After 1 h of accumulation, a 15-ml gas sample was taken of each chamber by a syringe, injected into 10-ml Vacutainers™, and returned to the laboratory for analysis (Burton et al., 2000).

Sampling of all soil attributes took place at two times, July 2000 and July 2001, to assay the effect of topography on the activity of denitrifiers in different humidity levels. The 2000 sampling occurred during a period of increased rainfall, while the 2001 sampling occurred following a period of decreased rainfall. Monthly precipitation of 133 mm and 26 mm was observed in July 2000 and July 2001, respectively, at the nearest weather station in the city of Brandon located 40 km southward from the site.

11.3.2 Laboratory Analyses

The soil moisture content was determined by drying 10-g soil subsamples at 105°C for 24 h. The bulk density was calculated from the moist weight of soil, water content, and core volume (Carter, 1993).

To allow direct comparison of microbial activity indices, analyses were conducted on a single subsample taken from each soil core within 1 h of collection.

The microbial biomass carbon content was determined using a fumigation-direct extraction technique (Carter, 1993). Two 15-g samples were weighed into square French bottles. One sample was extracted immediately using 30 ml 0.5 M K_2SO_4. The second sample was fumigated for 24 h under chloroform atmosphere and then extracted. Filtrate was analyzed for carbon using a Technicon Auto-analyzer.

The most probable number of denitrifiers was determined with the modified method of Tiedje (1994). Aliquots (0.5 ml) of serial dilutions from 10^{-3} to 10^{-6} were added to 4.5 ml of sterile nutrient broth in 10-ml Vacutainers^{TM}. They were incubated at 25°C for approximately 7 days. Denitrifier presence was determined by measuring N_2O accumulation in the headspace using a Varian 3800 gas chromatograph.

Measurements were made sequentially (microbial respiration rate, denitrification rate, and denitrifier enzyme activity) within 48 h of sampling.

To measure the microbial respiration rate, we used the modified method of Zibilske (1994). Soil (5 g) was incubated in 20-ml headspace vials at ambient temperature for 2 h. Then, the headspace was sampled, and CO_2 concentration was determined by a Varian 3800 gas chromatograph.

To measure the denitrification rate, we used the modified method of Beauchamp and Bergstrom (1993). Soil (5 g) was incubated at ambient temperature in 20-ml headspace vials containing atmospheric air and 10% acetylene. After 24 h, the headspace was analyzed for N_2O concentration using a Varian 3800 gas chromatograph. The acetylene blocks the conversion of N_2O to N_2, so the N_2O accumulation rate can be used as a proxy for the total denitrification ($N_2O + N_2$).

To measure the denitrifying enzyme activity, we applied the modified method of Beauchamp and Bergstrom (1993). Soil (5 g) was incubated at ambient temperature in 20-ml headspace vials with 4 ml of a buffer solution under a helium/acetylene atmosphere. The buffer solution consisted of 10 mM glucose, 10 mM KNO_3, and 50 mM K_2HPO_4. N_2O in the headspace was measured at 30-min intervals using a Varian 3800 gas chromatograph.

The N_2O flux samples collected *in situ* were analyzed using a Varian 3800 gas chromatograph.

11.3.3 Data Processing

11.3.3.1 *Topographic Modeling*

The irregular DEMs (Section 11.3.1) were converted into regular DEMs (Figs. 11.2a and 11.3a) by the Delaunay triangulation and a piecewise quadric polynomial interpolation (Agishtein and Migdal, 1991).

The DEM grid spacings were 15 m and 20 m for the Miniota and Minnedosa sites, respectively.

All recommended morphometric variables (Section 10.4) were derived from the regular DEMs (Figs. 11.2b–f and 11.3b–f). Local topographic attributes were calculated by the author's method (Section 4.2), and nonlocal ones were estimated by the Martz–de Jong method (Section 4.4). Each DTM obtained includes 2743 and 3193 points for the Miniota and Minnedosa sites, respectively. To deal with the large dynamic range of morphometric variables, we logarithmically transformed DTMs by Eq. (7.1) with $n = 4$.

Then we used the Delaunay triangulation and a piecewise quadric polynomial interpolation (Agishtein and Migdal, 1991) of these DTMs to determine the values of all derived morphometric attributes at the sampling points.

Digital terrain analysis was performed by the software LandLord (Appendix B).

11.3.3.2 Statistical Analysis

Tables 11.2 and 11.3 represent statistical characteristics of morphometric variables for the plots and the entire sites. Tables 11.4 and 11.5 represent statistical characteristics of soil properties for the plots. On the Miniota site, the sample sizes were 210 and 2743 for the plot and the entire site, respectively. On the Minnedosa site, the sample sizes were 40 and 3193 for the plot and the entire site, respectively.

To estimate the topographic representativeness of the plots relative to the entire sites (Tables 11.2 and 11.3), we applied the Kolmogorov–Smirnov two-sample test (Daniel, 2000, ch. 8). Based on the test results, there are no statistical differences between curvature distributions within the plots and the entire sites. However, distributions of elevation, slope gradient, specific catchment area, topographic and stream power indices within the plots are statistically different from those within the entire sites.

We also applied the Kolmogorov–Smirnov two-sample test to estimate statistical differences between two distributions of each soil property measured in different years (1997 and 1998 for the Miniota site; 2000 and 2001 for the Minnedosa site). Results are presented in Tables 11.4 and 11.5.

To evaluate the statistical relationships between soil properties and topographic attributes on the plots, we carried out multiple rank correlation analysis between all measured soil properties, on the one hand, and all morphometric variables, on the other. The sample sizes were 210 and 40 for the Miniota and Minnedosa sites, respectively. The results of correlation analysis are presented in Tables 11.6 and 11.7.

To describe the relationships between measured soil properties and topographic variables on the plots, the "best" combinations of these

TABLE 11.2 The Miniota Site: Density Traces and Statistics for Morphometric Variables at the Plot and the Entire Site

Density Trace		Plot	Site	Density Trace		Plot	Site
	z, m				G, °		
	min	507.73	506.31		min	0.07	0.03
	max	511.82	512.13		max	2.34	3.43
	$\bar{x}$	509.66	509.31		$\bar{x}$	1.02	0.96
	s	1.13	1.47		s	0.51	0.50
	Dn (P)	0.211 (0.00)			Dn (P)	0.082 (0.14)	
	SCA, ×10², m				H, ×10⁻², m⁻¹		
	min	0.00	0.00		min	−0.097	−0.250
	max	60.28	201.15		max	0.160	0.291
	$\bar{x}$	3.84	11.66		$\bar{x}$	0.003	0.000
	s	10.74	41.17		s	0.051	0.053
	Dn (P)	0.386 (0.00)			Dn (P)	0.053 (0.64)	
	k_{hr} ×10⁻², m⁻¹				k_v, ×10⁻², m⁻¹		
	min	−0.173	−0.296		min	−0.130	−0.288
	max	0.184	0.446		max	0.182	0.403
	$\bar{x}$	0.006	0.001		$\bar{x}$	0.000	−0.001
	s	0.066	0.066		s	0.055	0.066
	Dn (P)	0.056 (0.59)			Dn (P)	0.076 (0.21)	

11.3 MATERIALS AND METHODS

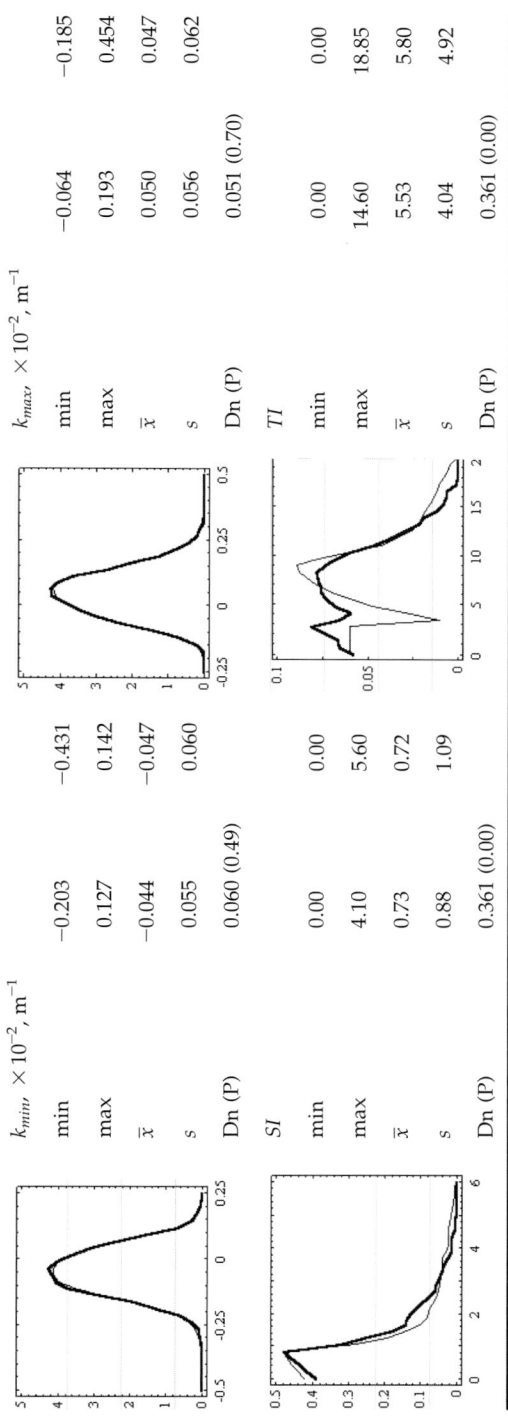

	$k_{min}, \times 10^{-2}, m^{-1}$		$k_{max}, \times 10^{-2}, m^{-1}$	
min	−0.203	−0.431	−0.064	−0.185
max	0.127	0.142	0.193	0.454
$\bar{x}$	−0.044	−0.047	0.050	0.047
s	0.055	0.060	0.056	0.062
Dn (P)	0.060 (0.49)		0.051 (0.70)	
SI			**TI**	
min	0.00	0.00	0.00	0.00
max	4.10	5.60	14.60	18.85
$\bar{x}$	0.73	0.72	5.53	5.80
s	0.88	1.09	4.04	4.92
Dn (P)	0.361 (0.00)		0.361 (0.00)	

1 Distribution density values are along Y-axis; morphometric variable values are along X-axis.
2 Heavy and thin lines describe morphometric variables at the plot and the entire site, respectively.
3 min—minimum, max—maximum, $\bar{x}$—average, s—standard deviation, Dn (P)—Kolmogorov-Smirnov statistics and significance level.
From (Florinsky, 2010, Table 3.4)

TABLE 11.3 The Minnedosa Site: Density Traces and Statistics for Morphometric Variables at the Plot and the Entire Site

Density Trace		Plot	Site	Density Trace		Plot	Site
z, m					G, °		
	min	580.89	575.13		min	0.3	0.0
	max	588.49	588.57		max	3.7	4.8
	$\bar{x}$	584.44	580.88		$\bar{x}$	1.7	1.3
	s	2.16	2.83		s	0.9	0.8
	Dn (P)	0.566 (0.00)			Dn (P)	0.280 (0.00)	
SCA, $\times 10^2$, m				H, $\times 10^{-2}$, m^{-1}			
	min	0.00	0.00		min	-0.08	-0.26
	max	17.54	89.00		max	0.14	0.22
	$\bar{x}$	2.09	9.17		$\bar{x}$	0.01	0.00
	s	4.70	22.97		s	0.05	0.06
	Dn (P)	0.333 (0.00)			Dn (P)	0.139 (0.43)	
k_{hr}, $\times 10^{-2}$, m^{-1}				k_v, $\times 10^{-2}$, m^{-1}			
	min	-0.09	-0.32		min	-0.14	-0.34
	max	0.19	0.40		max	0.22	0.29
	$\bar{x}$	0.01	0.00		$\bar{x}$	0.00	0.00
	s	0.06	0.07		s	0.08	0.08
	Dn (P)	0.143 (0.40)			Dn (P)	0.084 (0.94)	

11.3 MATERIALS AND METHODS

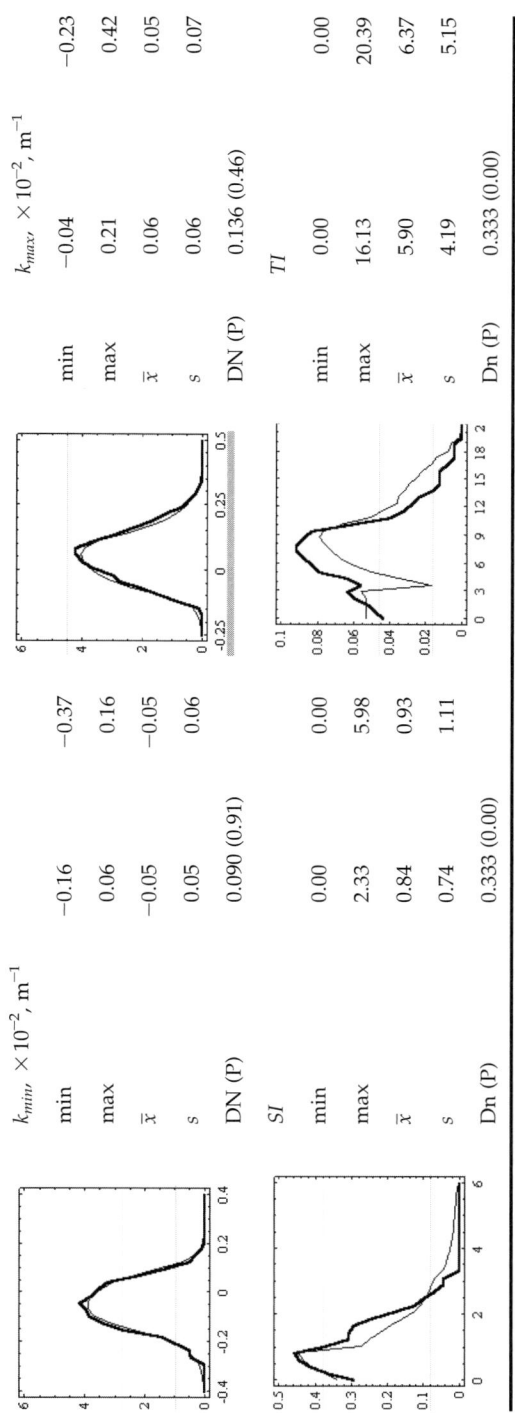

		$k_{min}, \times 10^{-2}, m^{-1}$			$k_{max}, \times 10^{-2}, m^{-1}$	
	min	−0.16	−0.37	min	−0.04	−0.23
	max	0.06	0.16	max	0.21	0.42
	$\bar{x}$	−0.05	−0.05	$\bar{x}$	0.06	0.05
	s	0.05	0.06	s	0.06	0.07
	DN (P)	0.090 (0.91)		DN (P)	0.136 (0.46)	
		SI			TI	
	min	0.00	0.00	min	0.00	0.00
	max	2.33	5.98	max	16.13	20.39
	$\bar{x}$	0.84	0.93	$\bar{x}$	5.90	6.37
	s	0.74	1.11	s	4.19	5.15
	Dn (P)	0.333 (0.00)		Dn (P)	0.333 (0.00)	

For explanation, see footnotes to Table 11.2.
From (Florinsky et al., 2004, Table 4)

TABLE 11.4 The Miniota Site: Density Traces and Statistics for the Soil Moisture Content (%)

Depth, m	Density Trace		May		Density Trace		July		Density Trace		August	
			1997	1998			1997	1998			1997	1998
0–0.3		min	12.46	10.78		min	11.87	17.95		min	7.18	12.65
		max	34.09	34.05		max	29.71	37.11		max	24.39	30.71
		$\bar{x}$	22.34	20.34		$\bar{x}$	20.16	26.70		$\bar{x}$	15.59	20.49
		s	3.88	3.59		s	3.00	3.76		s	3.13	3.03
		CV	17	18		CV	15	14		CV	20	15
		Dn (P)	0.256 (0.00)			Dn (P)	0.709 (0.00)			Dn (P)	0.600 (0.00)	
0.3–0.6		min	8.48	3.82		min	5.61	11.13		min	3.94	7.60
		max	39.86	25.20		max	24.83	42.26		max	18.16	28.73
		$\bar{x}$	17.29	13.64		$\bar{x}$	14.05	21.19		$\bar{x}$	10.23	16.75
		s	3.60	3.85		s	2.71	3.26		s	2.12	3.03
		CV	21	28		CV	19	15		CV	21	18
		Dn (P)	0.465 (0.00)			Dn (P)	0.833 (0.00)			Dn (P)	0.795 (0.00)	

11.3 MATERIALS AND METHODS

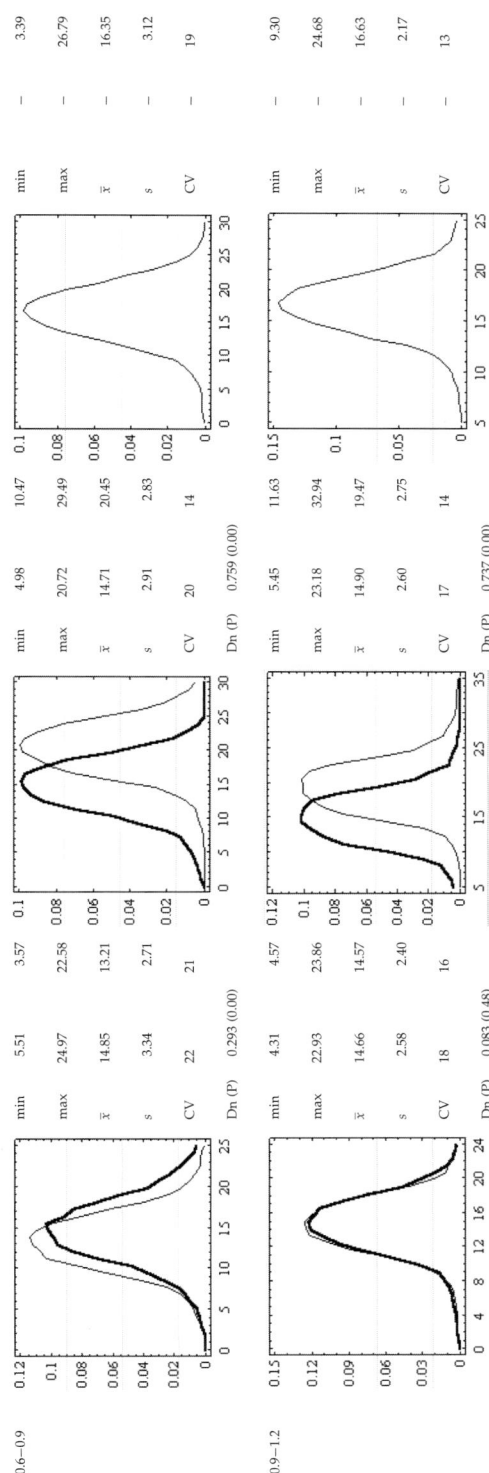

1 Distribution density values are along Y-axis; soil moisture values are along X-axis.
2 Heavy and thin lines describe soil moisture in 1997 and 1998, respectively.
3 CV—coefficient of variation. For other symbols, see footnote 3 to Table 11.2.
From (Florinsky, 2010, Table 3.6)

11. ANALYZING RELATIONSHIPS IN THE "TOPOGRAPHY–SOIL" SYSTEM

TABLE 11.5 The Minnedosa Site: Density Traces and Statistics for Soil Attributes

Density Trace		2000	2001	Density Trace			2000	2001
	Soil moisture content, %				Bulk density, g cm^{-3}			
	min	19.00	9.83			min	1.07	0.87
	max	57.00	52.27			max	1.51	1.41
	$\bar{x}$	33.35	23.24			$\bar{x}$	1.27	1.13
	s	7.16	7.60			s	0.12	0.14
	CV	21	33			CV	9	13
	Dn (P)	0.650 (0.00)				Dn (P)	0.475 (0.00)	
	Microbial biomass carbon content, $\mu g c g_{soil}^{-1}$				Denitrifier enzyme activity, $\mu g N g_{soil}^{-1} h^{-1}$			
	min	316.20	257.02			min	0.58	0.60
	max	1907.20	2053.82			max	6.75	6.24
	$\bar{x}$	926.88	803.89			$\bar{x}$	2.25	2.92
	s	364.71	361.80			s	1.36	1.56
	CV	39	45			CV	60	54
	Dn (P)	0.225 (0.26)				Dn (P)	0.300 (0.05)	

II. DIGITAL TERRAIN MODELING IN SOIL SCIENCE

11.3 MATERIALS AND METHODS

Most probable number, $\#_{organisms} g_{soil}^{-1}$

min	1056	141
max	293,919	3,004,034
$\bar{x}$	15,306	238,779
s	45,754	631,858
CV	299	265
Dn (P)	0.325 (0.03)	

Respiration rate, $\mu g_{CO_2} g_{soil}^{-1} h^{-1}$

min	0.27	0.14
max	9.21	5.04
$\bar{x}$	2.39	1.40
s	1.50	1.54
CV	63	110
Dn (P)	0.575 (0.00)	

Denitrification rate, $\mu g_N g_{soil}^{-1} h^{-1}$

min	0.00	0.00
max	0.32	0.18
$\bar{x}$	0.06	0.01
s	0.08	0.04
CV	136	276
Dn (P)	0.550 (0.00)	

N_2O flux, $ng_{N_2O} m^{-2} s^{-1}$

min	0.00	0.92
max	643.35	327.64
$\bar{x}$	41.99	31.51
s	113.35	58.60
CV	270	186
Dn (P)	0.138 (0.89)	

1 Distribution density values are along Y-axis; soil property values are along X-axis.
2 Heavy and thin lines describe samples collected in 2000 and 2001, respectively.
3 CV—coefficient of variation. For other symbols, see footnote 3 to Table 11.2.
From (Florinsky et al., 2009c, Table 3)

TABLE 11.6 The Miniota Site: Rank Correlations between the Soil Moisture Content and Morphometric Variables

Depth, m	Season	z	G	sinA	cosA	k_h	k_v	H	k_{min}	k_{max}	SCA	TI	SI
0–0.3	1997												
	May	−0.43	−0.38	—	0.17	−0.29	−0.41	−0.40	−0.35	−0.38	0.34	0.33	0.32
	July	−0.37	−0.39	—	0.27	−0.23	−0.37	−0.34	−0.25	−0.37	0.36	0.32	0.30
	August	−0.35	−0.25	—	0.14	−0.17	−0.24	−0.23	−0.21	−0.19	0.25	0.26	0.20
	1998												
	May	−0.44	−0.31	—	0.20	−0.29	−0.48	−0.44	−0.34	−0.46	0.46	0.46	0.42
	July	−0.29	−0.41	0.33	—	−0.23	−0.18	−0.26	−0.21	−0.24	0.29	0.30	0.22
	August	−0.36	−0.39	—	0.17	−0.23	−0.33	−0.31	−0.23	−0.36	0.33	0.30	0.24
0.3–0.6	1997												
	May	−0.23	−0.17	−0.19	—	—	—	—	—	—	—	—	—
	July	−0.26	−0.27	—	—	—	−0.21	—	—	—	0.19	—	0.14
	August	−0.19	−0.30	—	—	—	−0.20	−0.17	−0.15	−0.17	0.19	0.16	—
	1998												
	May	−0.38	−0.15	−0.25	—	—	−0.29	−0.22	−0.20	−0.19	0.21	0.17	0.21
	July	−0.23	−0.24	0.22	—	—	—	—	—	—	0.15	—	—
	August	−0.26	—	—	0.19	—	—	—	—	—	—	—	—

II. DIGITAL TERRAIN MODELING IN SOIL SCIENCE

0.6–0.9	1997											
	May	−0.28	−0.17	—	—	—	—	—	—	0.14	0.13	—
	July	−0.26	—	—	—	—	—	—	—	—	—	—
	1998											
	May	−0.33	—	−0.26	0.15	—	−0.15	—	−0.15	—	—	—
	July	−0.29	−0.16	—	0.19	—	—	—	—	—	—	—
	August	−0.30	—	—	0.14	—	−0.14	—	—	—	—	—
0.9–1.2	1997											
	May	−0.24	—	—	—	—	—	—	—	—	—	—
	July	−0.41	—	—	—	—	−0.16	−0.18	−0.21	0.20	0.17	0.20
	1998											
	May	−0.29	—	−0.21	—	—	−0.15	—	—	—	—	—
	July	−0.37	−0.15	0.17	0.19	−0.14	−0.17	−0.20	−0.24	—	—	—
	August	−0.21	—	—	0.20	—	—	—	—	—	—	—

1 $P \leq 0.05$ for statistically significant correlations; dashes are statistically insignificant correlations.
2 There were no significant correlations between soil properties and the following morphometric variables: k_{hv}, k_{zev}, E, M, K_r, K_{ar} and K.

From (Florinsky, 2010, Table 3.8)

TABLE 11.7 The Minnedosa Site: Rank Correlations between Soil Properties and Morphometric Variables in Wetter and Drier Conditions

Soil Property	Year	\multicolumn{10}{c}{Morphometric Variable}

Soil Property	Year	z	G	$\sin A$	k_v	H	k_{min}	k_{max}	SCA	TI	SI
Soil moisture content	2000	−0.51	—	—	−0.60	−0.48	−0.33	−0.43	0.42	0.53	0.34
	2001	−0.50	—	—	−0.37	−0.31	—	−0.33	—	—	—
Bulk density	2000	0.53	0.35	—	0.65	0.57	0.36	0.58	−0.47	−0.65	−0.49
	2001	—	—	—	—	—	—	—	—	—	—
Most probable number	2000	—	−0.41	−0.40	—	—	—	—	—	—	—
	2001	—	—	—	—	—	—	—	—	—	—
Microbial biomass carbon content	2000	−0.48	−0.30	—	−0.39	−0.32	—	−0.33	0.38	0.50	—
	2001	—	—	—	—	—	—	—	—	—	—
Denitrifier enzyme activity	2000	—	—	—	—	—	—	—	—	—	—
	2001	—	—	—	—	—	—	—	—	—	—
Denitrification rate	2000	—	—	—	—	—	—	—	0.52	0.46	0.40
	2001	—	—	—	—	—	—	—	—	—	—
Microbial respiration rate	2000	—	−0.40	—	—	—	—	—	—	—	—
	2001	—	—	—	—	—	—	—	—	—	—
N$_2$O flux	2000	—	—	—	−0.37	—	—	—	—	—	—
	2001	—	—	—	—	—	—	—	—	—	—

1 $P \leq 0.05$ for statistically significant correlations; dashes are statistically insignificant correlations.
2 There were no significant correlations between soil properties and the following variables: k_h, $k_{hе}$, k_{ve}, E, M, K_r, K_a, K, and $\cos A$.
From (Florinsky et al., 2009c, Table 5)

TABLE 11.8 The Miniota Site: Parameters and Statistics of the Regression Equation for the Dependence of the Soil Moisture Content (the depth 0–0.3 m, July 1997) on Morphometric Variables

Predictor	Estimate	Standard Error	95% Confidence Interval	
			Lower Limit	Lower Limit
Constant	250.02	89.11	74.31	425.73
z	−0.45	0.17	−0.79	−0.10
G	−2.41	0.34	−3.08	−1.74
$\sin A$	−0.79	0.24	−1.26	−0.32
$\cos A$	1.20	0.35	0.52	1.89
k_v	−22.11	4.62	−31.22	−13.01
k_{min}	16.82	5.16	6.65	27.00
$\ln SCA$	0.34	0.10	0.15	0.52

Source	Sum of Squares	Df	Mean Square	F-Ratio	P-Value
Model	854.65	7	122.09	24.08	0.00
Residual	1024.02	202	5.07		
Total (Corr.)	1878.67	209			
R^2	0.45				
Standard error of the estimate	2.25				
Durbin-Watson statistic	1.75				

From (Florinsky, 2010, Table 3.10)

variables were chosen by stepwise linear regression (Kleinbaum et al., 2008). The sample sizes were 210 and 40 for the Miniota and Minnedosa sites, respectively. Results of regression analysis are presented in Tables 11.8–11.10 (we show regression equations with $R^2 \geq 0.40$).

Using the regression equations (Tables 11.8–11.10) and DTMs (Figs. 11.2 and 11.3) inserted into the corresponding equations as predictors, we derived predictive digital models and maps of some soil properties (Fig. 11.4). These models include 2756 and 3193 points for the Miniota and Minnedosa sites, respectively.

Notice that we did not derive predictive maps for areas of the sites wherein values of morphometric predictors differ significantly from the

TABLE 11.9 The Miniota Site: Parameters and Statistics of the Regression Equation for the Dependence of the Soil Moisture Content (the depth 0–0.3 m, May 1998) on Morphometric Variables

Predictor	Estimate	Standard Error	95% Confidence Interval	
			Lower Limit	Upper Limit
Constant	319.80	96.71	129.12	510.49
z	−0.58	0.19	−0.96	−0.21
G	−2.56	0.39	−3.33	−1.79
$sinA$	−0.88	0.27	−1.42	−0.35
k_v	−33.87	5.29	−44.30	−23.44
k_{min}	25.68	5.94	13.97	37.38
$lnSCA$	0.55	0.11	0.33	0.76

Source	Sum of Squares	Df	Mean Square	F-Ratio	P-Value
Model	1326.28	6	221.05	32.91	0.00
Residual	1363.44	203	6.72		
Total (Corr.)	2689.72	209			
R^2		0.49			
Standard error of the estimate		2.59			
Durbin-Watson statistic		1.802			

From (Florinsky, 2010, Table 3.11).

dynamic ranges of the predictors within the plots (see Kolmogorov–Smirnov statistics in Tables 11.2 and 11.3). For the Miniota site, we did not perform predictive mapping for areas with values of $SCA > 6028$ m^2/m, $z < 507.7$ m, and $G > 2.5°$. For the Minnedosa site, we did not map areas with values of $SCA > 1754$ m^2/m, $z < 580.8$ m, and $G > 3.7°$. Hatching indicates areas omitted from the prediction (Fig. 11.4). Predictive patterns of soil properties (Fig. 11.4) generally resemble the structure of some morphometric maps (Figs. 11.2 and 11.3). Standard errors of the estimate may be used to assess roughly the prediction accuracy outside the plots (Tables 11.8–11.10).

Statistical analysis was carried out by Statgraphics Plus 3.0 (© Statistical Graphics Corp., 1994–1997). Predictive soil maps were derived using the software LandLord (Appendix B).

TABLE 11.10 The Minnedosa Site: Parameters and Statistics of the Regression Equation for the Dependence of the Microbial Biomass Carbon Content (July 2000) on Morphometric Variables

			95% Confidence Interval	
Predictor	Estimate	Standard Error	Lower Limit	Lower Limit
Constant	47,247.7	19,038.5	8309.6	86,185.8
z	−79.17	32.61	−145.85	−12.48
lnSCA	73.29	34.29	3.15	143.42

Source	Sum of Squares	Df	Mean Square	F-Ratio	P-Value
Model	2,285,180	2	1,142,590	12.14	0.00
Residual	2,730,100	29	94,141.3		
Total (Corr.)	5,015,280	31			
R^2			0.46		
Standard error of the estimate			306.83		
Durbin-Watson statistic			2.32		

From (Florinsky, 2010, Table 3.12)

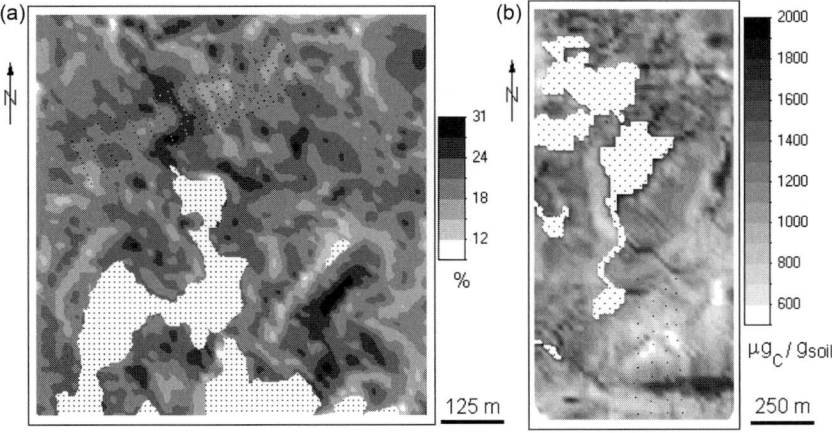

FIG. 11.4 Prediction of the spatial distribution of soil properties: (a) The Miniota site, soil moisture content (the depth 0–0.3 m, May 1998). From (Florinsky et al., 2009a, Fig. 4a). (b) The Minnedosa site, the microbial biomass carbon content (July 2000). Dots indicate soil-sampling sites. Hatching indicates depressions omitted from the prediction. *From (Florinsky et al., 2009c, Fig. 2a)*

11.4 RESULTS AND DISCUSSION

11.4.1 Variability in Relationships between Soil and Morphometric Variables

For the Miniota site, results of correlation analysis (Table 11.6) show that soil moisture at the depth 0–0.3 m was dependent on many topographic variables (except k_{he}, k_{ve}, E, M, K_r, K_a, and K). This was expected and supported by physical interpretations of topographic variables (Table 2.1) and results of previous research (see details in Chapter 8).

11.4.1.1 Temporal Variability

However, different correlation coefficients describe statistical relationships of the soil moisture with topography in different seasons (Table 11.6). For different seasons, we also obtained different regression equations of soil moisture content at the depth 0–0.3 m with $R^2 \geq 0.40$ (Tables 11.8 and 11.9).[1] This is clear evidence of temporal variability in "soil–topography" relationships.

The existence of this phenomenon was quantitatively established for soil moisture and for some dynamic soil properties, which depend on soil moisture, such as residual phosphate and potassium (Florinsky et al., 1999). This was supported by results from other research groups that analyzed the spatiotemporal regularities of the distribution of soil moisture (Western et al., 1999; Sulebak et al., 2000). Topography "controls" many soil properties because its characteristics determine the ways of the lateral migration of water and zones of water accumulation (Chapter 8). Therefore, although temporal variability was certainly established for soil moisture only, we believe that this regularity can be extended to other dynamic soil properties associated with soil moisture.

Soil properties are the integral result of various processes having (1) different typical times (that is, related to different temporal scales); and (2) different temporal variabilities (Stepanov et al., 1991; Targulian and Sokolova, 1996; Targulian and Krasilnikov, 2007). That is the reason relationships between soil and topographic attributes are temporally unstable. Indeed, tectonic and erosion processes change the land surface relatively slowly, so morphometric variables can be seen as temporally stable determinants of soil development. Other factors of soil formation (e.g., characteristics of plant cover) have higher temporal variability. This leads to temporal variability in a spatially distributed soil response, which can be observed as temporal variability in relationships between soil and topographic properties. The rate of this temporal

---

[1] Such results are acceptable because $R^2 > 0.70$ are unusual, and $R^2 \approx 0.50$ are common for quantitative soil spatial models (Beckett and Webster, 1971; Malone et al., 2009).

variability may be connected with the dynamic rate of a particular soil property. For example, relationships between topography and relatively static soil attributes, such as soil horizon thickness, may have a low temporal variability, that is, may remain unaltered over years.

11.4.1.2 Depth Variability

Absolute values of the correlation coefficients of the soil moisture content with a topographic variable decreased and, at times, became statistically insignificant with increasing the depth of soil layer (Table 11.6). This demonstrates a decrease of the topographic influence on soil moisture with the soil layer depth, at least for the conditions of the Miniota site.

The variability in relations between soil and topography depending on the soil layer depth (let us call this phenomenon "depth variability") may stem from the spatial variability in the characteristic decline of hydraulic conductivity with depth. If this decline were the same at all points in the landscape, we would observe equal correlations between soil and topographic attributes for all depths. The spatial variability of the decline in hydraulic conductivity with depth can be associated with spatial variability of pedogenetic processes, the existence of relict soil patterns, and random inclusions of sand or silt lenses in glacial till. The strongest dependence of soil properties on topography occurred at the 0–0.3 m depth within the Miniota site.

We suppose that in different landscapes, one may observe different depths of "effective soil layers" wherein relations of soil to topography are statistically significant. Moreover, more intricate situations may be found in some natural conditions and for some soil properties. In particular, topographic control of a soil property may increase with depth.

Indeed, the depth variability of "topography–soil" relationships was first quantitatively described analyzing soil moisture and some dynamic soil properties depending on soil moisture, such as residual phosphate and potassium (Florinsky et al., 1999). For residual phosphate, the highest correlations with morphometric variables were observed for the depth 0.3–0.6 m (negative correlations with curvatures and positive ones with nonlocal variables). Then, all correlations decreased and became insignificant with increases in the soil layer depth. For residual potassium, we observed a relatively stable negative correlation with G for all depths within the 1.2-m soil layer. Other morphometric variables did not influence the spatial distribution of residual potassium at 0.3–0.9 m depths. For the surficial layer (0–0.3 m), we observed slight positive correlations with nonlocal topographic attributes and slight negative ones with some curvatures. However, for the depth 0.9–1.2 m we found the inverse situation: there were slight negative correlations with nonlocal topographic attributes and slight positive ones with curvatures. The cause of such a behavior of residual potassium is not clear.

Later, Mitusov (2001) and Shary (2005) studied depth variability. They found that the dependence of soil properties on morphometric variables may both decrease and increase as the soil layer depth increases. In particular, for a cultivated field located in the southern Moscow Region, the dependence of available phosphorus (P_2O_5) and calcium and magnesium exchange cations on morphometric metrics of overland flows (i.e., k_h, k_v, k_{he}, and CA) decreases as the soil layer depth increases. More complex relationships were found between topography and potassium and sodium exchange cations. For potassium cations, as the soil layer depth increases, there is a decrease of correlations with local flow attributes (k_h and E), and an increase of those with nonlocal morphometric variables. In contrast, for sodium cations, as the soil layer depth increases, there is a decrease of correlations with nonlocal topographic variables, and an increase of those with local variables (k_h). The cause of such a behavior of potassium and sodium cations is not clear.

For "topography–NO_3^-" relationships, there were no significant correlations for the depth 0–0.2 m (Mitusov, 2001; Shary, 2005). This finding may be explained by a stronger impact of nontopographic factors (biota, etc.) on this nitrogen form in upper soil layers. For deeper soil layers, correlations of NO_3^- with all flow attributes decrease, while correlations with form attributes increase. Moreover, the spatial distribution of NO_3^- in the deepest soil layer is the reverse of that in the medium soil layer, which is demonstrated by the change of signs of correlation coefficients (as in the case with residual potassium - see above). Thus, a lateral redistribution of NO_3^- takes place during its vertical movement. The causes of this phenomenon are not understood. It is clear that the depth variability of "topography–soil" relationships invites further investigations.

Temporal and depth variabilities in relationships between soil and topography should be considered along with regional and scale variabilities in the topographic control of soil attributes. Regional variability refers to distinctions in the topographic control of soil properties under different natural conditions. Scale variability refers to the change in the character of soil–topography relationships under changes of a hierarchical level of the biogeocenosis and/or study scale (Chapter 9).

11.4.2 Topography and Denitrification

Different rainfall conditions in July 2000 and July 2001 (Section 11.3.1) resulted in different levels of soil water content in the landscape: the averages of the soil moisture content in the plot differed by 10% in these seasons (Table 11.5). In wetter and drier soil conditions, there were strong differences in relationships between soil environmental and microbial properties (Florinsky et al., 2004), as well as between soil properties and topographic attributes (Table 11.7).

11.4.2.1 Wetter Conditions

An earlier correlation analysis (Florinsky et al., 2004) demonstrated the significant dependence of all soil microbial variables, notably the microbial biomass carbon content and denitrification enzyme activity, on the soil moisture content in the wetter soil conditions of July 2000. The microbial biomass carbon content, most probable number, and N_2O flux also depended on the bulk density. These results are consistent with previous observations (Myrold and Tiedje, 1985; Groffman and Tiedje, 1989; Webster and Hopkins, 1996). Indeed, gravimetric moisture content and bulk density are indicative of aeration status in the soil. These parameters also influence the movement of water through soil and thereby the distribution of nitrogen and organic carbon, proximal regulators of denitrification. An increase of water content (water-filled pores) and/or bulk density (viz., a decreased total porosity) will result in a lower air-filled porosity and, therefore, a greater number of anaerobic sites in the soil that increase the suitability of the environment to support denitrification.

Note that a topographic influence on the spatial distribution of the soil organic carbon content has been previously demonstrated at this site (Bergstrom et al., 2001b). This was expected and stems from the spatial differentiation of organic matter and moistening according to the land surface morphology (Ototzky, 1901; Moore et al., 1993; Arrouays et al., 1998; Bell et al., 2000; Gessler et al., 2000; Chaplot et al., 2001; Florinsky et al., 1999, 2002; Mitusov, 2001; Shary, 2005).

There was a relatively strong influence of topography on the spatial distribution of soil moisture for the wetter soil conditions (Table 11.7). This was expected and supported by interpretations of topographic variables (Table 2.1) and previous results (Chapter 8). The soil moisture content had the highest values in such locations where k_v had negative values (concave profiles) and SCA had high values (draining of large upslope areas).

Relatively high correlations between the bulk density and topographic attributes (Table 11.7) resulted from its dependence on soil moisture, soil texture, and soil organic matter usually distributed according to the land surface morphology. Consequently, convex-profile upslopes ($k_v>0$, low values of SCA) had higher values of the bulk density than downslopes and depressions within the plot.

Almost all soil microbial indices, in one way or another, depended on topographic variables in the wetter soil conditions (Table 11.7). This was expected, as essential factors for denitrification—the soil moisture content, soil organic carbon content, and bulk density—depended on topographic attributes within the plot (see above). The denitrification rate depended on nonlocal and combined topographic variables: CA, TI, and SI (Table 11.7). The microbial biomass carbon content depended on both nonlocal and combined topographic variables (SCA and TI) and some local attributes (z, G, k_v, H, and k_{max}). Only local topographic variables influenced the N_2O flux, number of denitrifiers, and microbial respiration rate: they depended on k_v, G, and A (Table 11.7).

Thus, under wetter soil conditions, the spatial variability of the denitrification rate was mostly affected by redistribution and accumulation of soil moisture and soil organic matter due to their gain along a slope from its top to bottom, because of their additional amounts contributed from upslope (that is, according to the relative position of the point in the landscape). However, the N_2O emission, number of denitrifiers, and microbial respiration rate were affected by the distribution of soil water and organic matter according to the local slope geometry. Both groups of topographic factors (nonlocal and local) of spatial redistribution and accumulation of soil moisture and organic matter influenced the microbial biomass content. These results are generally consistent with previous observations that "hot spots" of denitrification are associated with downslope positions (Pennock et al., 1992; Van Kessel et al., 1993; Corre et al., 1996).

11.4.2.2 Drier Conditions

Based on the Kolmogorov–Smirnov statistics (Table 11.5), in the drier July 2001 there were marked decreases in the soil moisture content and bulk density, decreases in the denitrification and microbial respiration rates, a slight increase in the number of denitrifiers, and no significant changes in the microbial biomass carbon content, denitrification enzyme activity, or N_2O flux (Table 11.5).

Comparisons between the two sampling events demonstrated significant changes in the spatial differentiation of the soil properties. In the drier soil conditions, the coefficient of variation values showed pronounced increases in variation of the soil moisture content, denitrification and microbial respiration rates, and pronounced decrease in variation of the N_2O flux (Table 11.5). Correlations between indices of soil microbial activity and selected soil environmental properties became statistically insignificant in the drier soil conditions, except for the microbial biomass carbon content (Florinsky et al., 2004).

We observed a decrease of the topographic control of the soil moisture content in the drier July 2001, while associations of other soil properties with topographic attributes became insignificant (Table 11.7). This demonstrates a phenomenon of temporal variability in topographic control of dynamic soil properties (Section 11.4.1.1) (Florinsky et al., 1999, 2002).

From the observed distribution of the denitrification rate, it may be deduced that the denitrifier activity continued to persist under the drier soil conditions, but it was reduced and ceased to depend on the spatial distribution of soil moisture and thus land surface morphology. This likely reflects a transition of some critical level of the soil moisture status, and the ability of denitrifiers to be effective aerobic heterotrophs under aerobic conditions. The soil moisture status was still sufficient for the activity of these organisms, but was no longer a dominant force in influencing their spatial patterns.

11.4.2.3 Interpretations

It is possible to conclude that to keep the topographic control of the spatial distribution of denitrifiers and their activity, the landscape must contain some sufficient amount of soil water (more than some threshold). Physically, this idea seems reasonable as topographically controlled gravity-driven lateral transport of substances (e.g., nutrients) generally acts through the medium of gravimetric soil water. It is possible to generalize that the spatial distribution of soil dynamic properties depends on topographic variables only if soil moisture content is higher than some threshold value.

In drier and wetter soil conditions, the number of denitrifiers was different. It depended on topography in the wetter season, while the denitrifier enzyme activity was essentially the same on both sampling dates, ceasing to be affected by topography (Tables 11.5 and 11.7). This result reinforces the view that a direct relationship seldom exists between the number of denitrifiers and the amount of denitrifying enzyme in the soil (Parsons et al., 1991). This is because of the dual aerobic/anaerobic nature of the ecology and physiology of denitrifiers. The occurrence of denitrifying bacteria in any given habitat is primarily controlled by their ability to compete as heterotrophs rather than ability to denitrify (Groffman and Tiedje, 1989). The expression of denitrifying enzyme is just a response to anaerobic conditions reflecting soil aeration status.

Although the spatial distribution of denitrifiers was affected by topography in the wetter season, other groups of soil microbiota were probably more sensitive to the land surface morphology. This is reflected in the higher topographic effect on the microbial biomass carbon content compared with the most probable number (Table 11.7). Recall that microbial biomass carbon content measures the total microbial biomass in the soil, including both denitrifiers and other organisms (Table 11.1).

The lower topographic control of the spatial differentiation of the N_2O emission compared with other soil microbial variables in wetter soil conditions, as well as the disappearance of this control in drier soil conditions (Table 11.7) may reflect the high temporal and spatial variability of this attribute (Parsons et al., 1991) as well as the nature of N_2O production. Since N_2O production results from both autotrophic, aerobic processes (nitrification) and heterotrophic, anaerobic processes (denitrification) and is merely an intermediate in denitrification, it is not surprising that N_2O production and flux is highly variable and does not always reflect the environmental prerequisites of either of the microbial groups producing this gas.

The results obtained relate to the microbial activity in the upper soil layer wherein samples were collected and may vary with depth. This is caused by two factors:

- Soil microbial communities and enzyme activity change with depth (Zvyagintsev, 1994; Lehman, 2007).
- Topographic control of soil properties may both decrease and increase as the soil layer depth increases (Section 11.4.1.2) (Florinsky et al., 1999, 2002).

For the Minnedosa site, regression equations explained 46% of the variability of the microbial biomass carbon content in the wetter soil conditions (Table 11.10). Spatial heterogeneity of some other soil microbial attributes can be predicted using DTM-based regression, as demonstrated by the predictive maps of microbial biomass carbon content (Fig. 11.4b). However, DTM-based regression should be used with caution as different equations may be obtained in different seasons because:

- Temporal variability in the dependence of dynamic soil properties on topography (Section 11.4.1.1) (Florinsky et al., 1999, 2002)
- Seasonal dynamics of soil microbiota (Golovchenko and Polyanskaya, 1996)
- Soil microbial succession (Polyanskaya and Zvyagintsev, 1995)
- Spatiotemporal oscillations of soil microbial populations (Semenov et al., 1999) as well as other oscillating processes in soil (Florinsky et al., 2009b)

PART III

DIGITAL TERRAIN MODELING IN GEOLOGY

CHAPTER

12

Folds and Folding

OUTLINE

12.1 Introduction	223
12.2 Fold Geometry and Fold Classification	223
12.3 Predicting the Degree of Fold Deformation and Fracturing	225
12.4 Folding Models and the *Theorema Egregium*	226

12.1 INTRODUCTION

Folds of various hierarchical levels are the most abundant and studied geological features (Ramsay, 1967). In the context of fold studies, methods of digital terrain modeling (i.e., curvature derivation from DEMs of the land and stratigraphic surfaces) are applied to solve at least three problems (Bergbauer, 2007):

1. Analyze the geometry of folds and their classification.
2. Predict the degree of deformation or strain of folds, as well as fracture orientation and density of folded strata.
3. Analyze the evolution of folded strata and estimate the degree of their plasticity.

12.2 FOLD GEOMETRY AND FOLD CLASSIFICATION

Folds are characterized by the great diversity of geometric forms. This can be associated with regional lithological differences, the deformation degree of geological strata, stress distribution during deformation, and so on. Quantitative description of the fold geometry can

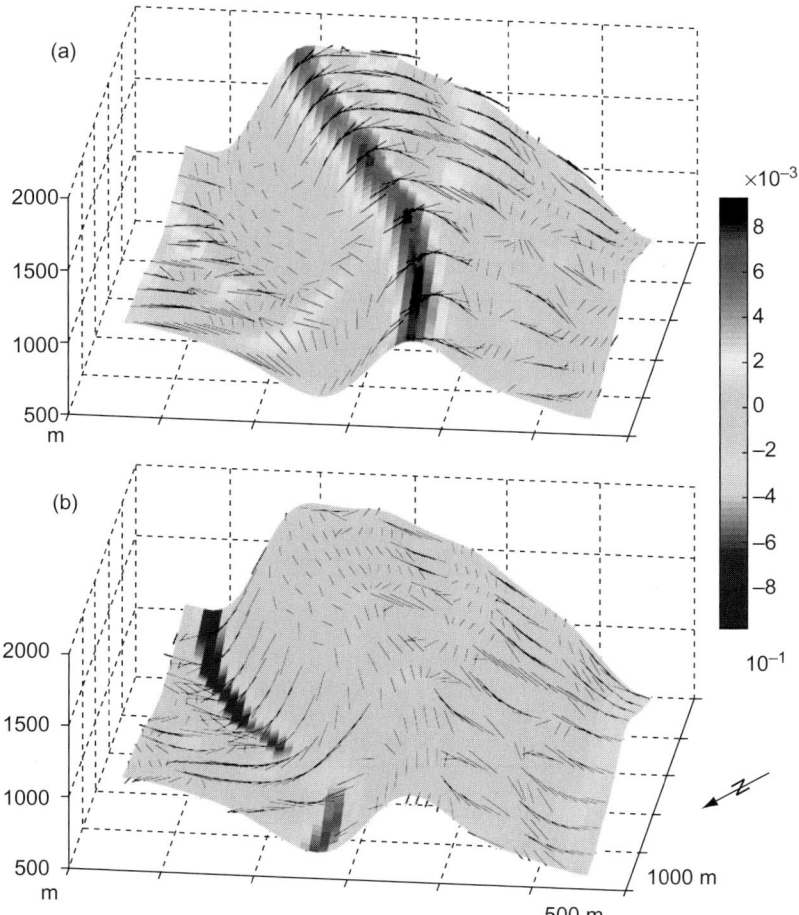

FIGURE 12.1 Magnitudes (colors) and axis directions (black tic marks) of the principal curvatures draped on the 3D model of the Sheep Mountain Anticline, Wyoming, USA: (a) maximal curvature, (b) minimal curvature. For the DEM description, see Fig. 6.7. *From (Mynatt et al., 2007a, Fig. 4), reproduced with permission.* For color version, please see page 24 in the Color Plate Section at the end of the book.

provide the key information necessary to analyze, reconstruct, and model the origin and evolution of folds (Bastida et al., 1999).

Bergbauer (2007, p. 185) correctly reasoned that for structural geology, "surface curvature is a unique descriptor of shape. Describing the geometry of a horizon quantitatively is an essential first step when attempting to compare physical and numerical models with natural surfaces." From the standpoint of terrain modeling, however, visualization of a form by itself is a standard procedure (Fig. 12.1). So, we do not discuss this issue in depth. Examples of visualization of the fold geometry with data on geological surface curvatures can be found elsewhere (Bergbauer, 2007; Hart and Sagan, 2007; Mynatt et al., 2007a).

Fold visualization combined with fold classification is a more complex task. Hudleston (1973) proposed quantitative classifications of folds. These were based on stereographic projections of bedding attitudes and cylindrical fold profiles. However, an assumption of cylindricity of folds was the main problem of such fold descriptions: it is obvious that a folded geological surface cannot be totally cylindrical.

Lisle (1994), Lisle and Robinson (1995), and Stewart and Podolski (1998) used the sign of the Gaussian curvature to distinguish three classes of folds: elliptical (domes or basins), hyperbolical (saddles), and cylindrical folds (Section 2.7.1). Roberts (2001) used a combination of the signs of the Gaussian and mean curvatures to distinguish six classes of folds (Section 2.7.1): domes ($K>0$, $H>0$), bowls ($K>0$, $H<0$), saddles ($K<0$), ridges ($K=0$, $H>0$), planes ($K=0$, $H=0$), and synforms ($K=0$, $H<0$). Within the framework of this approach, Lisle and Toimil (2007) suggested a generalized definition of a fold as a patch consisting of contiguous points with K and H of consistent sign. They distinguished four classes of folds: synclastic antiforms ($K>0$, $H>0$), anticlastic antiforms ($K<0$, $H>0$), anticlastic synforms ($K<0$, $H<0$), and synclastic synforms ($K>0$, $H<0$). For the continuous classification of folded surfaces, these authors applied the shape index (Koenderink and van Doorn, 1992) (Section 2.7.1). Mynatt et al. (2007a) elaborated the approaches of Roberts (2001) and Lisle and Toimil (2007): they proposed a classification including eight theoretically possible fold types (Fig. 2.11a), four of which correspond to "rare forms" (for details, see Sections 2.7.1 and 2.7.3). See Fig. 6.7 for an example of the classification of the geological surface by the signs of K and H.

We should note, however, that the landform classification by the signs of the Gaussian and mean curvatures is not complete and includes "rare forms." The Shary classification (Section 2.7.3) holds much promise.

12.3 PREDICTING THE DEGREE OF FOLD DEFORMATION AND FRACTURING

Assume there is an elastic thin beam. It is known that if the beam is bent, the shearing constraints at a point of the beam are proportional to its curvature (Timoshenko and Goodier, 1951). If these constraints reach some threshold, they are responsible for the rupture of the beam. Let us also consider a geological layer as an elastic beam. In this case, fractures may be typical for the layer portions where the shearing constraints are higher than the threshold (Samson and Mallet, 1997).

Murray (1968) was probably the first researcher to employ this idea and to link mathematically the curvature of a folded surface with the degree of fracture porosity and fracture permeability of folds, and so with the degree of potential oil- and gas-bearing of folds. He proposed that the product of the curvature and bed thickness can be used to estimate fracture porosity, while the third power of the product can be used to estimate fracture permeability. The computerization of the Murray approach to reveal areas of highest fracture intensity was based on the Laplacian derivation[1] from DEMs of stratigraphic surfaces, which were generated from structural contours (Muñoz-Espinoza, 1968; Pirson, 1970; Schultz-Ela and Yeh, 1992). Lisle (1994) argued that areas having high values of the Gaussian curvature correspond to zones of greater deformation and higher density of fractures. Fischer and Wilkerson (2000) proposed that joints formed by a deformation are parallel to axis directions of k_{min} normal sections at the given point of the geological surface.

However, Schultz-Ela and Yeh (1992) did not observe relationships between the high absolute values of the Laplacian and rock permeability (fracturing), which was estimated from the maximum annual gas productivity. Using results of detailed field studies, Fiore Allwardt et al. (2007, p. 419) concluded: "Fracture intensities do not correlate directly with curvature" because "curvature is only one mechanism for fracture development. Other mechanisms such as stretching, faulting, bedding-plane slip, the influence of preexisting fractures, and tectonic loading may contribute to fracture development and should be evaluated. ... A complete understanding of the heterogeneity in fracture intensities requires analysis beyond curvature calculations." Bergbauer (2007) came to a similar conclusion.

12.4 FOLDING MODELS AND THE *THEOREMA EGREGIUM*

In the *Theorema Egregium*[2] Gauss (1828)[3] demonstrated that the total (Gaussian) curvature is a bending invariant for the surface, if bending takes place without stretching, compression, or breaking of the surface. In other words, the Gaussian curvature does not change its values at each point of a surface during its deformations, if the surface was bent

---

[1]The Laplacian can be considered an approximation of mean curvature (Section A.3.4).

[2]This is Latin for "Remarkable Theorem."

[3]There is an English translation of this work (Gauss, 2009).

12.4 FOLDING MODELS AND THE THEOREMA EGREGIUM

and/or inclined without breaking, compression, or stretching. Changes in the Gaussian curvature value indicate compression or stretching of the surface in the vicinity of the given point.

In a pioneer work, Belonin and Zhukov (1968) utilized the *Theorema Egregium* to study the uplift evolution of an elevated block. These authors used DEMs of several conformable stratigraphic surfaces. To estimate the level, direction, and unevenness of stretching and bending of the block, they analyzed K as well as H, k_{min}, and k_{max} of these surfaces.

The *Theorema Egregium* was implemented to estimate "the degree of plasticity of the tectonics" (Samson and Mallet, 1997)—that is, to investigate the geometry and properties of oceanic lithospheric plates in subduction zones, in particular, whether or not the oceanic lithosphere maintains its mechanical integrity during subduction (Bevis, 1986; Cahill and Isacks, 1992; Ansell and Bannister, 1996; Nothard et al., 1996). However, DEMs of lithospheric plates were generated from data on earthquake hypocenter locations. It is clear that the accuracy of such DEMs and, hence, the certainty of conclusions about lithospheric plate plasticity are highly conjectural.

It would be interesting to apply this property of the Gaussian curvature to reconstruct paleotopography. Such a reconstruction can be best explained by the following example (Shary et al., 1991): Let us consider a plane, which is affected by two folding processes with different vergencies (Fig. 12.2a, b). As a result, a topographic surface arises with the pronounced regular cell-like structure; this structure can be seen on the elevation map (Fig. 12.2c). Assume that the plane is affected by two other folding processes with distinct vergencies (Fig. 12.2d, e). In this case, the processes are characterized by the same directions but shorter wavelengths as compared with the first case. As a result, there arises a topographic surface with the pronounced regular cell-like structure, but the cell size is less than it was in the first case (Fig. 12.2f). Now, assume that all four folding processes affect the plane. As a result, a topographic surface arises with a rather complex cell-like structure, which can, however, be seen on the elevation map (Fig. 12.2i).

Finally, assume that the surface (Fig. 12.2i) was inclined (Fig. 12.2g) and bent (Fig. 12.2h). The final topographic surface (Fig. 12.2j) obviously keeps some traces of the four folding processes, but they are hidden by the inclination and bending. On the final elevation map, it is impossible to see the cell-like structure, while a valley network is well marked.

Let us assume that (1) all surface deformations were free of breaks, stretching, and compression; and (2) there is an erosion-free condition. There is a need to reveal the hidden cell-like folding structure. Let us derive the Gaussian, horizontal, and vertical curvatures from the DEM of the topographic surface before (Fig. 12.2i) and after (Fig. 12.2j) its

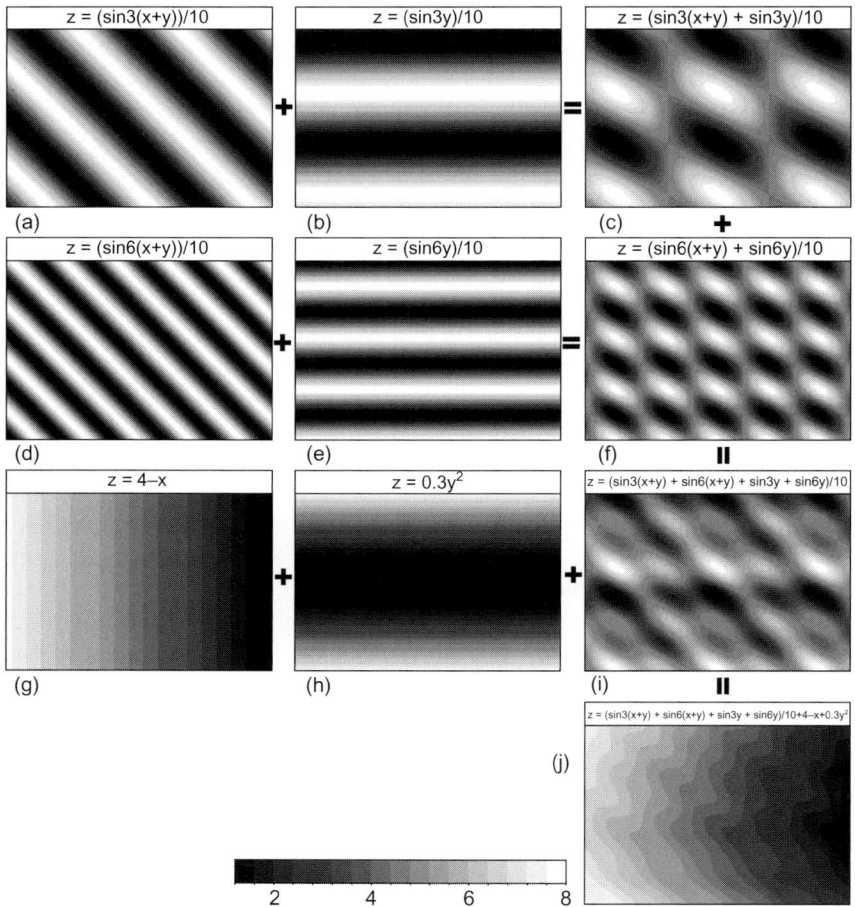

FIGURE 12.2 Modeling of the formation of the hidden complex folding (Shary et al., 1991). Two folding processes with different vergencies, (a) and (b), form a cell-like folding structure (c). Two other folding processes having distinct vergencies, (d) and (e), form the other cell-like folding structure (f). All four folding processes form the complex folding structure (i). Its inclination (g) and bending (h) form the final surface (j). "Elevation" maps of the surfaces are presented.

inclination and bending. One can see that there are distinct patterns on the maps of horizontal curvature of the surface before and after the inclination and bending (Fig. 12.3a). The same is true for the maps of vertical curvature (Fig. 12.3b). Moreover, there is no clear evidence of the cell-like structure (Fig. 12.2i) on the maps of these two curvatures. However, the desired cell-like structure can be seen on the maps of the Gaussian curvature (Fig. 12.3c). Notice that the maps of the Gaussian

12.4 FOLDING MODELS AND THE *THEOREMA EGREGIUM*

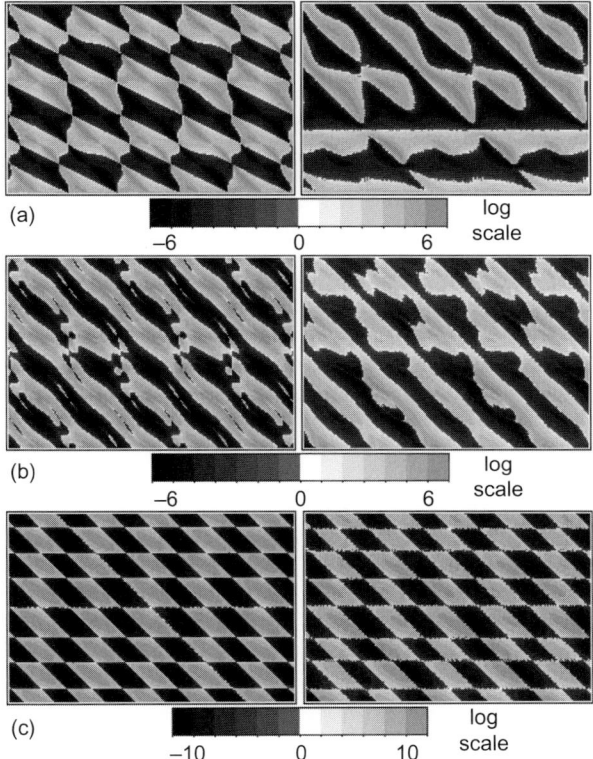

FIGURE 12.3 Revelation of the hidden complex folding structure with the *Theorema Egregium* (Shary et al., 1991): (a) horizontal curvature, (b) vertical curvature, (c) the Gaussian curvature. Left—curvatures of the surface with the hidden structure before the inclination and bending (Fig. 12.2i); right—curvatures of the surface with the hidden structure after the inclination and bending (Fig. 12.2j).

curvature of the surface before and after the inclination and bending have almost the same patterns (minor differences are caused by discretization errors).

Modeling was performed by the software LandLord (Appendix B).

We should finally note that it is difficult to use the *Theorema Egregium* to reconstruct paleotopography from DEMs of actual terrains. This is due to surface deformations caused by erosion, as well as tectonic stretching, compression, and faulting. This problem calls for further research.

CHAPTER

13

Lineaments and Faults

OUTLINE

13.1 Motivation	231
13.2 Theory	235
13.3 Method Validation	237
13.3.1 Materials and Data Processing	237
13.3.2 Results and Discussion	239
13.3.3 Strike, Dip, and Displacement Estimation	240
13.4 Two Case Studies	241
13.4.1 The Crimean Peninsula	241
13.4.2 The Kursk Nuclear Power Plant Area	248

13.1 MOTIVATION

It is well known that data on geological faults (Peive, 1956a; Slemmons and Depolo, 1986; Trifonov, 1995; Nikonov, 1995; Groshong, 2006) are required in neotectonic and seismic studies, ore and petroleum exploration, and so on. Revealing and monitoring local and regional faults are necessary in the development and operation of some potentially dangerous objects, such as nuclear power plants and hydroelectric power stations (Tabor, 1986; Chekunov et al., 1990; Poletaev et al., 1991). Indeed, construction and operation of such objects can cause changes in regional tectonic activity—for instance, tectonic movements along previously stable faults. It is obvious that this can affect the security of an object and its operation.

Faults are as a rule indicated by lineaments. The term *lineament* is commonly used in reference to any linear geological features of different origin, age, depth, and scale (Hobbs, 1904; Sonder, 1938; Hills, 1956;

Katterfeld and Charushin, 1973; Hodgson et al., 1976; O'Leary et al., 1976; Makarov, 1981; Trifonov et al., 1983; Wise et al., 1985; Kats et al., 1986; Bryukhanov and Mezhelovsky, 1987; Twidale and Bourne, 2007). Lineaments are usually associated with faults, linear zones of fracturing, bending deformation, and increased permeability of the crust, as well as linear chains of some geological features (i.e., laccolites, volcanoes, etc.). Lineaments are, as a rule, expressed in topography. They can be observed from air- and spacecrafts, and can also be recorded on aerial and satellite images, 3D seismic models, and thematic maps at a wide range of scales.

Geometric characteristics of linear topographic dislocations formed by vertical tectonic motions differ from those formed by horizontal tectonic movements (Ollier, 1981; Trifonov, 1983; Keller, 1986). Qualitative and quantitative signs of these differences can be used to recognize the origin (or type) of topographically expressed faults.

Qualitative approaches for revelation and classification of faults using topographic data have been repeatedly discussed (see reviews elsewhere—Trifonov, 1983; Slemmons and Depolo, 1986). One of the most popular approaches was visual analysis of topographic maps (Hobbs, 1904, 1911; Nezametdinova, 1970; Yabrova, 1981) and remotely sensed images (Wilson, 1941; Lattman, 1958; Skaryatin, 1973; Hodgson et al., 1976; Trifonov et al., 1983; Bryukhanov and Mezhelovsky, 1987). To recognize lineaments and ring structures visually, some researchers used hill-shading photographic images, which were obtained by taking pictures of 3D plastic terrain models illuminated at different angles (Hills, 1956; Wise, 1969; Saul, 1978; Wise et al., 1985). Analogue photogrammetric procedures were utilized to reveal and classify faults, as well as to measure their dip and strike angles (Vinogradova and Eremin, 1971).

To reveal and classify faults, geologists employed some standard techniques of digital terrain modeling, such as 3D topographic models dropped by aerial or satellite images or geological maps (Campagna and Levandowski, 1991; Morris, 1991; Kukowski et al., 2001, 2008; Ganas et al., 2005; Palyvos et al., 2006; Nazari et al., 2009) (Fig. 13.1), thalweg mapping (Eliason and Eliason, 1987; Thienssen et al., 1994; Jordan, 2003), as well as G and A mapping (Onorati et al., 1992; Steen et al., 1998; Kopp, 2000; Ganas et al., 2005; Kukowski et al., 2008). Lisle (1994) argued that high values of the Gaussian curvature may indicate fault zones. To reveal faults, Steen et al. (1998) approximated the derivation of vertical curvature by double derivation of slope gradients from DEMs of stratigraphic surfaces (viz., derivation of G from the digital model of G). To reveal lineaments, Wladis (1999) applied the geophysical method of second derivatives (Elkins, 1951) to DEMs. Maximal and minimal curvatures were also used to locate faults (Samson and Mallet, 1997; Bergbauer et al., 2003; Wynn and Stewart, 2003).

13.1 MOTIVATION

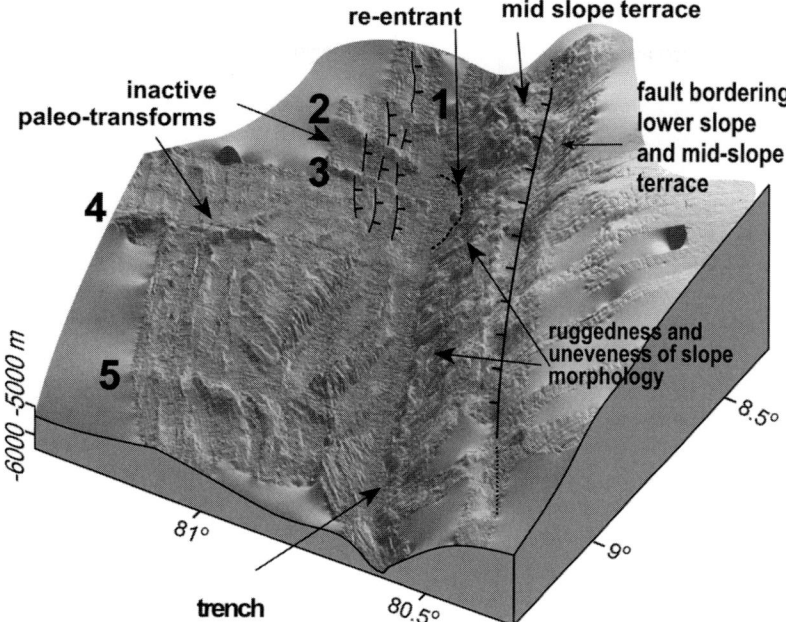

FIGURE 13.1 Expression of faults on a 3D topographic model, exemplified by the Nazca Plate (the trench and the Peruvian submarine continental slope). View from south; 1–5 are ridges. A DEM was derived from swath bathymetry data. *From (Kukowski et al., 2008, Fig. 4c), reproduced with permission.*

Hill shading (Section 2.5) is the most popular technique for revealing faults (Moore and Simpson, 1983; Schowengerdt and Glass, 1983; Zeilik et al., 1989; Onorati et al., 1992; Deffontaines et al., 1997; Chorowicz et al., 1998, 1999; Dhont et al., 1998a, 1998b; Adiyaman et al., 2001; Szynkaruk et al., 2004; Ganas et al., 2005; Dhont and Chorowicz, 2006; Masoud and Koike, 2006; Palyvos et al., 2006; Kukowski et al., 2008; Saadi et al., 2009) (Fig. 13.2). It has been noted that this technique is usually more effective than interpretation of remotely sensed images in terms of fault recognition (Schowengerdt and Glass, 1983). On hill-shading maps, one can find up to 90% of faults, which may be revealed by conventional geological and geophysical methods (Onorati et al., 1992).

In all these techniques, DEM processing allows enhancing "the contrast" of existing lineaments and faults, which are poorly expressed in topography. Finally revealing them is carried out by the human visual system. Moreover, the use of these geomorphometric approaches without ancillary geological data does not allow classification of revealed faults. This is mainly connected with difficulties in formalizing fault diagnostic signs. Detailed qualitative descriptions of the signs can be

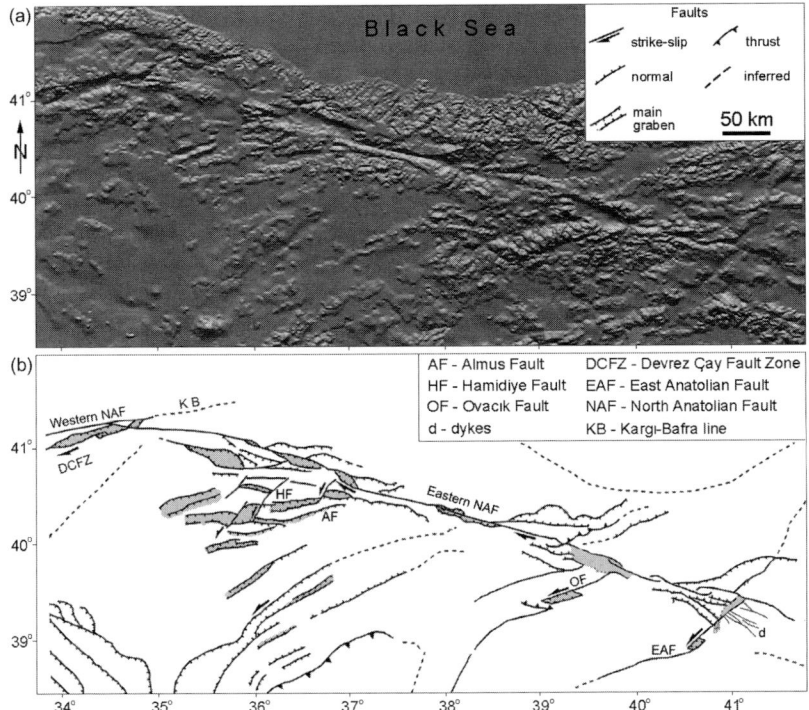

FIGURE 13.2 Fault recognition from a hill-shading map, exemplified by Central and Eastern Anatolia, Turkey: (a) Hill-shading map; illumination is from the north. A DEM was derived from the Digital Chart of the World (US DMA, 1991) by contour interpolation; DEM grid spacing is 300 m. (b) Fault patterns visually recognized on the hill-shading map. Gray areas are Neogene–Quaternary basins. *From (Chorowicz et al., 1999, Fig. 1b, c), reproduced with permission.*

found elsewhere (Ollier, 1981; Keller, 1986; Nikonov, 1995; Groshong, 2006). For instance, Dhont and Chorowicz (2006, p. 36) described the following principles to reveal and classify faults using hill-shading maps:

> Neotectonic faults can be identified by (1) their morphology, forming asymmetric ranges with one side corresponding to breaks in slope or scarps, (2) the displacement of late Neogene volcanic boundaries, structural or erosional surfaces, and (3) the occurrence of straight lines of several tens of kilometers in length. ... Strike-slip faults have rectilinear traces and they locally bound push-up hills or extensional basins at step-over or bends of the fault trace. They can be associated with typical patterns such as tail-crack or horse-tail structures at fault ends. Reverse faults have sinuous traces and they are associated with half-cylindrical-shaped hills of the uplifted blocks due to drag folds deforming ancient planar erosion surface in the hanging wall. Normal faults are recognized by the following geomorphic

characters: (1) they generally have a widely arched trace, concave (mainly) or convex toward the footwall, in contrast to the strike-slip faults whose trace is generally straighter; (2) they bound tilted plateaus (tilted blocks); (3) as is also the case for the strike-slip faults, they are not related to half-cylindrical-shaped hills corresponding to recent drag folds, which accompany active reverse faulting.

Revealing topographically expressed lineaments can be done by k_h and k_v calculation and mapping; k_h and k_v values should be subdivided into two levels relative to the zero value (Florinsky, 1992). It has been observed that lineaments recorded on k_h and k_v maps have essentially distinct statistical properties (i.e., orientation and density). Geomorphologically, this can be explained by the fact that k_h mapping reveals predominantly valley and crest spurs, while k_v mapping reveals mainly terraces (Evans, 1980). P. A. Shary (1992, personal communication) supposed that lineaments displayed on k_h and k_v maps may be associated with faults generated by horizontal and vertical tectonic motions, correspondingly.

In this chapter, we describe a method to reveal and classify topographically expressed lineaments. The method is based on the derivation of horizontal and vertical curvatures from DEMs of the land or stratigraphic surfaces (Florinsky, 1996). The method is illustrated by two case studies: (1) the Crimean Peninsula and the adjacent sea bottom, a seismically active, mountainous territory (Florinsky, 1996); and (2) the Kursk nuclear power plant area, a platform plain region (Florinsky et al., 1995).

13.2 THEORY

Consider an arbitrary surface (Fig. 13.3a). Recall that k_h and k_v are positive for convex areas, while negative for concave ones.

Suppose a dip-slip (or reverse) fault occurs in the zone of k_v positive values (Fig. 13.3b). k_h and k_v values in the deformation zone will change; k_v will take negative values along the fault line $x-x$. Let us subdivide k_h and k_v values into two levels relative to the zero value, and display areas with k_h and k_v positive values in white while displaying areas with k_h and k_v negative values in black. In this case, one will be able to see an indicator of the dip-slip (or reverse) fault as a black lineament on a white background on the k_v map. A similar lineament will be seen on the k_v map if the surface was plane before the vertical motion. If k_v values were negative before the movement, the k_v sign will also change along the fault line. In this case, the dip-slip fault will be indicated by a white lineament on a black background on the k_v map. A lineament consisting of black and white spots will be recorded on the k_v map after the vertical movement if the surface has a complex shape.

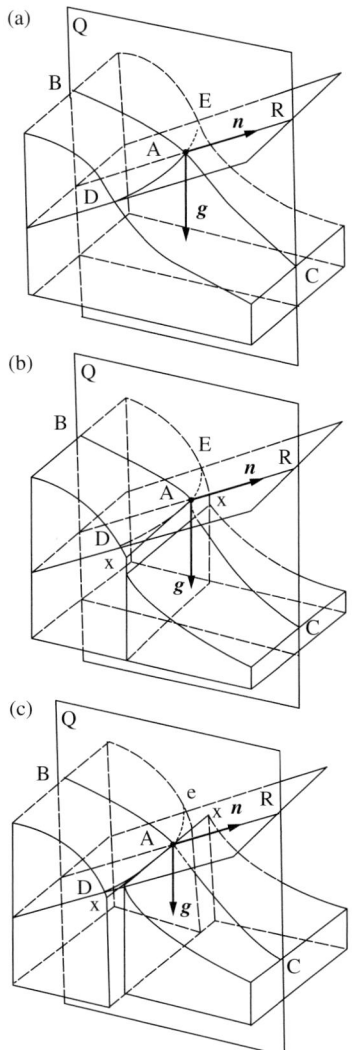

FIGURE 13.3 Block diagrams illustrating sign changes of horizontal and vertical curvatures after tectonic motions: (a) the initial surface; (b) the surface with a dip-slip fault after a vertical displacement; (c) the surface with a strike-slip fault after a horizontal displacement. k_v is the curvature of the normal section BAC formed by the intersection of the surface with the plane Q, which contains the gravitational acceleration vector g at the given point A. k_h is the curvature of the normal section DAE formed by the intersection of the surface with the plane R, which is orthogonal to the normal vertical section BAC at the point A. n is the external normal to the surface at the point A; x—x are fault lines. *From (Florinsky, 1996, Fig. 1).*

In like manner, a lineament indicating a thrust will be revealed on the k_v map, since thrusting, as a rule, produces a scarp. However, lineaments indicating dip-slip, reverse, and thrust faults will not be recorded on the k_h map, because changes of the k_h sign along fault lines will be random rather than systematic (some nonlinear changes in patterns will arise on the k_h map anyway).

Suppose a strike-slip fault occurs in the zone of k_h positive values (Fig. 13.3c). k_h and k_v values will also change in the deformation zone. However, the k_h will take negative values along the length of the fault line $x-x$, while changes of the k_v sign will be random rather than systematic. Therefore, the following lineaments indicating horizontal movement traces will be recorded on the k_h map: (1) a black lineament on a white background for the given case and planar surface before deformation; (2) a white lineament on a black background for the surface with negative k_h values; and (3) a lineament consisting of white and black spots for the complex surface. Some nonlinear traces of horizontal movements will be recorded on the k_v map.

After an oblique-slip and a gaping faults formation, both curvatures ought to change signs systematically along fault lines. Therefore, we can expect that lineaments indicating these faults will be recorded on maps of both k_h and k_v.

There is a need to use ancillary geological, geophysical, and geomorphic data to distinguish and/or separate: (1) lineaments of nontectonic (e.g., erosion) origin; (2) lineaments associated with flexures and some folds; (3) dip-slip, reverse, and thrust faults equally revealed by k_v; and (4) oblique-slip and gaping faults revealed by k_h in similar manner. Notice that for platform areas, the direct association between lineaments and faults may be conjectural. In such cases, lineaments on k_h and k_v maps can be interpreted as linear zones of concentration of bending deformations, as follows:

- Lineaments revealed on k_h maps are predominantly associated with zones of horizontal bending deformations.
- Lineaments recorded on k_v maps mostly relate to zones of vertical bending deformations (e.g., flexures).
- Lineaments revealed on both k_h and k_v maps can indicate zones of bending deformations formed by both vertical and horizontal movements.

13.3 METHOD VALIDATION

13.3.1 Materials and Data Processing

The method validation was performed using an artificial site. The site (Fig. 13.4a) measures 60×60 m; the maximum elevation difference is about 7.5 m. It includes an east-striking valley and two adjacent crests. An irregular DEM includes 129 points. The following seven simple typical faults (Gzovsky, 1954) were modeled by deformations of the initial irregular DEM:

1. A vertical dip-slip fault with 1-m displacement (Fig. 13.4b)
2. A low-angle dip-slip fault with 1-m displacement and 30° dip (Fig. 13.4c)

FIGURE 13.4 Modeling faults on the artificial DEM: (a) the initial surface; (b) vertical dip-slip fault; (c) low-angle dip-slip fault; (d) strike-slip fault; (e) oblique-slip fault; (f) overthrust, 5-m displacement; (g) overthrust, 15-m displacement; (h) gaping fault. Upper are elevations, middle are horizontal curvatures, lower are vertical curvatures; x—x are fault axes. *From (Florinsky, 1996, Fig. 2).*

3. A strike-slip fault with 3.5-m displacement (Fig. 13.4d)
4. An oblique-slip fault with 3.5-m horizontal and 1-m vertical displacements (Fig. 13.4e)
5. An overthrust with 5-m displacement (Fig. 13.4f)
6. An overthrust with 15-m displacement (Fig. 13.4g)
7. A gaping fault with a trench of 1-m width and 0.2-m depth (Fig. 13.4h)

Eight irregular DEMs of the initial and deformed surfaces were converted into regular ones by the Delaunay triangulation and a piecewise quadric polynomial interpolation (Agishtein and Migdal, 1991). The grid spacing was 2 m. Digital models of k_h and k_v (Fig. 13.4) were derived from all regular DEMs by the Evans–Young method (Section 4.1). Data processing was performed by the software LandLord (Appendix B).

13.3.2 Results and Discussion

Visual analysis of the maps obtained (Fig. 13.4) confirms the general correctness of the method. Indeed, for the dip-slip fault models (Fig. 13.4b, c), k_v maps display clear lineaments passing strictly along the fault axes. k_h mapping also revealed some traces of the vertical motion but no lineaments along the fault axes. The lineament corresponding to the vertical dip-slip fault (Fig. 13.4b) does not essentially differ from the lineament corresponding to the low-angle dip-slip fault (Fig. 13.4c).

For the strike-slip fault model (Fig. 13.4d), k_h map reveals a small lineament in the lower part of the map. It strictly passes along the fault axis. The k_v map does not contain lineaments, although strike-slip fault traces can be found: these are the break of the valley and the displacement of its eastern portion to the north.

For the oblique-slip fault model (Fig. 13.4e), the clearly expressed lineament is revealed on the k_v map. Traces of vertical and horizontal motions are recorded by k_h mapping, but there are no lineaments along the fault axis. This fact contradicts the theoretical expectation and may be connected with drawbacks of the oblique-slip fault model.

For the thrust models (Fig. 13.4f, g), clear lineaments passing along the fault axes are revealed by k_v mapping. Some overthrust traces are also recorded on k_h maps, but they are not regular. For the gaping fault model (Fig. 13.4h), lineaments are recorded on the k_h and k_v maps.

To validate the method, we applied a very simple DEM and modeled single simple faults. It is obvious that effectiveness and impartiality of fault revealing and classification will be lower for DEMs of actual terrains with many faults of different strikes and nontectonic lineaments.

13.3.3 Strike, Dip, and Displacement Estimation

DEMs are widely utilized to measure the dip and strike of faults and other geological contacts (Chorowicz et al., 1991; Morris, 1991; Koike et al., 1998) (Fig. 13.5a): For a locally planar geological contact, three points belonging to the contact define a local approximation of the contact surface between two lithologic units. If these points also belong to a DEM grid, one can estimate the following characteristics from the point coordinates: (1) the strike, the azimuth of the intersection of the contact surface with a horizontal plane; (2) the dip direction, that is, the azimuth of local slope lines; and (3) the dip, the angle between local slope lines and a horizontal plane (Chorowicz et al., 1991).

For a simple surface with a single fault (Fig. 13.4), DTM-based quantitative estimation of a fault displacement may be done as follows. To

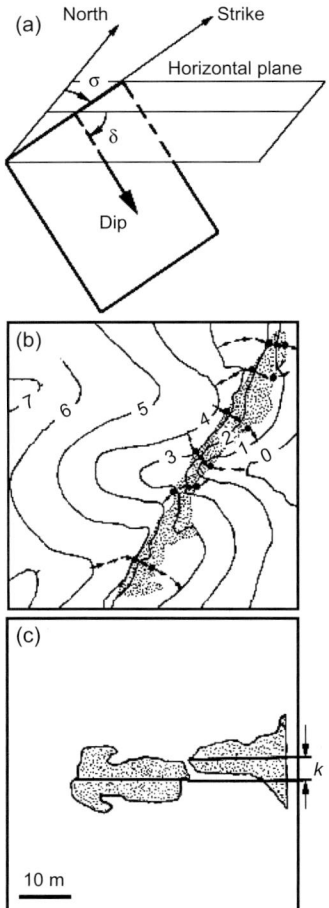

FIGURE 13.5 Quantitative estimation of some properties of faults: (a) Estimation of the dip (δ) and strike (σ) of a fault or geological contact. *From (Chorowicz et al., 1991, Fig. 2), reproduced with permission.* (b) Estimation of the vertical displacement modeled in Fig. 13.4b; the lineament is hachured; arrows are flow lines; dots are intersections of flow lines with the lineament. (c) Estimation of the horizontal displacement modeled in Fig. 13.4d; parts of the valley are hachured, k is the horizontal displacement. *From (Florinsky, 1996, Fig. 6).*

estimate a vertical displacement, one may use an average value of elevation differences in points where flow lines intersect a lineament displayed on a k_v map (Fig. 13.5b). The estimation accuracy obviously depends on the erosion intensity, the DEM accuracy and resolution, and the number of measurements along flow lines. A horizontal displacement of a strike-slip fault may be estimated analyzing relative displacements of valley portions displayed on a k_v map along a fault line (Fig. 13.5c). However, displacement estimation will be complicated in the case of an actual terrain having complex topography and tectonics.

13.4 TWO CASE STUDIES

The method is exemplified by a part of the Crimean Peninsula and the adjacent sea bottom, as well as the Kursk nuclear power plant area. The study regions are representative of seismically active, mountainous terrains and platform plain territories, respectively.

13.4.1 The Crimean Peninsula

The part of the Crimean Peninsula and the adjacent sea bottom (Fig. 13.6) measures 210 × 132 km. This territory was selected for two reasons. First, it has been much studied. A lot of geological, geophysical, and satellite data are available that can be used to test results of lineament revealing and classification. Second, topographic (Fig. 13.7) and geological diversity of the region allows one to demonstrate the possibilities of the method in different geomorphic and geological conditions.

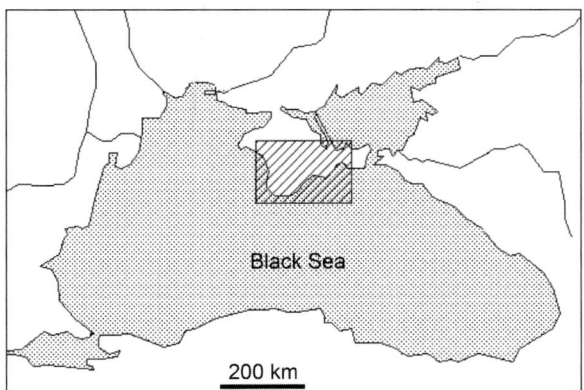

FIGURE 13.6 Geographical location of the studied portion of the Crimean Peninsula and the adjacent sea bottom. *From (Florinsky, 2010, Fig. 4.4).*

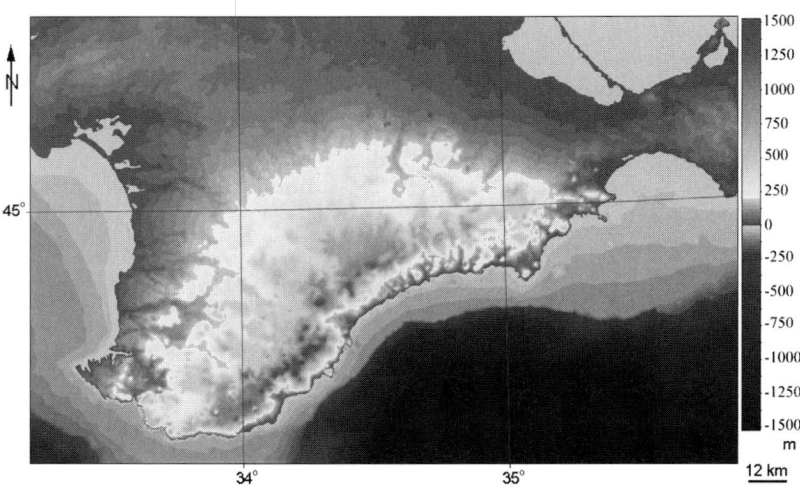

FIGURE 13.7 The Crimea and the adjacent sea bottom: elevation. *From (Florinsky, 2010, Fig. 4.5)*. For color version, please see page 25 in the Color Plate Section at the end of the book.

13.4.1.1 The Geological Setting

Rather complicated topography and geological setting are typical for this territory due to an interaction of the Precambrian East European Platform and the Alpine–Himalayan orogenic belt. It is customary to distinguish three main parts of the territory: (1) the Epi-Paleozoic Scythian Plate correlating with the Crimean Plain; (2) the Crimean Meganticlinorium topographically expressed as the Crimean Mountains; and (3) an adjacent foredeep system topographically related to the continental shelf and slope of the Black Sea Basin. One can distinguish several elevated blocks of the Proterozoic folded basement covered by Miocene and Pliocene sediments in the Scythian Plate. The central portion of the meganticlinorium—the Main Ridge of the Crimean Mountains—is mainly comprised of Tauric series and Middle and Late Jurassic rocks. The Main Ridge and piedmonts include several anticlinoria and synclinoria covered by Early Cretaceous, Paleogene, and Neogene limestones, clays, and marls (Muratov, 1969).

Two systems of transregional deep mantle fault zones intersect the Crimean Peninsula (Fig. 13.8a) (Chekunov et al., 1965; Kovalevsky, 1965; Pustovitenko and Trostnikov, 1977):

1. The pre-Riphean, approximately north-striking fault zones of the southern part of the East European Platform crossing the Ukrainian Shield and Scythian Plate, and extending southward to the Anatolian Peninsula.
2. The Paleozoic, approximately east-striking fault zones separating the Crimean Mountains from the Crimean Plain and the Black Sea Basin.

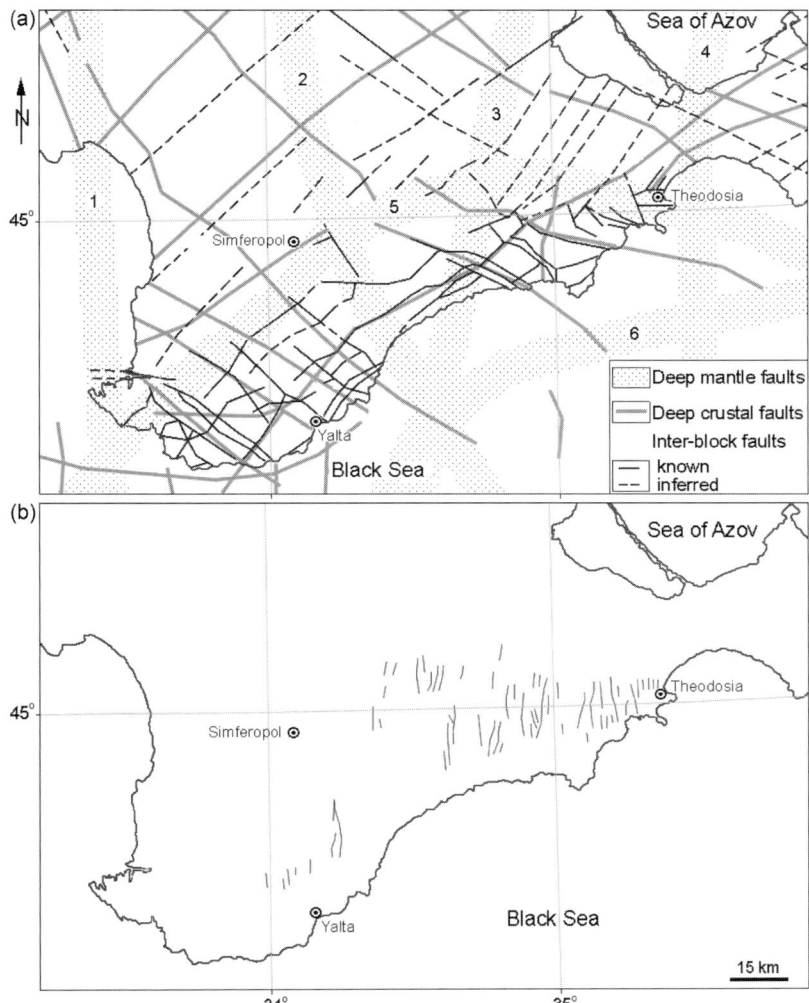

FIGURE 13.8 The Crimea and the adjacent sea bottom, known faults and lineaments: (a) Deep crustal faults (Borisenko, 1986) and deep mantle fault zones (Chekunov et al., 1965; Kovalevsky, 1965; Lebedev and Orovetsky, 1966; Pustovitenko and Trostnikov, 1977): 1—the Eupatoria–Skadovsk zone, 2—the Salgir–Oktyabrskoe zone, 3—the Orekhovo–Pavlograd zone, 4—the Korsak–Theodosia zone, 5—the Foothill Crimean–Caucasian zone, 6—the Central Crimean–Caucasian zone. (b) North-striking faults in the southeastern Crimea (Muratov, 1969). (c) Main lineaments from satellite images (Kats et al., 1981). *From (Florinsky, 2010, Fig. 4.6).*

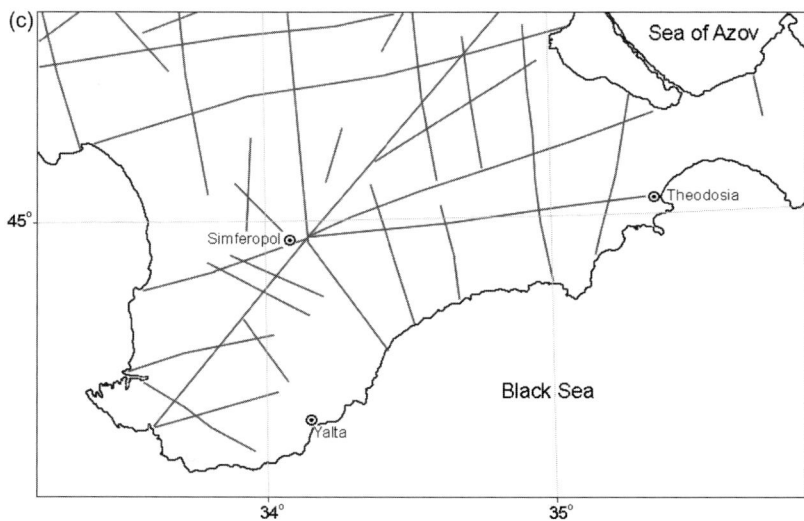

FIGURE 13.8 (Continued).

These fault zones range in width from 15 to 20 km and include numerous smaller faults (Pustovitenko and Trostnikov, 1977). The latest reactivation of the Eupatoria−Skadovsk, Salgir−Oktyabrskoe, and Korsak−Theodosia fault zones occurred in the Mesozoic time, whereas the latest reactivation of the Orekhovo−Pavlograd, Foothill Crimean−Caucasian, and Central Crimean−Caucasian fault zones took place during the neotectonic period. Also, a network of deep crustal faults (Fig. 13.8a) was activated during the neotectonic period (Borisenko, 1986).

Spatial orientation of regional and local faults (Fig. 13.8a, b) is generally controlled by deep crustal faults. However, different authors presented different systems of regional and local lineaments and faults (Shalimov, 1966; Muratov, 1969; Rastsvetaev, 1977; Kats et al., 1981; Borisenko, 1986; Yudin and Gerasimov, 2001; Wolfman et al., 2008; Verkhovtsev, 2008). For example, Borisenko (1986) proposed that there are two diagonal systems of faults in the Crimean Mountains: (1) approximately northeast-striking normal and reverse dip-slip faults with vertical displacements of up to several hundred meters; and (2) approximately northwest-striking oblique- and strike-slip faults with horizontal displacements of up to several hundred meters. These faults probably originated in the Triassic−early Jurassic time. Being reactivated several times, they are currently active.

Rastsvetaev (1977, p. 99) suggested that "there dominates approximately north-striking faults in the Crimean Mountains, which are up to several tens of kilometers long. There are horizontal sinistral displacements of geological contacts and portions of large folds along most faults of this system... Horizontal displacements are not less than 3–5 km along each fault... Available data allow us to characterize approximately north-striking faults of the Crimean Mountains as a system of sinistral oblique-slip faults with the dominant horizontal displacement." Some of these structures are represented in Fig. 13.8b. This author also described two other groups of faults in the Crimean Mountains: (1) northeast-striking thrusts 10–15 km long, with up to 3-km horizontal displacements and 30°–45° dips; and (2) dextral northwest-striking strike-slip faults with steep dips (Rastsvetaev, 1977).

13.4.1.2 Materials and Data Processing

We used an irregular DEM (Fig. 13.7) (Florinsky, 1992) generated from topographic maps (Central Board of Geodesy and Cartography, 1953; USSR General Headquarters, 1986). The irregular DEM includes 11,936 points. It was converted into the regular DEM by the Delaunay triangulation and a piecewise quadric polynomial interpolation (Agishtein and Migdal, 1991). The grid spacing was 500 m. Digital models of k_h and k_v (Fig. 13.9) were derived from the regular DEM by the Evans–Young method (Section 4.1); the grid spacing was 3000 m. Data processing was performed by the software LandLord (Appendix B).

The map of revealed and classified lineaments (Fig. 13.10) was produced by a visual analysis of the k_h and k_v maps (Fig. 13.9); the skeleton of linear patterns was constructed manually. To estimate the effectiveness of the method, we compared this map (Fig. 13.10) and available geological data (Fig. 13.8).

13.4.1.3 Results and Discussion

k_h mapping allowed us to reveal (1) a system of approximately northwest-striking lineaments in the eastern and central parts of the region; (2) a system of approximately east-striking lineaments in the western part of the region; (3) some approximately northeast-striking linear structures in the northern part of the study area; and (4) some approximately northwest-striking lineaments in the southern part of the region (Fig. 13.9a). We believe that these lineaments are associated with strike-slip faults (Fig. 13.10).

k_v mapping allowed us to reveal (1) a system of approximately west-striking lineaments in the east and central part of the study area; (2) a system of approximately north-striking lineaments mainly in the west part of the region; and (3) some approximately northwest-striking linear structures mostly in the southern and the northern part of the study area (Fig. 13.9b). We interpreted these lineaments as dip-slip faults and thrusts (Fig. 13.10).

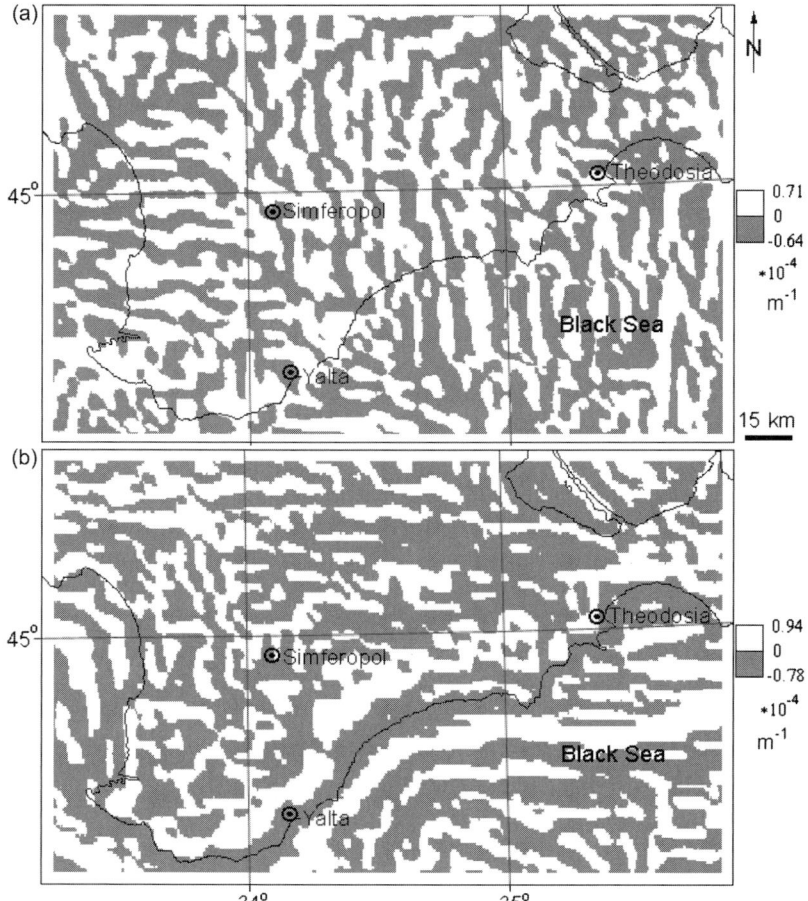

FIGURE 13.9 The Crimea and the adjacent sea bottom: (a) horizontal curvature, (b) vertical curvature. *From (Florinsky, 1996, Fig. 5a, b).*

Lineaments revealed on both k_h and k_v maps (Fig. 13.9) are interpreted as oblique-slip faults (Fig. 13.10).

The map of classified lineaments (Fig. 13.10) displays a complex spatial distribution of these structures. As a rule, lineaments of the same origin unite into systems, whereas lineaments associated with dip-slip faults stretch across lineaments relating to strike-slip faults. There are some complex faults, which include dip-slip, strike-slip, thrust, and oblique-slip offsets. The system of approximately north-striking strike-slip faults passes through areas of different geological origin (i.e., the Scythian Plate and the Mountain Crimean meganticlinorium).

A comparison of the lineament map (Fig. 13.10) and fault maps based on geological and geophysical data (Fig. 13.8a) demonstrated that

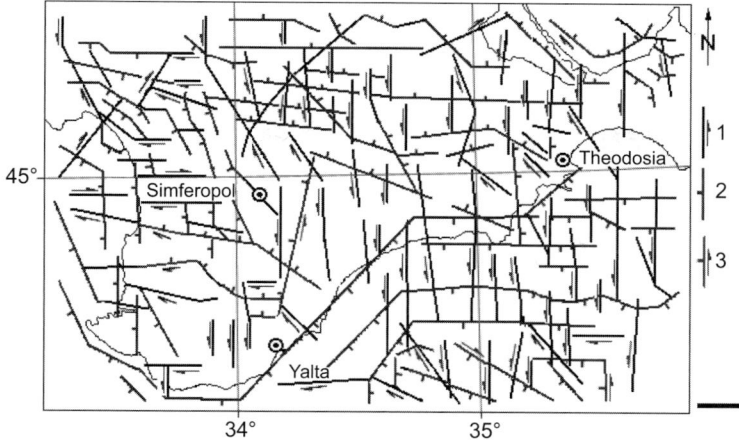

FIGURE 13.10 The Crimea and the adjacent sea bottom: classification of lineaments revealed by horizontal and vertical curvature mapping. Lineaments associated with strike-slip faults (1), dip-slip faults and thrusts (2), and oblique-slip faults (3). *From (Florinsky, 1996, Fig. 5c).*

a part of revealed lineaments correlates with known faults. For example, the lineament map reflects the Korsak–Theodosia, Salgir–Oktyabrskoe, Eupatoria–Skadovsk, and Central Crimean–Caucasian deep mantle fault zones (#1, 2, 4, and 6, respectively, on Fig. 13.8a). However, most of the revealed lineaments do not correlate with previously mapped faults. Generally, we expected such a result: it is well-known that lineaments detected on topographic maps and satellite images do not necessary correlate with faults established by conventional geological methods and displayed on geological maps (Makarov, 1981; Trifonov et al., 1983; Bryukhanov and Mezhelovsky, 1987). In this connection, it is interesting that the k_h and k_v maps (Fig. 13.9) show most lineaments recognized from satellite images (Kats et al., 1981) (Fig. 13.8c). The latter authors related approximately north-striking lineaments to the north-striking deep mantle faults (Fig. 13.8a).

Most of the revealed lineaments relate to the main groups of known regional faults (Section 13.4.1.1). In particular, the system of approximately north-striking lineaments, which are from our point of view associated with strike-slip faults (Fig. 13.10), relates to the system of sinistral oblique-slip faults with the dominant horizontal displacement described by Rastsvetaev (1977). Northwest-striking lineaments, which from our standpoint indicate strike-slip, dip-slip, and oblique-slip faults (Fig. 13.10), relate to groups of such faults described by both Rastsvetaev (1977) and Borisenko (1986).

The lineament map (Fig. 13.10) has somewhat subjective nature. First, we generally drew fault lines as medians of linear patterns recorded on the k_h and k_v maps. Second, a visual analysis of these maps may result in the loss of some lineaments. Third, the lineament map corresponds to the only DEM grid spacing. One may obtain maps containing more or fewer lineaments using smaller or larger grid spacings, respectively. To reveal and classify virtually all topographically expressed lineaments, as well as to grade them into transregional, regional, and local scales, it is necessary to handle a series of k_h and k_v maps with different resolutions.

13.4.2 The Kursk Nuclear Power Plant Area

In this section, we present results of revealing and interpreting lineaments for a territory near the Kursk nuclear power plant (Fig. 13.11) (Florinsky et al., 1995). The area measures 68 × 48 km.

13.4.2.1 *The Geological Setting*

The territory is composed by Archean, Proterozoic, Devonian, Carboniferous, Jurassic, Cretaceous, Paleogene, Neogene, and Quaternary rocks (Dubyansky and Khakman, 1949; Nalivkin, 1970, 1974; Myrzin et al., 1988; Poletaev et al., 1992). Precambrian rocks (crystalline and foliated schists, marbles, dolomites, quartzites, etc.) composing the crystalline basement are highly metamorphosed and dislocated (Fig. 13.12a). The basement includes Archean blocks, horst-anticlinoria,

FIGURE 13.11 Geographical location of the Kursk nuclear power plant (KNPP) area. *From (Florinsky, 2010, Fig. 4.9).*

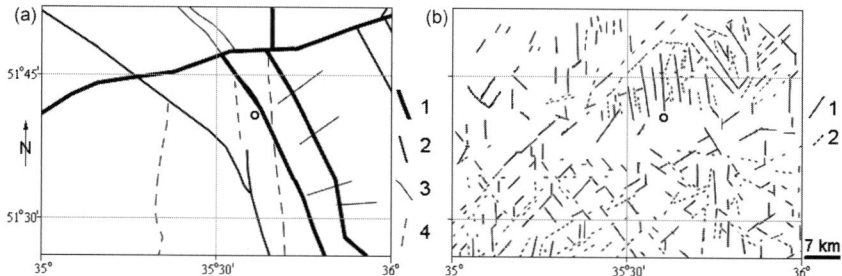

FIGURE 13.12 The Kursk nuclear power plant area, known faults and lineaments: (a) faults revealed by geophysical data (1—deep crustal faults, 2—regional faults, 3—other faults) and satellite images (4) (Myrzin et al., 1988); (b) lineaments revealed by topographic maps and satellite images: 1—true, 2—inferred (Poletaev et al., 1992). The dot in the circle shows the KNPP location. *From (Florinsky, 2010, Fig. 4.10).*

and synclinoria, which are extended to the northwest and separated by faults of different hierarchical levels. Devonian sediments are inclined to the northwest, while Carboniferous, Mesozoic, and Cenozoic sediments are inclined to the southwest.

The depth of the top of the crystalline basement ranges from −300 m to 50 m above sea level. Poletaev et al. (1992) argued that the key structural elements of the basement (e.g., faults) may be expressed as lineaments on the land surface. They should reflect both the northwestern strike of the main folding structures and faults, and the northeastern strike of transverse faults. In particular, these authors considered rectilinear, northwest-striking coasts of pre-Mesozoic sea bays and northeast-striking coasts of Paleogene ones as reactivated basement faults. Poletaev et al. (1992) also supposed that approximately north-striking linear elements of the modern topography may reflect both portions of the pre-Bajocian north-striking paleodrainage network (Myrzin et al., 1988) and some elements of the deep structure of the crust also associated with approximately north-striking faults of the Precambrian basement.

13.4.2.2 *Materials and Data Processing*

As initial data, we used a topographic map (USSR General Headquarters, 1981) and geological and geophysical cartographic materials, scale 1 : 200,000 (Myrzin et al., 1988). Using a digitizer, we produced irregular DEMs of the land surface (Fig. 13.13a), the top of the Cretaceous deposits (Fig. 13.13b), the top of the Cenomanian deposits (Fig. 13.13c), and the top of the crystalline basement (Fig. 13.13d). The irregular DEMs included 46,694, 941, 1002, and 1308 points, respectively.

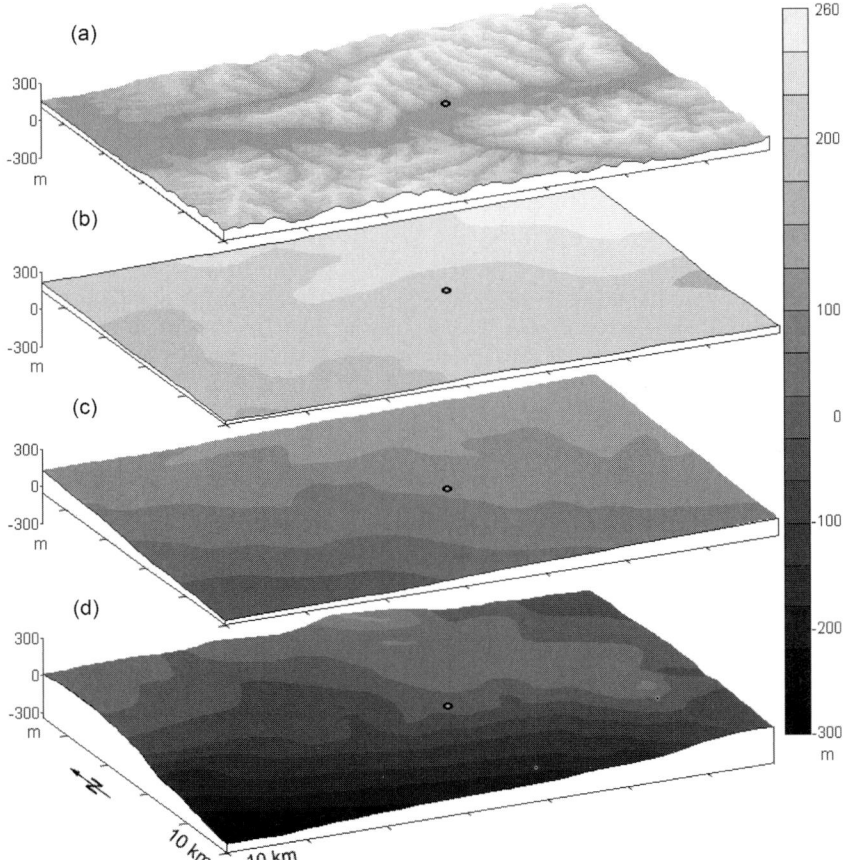

FIGURE 13.13 The Kursk nuclear power plant area, elevation: (a) the land surface, (b) the top of the Cretaceous deposits, (c) the top of the Cenomanian deposits, (d) the top of the crystalline basement. All elevations are above sea level; 30^x exaggeration of the vertical scale. The dot in the circle shows the KNPP location and its projection on the top surfaces of stratigraphic horizons. *From (Florinsky, 2010, Fig. 4.11).*

The irregular DEMs were converted into regular ones by the Delaunay triangulation and a piecewise quadric polynomial interpolation (Agishtein and Migdal, 1991). Digital models of k_h and k_v (Fig. 13.14) were derived from the regular DEMs by the Evans–Young method (Section 4.1); the grid spacing was 1500 m. It is obvious that all land surface DTMs have higher accuracy and actual resolution than DTMs of stratigraphic surfaces.

Maps of revealed and classified lineaments (Fig. 13.15) were produced by visual analysis of the k_h and k_v maps (Fig. 13.14); skeletons of linear patterns were constructed manually.

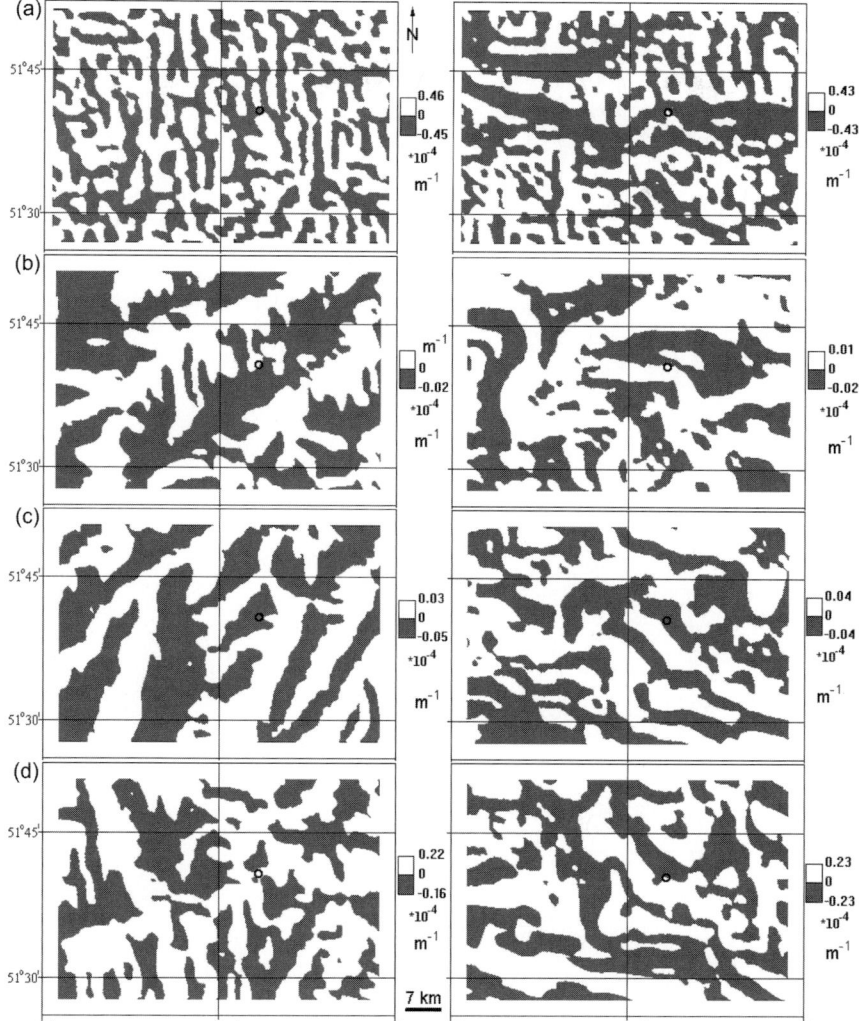

FIGURE 13.14 The Kursk nuclear power plant area, horizontal curvature (left) and vertical curvature (right): (a) the land surface, (b) the top of the Cretaceous deposits, (c) the top of the Cenomanian deposits, (d) the top of the crystalline basement. The dot in the circle shows the KNPP location. *From (Florinsky, 2010, Fig. 4.12).*

13.4.2.3 Results and Discussion

We are concerned with a platform territory, and so it is reasonable to consider lineaments as zones of concentration of bending deformations rather than faults (V.G. Trifonov, 2010, personal communication).

Lineaments probably associated with zones of horizontal bending deformations were detected on all surfaces. For the top of the crystalline

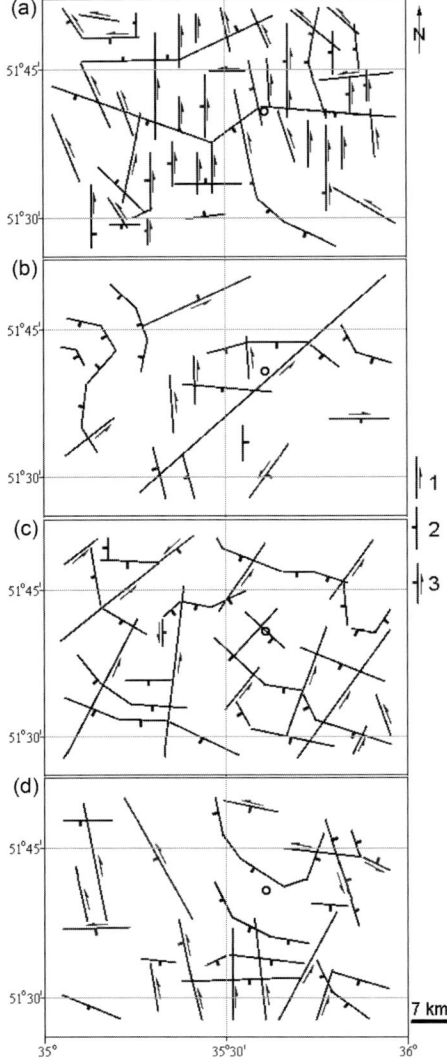

FIGURE 13.15 The Kursk nuclear power plant area, classification of lineaments revealed on the maps of horizontal and vertical curvatures: (a) the land surface, (b) the top of the Cretaceous deposits, (c) the top of the Cenomanian deposits, (d) the top of the crystalline basement. Lineaments associated with zones of horizontal bending deformations (1), flexures (2), and zones of bending deformations formed by both vertical and horizontal movements (3). Directions of horizontal motions are arbitrary. The dot in the circle shows the KNPP location. *From (Florinsky, 2010, Fig. 4.13).*

basement, these are predominantly north-striking structures (Fig. 13.15d). For the tops of the Cenomanian and Cretaceous deposits, there are generally northeast-striking structures, but there are also minor north- and northwest-striking ones (Fig. 13.15b, c). For the land surface, there are predominantly north-striking structures (Fig. 13.15a). On all surfaces, lineaments associated with flexures generally stretch across lineaments, which indicate zones of horizontal bending deformations (Fig. 13.15).

As for the Crimea (Section 13.4.1.3), revealed lineaments do not relate to the basement faults recognized by geological and geophysical techniques (Fig. 13.12a). However, one can see a similarity of the two lineament maps of the land surface derived from satellite images (Fig. 13.12b) and morphometric maps (Fig. 13.15a). Moreover, our results are consistent with the conclusion by Poletaev et al. (1992, p. 34) that "landscape elements of the Kursk Nuclear Power Plant area can reflect northwest- and northeast-striking folds and faults of the basement, as well as north-striking elements of the deep structure and paleotopography (the pre-Bajocian drainage network), and coasts of pre-Mesozoic and Paleogene sea bays, which have northwestern and northeastern orientations".

CHAPTER

14

Accumulation Zones and Fault Intersections

OUTLINE

14.1 Motivation	255
14.2 Study Area	257
14.3 Materials and Methods	258
14.4 Results and Discussion	261

14.1 MOTIVATION

In many ways topography controls lateral transport of substances, viz. gravity-driven migration and accumulation of water as well as mineral and organic materials over the land surface and in the soil (Huggett and Cheesman, 2002). At the same time, there are strong associations between topography and tectonics (Penck, 1924; Meshcheryakov, 1965; Ollier, 1981; Ufimtsev, 1988; Scheidegger, 2004). In particular, valley networks, determining principal routes of overland flows, can often be associated with networks of faults and fracturing zones of different scales (Zernitz, 1931; Filosofov, 1960; Gerasimov and Korzhuev, 1979; Hantke and Scheidegger, 1999; Gudmundsson, 2000; Twidale, 2004). In turn, faults and fracturing zones serve as pathways for upward transport of deep groundwaters, fluids, and gases to the Earth's surface (Peive, 1956b; Kerrich, 1986; Gudmundsson, 2000) and, hence, determine areas of soil geochemical anomalies (Kovda, 1973, pp. 214–221; Kasimov et al., 1978; Kasimov, 1980). Therefore, some migration routes of surface flows can coincide with some surface discharge areas of deep flows. Study of such associations can be useful in gaining a better understanding of relationships between (1) geological structures and landforms; and (2) endo- and exogenous processes influencing the evolution of the biosphere.

Gravity-driven overland and intrasoil transport can be interpreted in terms of flow divergence/convergence and flow deceleration/acceleration determined by horizontal and vertical curvatures, respectively (Table 2.1). Flow convergence and deceleration result in accumulation of substances at soil caused by the slowing down or termination of overland and intrasoil transport. At different scales, the intensity and direction of the transport of substances in many respects depend on the spatial distribution of three types of landforms (Fig. 2.11b) (Section 2.7.2):

1. *Relative accumulation zones.* There are concurrent convergence and deceleration of flows ($k_h < 0$ with $k_v < 0$);
2. *Relative dissipation zones.* There are concurrent divergence and acceleration of flows ($k_h > 0$ with $k_v > 0$);
3. *Transit zones.* There is no concurrent action of "unidirectional" processes (k_h and k_v have different signs or are zero).

Qualitatively, the role of relative accumulation zones in the formation of soil properties has been repeatedly discussed. Accumulation zones are usually closed depressions of different scales; soil moisture content can be highly increased in such depressions (Sibirtsev, 1899, p. 134; Ponagaibo, 1915; Neustruev, 1927; Rode, 1953; Fedoseev, 1959). In arid and semiarid regions, there occur buildup of salts, secondary salinization of soils, and groundwater salinization in these zones at macro- and mesotopographic scales (Kovda, 1946).[1]

Quantitatively, it was demonstrated that the wettest soil areas and saturation zones are typical for relative accumulation zones (O'Loughlin, 1981; Feranec et al., 1991). These may initiate overflows in accumulation zones (Wood et al., 1990). Landsliding (Lanyon and Hall, 1983), soil gleying, maximum thickness of the A horizon, and maximum depth to calcium carbonate (Pennock et al., 1987) are seen in these zones due to increased water content in soils and grounds.

Fault intersections of different scales are important features of the geological environment (Poletaev, 1992). Increased fracturing, porosity, and permeability of rocks characterize them. It is established that fault intersections are often marked by increased seismicity (Gelfand et al., 1972; Kozlov, 1991; Bhatia et al., 1992; Balassanian, 2005; Gorshkov and Solov'ev, 2009), high discharges of springs and boreholes (Lattman and Parizek, 1964; Morozov et al., 1988; Babiker and Gudmundsson, 2004), development of swamps (Garetsky et al., 1983), abnormal structure of

---

[1]There is an English translation of this book (Kovda, 1971).

soil horizons and composition of soil cover (Romashkevich et al., 1997), activization of erosion, karstification, and landsliding (Korobeynik et al., 1982; Glasko and Rantsman, 1996; Karakhanian, 1981). Fault intersections can control intensive magmatism and volcanism (Poletaev, 1992), ore fields and deposits (Kutina, 1974; Wertz, 1974; Volchanskaya, 1981; Trifonov et al., 1983; O'Driscoll, 1986; Lopatin, 2002), as well as oil and gas fields (Guberman et al., 1997). These do not imply that all the phenomena are observed in each fault intersection; this depends on the geological background and other environmental factors.

Garetsky et al. (1983) were probably the first to establish an association of some topographic depressions with fault intersections by visual interpretation of satellite images. Analyzing geological and remotely sensed data, Poletaev (1992) found qualitatively that fault intersections are, as a rule, expressed by topographic depressions. Florinsky (1993) proved quantitatively that topographically expressed accumulation zones, as a rule, coincide with sites of fault intersections (at least, in tectonic terrains).

Indeed, k_h and k_v mapping reveals two groups of topographically expressed lineaments (Florinsky, 1992). Structures of the first group correspond to convergence areas ($k_h < 0$), whereas lineaments of the second group relate to deceleration areas ($k_v < 0$). These lineaments indicate predominantly strike-slip and dip-slip faults, respectively (Florinsky, 1996) (Chapter 13). It is clear that an intersection of two lineaments or faults of different groups relates to a topographically expressed accumulation zone, by its definition (see above). Therefore, values $k_h < 0$ with $k_v < 0$ determine fault intersections, values $k_h > 0$ with $k_v > 0$ (dissipation zones) determine blocks between faults, while other combinations of k_h and k_v values (transit zones) determine fault segments outside fault intersections.[2]

In this chapter, we analyze relationships of some natural phenomena *a priori* associated with fault intersections (i.e., sites of intensive rock fracturing and sites of springs/boreholes with abnormally high discharges) with zones of flow accumulation, transit, and dissipation (Florinsky, 2000).

14.2 STUDY AREA

We studied a part of the Crimean Peninsula (Fig. 13.6). For geological setting, see Section 13.4.1.1.

---

[2] We do not advocate that geological structures control all accumulation zones. Some of them may be formed by exogenous processes, such as erosion.

14.3 MATERIALS AND METHODS

We employed the DEM of the part of the Crimean Peninsula and the adjacent sea bottom (Fig. 13.7). For description of the DEM and calculation of k_h and k_v digital models (Fig. 13.9), see Section 13.4.1.2. A map of accumulation, transit, and dissipation zones (Fig. 14.1) was derived combining k_h and k_v models.

For the analysis, we selected the following natural phenomena, which can be *a priori* associated with fault intersections:

1. Springs and boreholes with abnormally high discharges (Fig. 14.2) (Morozov et al., 1988).
2. Intensive rock fracturing, wherein discharges of deep thermal groundwaters occur (Fig. 14.2) (Shtengelov, 1978, 1982; Korobeynik et al., 1982)

Sites of springs and boreholes with abnormally high discharges (Fig. 14.2) are controlled by intersections of local northwest- and approximately west-striking faults, with transregional approximately north-striking fault zones (Morozov et al., 1988). These authors did not publish quantitative data on all discharges. An example is Kara Sou Bashi, the greatest Crimean spring located at one of the fault intersections. Its mean annual discharge is about 1450 L/s.

Sites of intensive and highly intensive rock fracturing (Fig. 14.2) were recognized by a gammametric technique based on a relationship between rock radioactivity and degree of rock fracturing (Shtengelov, 1978, 1982).

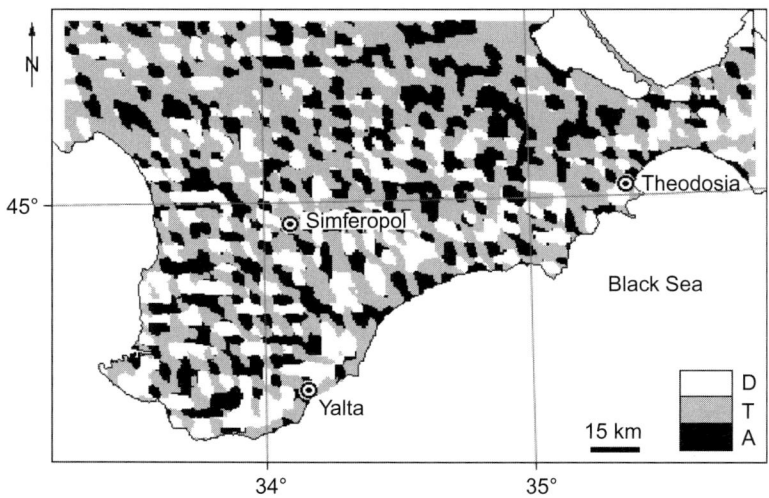

FIGURE 14.1 The Crimea: topographically expressed zones of dissipation (D), transit (T), and accumulation (A). *From (Florinsky, 2000, Fig. 6).*

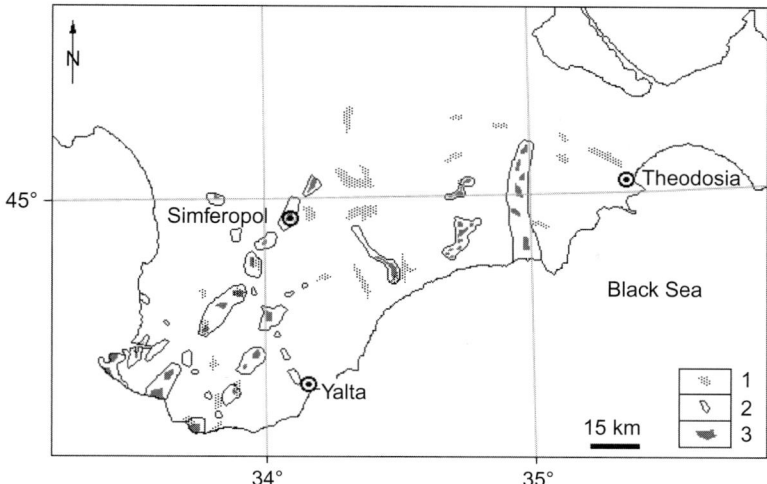

FIGURE 14.2 The Crimea: sites of springs and boreholes with abnormally high discharges (1) (Morozov et al., 1988); sites of intensive (2) and highly intensive (3) rock fracturing (Shtengelov, 1978, 1982; Korobeynik et al., 1982). *From (Florinsky, 2000, Fig. 4)*.

For lithologically homogeneous rocks, the greater the rock fracturing, the less the gamma activity. This is caused by increased leakage of fine materials, the major contributor to gamma activity for most sedimentary and metamorphic rocks. Shtengelov (1978, 1982) and Korobeynik et al. (1982) did not publish quantitative data on the background gamma activity and gamma activity for sites of intensive rock fracturing.

To estimate the association of sites of intensive/highly intensive rock fracturing and sites of springs/boreholes having abnormally high discharges with accumulation, transit, and dissipation zones, we estimated the following parameters (Tables 14.1 and 14.2):

1. Total areas of accumulation, transit, and dissipation zones S_a, S_t, and S_d.
2. Proportions of the total area of the region S_Σ corresponding to the total areas of accumulation, transit, and dissipation zones (E_a, E_t, and E_d, respectively): $E_i = S_i/S_\Sigma$, where i denotes a, t, and d.
3. Weights of accumulation, transit, and dissipation zones W_a, W_t, and W_d: $W_i = \frac{4}{3} - E_i$.
4. A total area of a particular phenomenon under study U_Σ and its total areas located in accumulation, transit, and dissipation zones U_a, U_t, and U_d.
5. Proportions of the total area of a particular phenomenon corresponding to its total areas located in accumulation, transit, and dissipation zones (P_a, P_t and P_d, correspondingly): $P_i = U_i/U_\Sigma$.

TABLE 14.1 Parameters of Topographically Expressed Accumulation, Transit, and Dissipation Zones within the Study Region

Topographic Zone	Parameter	Value
Accumulation	Total area, S_a (km^2)	6588
	Proportion of the total area, E_a	0.25
	Weight, W_a	1.09
Transit	Total area, S_t (km^2)	14,751
	Proportion of the total area, E_t	0.55
	Weight, W_t	0.78
Dissipation	Total area, S_d (km^2)	5364
	Proportion of the total area, E_d	0.20
	Weight, W_d	1.13

The total area of the study area $S_\Sigma = 26{,}703$ km^2.
From (Florinsky, 2000, Table 1).

TABLE 14.2 Parameters for the Association of the Phenomena under Study with Topographically Expressed Accumulation (A), Transit (T), and Dissipation (D) Zones

Phenomenon	Total Area, U_Σ (km^2)	Parameter	Topographic Zone		
			A	T	D
Sites of springs and boreholes with abnormally high discharges	301	Total area, U_i (km^2)	204	90	7
		Proportion of the total area, P_i	0.68	0.30	0.02
		Association coefficient, R_i^{as}	0.74	0.23	0.03
Sites of intensive rock fracturing	1006	Total area, U_i (km^2)	492	382	132
		Proportion of the total area, P_i	0.49	0.38	0.13
		Association coefficient, R_i^{as}	0.54	0.31	0.15
Sites of highly intensive rock fracturing	134	Total area, U_i (km^2)	95	25	14
		Proportion of the total area, P_i	0.71	0.19	0.10
		Association coefficient, R_i^{as}	0.74	0.15	0.11

From (Florinsky, 2000, Table 2).

Finally, we applied "an association coefficient" R_a^{as}, R_t^{as}, and R_d^{as} (Table 14.2) to characterize the association of a particular phenomenon with accumulation, transit, and dissipation zones, respectively: $R_i^{as} = W_i P_i$. For each phenomenon, R_i^{as} was normalized: $R_d^{as} + R_t^{as} + R_a^{as} = 1$.

Data processing was performed by the software LandLord (Appendix B).

14.4 RESULTS AND DISCUSSION

The map of accumulation, transit, and dissipation zones (Fig. 14.1) shows a regular, cell-like spatial distribution of these landforms that is probably connected with intersections of north-, east-, northwest-, and northeast-striking regional faults (Florinsky, 1996) (Section 13.4.1.1).

Sites of springs/boreholes with abnormally high discharges and sites of highly intensive rock fracturing are closely associated with accumulation zones: the coefficient is 0.74 (Table 14.2 and Fig. 14.3). This is no surprise: as accumulation zones are connected with fault intersections, so phenomena located at fault intersections should correlate strongly with accumulation zones. Sites of intensive rock fracturing are associated with accumulation zones to a lesser extent: the coefficient is 0.54 (Table 14.2 and Fig. 14.3).

The association of intensive rock fracturing sites with transit zones is described by the coefficient of 0.31 (Table 14.2 and Fig. 14.3). This is also not surprising, since transit zones are connected with fault segments outside intersections (Section 14.1). At these segments, fracturing is less than at fault intersections, so transit zones can control lesser parts of phenomena typical for accumulation zones. Indeed, the associations of sites of springs/boreholes with abnormally high discharges and sites of highly intensive rock fracturing with transit zones are

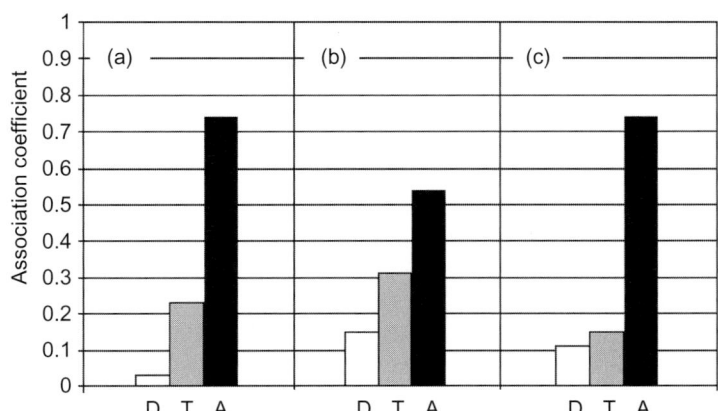

FIGURE 14.3 The Crimea: the association of the phenomena under study with topographically expressed zones of dissipation (D), transit (T), and accumulation (A): (a) sites of springs and boreholes with abnormally high discharges; (b) sites of intensive and (c) highly intensive rock fracturing. *From (Florinsky, 2000, Fig. 7).*

described by the coefficients of 0.23 and 0.15, respectively (Table 14.2 and Fig. 14.3).

The smallest parts of the phenomena under study are associated with dissipation zones: the coefficients range from 0.03 for abnormally high discharge sites to 0.15 for intensive rock-fracturing sites (Table 14.2 and Fig. 14.3). We anticipate that these are artifacts caused by data errors, because rock fracturing is generally not typical for dissipation zones (blocks between faults).

It is clear that numerical results of the analysis (Table 14.2) are area-specific. Associations concerned should depend on geological and geomorphic conditions, as well as the study scale.

Topographically expressed accumulation zones are closely associated with sites of abnormally high discharges and highly intensive rock fracturing (Table 14.2 and Fig. 14.3), which control upward transport of deep groundwaters. In accumulation zones, increased soil moisture can therefore be controlled by increased upwelling of deep groundwaters coincident with accumulation of water from overland and intrasoil lateral flows. This statement is consistent with the established facts:

- Increased discharges of springs and boreholes are typical for both accumulation zones (Lukin, 1987) and fault intersections (Lattman and Parizek, 1964; Morozov et al., 1988; Babiker and Gudmundsson, 2004).
- Increased soil and ground moisture is observed in both accumulation zones (Rode, 1953; Feranec et al., 1991) and fault intersections (Garetsky et al., 1983).
- Increased soil and ground moisture leads to landslide development in both accumulation zones (Lanyon and Hall, 1983) and fault intersections (Karakhanian, 1981);
- Soil development in depressions at fault zones is affected by upwelling of saline and fresh groundwater, as well as overland flows carrying dissolved materials. This results in formation of salt-affected and hydromorphic soils (Kasimov et al., 1978).

Thus, accumulation zones/fault intersections are areas of contact and interaction between overland and deep substance flows. The association of accumulation zones with fault intersections is probably caused by increased rock fracturing typical for sites of fault intersection. They have a higher degree of fracturing as compared to faults outside intersections. Therefore, rocks in fault intersections are more prone to weathering, erosion, and other processes leading to the occurrence and development of depressions. An increased soil and ground moisture in accumulation zones may increase erosion that, in turn, may enhance the topographic expression of these areas.

CHAPTER 15

Global Topography and Tectonic Structures

OUTLINE

15.1 Motivation	263
15.2 Materials and Data Processing	266
15.3 Results and Discussion	269
15.3.1 General Interpretation	269
15.3.2 Global Helical Structures	277

15.1 MOTIVATION

As mentioned in Section 13.1, faults and fracture zones are expressed as lineaments on the Earth's surface. Geometrically, one can treat lineaments as planar straight lines at a regional scale and spatial curves at continental and global scales.

Much attention has been paid to planetary systems of lineaments in global tectonic studies and modeling. Three groups of investigations have been presented:

1. Detection of regularities in the global distribution and orientation of lineaments by analyzing physiographic and geological maps and then developing a model for the regularities observed (Chebanenko, 1963; Moody, 1966; Katterfeld and Charushin, 1973; Besprozvanny et al., 1994).
2. Development of a physical-mathematical model of a global tectonic process causing an ideal planetary network of lineaments and then comparison of the ideal and actual lineament networks

(Weinberg, 1934b; Vening Meinesz, 1947; Dolitsky and Kiyko, 1963; Chebanenko and Fedorin, 1983).
3. Laboratory simulation of a global lineament network using rotatable spheres (Knetsch, 1965; Cherednichenko et al., 1966). In this case, the origin of global lineament systems was usually associated with rotational forces.

Rance (1968) developed a physical-mathematical model of the torsional deformation of a sphere. The torsion was attributed to an action of mantle convection currents on the crust. According to the model, there are two systems of traces of torsional failure surfaces on the surface of the sphere: shear fractures and cleavage cracks (Rance, 1967). Geometrically, traces of torsional deformation constitute two systems of double helices encircling the sphere from pole to pole (Fig. 15.1a). The traces vary in inclination at the equator: a system of double helices tracing shear fractures has the inclination of $15°-18°$ and $165°-162°$, and another system of double helices tracing cleavage cracks has the inclination of $56°-62°$ and $124°-118°$. A search for actual global helical tectonic features resulted in the detection of several relatively small lineaments referring to faults, trenches, ridges, fracture zones, and seamount chains in basins of the Pacific and Indian Oceans (Rance, 1967, 1969).

Using physiographic maps, O'Driscoll (1980) visually detected two global topographically and tectonically expressed double helical zones. The zones have the same inclination at the equator: about $32°$ and $160°$ (Fig. 15.1b). O'Driscoll (1980) believed that these are fundamental structural belts governing the global deformation network and planetary evolution. Volkov (1995) reported six global double-helical structures that were also detected by a visual analysis of physiographic maps. At the equator, three of them were inclined at about $12°$ and $168°$, and the other three structures were inclined at about $22°$ and $158°$ (Fig. 15.1c). Volkov (1995) presumed that these are ancient traces of tidal effects within the Earth−Moon resonance system.

Although much attention has been paid to global lineaments, their existence is still questionable. This is because of:

- The qualitative character of topographic, physiographic, and geological maps analyzed in previous works
- Inaccurate presentation of seafloor bathymetry on those maps produced before reasonably accurate bathymetric data became available
- The impossibility of considering all natural conditions in a mathematical model
- Obvious differences between the Earth's rotation and its laboratory simulation

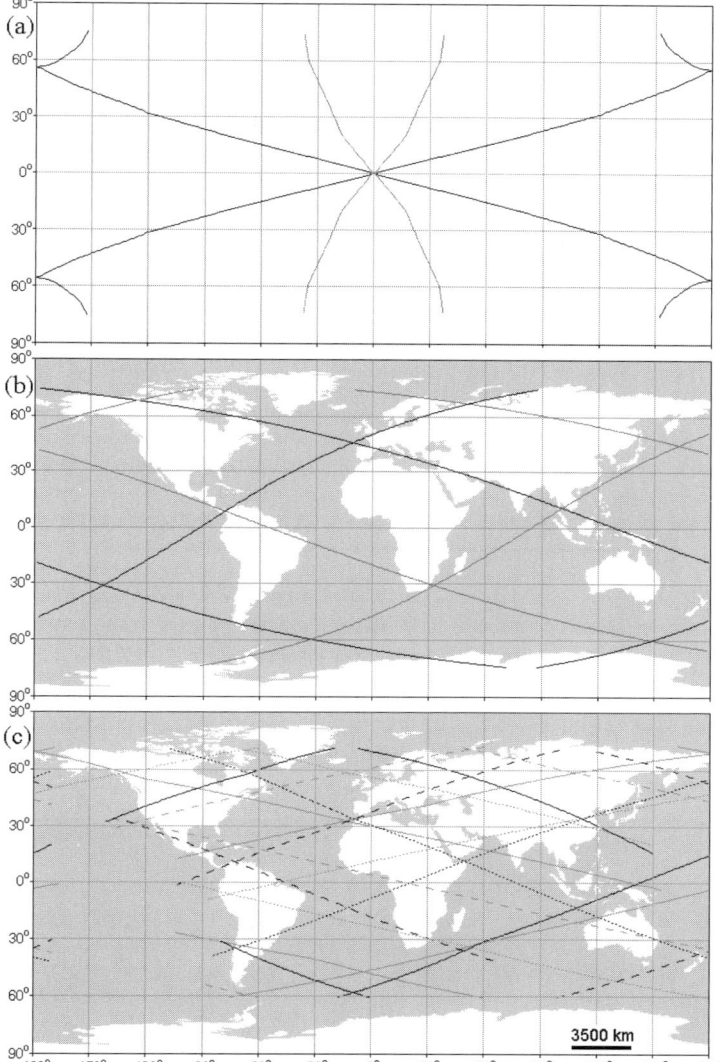

FIGURE 15.1 The Earth, global helices: (a) Theoretical traces of torsional deformation (Rance, 1967): shear fractures (black) and cleavage cracks (grey). (b) Axes of double helical zones (O'Driscoll, 1980). (c) Six double helices (Volkov, 1995): lines of different styles show different structures. *From (Florinsky, 2008a, Fig. 1), © 2008 Springer-Verlag Berlin Heidelberg, reproduced with kind permission from Springer Science + Business Media.*

- A basic conflict between the plate tectonic theory and the possibility of the existence of global topographic and tectonic structures

Technical flaws can be obviated using quantitative descriptions of global topography (a global DEM), and methods of digital terrain modeling. Indeed, topography, resulting from the interaction of endogenous and exogenous geophysical processes of different spatial and temporal scales, carries information on both surface processes and tectonic features. Thus, if global helical structures really exist, there is a good chance that they are expressed on the topographic surface.

In this chapter, we present results of digital terrain modeling of the Earth and some other celestial bodies of the solar system at the planetary scale to detect global helical structures to support or refute the hypotheses indicated above (Florinsky, 2008a, 2008b).

15.2 MATERIALS AND DATA PROCESSING

The study was based on a 30′-gridded global DEM (Fig. 15.2) assembled from several sources. Elevations of the land topography were derived from GLOBE, the 30″-gridded global DEM (GLOBE Task Team and others, 1999). Most of the seafloor topography was taken from ETOPO2, the 2′-gridded global DEM (National Geophysical Data Center, 2001). Bathymetry of the Antarctic Continental Shelf, the

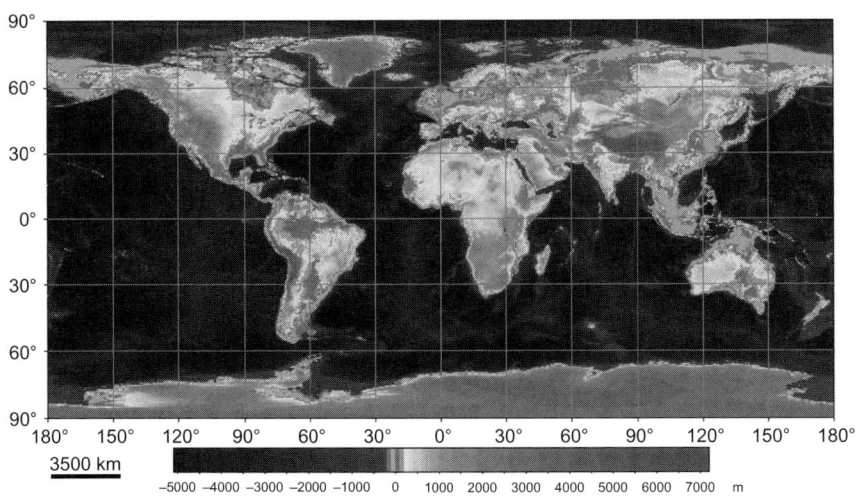

FIGURE 15.2 The Earth, elevation. *From (Florinsky, 2008c, Fig. 1).* (See page 26 in Color Plate Section)

Caspian Sea, and some large lakes was digitized from topographic maps. The DEM includes 260,281 points (the matrix 721 × 361). For Antarctica and Greenland, GLOBE includes elevations of ice sheets rather than subglacial topography (Hastings and Dunbar, 1998). These areas were included in the DEM to retain a united configuration of data.

We selected Mars, Venus, and the Moon for a comparative analysis. First, these celestial bodies are most similar to the Earth with respect to evolution and shape. Second, there are global DEMs for Mars, Venus, and the Moon. A DEM of Mars (Fig. 15.3a) was obtained from the Mars Orbiter Laser Altimeter Mission Experiment Gridded Data Record archive (Smith, 2003) of the Mars Global Surveyor mission (Albee, 2000). A DEM of Venus (Fig. 15.3b) was compiled using data from two archives of the Magellan mission (Saunders and Pettengill, 1991), viz. Global Topography Data Record (Ford, 1992) and Magellan Spherical Harmonic Models and Digital Maps (Sjogren, 1997). A DEM of the Moon (Fig. 15.3c) was extracted from the Clementine Gravity and Topography Data archive (Zuber, 1996) of the Clementine mission (Nozette et al., 1994). Each DEM was based on the global spheroidal equal angular grid with the grid spacing of 30′ including 260,281 points (the matrix 721 × 361). Martian and Lunar elevations were presented with reference to related geoids (Zuber et al., 1994; Smith et al., 1999), while Venusian elevations were presented with reference to the mean planetary radius (Rappaport et al., 1999).

To suppress high-frequency noise, the initial DEMs were smoothed using the 3 × 3 moving window. Our experiments demonstrated that one to three smoothing iterations should be done to obtain readable global morphometric maps of the celestial bodies.

For all four bodies, we derived models of all local morphometric attributes (Table 2.1) by the author's method for spheroidal equal angular grids (Section 4.3). Models of catchment and dispersive areas (Table 2.1) were calculated by the Martz–de Jong method adopted to spheroidal equal angular grids (Section 4.4). All topographic attributes were derived from smoothed DEMs. The global DEMs were processed as virtually closed spheroidal matrices of elevations. Each global spheroidal DTM included 260,281 points (the matrix 721 × 361); the grid spacing was 30′.

To estimate linear sizes of spheroidal trapezoidal windows in DTM calculation and smoothing (Section 4.3.3), standard values of the major and minor semiaxes of the Krasovsky ellipsoid (6,378,245 m and 6,356,863 m) and the Martian ellipsoid (3,396,190 m and 3,376,200 m) were used for the Earth and Mars, correspondingly. Venus and the Moon were considered as spheres of 6,051,848 m and 1,738,000 m radii, respectively.

268   15. GLOBAL TOPOGRAPHY AND TECTONIC STRUCTURES

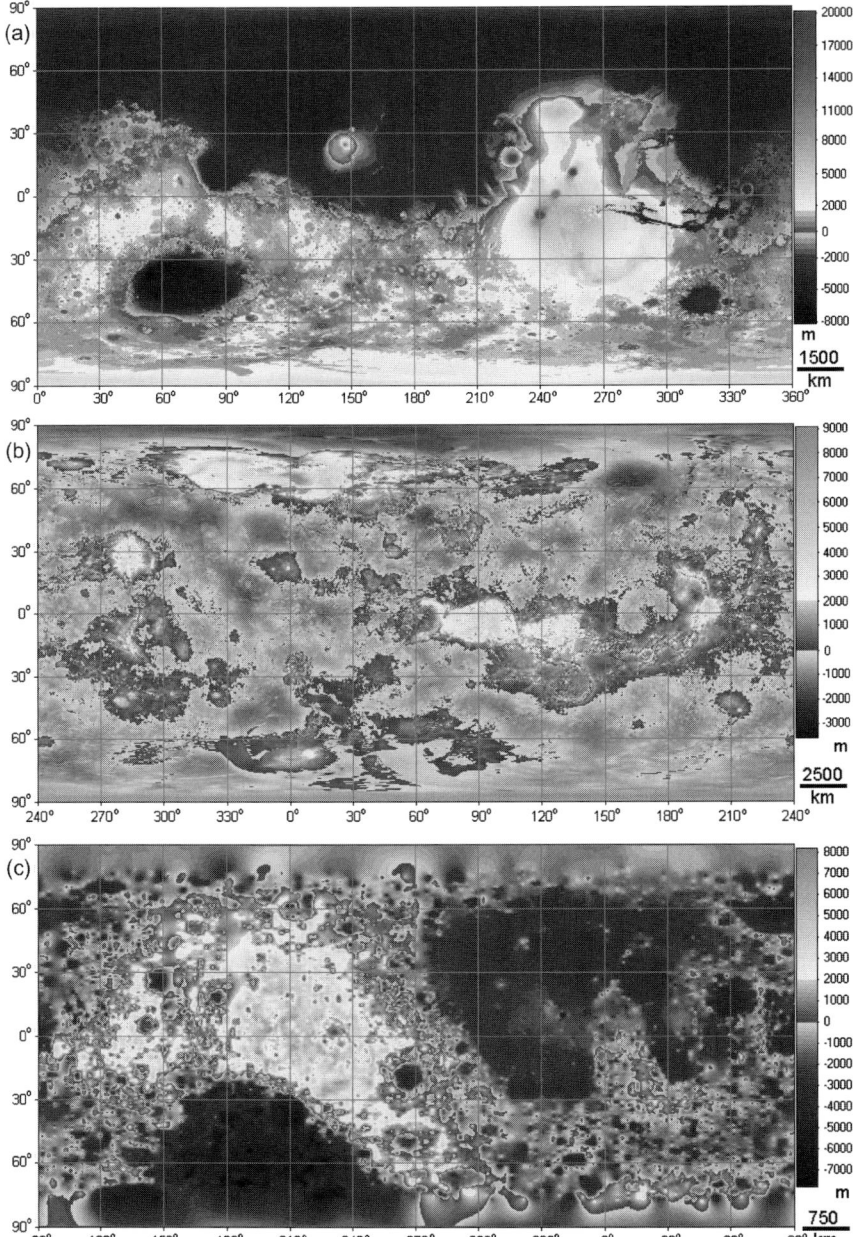

FIGURE 15.3 Elevation: (a) Mars, (b) Venus, (c) the Moon. *From (Florinsky, 2010, Fig. 4.19).* For color version, please see page 27 in the Color Plate Section at the end of the book.

We derived small-scale global maps of all calculated topographic attributes (Figs. 15.4–15.9). To deal with the large dynamic range of morphometric variables, we logarithmically transformed their digital models by Eq. (7.1) with $n = 9$. We also derived "binary" maps of CA and DA subdividing their values into two levels (Figs. 15.10 and 15.11). Data processing was performed by the software LandLord (Appendix B).

The maps obtained were visually examined. Attention was paid to lineaments running over the entire globe or a hemisphere. Contrary to many lineaments of regional and continental scales, global lineaments are not expressed as uninterrupted linear image patterns or their sequences. A global lineament may be visually detected due to traits of the image texture strung out along some line running over the Earth.

15.3 RESULTS AND DISCUSSION

15.3.1 General Interpretation

For the same angular resolution of 30′, DTMs and morphometric maps of the Earth, Mars, Venus, and the Moon have distinct linear resolution (around 55.2 km, 29.5 km, 52.8 km, and 15.2 km on the equator, respectively) and scales. It is recommended that the same angular resolution be used in comparative cartographic analysis of different-sized celestial bodies: the ratio of map scales of celestial bodies should be equal to the ratio of their sizes (Burba, 1984). For example, a Venus map at the scale 1 : 1,000,000 approximately corresponds to a Mars map at the scale of 1 : 500,000 and a Moon map at the scale 1 : 250,000.

Global maps of morphometric variables represent peculiarities of megatopography in different ways, according to the physical and mathematical sense of a particular variable. Horizontal curvature delineates areas of flow divergence and convergence (positive and negative values, respectively). These areas relate to spurs of valleys and ridges (blue and yellow patterns on the k_h maps, respectively), which form so-called flow structures. For the Earth, they are most pronounced in ocean basins (Fig. 15.4a). On the k_h map of Mars (Fig. 15.5a), one can see flow structures, probably of lava origin, beginning on slopes of Alba Patera and forming a huge fan in the North Polar Basin (30°–75°N 200°–310°E). There is a system of flow structures incoming to Utopia Planitia from Nilosyrtis and Protonilus Mensae and Elysium Planitia and Mons (5°–70°N 75°–150°E). On Venus, flow structures appear slightly at the global scale (Fig. 15.5b). For example, one can see them on Beta Regio slopes (15°–45°N 270°–300°E). For the Moon, the horizontal curvature represents cell-like patterns (Fig. 15.5c) resulting from a predominance of craters at the global scale.

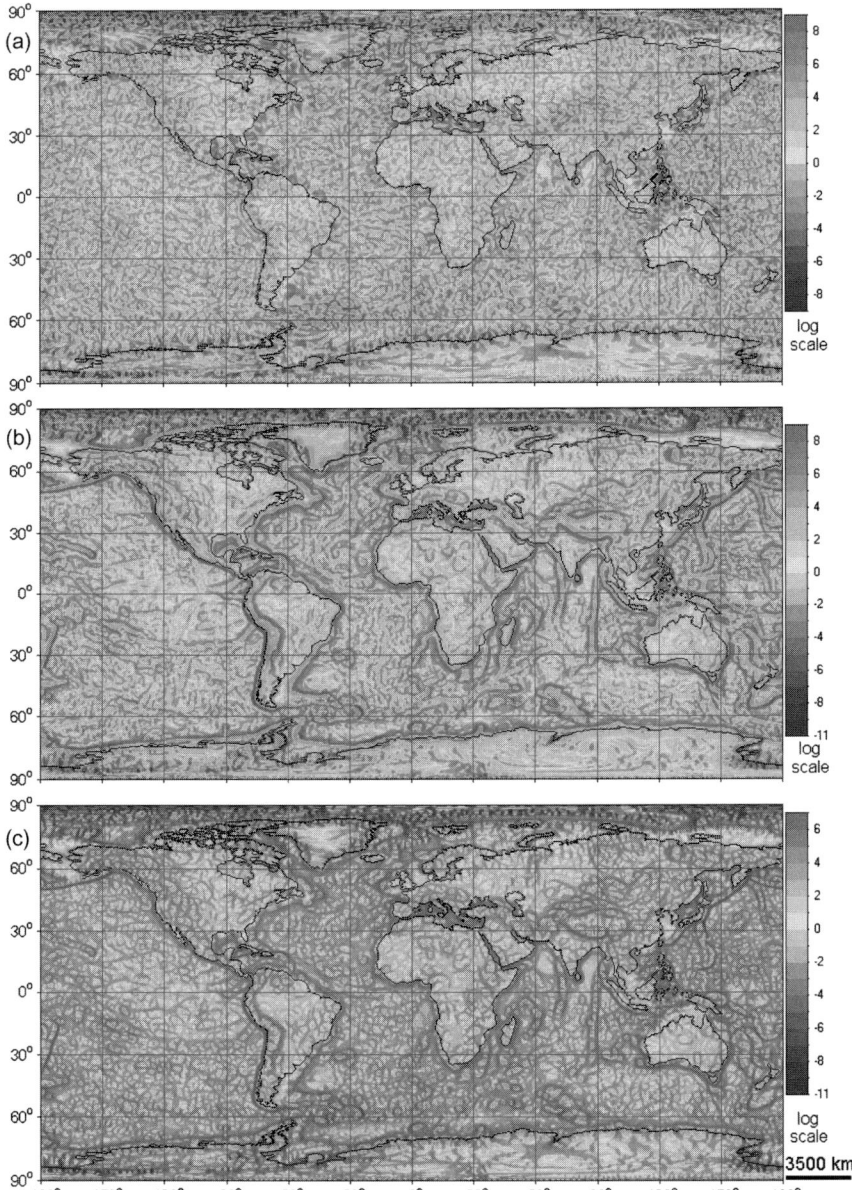

FIGURE 15.4 The Earth, global morphometric maps derived from the three times smoothed DEM: (a) horizontal curvature, (b) vertical curvature, (c) minimal curvature. *From (Florinsky, 2008c, Fig. 2a, b, c).* For color version, please see page 28 in the Color Plate Section at the end of the book.

15.3 RESULTS AND DISCUSSION 271

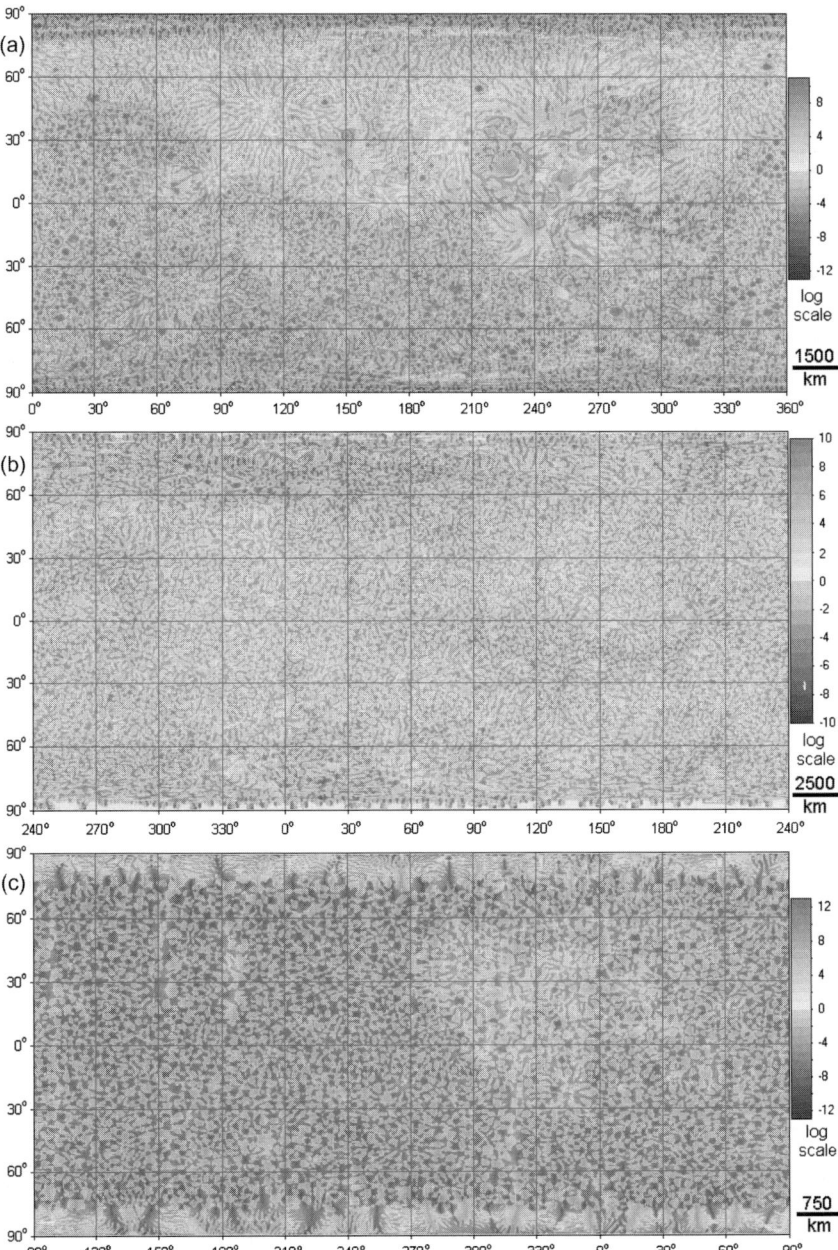

FIGURE 15.5 Horizontal curvature derived from the two times smoothed DEMs: (a) Mars, (b) Venus, (c) the Moon. *From (Florinsky, 2010, Fig. 4.21).* For color version, please see page 29 in the Color Plate Section at the end of the book.

Vertical curvature is a measure of relative acceleration and deceleration of flows (positive and negative values, respectively). Among other features, the k_v map of the Earth shows "mega-scarps," such as edges of continents and mountains (Fig. 15.4b). On the k_v map of Mars (Fig. 15.6a), one can see boundaries of Hellas Planitia (30°−50°S 50°−90°E), Isidis Planitia (10°−25°N 75°−100°E), Valles Marineris (10°−20°S 270°−335°E), foothills of Olympus Mons (15°−20°N 220°−230°E), Alba Patera (30°−50°N, 225°−265°E), and so on. Artemis Chasma (30°−45°S 120°−145°E) and some other surface features are pronounced on the k_v map of Venus (Fig. 15.6b). On the k_v map of the Moon (Fig. 15.6c), one can see well-marked boundaries of Mare Serenitatis (15°−40°N 10°−30°E), Mare Crisium (10°−20°N 50°−70°E), and so on.

Catchment area measures an upslope area potentially drained through a given point on the topographic surface (Table 2.1). For the Earth at the global scale (Fig. 15.7a), low values of CA delineate land and ocean ridges as black lines (e.g., the Andes, Alps, mid-ocean ridges). High values of CA show land valleys and ocean canyons as white lines, as well as land depressions and ocean basins as light areas (e.g., the Mediterranean Sea, Gulf of Mexico, Angola Basin). On the CA map of Mars (Fig. 15.8a), one can see the planetary network of valleys and canyons, as well as a large feature, Solis Planum (15°−30°S 270°−290°E) and a plethora of smaller depressions, predominantly craters. There are boundaries of Mare Orientalis (15°−25°S 255°−275°E), Mare Nubium (20°−25°S 340°−350°E), and so on, on the CA map of the Moon (Fig. 15.8c).

Dispersive area measures a downslope area potentially exposed by slope lines passing through the given point on the topographic surface (Table 2.1). For the Earth at the global scale, high values of DA delineate mountains and highlands as light areas (e.g., Himalayas, Urals, Ethiopian Highlands), as well as land and ocean ridges as light lines (Fig. 15.7b). The planetary ridge network can be observed on the DA map of Mars (Fig. 15.9a). One can also see Alba Patera (30°−50°N 230°−260°E), Tharsis Montes (15°N−15°S 230°−260°E), and other surface features.

Global morphometric maps of the Earth are complementary to geomorphometric atlases (Guth, 2007). They can be integrated with digital geological and geomorphological globes (Ryakhovsky et al., 2003; Tooth, 2006; Bernardin et al., 2011) to solve various tectonic and geophysical problems of the planetary scale. The existing geological maps of Mars (Scott and Carr, 1978), Venus (Ivanov, 2008), and the Moon (USGS, 1972) can also be integrated with the morphometric maps of these celestial bodies. They may be useful in solving various tasks of planetary science, such as refining shapes of topographic and geological

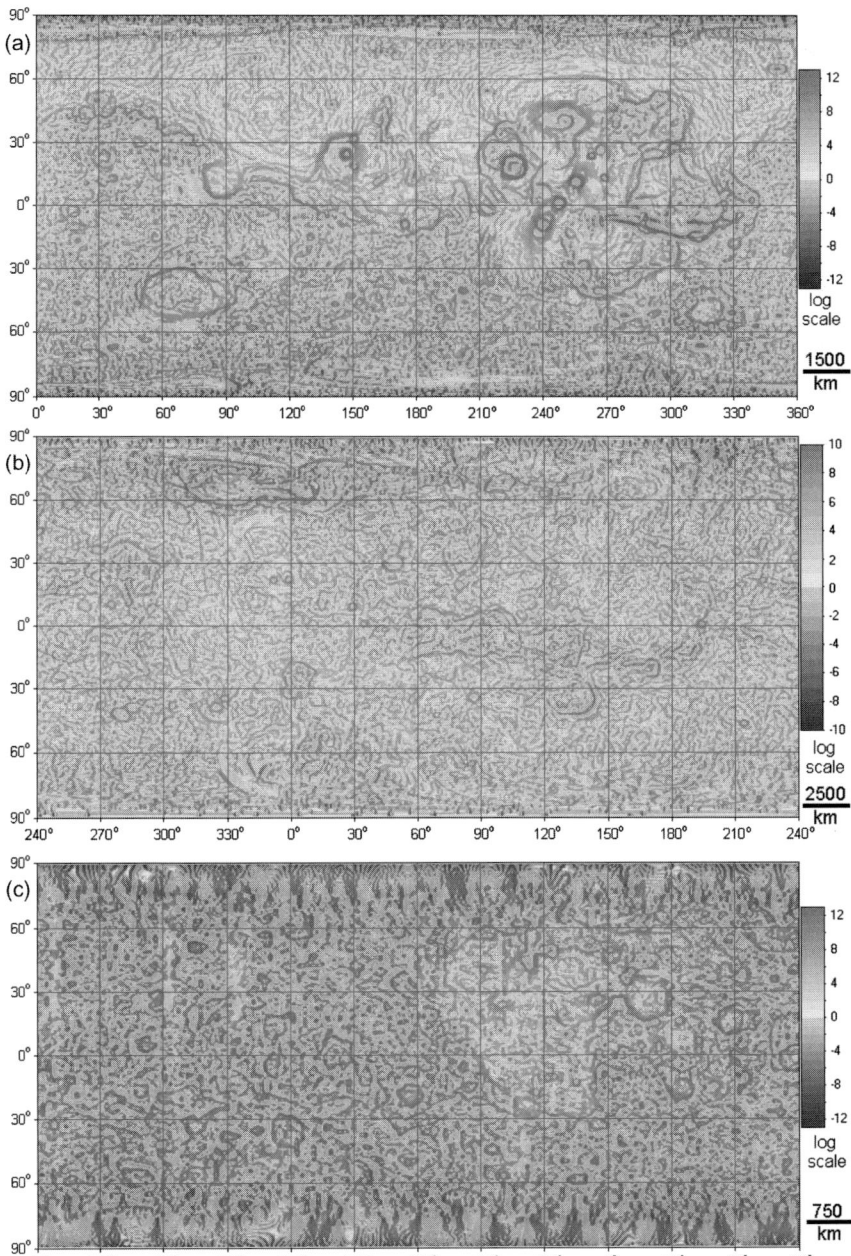

FIGURE 15.6 Vertical curvature derived from the two times smoothed DEMs: (a) Mars, (b) Venus, (c) the Moon. *From (Florinsky, 2010, Fig. 4.22).* For color version, please see page 30 in the Color Plate Section at the end of the book.

274   15. GLOBAL TOPOGRAPHY AND TECTONIC STRUCTURES

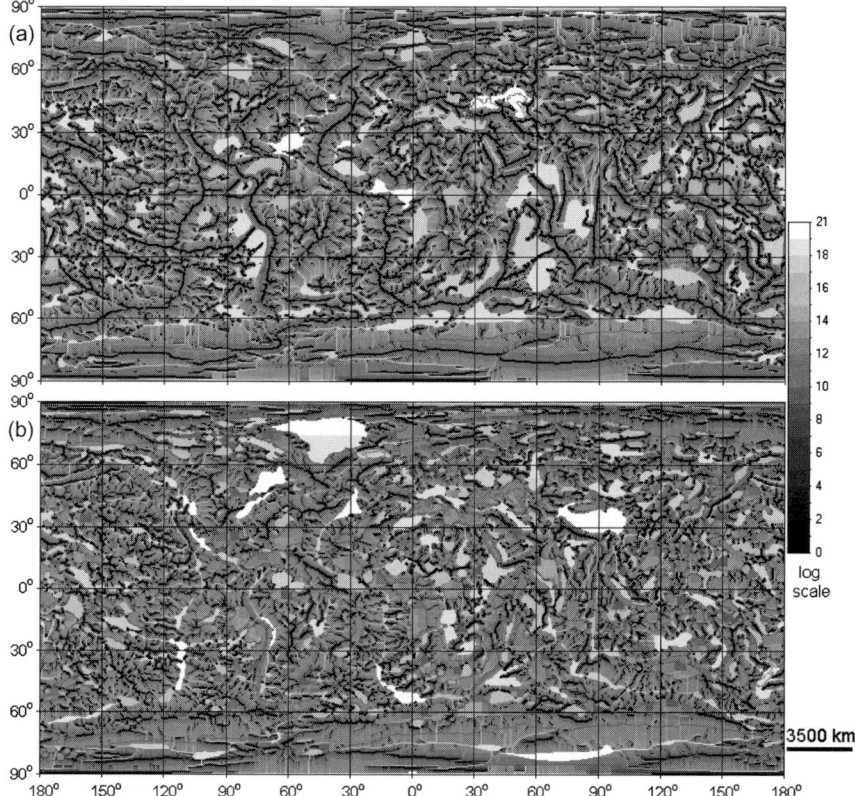

FIGURE 15.7 The Earth, global morphometric maps derived from the three times smoothed DEM: (a) specific catchment area, (b) specific dispersive area. *From (Florinsky, 2008c, Fig. 3).* For color version, please see page 31 in the Color Plate Section at the end of the book.

structures, describing them quantitatively, and analyzing their spatial distribution over the planetary surface.

Recognizable artifacts are typical for polar regions on all maps, especially for the Moon[1] (Figs. 15.4—15.9). They were caused by a relatively low accuracy of the description of these areas in the databases used. Moreover, systematic errors of the Clementine Gravity and Topography Data led to several meridian artifacts, such as stripes up to 5° wide (80°S—80°N 155°E, 10°—60°N 185°E, 15°—80°N 280°E, etc.) on the maps

---

[1] It seems that the newest high-resolution DEM of the Moon generated from the Lunar Orbiter Laser Altimeter (LOLA) data (Smith D.E. et al., 2010) will allow production of much better morphometric maps. However, the LOLA DEM was not yet publicly available at the time of writing of this book.

15.3 RESULTS AND DISCUSSION 275

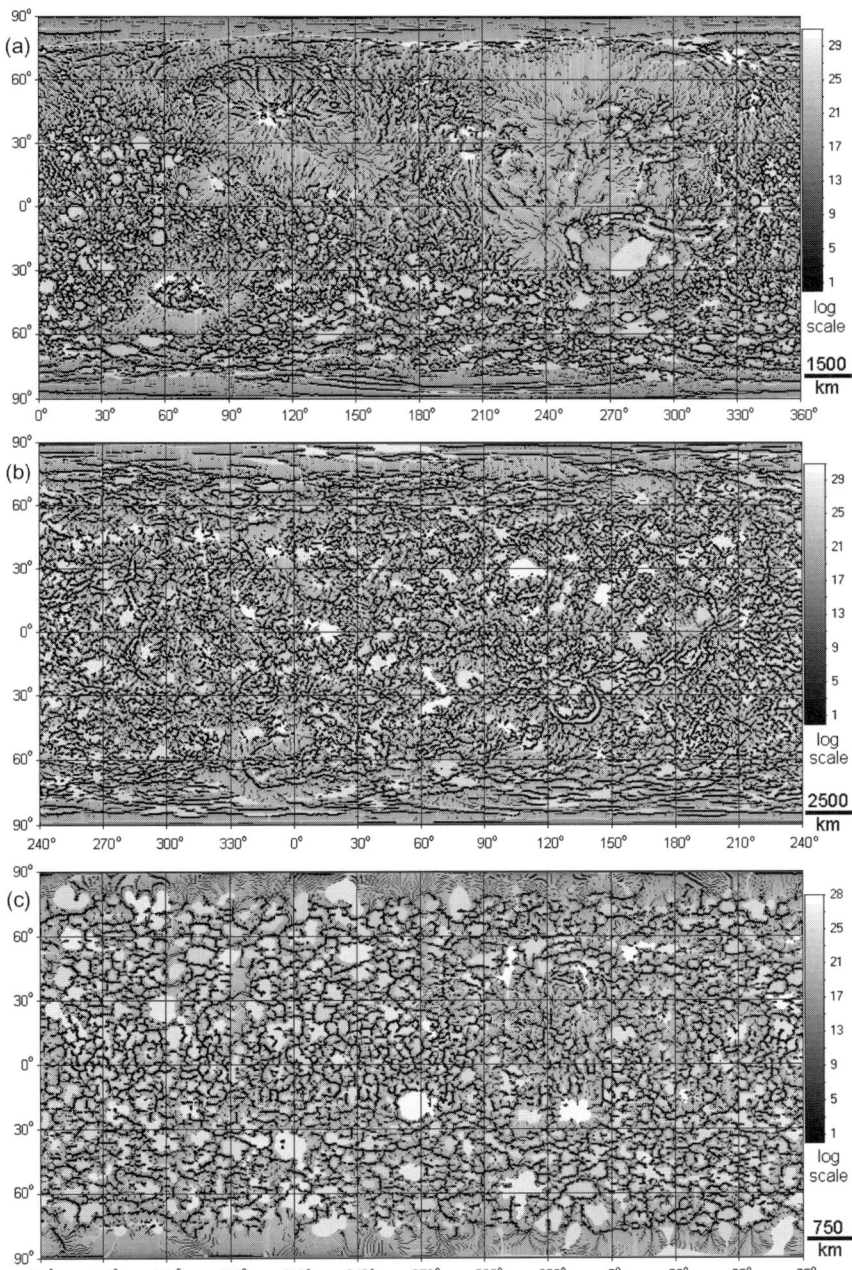

FIGURE 15.8 Catchment area derived from the smoothed DEMs: (a) Mars, (b) Venus, (c) the Moon. *From (Florinsky, 2008b, Fig. 8.4), © 2008 Springer-Verlag Berlin Heidelberg, reproduced with kind permission from Springer Science + Business Media.*

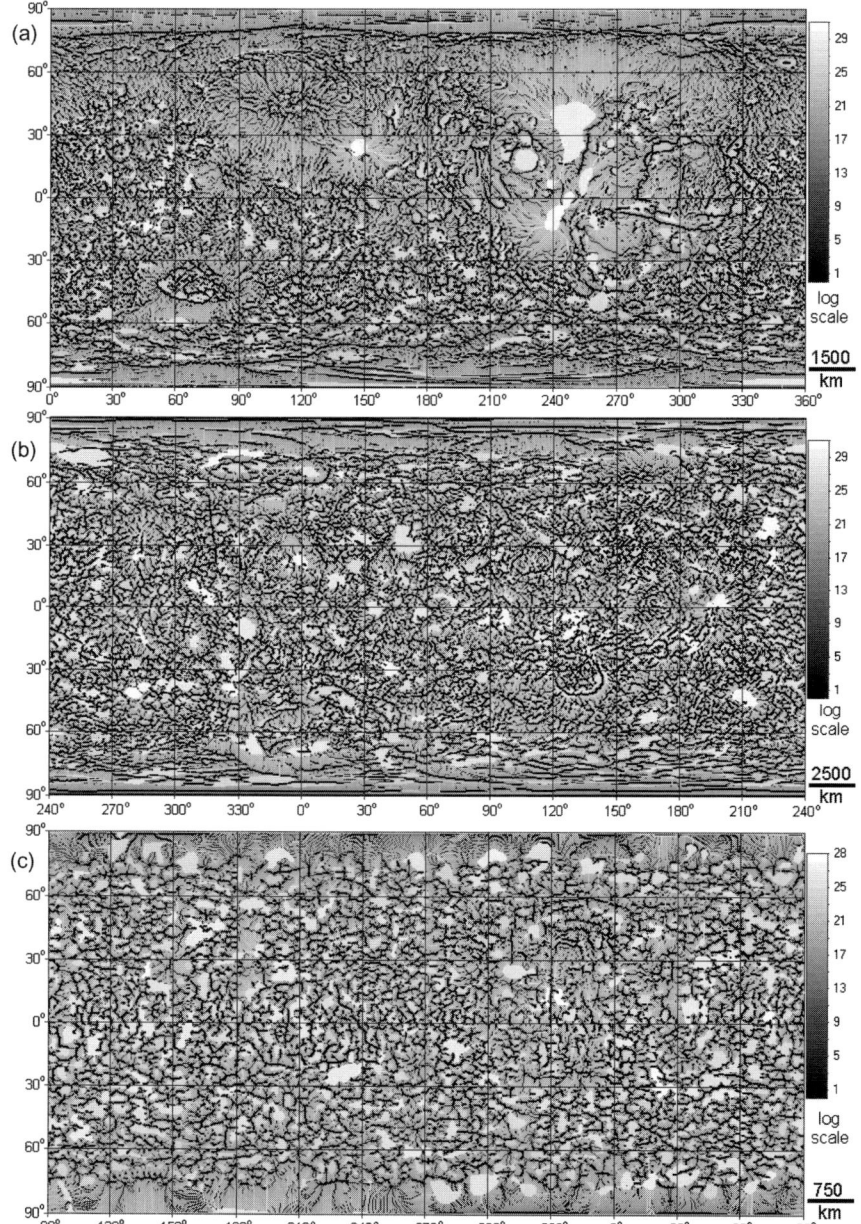

FIGURE 15.9 Dispersive area derived from the smoothed DEMs: (a) Mars, (b) Venus, (c) the Moon. *From (Florinsky, 2008b, Fig. 8.5), © 2008 Springer-Verlag Berlin Heidelberg, reproduced with kind permission from Springer Science + Business Media.*

of elevation, k_h, k_v, and DA of the Moon (Figs. 15.3c, 15.5c, 15.6c, and 15.9c). The Venus DEM was compiled combining the Global Topography Data Record and Magellan Spherical Harmonic Models and Digital Maps. These archives have different resolutions and accuracies. As a result, one can see several linear artifacts—marks of this combination (e.g., 15°–55°N 325°E, 60°–80°S 325°–355°E) on the maps of elevation, k_v, CA, and DA of Venus (Figs. 15.3b, 15.6b, 15.8b, and 15.9b).

There are also computational artifacts on the maps of CA and DA expressed as straight parallel lines, predominantly in polar regions (Figs. 15.7–15.9). These are well-known artifacts of single-flow direction algorithms common for flat slopes (Liang and Mackay, 2000).

15.3.2 Global Helical Structures

Subdivision of CA values into two levels allowed display of ridge networks of the Earth (Fig. 15.10), Mars (Fig. 15.11a), Venus (Fig. 15.11b), and the Moon (Fig. 15.11c). The greater the number of DEM smoothing, the more generalized picture of the network is mapped (Fig. 15.10).

It was observed that maps of a (specific) catchment area with values subdivided into two levels were best suited to detect global lineaments. Analysis of the SCA binary maps of the Earth (Fig. 15.12) revealed five systems of double-helical-like tectonic structures encircling the planet from pole to pole (Fig. 15.13). It is clear that the structures revealed are helical zones many kilometers wide rather than simply lines. Each helical zone transgresses plate boundaries and regions dissimilar in respect to their tectonic origin, rock composition, and age. Each double helix was named after the area(s) of intersection(s) of its arms (Table 15.1). Arms running clockwise upward and counterclockwise upward (dextral and sinistral helices) were called right and left arms, respectively.

For Mars and Venus, several global lineaments were also detected by the visual analysis of the SCA binary maps (Fig. 15.14). The global helical structures encircle these planets from pole to pole.

The global lineaments revealed cannot be artifacts due to DEM errors, treatment, or grid geometry (Section 5.7). First, DEM noise and errors usually have a spatially random distribution. Second, smoothing and derivation of topographic variables were carried out using local filters (3×3 moving windows). Third, the grid geometry may amplify its own preferential directions: orthogonal (north-south, east-west) and diagonal (northeast-southwest, northwest-southeast). However, the structures detected have (1) the global character relative to the DEMs, and (2) directions distinct from orthogonal and diagonal ones (Figs. 15.13 and 15.14).

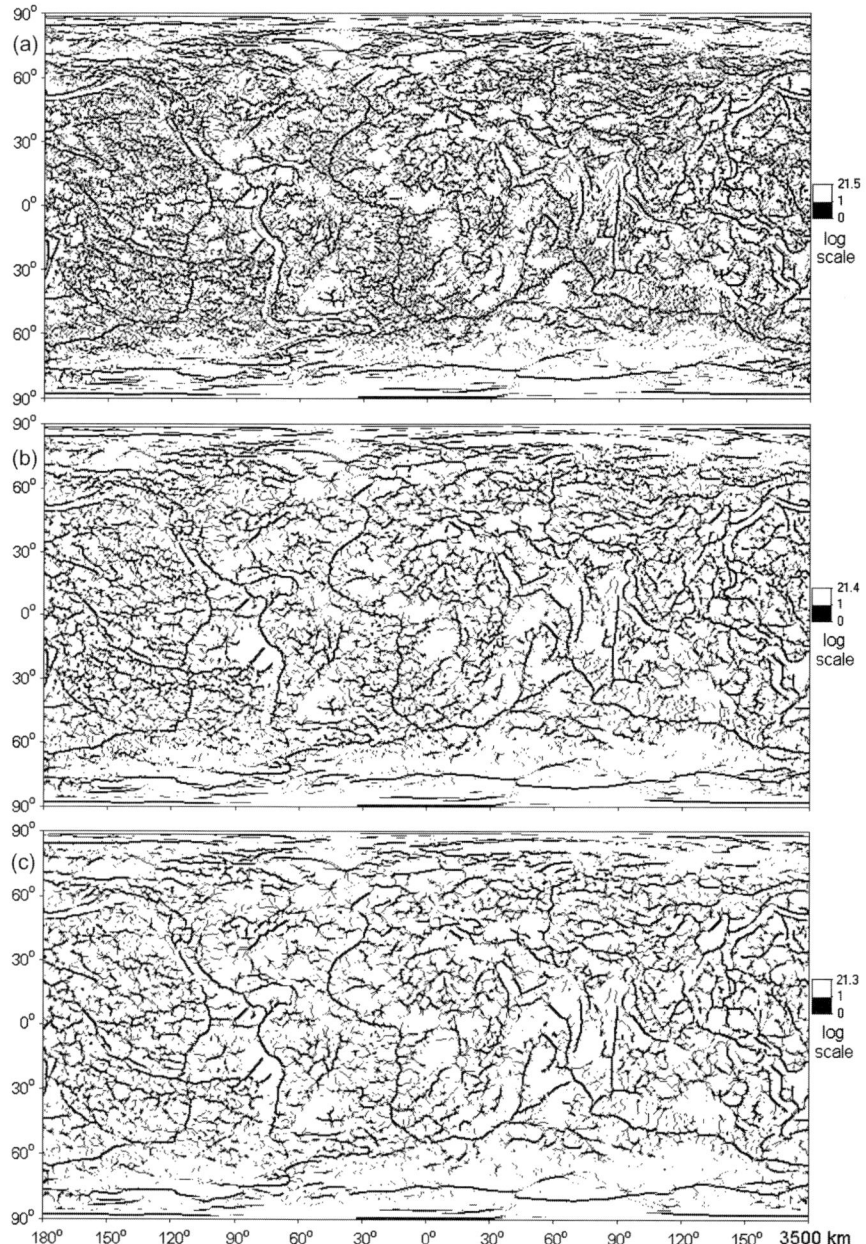

FIGURE 15.10 The Earth, global binary maps of specific catchment area: (a) smoothed DEM; (b) two times smoothed DEM; (c) three times smoothed DEM. *From (Florinsky, 2008a, Fig. 3), © 2008 Springer-Verlag Berlin Heidelberg, reproduced with kind permission from Springer Science + Business Media.*

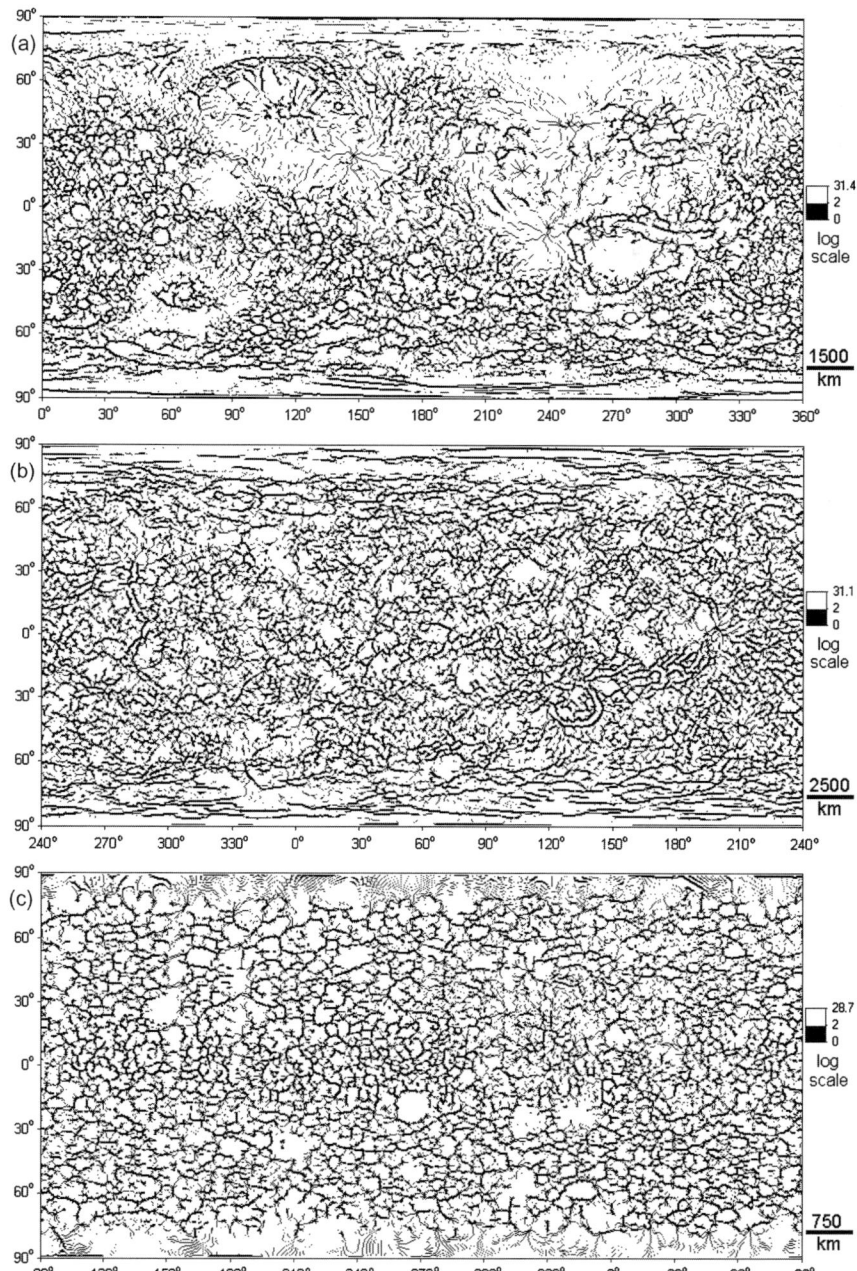

FIGURE 15.11 Global binary maps of the catchment area: (a) Mars, (b) Venus, (c) the Moon. For Mars and Venus, the two times smoothed DEMs were processed; for the Moon, the smoothed DEM was used. *From (Florinsky, 2008b, Fig. 8.7), © 2008 Springer-Verlag Berlin Heidelberg, reproduced with kind permission from Springer Science + Business Media.*

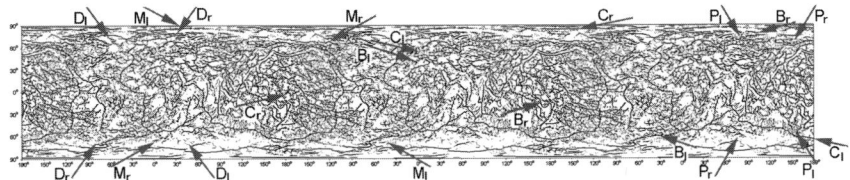

FIGURE 15.12 The Earth, a scheme for visual detection of global helical structures on the threefold map of specific catchment area derived from the two times smoothed DEM (Fig. 15.10b). Arrows show positions of helical arms. Structures (Table 15.1): C—Caucasus−Clipperton, B—Biscay−Santa Cruz, M—Marcus, D—Dakar, P—Palawan. Subscripts: r—right arm, l—left arm. *From (Florinsky, 2008a, Fig. 4), © 2008 Springer-Verlag Berlin Heidelberg, reproduced with kind permission from Springer Science + Business Media.*

Although visual analysis may cause artifacts, it is questionable whether an application of automated tools (e.g., Fukue et al., 1981; Takahashi, 1981; Zlatopolsky, 1992) may reveal radically different lineaments. Indeed, experiments demonstrated that the ideal observer (Swets, 1961) and the human visual system have almost similar capabilities to recognize geometrical patterns on noisy binary images (Krasilnikov et al., 2000). Numerous geological studies have demonstrated that visual analysis of maps and remotely sensed images can be successfully used to detect lineaments (Trifonov et al., 1983; Kats et al., 1986; Bryukhanov and Mezhelovsky, 1987; Kozlov, 1991).

The artifacts described in Section 15.3.1 did not influence the detection of global helical structures because the artifacts are located only in the polar regions.

The Caucasus−Clipperton double helix (Fig. 15.13a) coincides with one of the structures reported by Volkov (1995) (Fig. 15.1c). The left arm of the Biscay−Santa Cruz structure (Fig. 15.13b) partly agrees with the left arm of one of the helices detected by O'Driscoll (1980) (Fig. 15.1b). A comparison of inclination angles of theoretical traces of torsional deformation (Section 15.1 and Fig. 15.1a) and that of the double helices revealed (Table 15.1) shows that the Caucasus−Clipperton and Biscay−Santa Cruz structures can be assigned to traces of shear fractures, whereas the Dakar and Palawan structures can be assigned to traces of cleavage cracks. The mean deviation of the inclination angles of the structures from the theoretical values is 2.8°. There are, however, disagreements with the Rance model, such as: (1) arms of each double helix meet off the equator; and (2) there is the Marcus structure with the "abnormal" inclination. These disagreements may be attributed to (a) the deviation of the Earth's shape from a sphere, which was used in the model; and (b) the impossibility of considering all natural factors in mathematical models.

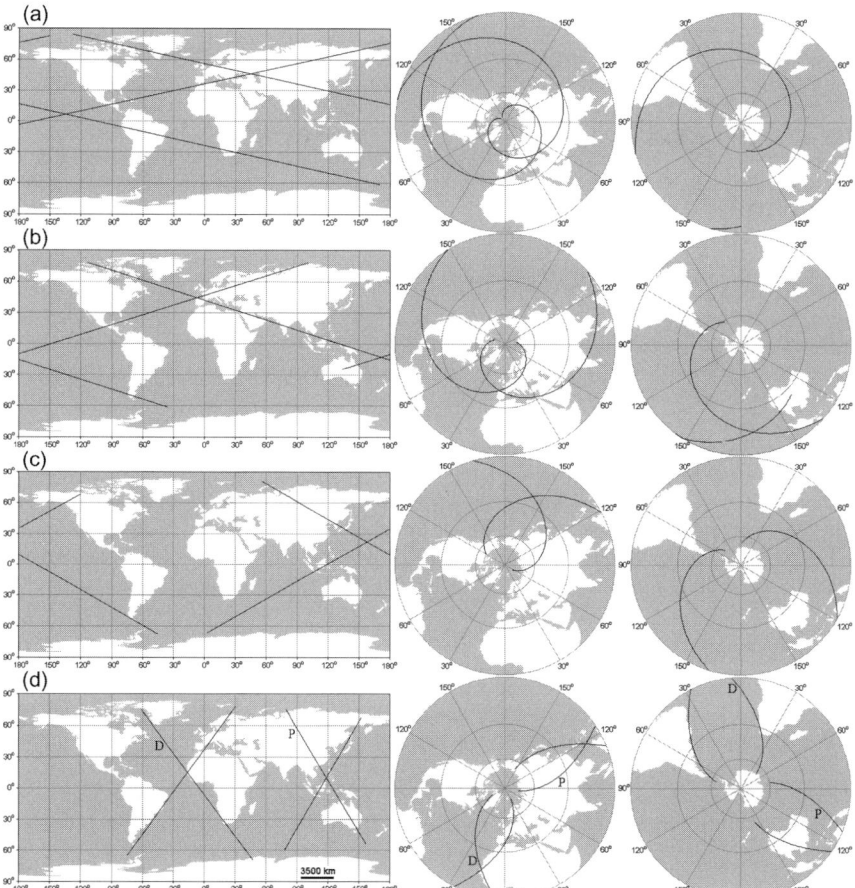

FIGURE 15.13 The Earth, global double-helical structures: (a) Caucasus−Clipperton, (b) Biscay−Santa Cruz, (c) Marcus, (d) Dakar (D) and Palawan (P). The Plate Carrée projection (left), and polar stereographic projections for the Northern (center) and Southern (right) hemispheres. *From (Florinsky, 2008a, Fig. 5), © 2008 Springer-Verlag Berlin Heidelberg, reproduced with kind permission from Springer Science + Business Media.*

Thus, of the five double-helical structures detected, four have the inclination angles fitting the theoretical inclination angles for traces of torsional deformation of the sphere. This suggests that one may consider topographically expressed helical structures as fracture traces of global torsion stresses and deformations. They may be caused by the rotational unevenness of the layers of the Earth's spheroid, which constitute a rheologically and lithologically nonuniform body.

The helical structures are expressed in modern topography; each of the structures crosses adjacent tectonic plates. Therefore, the helical

TABLE 15.1 Parameters of the Global Topographic Helices of the Earth

Structure	Left Arm Lengths (km)	Left Arm Inclination at the Equator (°)	Right Arm Lengths (km)	Right Arm Inclination at the Equator (°)	Geographical Coordinates of the Arm Intersection(s)
Caucasus–Clipperton	55,800	167.5	31,500	12.5	46.4°N 44.81°E, 5.9°N 134.7°W
Biscay–Santa Cruz	39,600	162.2	29,800	17.5	44.4°N 7.3°W, 12.9°S 171.4°E
Marcus	26,500	150.6	24,900	29.7	21.4°N 157.5°E
Dakar	17,700	126.9	17,200	53.3	14.9°N 16.0°W
Palawan	15,400	121.3	15,300	59.5	9.9°N 119.1°E

From (Florinsky, 2008a, Table 1), © 2008 Springer-Verlag Berlin Heidelberg, reproduced with kind permission from Springer Science + Business Media.

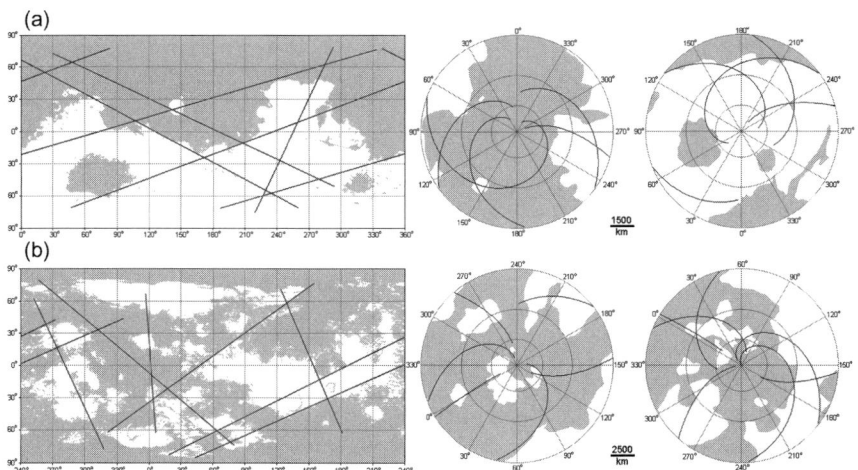

FIGURE 15.14 Global helical structures: (a) Mars, (b) Venus. The Plate Carrée projection (left), and polar stereographic projections for the Northern (center) and Southern (right) hemispheres. The gray color indicates areas located below datum. From (Florinsky, 2008b, Fig. 8.9), © 2008 Springer-Verlag Berlin Heidelberg, reproduced with kind permission from Springer Science + Business Media.

structures could have occurred in the Cenozoic only if the plate tectonic hypothesis is true. If one rejects the hypothesis, the structures could have arisen in earlier geological periods.

Notice that some spiral structures have been documented for the Martian polar caps (Howard, 1978; Weijermars, 1985/86; Fishbaugh and Head, 2001) (Fig. 15.15a). These are troughs and scarps 5–30 km wide and up to several hundreds kilometers long. Trough depths are up to 500–1000 m on polar cap edges and 100–200 m near poles. There are several models for the origin of Martian polar spiral structures. They

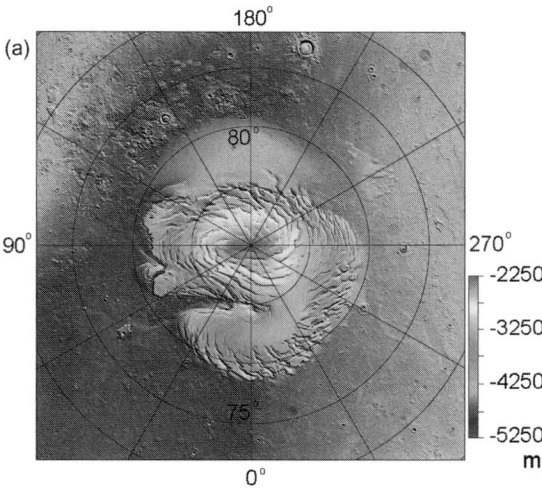
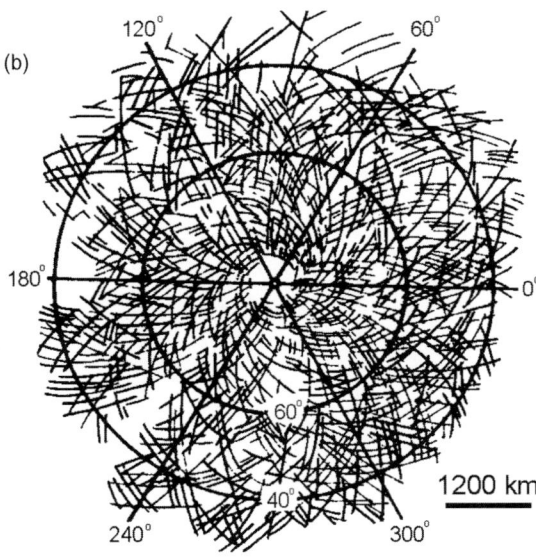

FIGURE 15.15 Helical structures of Mars and Venus detected on satellite data: (a) Structures of the Northern polar cap of Mars, from the Mars Orbiter Laser Altimeter DEM. *From (Fishbaugh and Head, 2001, Fig. 3-top), reproduced with permission.* (b) Structures of the Northern Hemisphere of Venus, from Venera-15 and -16 radar images. *From (Slyuta et al., 1989, Fig. 2), © 1990 Plenum Publishing Corporation, reproduced with kind permission from Springer Science + Business Media.* For color version, please see page 32 in the Color Plate Section at the end of the book.

are considered as a result of ice elastic deformations and sublimation, as well as the influence of katabatic wind flows affected by Coriolis forces (Howard, 1978; Weijermars, 1985/86; Fisher, 1993).

For the Northern Hemisphere of Venus, Slyuta et al. (1989) discovered a dense, regular network of dextral and sinistral spiral structures (Fig. 15.15b). They are topographically expressed as troughs, scarps, and depressions. These authors believed that strong rotational forces had formed the network during the deceleration of Venus's rotation. They suggested that the helical network is a relict feature, an "imprint" of ancient rotational stress fields, because the current rotational velocity of Venus is quite slow.

These facts on the Martian and Venusian spiral structures are circumstantial evidence that (1) global helical structures of the Earth are not artifacts; and (2) their origin may be associated with planetary rotational (torsional) stresses.

CHAPTER 16

Synthesis

In the last two decades, great progress has been made in digital terrain analysis: (1) a physical and mathematical theory of the topographic surface has been developed; (2) effective algorithms to derive DTMs have evolved; (3) large-scale and detailed DEMs have become widely available owing to advances in kinematic GPS survey and LiDAR aerial survey; and (4) global spheroidal medium- and small-scale DEMs have been produced from SRTM and ASTER data. These factors, as well as reproducibility, relative simplicity, and flexibility of digital terrain modeling methods, determine their practical potential for soil and geological research.

It should be realized, however, that the governing factor for the evolution of digital terrain analysis is the advancement in the theory of the topographic surface (Chapter 2 and Appendix A), which lays a rigorous physical and mathematical foundation for both computation algorithms and applied issues of topographic modeling. Production of increasingly accurate and detailed DEMs is of secondary importance for the scientific evolution of geomorphometry.

In the last three decades, effective algorithms have been developed for computing local and nonlocal topographic variables as well as structural lines (Chapter 4). When deciding on a particular calculation method, one of the key criteria is the algorithm's sensitivity to DEM high-frequency noise and the possibility of its suppression. This is because any DEM contains errors (Chapter 5), but filtering techniques (Chapter 6) are unable to remove all the errors from DEMs. All calculation methods of geomorphometry more or less increase the manifestation of high-frequency noise in models of topographic variables that may limit the possibility to use and interpret such models.

One other important factor influences the selection of a particular calculation algorithm. This is the type of DEM grid (Section 3.2). It is important to realize that widespread calculation methods to process DEMs on a plane square grid cannot be applied to handle DEMs on a spheroidal equal angular grid. This is associated with principal differences in the geometry of these grid types.

The author developed two calculation methods based on the approximation of partial derivatives of elevation by finite differences (Sections 4.2 and 4.3). These methods permit derivation of digital models of local morphometric attributes on plane square grids and spheroidal equal angular grids of any linear or angular resolution.

The method for deriving local topographic variables on plane square grids (Section 4.2) is based on the approximation of the third-order polynomial to the 5×5 moving window by the least-squares approach. Compared to conventional techniques, this method has higher calculation accuracy and stronger suppression of high-frequency noise. The method can be utilized in soil and geological studies at field, watershed, and regional scales.

The method for deriving local morphometric variables on spheroidal equal angular grids (Section 4.3) is based on the approximation of the second-order polynomial to the 3×3 moving window by the least-squares approach. The method is intended for geological studies at regional, continental, and global scales, as well as regional soil research.

These methods extend the capabilities of digital terrain analysis in soil science and geology. In particular, they can be applied to DEMs produced by any technique. This allows modeling and analysis of topography of any hierarchical level in soil and geological studies of any scale.

To accomplish a DTM-based study successfully, proper value of the DTM grid spacing must be selected. Its minimal value is determined by the fundamental sampling theorem and its three sequences (Section 3.3). Ignoring this rule inevitably leads to the production of artifacts in DTMs, which reflect properties of an interpolator rather than the topographic surface. Such artifacts hamper the performance of DTM-based soil and geological studies. Relationships in the "soil–topography" system are manifested not at any scale level, but at scale range intervals unique for each landscape and each soil property. This dictates a need to estimate the adequate resolution of DTMs to be used in a soil study (Chapter 9).

Use of digital terrain analysis in soil research stems from the fact that topography is one of the soil-forming factors (Chapter 8). Dozens of methods can be used to analyze, model, and map the spatial distribution of soil properties using DTMs and various mathematical tools (Chapter 10). Although spatial prediction of soil properties can be performed by many mathematical approaches, the success of such modeling depends primarily on the correct selection of predictors for a particular soil property. Morphometric attributes can be considered the most useful predictors.

Geology-oriented digital terrain modeling is based on the fact that topography is the most observable indicator of geological structure in the Earth's crust. One can use this feature of topography at any scale. Moreover, it was recently demonstrated that differential geometry and topographic modeling can be employed as a basis for mathematization of structural geology (Pollard and Fletcher, 2005) (Chapters 12 and 13).

Digital terrain modeling extends the possibilities of soil science and geology. For instance, the author obtained fundamental results using the methods described in this book:

- For Chernozemic soils of agrolandscapes in the boreal zone of North America, the following regularities were established and quantitatively described: (1) spatial distribution of soil dynamic properties depends on topography only if soil moisture content is higher than some threshold value; (2) dependence of soil dynamic properties on morphometric variables may both decrease and increase as the soil layer depth increases; (3) there exists temporal variability in relationships between spatial distribution of soil dynamic properties and morphometric attributes (Chapter 11).
- For tectonic terrains, it is established that topographically expressed zones of flow accumulation, as a rule, coincide with sites of fault intersections owing to increased rock fracturing. Topographically expressed accumulation zones are areas of contact and substance exchange between overland/intrasoil lateral water flows and upward groundwater flows (Chapter 14).
- Using quantitative data and numerical methods, we confirmed the hypothesis that there exist topographically expressed, global double-helical structures, which are apparently associated with traces of the rotational stresses of the Earth's crust (Chapter 15).

I hope that this book provides insight into the theory, methods, and applications of digital terrain analysis, as well as further understanding of relationships between topography, soil, and geology.

Appendix A
The Mathematical Basis of Local Morphometric Variables

Peter A. Shary

Institute of Physical, Chemical, and Biological Problems in Soil Science,
Russian Academy of Sciences, Pushchino, Moscow Region, 142290,
Russia; email: p_shary@mail.ru

OUTLINE

A.1 Gradient, flow lines, and special points	289
A.2 Aspect and insolation	294
A.3 Curvatures	297
A.4 Generating function	312

A.1 GRADIENT, FLOW LINES, AND SPECIAL POINTS

A smooth model of the land surface is used in this Appendix. A smooth surface S is described as a continuously differentiable function $z = f(x, y)$ representing elevation z as a function of planar coordinates x and y. It is also assumed that the gravitational field is uniform,[1] with the gravitational acceleration vector, $\mathbf{g}$, parallel to the z-axis and directed to the decrease of elevation—that is, $\mathbf{g} = -g\mathbf{k}$, where g is the gravitational acceleration constant, $\mathbf{i}$, $\mathbf{j}$, and $\mathbf{k}$ represent unit vectors in the directions of coordinate axes x, y, and z, respectively.

A.1.1 Gradient

There are two components of $\mathbf{g}$: tangential, $\mathbf{g}_t$, and normal, $\mathbf{g}_n$, ones (Fig. A.1). Substance flows are driven along the surface S by gravity governed by the tangential component, $\mathbf{g}_t$.

---
[1] This assumption is real for sufficiently small portions of the geoid, for which the equipotential surface can be considered as a plane.

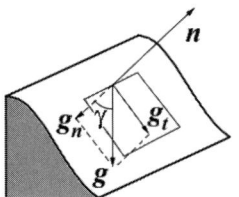

FIGURE A.1 Tangential, g_t, and normal, g_n, components of the gravitational vector g.

Let us find an expression for g_t. n is a unit external normal vector to S at a point R_0 on S. Then an equation of a tangent plane to S is known from linear algebra:

$$(R - R_0, n) = 0, \tag{A.1}$$

where parentheses denote a scalar product. The end of the vector g beginning at the point R_0 is $R_0 + g$. The distance from the point $R_0 + g$ to the tangent plane is $|(R_0 + g - R_0, n)| = |(n,g)|$. It follows from $g = g_t + g_n$ that $g_t = g - g_n$, but $|g_n|$ is the distance $|(n, g)| = -(n, g)$. The direction of g_n coincides with that of the unit vector $-n$, so that $g_n = n(n, g)$, and finally

$$g_t = g - n(n, g). \tag{A.2}$$

Note that the inequality

$$(n, g) < 0 \tag{A.3}$$

means that the angle between n and g is greater than 90°. This describes a topological restriction (Shary, 1995) excluding caves, infinitely steep slopes, and so on.

Now introduce notation of partial derivatives:

$$r = \frac{\partial^2 z}{\partial x^2}, \quad t = \frac{\partial^2 z}{\partial y^2}, \quad s = \frac{\partial^2 z}{\partial x \partial y}, \quad p = \frac{\partial z}{\partial x}, \quad q = \frac{\partial z}{\partial y}. \tag{A.4}$$

In a uniform field $g = -gk$, considering the expression for the external normal to the surface S (e.g., Zevenbergen and Thorne, 1987)

$$n = \frac{-pi - qj + k}{\sqrt{1 + p^2 + q^2}}, \tag{A.5}$$

one ensures that the inequality (Eq. A.3) is always valid for $f(x, y)$:

$$(n, g) = -\frac{g}{\sqrt{1 + p^2 + q^2}} < 0. \tag{A.6}$$

It follows from $g = -gk$ and (Eqs. A.2 and A.5) that

$$g_t = -g \frac{pi + qj + (p^2 + q^2)k}{1 + p^2 + q^2}. \tag{A.7}$$

Note also that in a uniform field

$$\cos \gamma = \left(-\mathbf{n}, \frac{\mathbf{g}}{g}\right) = \frac{1}{\sqrt{1 + p^2 + q^2}}, \qquad (A.8)$$

$$\sin \gamma = \sqrt{\frac{p^2 + q^2}{1 + p^2 + q^2}}, \qquad (A.9)$$

$$\tan \gamma = \sqrt{p^2 + q^2}, \qquad (A.10)$$

where γ is the angle between $\mathbf{k}$ and $\mathbf{n}$ (Fig. A.1). This angle is slope steepness, or gradient. Dimensionless gradient, $\overline{G}$, is

$$\overline{G} = \tan \gamma = \sqrt{p^2 + q^2}; \qquad (A.11)$$

gradient in degrees, G, is

$$G = \arctan \sqrt{p^2 + q^2}. \qquad (A.12)$$

It follows from Eqs. (A.7) and (A.9) that

$$|\mathbf{g}_t| = g \sin \gamma, \qquad (A.13)$$

that is, the tangential component of the gravitational force is proportional to gradient factor (or, slope factor), GF, rather than to gradient (Strahler, 1952):

$$GF = \frac{|\mathbf{g}_t|}{g} = \sin \gamma = \frac{\sqrt{p^2 + q^2}}{\sqrt{1 + p^2 + q^2}}. \qquad (A.14)$$

One obtains from Eqs. (A.14) and (A.11) that

$$GF = \frac{\overline{G}}{\sqrt{1 + \overline{G}^2}}. \qquad (A.15)$$

A.1.2 Special Points

In a uniform field, a contour line is a set of points of intersection of the topographic surface S and a horizontal plane $z =$ const; that is, it is described by the equation

$$z(x,y) = \text{const}, \qquad (A.16)$$

and it is a plane curve. The sufficient conditions for a contour line (Eq. A.16) to be a differentiable curve may be obtained from theorems of implicit functions (Nikolsky, 1977a, § 7.16). These conditions (namely, p and q are not equal to zero simultaneously) may be written as follows:

$$\overline{G} = \sqrt{p^2 + q^2} > 0. \qquad (A.17)$$

Here a curve is a contour line portion, for which a one-to-one and unique solution of Eq. (A.16) does exist in the form of $y = y(x)$ or $x = x(y)$ in a vicinity of each point.

Denote special points as a set O of points (x,y) for which

$$\overline{G} = \sqrt{p^2 + q^2} = 0. \qquad (A.18)$$

The inequality (Eq. A.17) is valid for a set N of nonspecial points. A contour line is a curve at nonspecial points. At special points, a contour line may not be a curve in the usual sense defined above. For example, this may be an isolated point (a top of a hill, or a bottom of a pit), an absolutely horizontal area, or an edge point of a contour line.[2] It is shown below that not all local morphometric variables can be defined in special points.

Are special points seldom or frequently distributed in space? Let P be an opened rectangle with edges parallel to the x- and y-axes, and $\overline{P}$ is its closure. Let also Q be an opened rectangular parallelepiped with faces parallel to coordinate axes such that Q contains $\overline{P}$. Now denote a set of all nonspecial points of Q as N_Q, and let $\overline{O} = \overline{P}\backslash N_Q$ be the set of all special points of $\overline{P}$. The following theorem is valid:

THEOREM 1.

The set N of nonspecial points is open, and the set $\overline{O}$ of special points is closed (Shary, 1991).

THE PROOF.

The proof of the first part follows from continuity of the function $\overline{G} = \sqrt{p^2 + q^2}$. If a point $(x,y) \in Q$ is nonspecial (i.e., $\overline{G} > 0$), then all points in its vicinity belong to N_Q, so that N_Q consists of internal points; that is, it is an open set. The intersection of N_Q and P is that of two open sets; consequently it is an open set. On the other hand, it is a subset of the nonspecial points of P, that is, of N. The proof of the second part follows from the definition of $\overline{O}$ as a difference between the closed set $\overline{P}$ and the open set N_Q. Consequently, $\overline{O}$ is a closed set. The theorem is proven.

---

[2]Sard (1942) suggested a mathematical description of contour line behavior near special points.

As $\overline{O}$ is the closed set, special points are usually sparsely distributed in space. If flat areas occupy more than a small proportion of a DEM, this is probably due to rounding of elevation, or excessive use of interpolation.

A.1.3 Slope Lines and Flow Lines

A slope line, or gradient line, is a curve (defined on N) on the surface S, in each point of which the tangential direction coincides with that of the tangential component of the gravitational force g_t (Cayley, 1859). A flow line is a projection of a gradient line to a horizontal plane. In special points, $g_t = 0$ (because $p = q = 0$) and the direction of g_t is not defined. Gradient lines are directed downslope because the z-component of g_t is negative for N.

The unit tangent vector to the gradient line is

$$r'_\eta = \frac{dx}{d\eta}\mathbf{i} + \frac{dy}{d\eta}\mathbf{j} + \frac{dz}{d\eta}\mathbf{k}, \qquad (A.19)$$

where η is a gradient line length, and

$$\left(\frac{dx}{d\eta}\right)^2 + \left(\frac{dy}{d\eta}\right)^2 + \left(\frac{dz}{d\eta}\right)^2 = 1. \qquad (A.20)$$

Since $|g_t| = g\sin\gamma$, one obtains from Eq. (A.7) that for N

$$r'_\eta = -\frac{p\mathbf{i} + q\mathbf{j} + (p^2 + q^2)\mathbf{k}}{\sqrt{(p^2 + q^2)(1 + p^2 + q^2)}}. \qquad (A.21)$$

THEOREM 2.

For the set N of nonspecial points, any contour line is orthogonal to any gradient line at the point of intersection. For N, contours and flow lines are also orthogonal (Shary, 1991).

THE PROOF.

For the first part of the theorem, it is sufficient to prove that for N

$$(r'_\eta, r'_\xi) = 0, \qquad (A.22)$$

where ξ is a contour line length, and r'_ξ is a unit tangent vector to the contour line. Taking the derivative of a contour line (Eq. A.16) on ξ and noting that, for a plane curve, the formula (Eq. A.20) is

$$\left(\frac{dx}{d\xi}\right)^2 + \left(\frac{dy}{d\xi}\right)^2 = 1, \tag{A.23}$$

one obtains

$$p\frac{dx}{d\xi} + q\frac{dy}{d\xi} = 0. \tag{A.24}$$

It follows from here that for N

$$r'_\xi = \pm \frac{-q\mathbf{i} + p\mathbf{j}}{\sqrt{p^2 + q^2}}. \tag{A.25}$$

The sign choice is defined by the contour line orientation. Therefore, Eq. (A.22) is proven, as is the first part of the theorem.

Using expression (Eq. A.21) and designating a flow line length as σ, one obtains an expression for a tangent vector r'_σ to a flow line (a plane curve):

$$r'_\sigma = -\frac{p\mathbf{i} + q\mathbf{j}}{\sqrt{p^2 + q^2}} \tag{A.26}$$

for N. It follows from Eqs. (A.25) and (A.26) that for N

$$(r'_\sigma, r'_\xi) = 0. \tag{A.27}$$

Therefore, flow lines are orthogonal to contour lines. The theorem is proven.

Now demonstrate that a direction of the maximal decrease of elevation is a flow line direction for each nonspecial point. Indeed, for N the directional derivative along a unit vector $\mathbf{m}$ is

$$\left|\frac{\partial z}{\partial \mathbf{m}}\right| = |(\nabla z, \mathbf{m})| \leq |\nabla z| = \sqrt{p^2 + q^2}, \tag{A.28}$$

where ∇z is a two-dimensional vector of gradient, $\nabla z = p\mathbf{i} + q\mathbf{j}$. It follows that the direction of $-\nabla z$ gives the maximal decrease of elevation among all possible directions. However, this is the direction of r'_σ along a flow line. Notice that the quantitative measure of this elevation decrease is the dimensionless gradient (Eq. A.11).

A.2 ASPECT AND INSOLATION

A.2.1 Aspect

The negative two-dimensional vector of gradient, $-\nabla z$, defines both slope steepness, G, and slope aspect, A. Aspect is defined as an angle between the northern direction $\mathbf{j}$ and the direction of $-\nabla z$ counted clockwise, from 0 to 360° (vectors of coordinate axes i, j, and k are directed to the East, North, and upward, respectively).

A unit tangent vector to a flow line, $\frac{-\nabla z}{|\nabla z|}$, is r'_σ. For N, a scalar product (r'_σ, j) is a cosine of the angle A between r'_σ and j:

$$\cos A = (r'_\sigma, j) = \frac{-q}{\sqrt{p^2 + q^2}}. \tag{A.29}$$

Considering that the inequality $q > 0$ refers to southern slopes ($90 < A \leq 180°$ if $p \leq 0$, $180 < A < 270°$ if $p > 0$), $A = 90°$ ($p < 0$) or $A = 270°$ ($p > 0$) if $q = 0$, and the inequality $q < 0$ refers to northern slopes ($0 < A < 90°$ if $p < 0$, $270 < A < 360°$ if $p > 0$), for N one obtains:

$$A = 0 + \frac{180}{\pi} \arccos\left(\frac{-q}{\sqrt{p^2 + q^2}}\right) \quad \text{for} \quad p \leq 0, \; q < 0, \tag{A.30}$$

$$A = 90 + \frac{180}{\pi} \arccos\left(\frac{-q}{\sqrt{p^2 + q^2}}\right) \quad \text{for} \quad p \leq 0, \; q \geq 0, \tag{A.31}$$

$$A = 180 + \frac{180}{\pi} \arccos\left(\frac{-q}{\sqrt{p^2 + q^2}}\right) \quad \text{for} \quad p > 0, \; q \geq 0, \tag{A.32}$$

$$A = 270 + \frac{180}{\pi} \arccos\left(\frac{-q}{\sqrt{p^2 + q^2}}\right) \quad \text{for} \quad p > 0, \; q < 0, \tag{A.33}$$

where arccosine is in radians.

Now one may use the function

$$\text{sign}(x) = \begin{cases} 1 & \text{for} \quad x > 0 \\ 0 & \text{for} \quad x = 0, \\ -1 & \text{for} \quad x < 0 \end{cases} \tag{A.34}$$

and transform formulas (Eqs. A.30 to A.33) into a single formula valid for N:

$$A = -90\left[1 - \text{sign}(q)\right](1 - |\text{sign}(p)|) + 180\left[1 + \text{sign}(p)\right]$$
$$- \frac{180}{\pi} \text{sign}(p) \arccos\left(\frac{-q}{\sqrt{p^2 + q^2}}\right). \tag{A.35}$$

For N, a sine of A may be deduced as a z-component of the vector product $[r'_\sigma, j]$:

$$[r'_\sigma, j]_z = \begin{vmatrix} i & j & k \\ \frac{-p}{\sqrt{p^2+q^2}} & \frac{-q}{\sqrt{p^2+q^2}} & 0 \\ 0 & 1 & 0 \end{vmatrix}_z = \frac{-p}{\sqrt{p^2+q^2}}, \tag{A.36}$$

so that

$$\sin A = [\mathbf{r}'_\sigma, \mathbf{j}]_z = \frac{-p}{\sqrt{p^2 + q^2}}. \tag{A.37}$$

A.2.2 Insolation

Insolation, $I(\theta,\psi)$, is defined as a proportion (in %) of maximal direct solar irradiation at the Sun's angular position determined by two angles: solar azimuth, θ, and solar elevation, ψ (Shary et al., 2005) (Fig. 2.8).

Solar rays may be considered as parallel, so a unit vector to the Sun from a given point on the surface S is

$$\mathbf{m} = \mathbf{i} \sin \theta \cos \psi + \mathbf{j} \cos \theta \cos \psi + \mathbf{k} \sin \psi. \tag{A.38}$$

The proportion (from maximal one) of solar irradiation intensity is a cosine of the angle between $\mathbf{m}$ and the unit normal vector $\mathbf{n}$ to S; that is, it is $(\mathbf{m},\mathbf{n})$ if this angle is smaller than 90°, and it is null in the opposite case (a shady side of a hill). The inequality $(\mathbf{m}, \mathbf{n}) > 0$ corresponds to angles smaller than 90°, so this proportion is

$$\frac{1}{2}(\mathbf{m}, \mathbf{n})[1 + \text{sign}(\mathbf{m}, \mathbf{n})]. \tag{A.39}$$

It follows from here that insolation is

$$I(\theta, \psi) = 50(\mathbf{m}, \mathbf{n})[1 + \text{sign}(\mathbf{m}, \mathbf{n})]. \tag{A.40}$$

Now one obtains from Eqs. (A.38) and (A.5) that

$$(\mathbf{m}, \mathbf{n}) = \frac{\sin \psi - \cos \psi (p \sin \theta + q \cos \theta)}{\sqrt{1 + p^2 + q^2}}. \tag{A.41}$$

Substituting this into Eq. (A.40), one obtains the final expression:

$$I(\theta, \psi) = 50 \frac{\{1 + \text{sign}[\sin\psi - \cos\psi(p\sin\theta + q\cos\theta)]\}[\sin\psi - \cos\psi(p\sin\theta + q\cos\theta)]}{\sqrt{1 + p^2 + q^2}}. \tag{A.42}$$

$I(\theta,\psi)$ can be defined at each point of the surface.

Notice that insolation is defined here as a local morphometric variable, so that hills may have shady sides, but do not produce shadows themselves, because cast shadow is a nonlocal concept. The intensity of direct solar irradiation to a surface perpendicular to solar rays during a sunny day is $\tau = 1.25$ kW/m^2. Thus, insolation can also be calculated in energy units using the following expression:

$$\tilde{I}(\theta, \psi) = \frac{\tau I(\theta, \psi)}{100}. \tag{A.43}$$

A.3 CURVATURES

To have a flexible mathematical tool for deducing curvatures, one may use a general formula of curvature of a smooth (not necessarily plane) curve (Nikolsky, 1977a, § 6.9):

$$|r''_\lambda| = \sqrt{\left(\frac{d^2x}{d\lambda^2}\right)^2 + \left(\frac{d^2y}{d\lambda^2}\right)^2 + \left(\frac{d^2z}{d\lambda^2}\right)^2}, \tag{A.44}$$

where λ is the length of a curve, r''_λ is the second derivative of the vector r by λ.

A.3.1 Plan Curvature

Plan curvature, k_p, is the curvature of a contour line (Evans, 1972; Krcho, 1973). Substituting a contour line length ξ (instead of λ) to the general formula (Eq. A.44), one may find the derivatives needed for this formula. It follows from Eq. (A.25) that for N

$$\frac{dx}{d\xi} = \pm \frac{-q}{\sqrt{p^2 + q^2}}, \quad \frac{dy}{d\xi} = \pm \frac{p}{\sqrt{p^2 + q^2}}, \tag{A.45}$$

also

$$\frac{dp}{d\xi} = r\frac{dx}{d\xi} + s\frac{dy}{d\xi}, \quad \frac{dq}{d\xi} = s\frac{dx}{d\xi} + t\frac{dy}{d\xi}. \tag{A.46}$$

Thus, for N

$$\frac{d^2x}{d\xi^2} = -\frac{p}{(p^2 + q^2)^2}(q^2 r - 2pqs + p^2 t), \tag{A.47}$$

$$\frac{d^2y}{d\xi^2} = -\frac{q}{(p^2 + q^2)^2}(q^2 r - 2pqs + p^2 t). \tag{A.48}$$

The vector to be found

$$r''_\xi = \frac{d^2x}{d\xi^2}i + \frac{d^2y}{d\xi^2}j \tag{A.49}$$

is therefore

$$r''_\xi = -k_p r'_\sigma \qquad (A.50)$$

for N, where

$$k_p = -\frac{q^2 r - 2pqs + p^2 t}{\sqrt{(p+q)^3}} \qquad (A.51)$$

for N, r'_σ is the unit tangent vector to flow line as defined by Eq. (A.26).

Let us prove that plan curvature k_p is a quantitative measure of flow line divergence and convergence. First, elucidate the meaning of divergence $\text{div}(r'_\sigma)$ of the unit tangent vector r'_σ to a flow line using the Gauss–Ostrogradsky theorem (Nikolsky, 1977b, § 13.10). Let an arbitrary but fixed point (x,y) belong to N and be surrounded by a small circle T_ε of radius $\varepsilon > 0$, which also belongs to N (this is always possible, because N is an open set — see Theorem 1), with an external normal n' to positively oriented boundary Γ_ε of the circle T_ε. According to the two-dimensional case of the Gauss–Ostrogradsky theorem,

$$\iint_{T_\varepsilon} \text{div}(r'_\sigma)\,dxdy = \int_{\Gamma_\varepsilon} (r'_\sigma, n')\,d\xi, \qquad (A.52)$$

where $d\xi$ is the differential of Γ_ε. Using a theorem on averages (Nikolsky, 1977a, § 5.8) and taking a limit, one finds that

$$\text{div}(r'_\sigma) = \lim_{\varepsilon \to 0, \varepsilon > 0} \left(\frac{1}{\pi\varepsilon^2}\right) \int_{\Gamma_\varepsilon} (r'_\sigma, n')\,d\xi, \qquad (A.53)$$

where the integral is the flow of a unit vector r'_σ through Γ_ε. This means that $\text{div}(r'_\sigma)$ is a quantitative measure of flow lines divergence.

THEOREM 3.

Divergence of the unit tangent vector r'_σ to a flow line is equal to k_p (Shary, 1995).

THE PROOF.

Calculating $\text{div}(r'_\sigma)$ with Eq. (A.26), one obtains for N

$$\text{div}(r'_\sigma) = \frac{\partial\left(\frac{-p}{\sqrt{p^2+q^2}}\right)}{\partial x} + \frac{\partial\left(\frac{-q}{\sqrt{p^2+q^2}}\right)}{\partial y} = k_p. \qquad (A.54)$$

Therefore, plan curvature is a quantitative measure of flow line divergence for N. The theorem is proven.

Flow lines converge where $k_p < 0$ and diverge where $k_p > 0$; $k_p = 0$ refers to parallel flow lines.

A.3.2 Horizontal Curvature

Horizontal or tangential curvature, k_h, is the curvature of the normal section tangential to a contour line (Krcho, 1983; Shary, 1991; Mitášová and Hofierka, 1993). To deduce the formula of k_h, one may use the Meusnier theorem (Nikolsky, 1977a, § 7.24):

$$k_0 = k \cos \theta, \tag{A.55}$$

where k_0 is a curvature of a normal section, passing through R and tangential to Γ; k is a curvature $k = |r''_\lambda|$ of a curve Γ, belonging to S and passing through a point R on S; and θ is the angle between vectors $\nu = \frac{r''_\lambda}{|r''_\lambda|}$ and n, which is orthogonal to S at the point R, $\cos \theta = (n, \nu)$. It follows from here that the Meusnier theorem may be rewritten as

$$k_0 = (n, r''_\lambda). \tag{A.56}$$

Considering a contour line as the curve Γ (so that $\lambda = \xi$) and using the formulas of n and $\sin\gamma$ (Eqs. A.5 and A.9), one obtains for N

$$k_h = k_p \sin \gamma, \tag{A.57}$$

where $0 < \gamma < \pi/2$, and consequently

$$k_h = -\frac{q^2 r - 2pqs + p^2 t}{(p^2 + q^2)\sqrt{1 + p^2 + q^2}} \tag{A.58}$$

for N. Notice that k_p and k_h have the same signs at nonspecial points, as seen from Eqs. (A.51) and (A.58). Thus, flow lines converge where $k_h < 0$ and diverge where $k_h > 0$; $k_h = 0$ refers to parallel flow lines.

A.3.3 Vertical Curvature

Vertical or profile curvature, k_v, is the curvature of a normal section, which has a mutual tangent line with a gradient line (Aandahl, 1948; Krcho, 1973; Young, 1978). According to Theorem 2, contour lines and gradient lines are orthogonal. So, one may use the Euler theorem (Nikolsky, 1977a, § 7.24) stating that mean curvature, H, is equal to half the sum of curvatures of two mutually orthogonal normal sections (Gauss, 1828).[3] In part,

$$H = \frac{1}{2}(k_h + k_v) \tag{A.59}$$

---

[3]There is an English translation of this work (Gauss, 2009).

for N. The formula for mean curvature is as follows (Young, 1805; Gauss, 1828):

$$H = -\frac{(1+q^2)r - 2pqs + (1+p^2)t}{2\sqrt{(1+p^2+q^2)^3}}.\tag{A.60}$$

This is valid for each point of the topographic surface. Substituting formulas of k_h and H (Eqs. A.58 and A.60) to Eq. (A.59), one obtains for N:

$$k_v = -\frac{p^2 r + 2pqs + q^2 t}{(p^2+q^2)\sqrt{(1+p^2+q^2)^3}}.\tag{A.61}$$

THEOREM 4.

For N, vertical curvature k_v is a derivative of gradient factor GF by a flow line length σ (Shary, 1995; Shary et al., 2002b).

THE PROOF.

For N, the first derivative of elevation z by a flow line length σ is

$$\frac{dz}{d\sigma} = p\frac{dx}{d\sigma} + q\frac{dy}{d\sigma} = -\sqrt{p^2+q^2} = -\overline{G},\tag{A.62}$$

where expressions for $\frac{dx}{d\sigma}$ and $\frac{dy}{d\sigma}$ were taken from Eq. (A.26). For N, the second derivative of elevation is

$$\frac{d^2 z}{d^2 \sigma} = -\frac{d\overline{G}}{d\sigma} = -\frac{d\sqrt{p^2+q^2}}{d\sigma} = -k_v\sqrt{(1+p^2+q^2)^3},\tag{A.63}$$

where it was assumed that

$$\frac{dp}{d\sigma} = \frac{dp}{dx}\frac{dx}{d\sigma} + \frac{dp}{dy}\frac{dy}{d\sigma} = r\frac{dx}{d\sigma} + s\frac{dy}{d\sigma}, \quad \frac{dq}{d\sigma} = \frac{dq}{dx}\frac{dx}{d\sigma} + \frac{dq}{dy}\frac{dy}{d\sigma} = s\frac{dx}{d\sigma} + t\frac{dy}{d\sigma},\tag{A.64}$$

and formulas of $\frac{dx}{d\sigma}$ and $\frac{dy}{d\sigma}$ were taken from Eq. (A.26). It follows from Eq. (A.63) that

$$k_v = \frac{\frac{d\overline{G}}{d\sigma}}{\sqrt{(1+\overline{G}^2)^3}}\tag{A.65}$$

for N. Considering Eq. (A.15), one obtains that

$$\frac{d(GF)}{d\sigma} = \frac{\frac{d\overline{G}}{d\sigma}\left(\sqrt{1+\overline{G}^2} - \frac{\overline{G}^2}{\sqrt{1+\overline{G}^2}}\right)}{1+\overline{G}^2} = \frac{\frac{d\overline{G}}{d\sigma}}{\sqrt{(1+\overline{G}^2)^3}} = k_v \qquad (A.66)$$

for N; that is, the derivative of gradient factor by the flow line length is equal to vertical curvature. The theorem is proven.

Negative values of the derivative $\frac{d(GF)}{d\sigma}$ mean relative deceleration of gravity-driven substance flows: the velocity of upslope particles is greater than that of downslope ones, resulting in their accumulation. Theorem 4 states that such areas are located in concave slope profiles.

A.3.4 Laplacian

The Laplacian of elevation, Δz, is defined (Nikolsky, 1977a, § 7.26) as

$$\Delta z = \left(\frac{\partial^2 z}{\partial x^2}\right)^2 + \left(\frac{\partial^2 z}{\partial y^2}\right)^2 = r + t. \qquad (A.67)$$

THEOREM 5.
$\Delta z \approx -2H$ as $p, q \to 0$ (Shary et al., 2002b).

THE PROOF.
The limit value of H as $p, q \to 0$ can be found from Eq. (A.60); it is equal to $-0.5(r+t)$. Therefore, $\Delta z = r + t \approx -2H$ as $p, q \to 0$. This proves the theorem.

Gradient $\overline{G} = \sqrt{p^2 + q^2}$ also tends to zero as $p, q \to 0$. Thus, it follows from this theorem that Δz is close to $-2H$ in gently sloping terrains.

A.3.5 Total Gaussian Curvature

The total Gaussian curvature, K, is defined by Gauss (1828) as the product of maximal and minimal curvatures of normal sections at a given point

$$K = k_{min} k_{max}, \qquad (A.68)$$

and its formula deduced by Gauss (1828) is as follows:

$$K = \frac{rt - s^2}{(1 + p^2 + q^2)^2}. \tag{A.69}$$

According to the Euler theorem (Nikolsky, 1977a, § 7.24), mean curvature is

$$H = \frac{1}{2}(k_{min} + k_{max}). \tag{A.70}$$

From Eqs. (A.68) and (A.70) one obtains that

$$k_{max}, k_{min} = H \pm \sqrt{H^2 - K}, \tag{A.71}$$

where the sign " $+$ " refers to k_{max}, and the sign " $-$ " refers to k_{min}.

A.3.6 Unsphericity Curvature

Unsphericity curvature, M, is defined by Shary (1995) as

$$M = \frac{1}{2}(k_{max} - k_{min}). \tag{A.72}$$

It follows from Eqs. (A.71) and (A.72) that

$$M = \sqrt{H^2 - K}. \tag{A.73}$$

To obtain an explicit formula for M, one should transform $H^2 - K$ to a sum of squares. This may be done using formulas of K and H (Eqs. A.69 and A.60):

$$M = \sqrt{\frac{[(1+q^2)r - 2pqs + (1+p^2)t]^2}{4(1+p^2+q^2)^3} - \frac{rt - s^2}{(1+p^2+q^2)^2}}$$

$$= \sqrt{\frac{[(1+q^2)r - 2pqs + (1+p^2)t]^2 - 4(1+p^2+q^2)(rt - s^2)}{4(1+p^2+q^2)^3}}. \tag{A.74}$$

Denote the following expression as α

$$\alpha = [(1+q^2)r - 2pqs + (1+p^2)t]^2 - 4(1+p^2+q^2)(rt - s^2). \tag{A.75}$$

Now Eq. (A.74) may be rewritten as

$$M = \sqrt{\frac{\alpha}{4(1+p^2+q^2)^3}}. \tag{A.76}$$

Transform α to the form:

$$\alpha = (1+q^2)^2 r^2 + 4(1+p^2)(1+q^2)s^2 + (1+p^2)^2 t^2 - 4pq(1+q^2)sr$$
$$- 4pq(1+p^2)st + 2(p^2q^2 - 1 - p^2 - q^2)rt. \tag{A.77}$$

A.3 CURVATURES

This is a second-order polynomial with respect to r, s, and t of the form

$$\alpha = a^2 r^2 + b^2 s^2 + c^2 t^2 + 2drs + 2est + 2frt, \tag{A.78}$$

where a^2, b^2, c^2, d, e, and f are

$$a^2 = (1+q^2)^2, \quad b^2 = 4(1+p^2)(1+q^2), \quad c^2 = (1+p^2)^2$$
$$d = -2pq(1+q^2), \quad e = -2pq(1+p^2), \quad f = p^2 q^2 - 1 - p^2 - q^2. \tag{A.79}$$

α may be represented as a sum of squares:

$$\alpha = (Ar + Bt)^2 + (Cr + Ds + Et)^2 = (A^2 + C^2)r^2 + D^2 s^2 + (B^2 + E^2)t^2$$
$$+ 2CDrs + 2DEst + 2(AB + CE)rt \tag{A.80}$$

with unknown coefficients A, B, C, D, and E, which are to be determined. Equations for their determination may be found from comparison of the last formula and Eq. (A.77):

$$A^2 + C^2 = a^2, \quad CD = d, \quad D^2 = b^2, \quad DE = e, \quad B^2 + E^2 = c^2, \quad AB + CE = f. \tag{A.81}$$

Note that there are six equations and five unknown variables (A, B, C, D, and E), so that one of the equations should be automatically satisfied. One may express A, B, C, D, and E through a^2, b^2, c^2, d, e, and f using Eq. (A.78):

$$D = d, \; C = \frac{d}{b}, \; E = \frac{e}{b}, \; A = \sqrt{a^2 - \frac{d^2}{b^2}} = \sqrt{\frac{a^2 b^2 - d^2}{b^2}}, \; B = \sqrt{c^2 - \frac{e^2}{b^2}} = \sqrt{\frac{b^2 c^2 - e^2}{b^2}}, \tag{A.82}$$

and substitute here expressions for a^2, b^2, c^2, d, e, and f from Eq. (A.79). Note that the choice of sign for b is not unique. Choosing

$$b = -2\sqrt{(1+p^2)(1+q^2)}, \tag{A.83}$$

one obtains

$$D = b = -2\sqrt{(1+p^2)(1+q^2)}, \tag{A.84}$$

$$C = \frac{d}{b} = \frac{-2pq(1+q^2)}{-2\sqrt{(1+p^2)(1+q^2)}} = \frac{pq\sqrt{1+q^2}}{\sqrt{1+p^2}}, \tag{A.85}$$

$$E = \frac{e}{b} = \frac{-2pq(1+p^2)}{-2\sqrt{(1+p^2)(1+q^2)}} = \frac{pq\sqrt{1+p^2}}{\sqrt{1+q^2}}, \tag{A.86}$$

$$A = \sqrt{\frac{4(1+q^2)^3(1+p^2) - 4p^2q^2(1+q^2)^2}{4(1+p^2)^3(1+q^2)}} = \sqrt{\frac{(1+q^2)(1+p^2+q^2)}{(1+p^2)}}, \quad (A.87)$$

$$B = -\sqrt{\frac{4(1+p^2)^3(1+q^2) - 4p^2q^2(1+p^2)^2}{4(1+p^2)(1+q^2)}} = \sqrt{\frac{(1+p^2)(1+p^2+q^2)}{(1+q^2)}}. \quad (A.88)$$

Now calculate:

$$AB + CE = -(1+p^2+q^2) + p^2q^2 = p^2q^2 - 1 - p^2 - q^2 = f, \quad (A.89)$$

that is, the last equation in Eq. (A.81) is automatically satisfied, and therefore the polynomial α may be represented as a sum of squares (Eq. A.80).

Substituting the expressions for A, B, C, D, and E (Eqs. A.84 to A.88) into the formula (Eq. A.80) for α, then substituting α into the formula (Eq. A.76) for unsphericity, and performing corresponding transformations, one obtains a formula convenient for calculations:

$$M = \sqrt{\frac{\left(r\sqrt{\frac{1+q^2}{1+p^2}} - t\sqrt{\frac{1+p^2}{1+q^2}}\right)^2 (1+p^2+q^2) + \left(pqr\sqrt{\frac{1+q^2}{1+p^2}} - 2s\sqrt{(1+q^2)(1+p^2)} + pqt\sqrt{\frac{1+p^2}{1+q^2}}\right)^2}{4(1+p^2+q^2)^3}}.$$

(A.90)

A.3.7 Rotor

The term *rotor* is defined as the flow line curvature (Shary, 1991). To deduce its formula, one may determine a flow line curvature $|r''_\sigma|$ and compare it to $|\mathrm{rot}(r'_\sigma)|$, where r'_σ is a unit tangent vector to a flow line, as defined by Eq. (A.26). It is seen from Eq. (A.26) that

$$\frac{dx}{d\sigma} = -\frac{p}{\sqrt{p^2+q^2}}, \quad \frac{dy}{d\sigma} = -\frac{q}{\sqrt{p^2+q^2}} \quad (A.91)$$

for N. Therefore,

$$\frac{dp}{d\sigma} = r\frac{dx}{d\sigma} + s\frac{dy}{d\sigma} = -\frac{pr+qs}{\sqrt{p^2+q^2}}, \quad (A.92)$$

$$\frac{dq}{d\sigma} = s\frac{dx}{d\sigma} + t\frac{dy}{d\sigma} = -\frac{ps+qt}{\sqrt{p^2+q^2}} \quad (A.93)$$

for N. So,

$$\frac{d^2x}{d\sigma^2} = -\frac{\left[-(pr+qs) - \frac{p\left(p\frac{dp}{d\sigma} + q\frac{dq}{d\sigma}\right)}{\sqrt{p^2+q^2}}\right]}{p^2+q^2}$$

$$= -\frac{-\{-(pr+qs)(p^2+q^2)+p[p(pr+qs)+q(ps+qt)]\}}{(p^2+q^2)^2}$$

$$= \frac{p^3r+p^2qs+pq^2r+q^3s-p^3r-p^2qs-p^2qs-pq^2t}{(p^2+q^2)^2}$$

$$= \frac{pq^2(r-t)-qs(p^2+q^2)}{(p^2+q^2)^2} = \left(\frac{-q}{\sqrt{p^2+q^2}}\right)\frac{(p^2-q^2)s-pq(r-t)}{\sqrt{(p^2+q^2)^3}}, \quad (A.94)$$

$$\frac{d^2y}{d\sigma^2} = -\frac{-\left[-(ps+qt)-\frac{q\left(p\frac{dp}{d\sigma}+q\frac{dq}{d\sigma}\right)}{\sqrt{p^2+q^2}}\right]}{p^2+q^2}$$

$$= -\frac{-\{-(ps+qt)(p^2+q^2)+q[p(pr+qs)+q(ps+qt)]\}}{(p^2+q^2)^2}$$

$$= \frac{p^3s+p^2qt+pq^2s+q^3t-p^2qr-pq^2s-pq^2s-q^3t}{(p^2+q^2)^2}$$

$$= \frac{ps(p^2-q^2)-p^2q(r-t)}{(p^2+q^2)^2} = \left(\frac{p}{\sqrt{p^2+q^2}}\right)\frac{(p^2-q^2)s-pq(r-t)}{\sqrt{(p^2+q^2)^3}} \quad (A.95)$$

for N. Comparing Eqs. (A.94), (A.95), and (A.25), one finds that

$$\mathbf{r}''_\sigma = \pm\frac{(p^2-q^2)s-pq(r-t)}{\sqrt{(p^2+q^2)^3}}\mathbf{r}''_\xi \quad (A.96)$$

for N. The choice of sign is defined by the contour line orientation. It follows from here that flow line curvature is

$$|\mathbf{r}''_\sigma| = \left|\frac{(p^2-q^2)s-pq(r-t)}{\sqrt{(p^2+q^2)^3}}\right| \quad (A.97)$$

for N, where the sign of rotor is not taken into account.

Rotor is also a vector with components

$$\mathrm{rot}(\mathbf{a}) = \begin{vmatrix} \mathbf{i} & \mathbf{j} & \mathbf{k} \\ \frac{\partial}{\partial x} & \frac{\partial}{\partial y} & \frac{\partial}{\partial z} \\ a_x & a_y & a_z \end{vmatrix}. \quad (A.98)$$

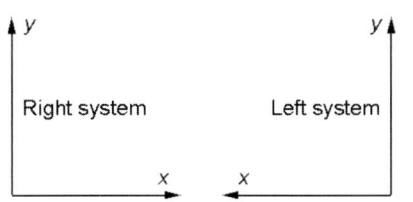

FIGURE A.2 Right- and left-handed coordinate systems.

Substituting here components of r'_σ (Eq. A.26) instead of a_x, a_y, and a_z, and considering their independence of z, one obtains for N:

$$\text{rot}(r'_\sigma) = -\left\{\frac{\partial}{\partial x}\left(\frac{q}{\sqrt{p^2+q^2}}\right) - \frac{\partial}{\partial y}\left(\frac{p}{\sqrt{p^2+q^2}}\right)\right\}k$$

$$= -\frac{s(p^2+q^2) - q(pr+qs) - s(p^2+q^2) + p(ps+qt)}{\sqrt{(p^2+q^2)^3}}k$$

$$= -\frac{(p^2-q^2)s - pq(r-t)}{\sqrt{(p^2+q^2)^3}}k. \tag{A.99}$$

Comparing this result and Eq. (A.97), one sees that for N

$$\text{rot}(r'_\sigma) = \pm |r''_\sigma|k. \tag{A.100}$$

According to the meaning of rotor, in a right-handed coordinate system (Fig. A.2) the flow line turns clockwise when the factor by which k is multiplied is negative. For this reason, the choice of sign for scalar rotor *rot* is opposite to the z-component of $\text{rot}(r'_\sigma)$. Therefore, one obtains that the flow line turns clockwise if *rot* is positive, while it turns counterclockwise in the opposite case. Finally,

$$rot = \frac{(p^2-q^2)s - pq(r-t)}{\sqrt{(p^2+q^2)^3}} \tag{A.101}$$

for N. Nevertheless, $\text{rot}(r'_\sigma)$ is an axial vector; therefore a negative sign for *rot* corresponds to a clockwise turn of the flow line in a left-handed coordinate system only (Fig. A.2), which is usually not used in geosciences.

A.3.8 Difference Curvature

Difference curvature, E, is defined by Shary (1995) as

$$E = \frac{1}{2}(k_v - k_h) \tag{A.102}$$

for N. Using formula (Eq. A.59), one obtains that $k_v = 2H - k_h$ and $E = H - k_h$. It follows from here and from Eqs. (A.60) and (A.58) that for N

$$E = \frac{q^2 r - 2pqs + p^2 t}{(p^2 + q^2)\sqrt{1 + p^2 + q^2}} - \frac{(1 + q^2)r - 2pqs + (1 + p^2)t}{2\sqrt{(1 + p^2 + q^2)^3}}. \qquad (A.103)$$

It follows from Eqs. (A.102) and (A.59) that

$$k_v = H + E, \qquad (A.104)$$

$$k_h = H - E \qquad (A.105)$$

for N. It follows from Eqs. (A.70) and (A.72) that

$$k_{max} = H + M, \qquad (A.106)$$

$$k_{min} = H - M. \qquad (A.107)$$

According to Eqs. (A.68), (A.106), and (A.107),

$$K = H^2 - M^2. \qquad (A.108)$$

Let us consider an interrelationship between difference curvature, E, and unsphericity, M. According to the Euler theorem (Nikolsky, 1977a, § 7.24), the curvature k of a normal section turned to angle φ from the principal normal section, which has curvature k_{max}, is a function of k_{max}, k_{min}, and φ:

$$k(\varphi) = k_{max} \cos^2 \varphi + k_{min} \sin^2 \varphi. \qquad (A.109)$$

Let θ be the smallest angle between a vertical plane of the normal section with curvature k_v (to which the gradient line is tangential). As Theorem 2 states, the plane of the normal section with the curvature k_h is orthogonal to the plane of the normal section with the curvature k_v; that is, $k_h = k\,(\theta + \pi/2)$. It follows from here and from the Euler theorem (Nikolsky, 1977a, § 7.24) that

$$k_v = k_{max} \cos^2 \theta + k_{min} \sin^2 \theta, \qquad (A.110)$$

$$k_h = k_{max} \cos^2\left(\theta + \frac{\pi}{2}\right) + k_{min} \sin^2\left(\theta + \frac{\pi}{2}\right), \qquad (A.111)$$

One may transform the last formula as follows:

$$k_h = k_{max} \sin^2 \theta + k_{min} \cos^2 \theta. \qquad (A.112)$$

Subtracting Eq. (A.112) from Eq. (A.110), one obtains

$$k_v - k_h = k_{max}(\cos^2 \theta - \sin^2 \theta) + k_{min}(\sin^2 \theta - \cos^2 \theta)$$

$$= (k_{max} - k_{min})(\cos^2 \theta - \sin^2 \theta) = (k_{max} - k_{min}) \cos(2\theta). \qquad (A.113)$$

It follows from here and from the definitions (Eqs. A.72 and A.102) that for N

$$E = M\cos(2\theta), \qquad (A.114)$$

$$|E| \leq M. \qquad (A.115)$$

A.3.9 Horizontal and Vertical Excess Curvatures

Shary (1995) defines horizontal and vertical excess curvatures, k_{he} and k_{ve}, as

$$k_{he} = k_h - k_{\min}, \qquad (A.116)$$

$$k_{ve} = k_v - k_{\min} \qquad (A.117)$$

for N. These curvatures are nonnegative, because $k_{\min}$ is not greater than a curvature of any other normal section at the same point of the surface S. Using these definitions and the formulas (Eqs. A.104 to A.107), one obtains that for N

$$k_{ve} = M + E, \qquad (A.118)$$

$$k_{he} = M - E. \qquad (A.119)$$

A.3.10 Total Ring Curvature

Total ring curvature, K_r, is defined as the product of k_{he} by k_{ve} (Shary, 1995). It follows from Eqs. (A.118) and (A.119) that

$$K_r = k_{ve}k_{he} = (M+E)(M-E) = M^2 - E^2 \qquad (A.120)$$

for N. Using this formula together with Eqs. (A.73) and (A.105), one finds that

$$K_r = 2Hk_h - k_h^2 - K \qquad (A.121)$$

for N, and substituting formulas (Eqs. A.58, A.60, and A.69) one obtains:

$$K_r = \frac{[(1+q^2)r - 2pqs + (1+p^2)t]\,(q^2r - 2pqs + p^2t)}{(p^2+q^2)(1+p^2+q^2)^2}$$

$$- \frac{(q^2r - 2pqs + p^2t)^2}{(p^2+q^2)^2(1+p^2+q^2)} - \frac{rt - s^2}{(1+p^2+q^2)^2} \qquad (A.122)$$

for N. Let $k'_h = q^2 r - 2pqs + p^2 t$, then the numerator of K_r is

$$(p^2 + q^2)(r+t)k'_h + (p^2 + q^2)(k'_h)^2 - (k'_h)^2 - (p^2 + q^2)(k'_h)^2 - (p^2 + q^2)^2(rt - s^2)$$
$$= (p^2 + q^2)(r+t)(q^2 r - 2pqs + p^2 t) - (q^2 r - 2pqs + p^2 t)^2 - (p^2 + q^2)^2(rt - s^2)$$
$$= (q^2 r - 2pqs + p^2 t)[(p^2 + q^2)(r+t) - q^2 r + 2pqs - p^2 t] - (p^2 + q^2)^2(rt - s^2)$$
$$= (q^2 r - 2pqs + p^2 t)(p^2 r + 2pqs + q^2 t) - (p^2 + q^2)^2(rt - s^2)$$
$$= p^2 q^2 r^2 - 2p^3 qsr + p^4 rt + 2pq^3 sr - 4p^2 q^2 s^2 + 2p^3 qst + q^4 rt - 2pq^3 st$$
$$\quad + p^2 q^2 t^2 - (p^4 + 2p^2 q^2 + q^4)(rt - s^2)$$
$$= p^2 q^2 r^2 - 2pq(p^2 - q^2)sr + (p^4 + q^4)rt + 2pq(p^2 - q^2)st - 4p^2 q^2 s^2 + p^2 q^2 t^2$$
$$\quad - (p^4 + 2p^2 q^2 + q^4)rt + (p^4 + 2p^2 q^2 + q^4)s^2$$
$$= p^2 q^2 r^2 - 2pq(p^2 - q^2)sr - 2p^2 q^2 rt + 2pq(p^2 - q^2)st + (p^2 - q^2)^2 s^2 + p^2 q^2 t^2$$
$$= p^2 q^2 (r - t)^2 - 2pq(p^2 - q^2)s(r - t) + (p^2 - q^2)^2 s^2$$
$$= \left[(p^2 - q^2)s - pq(r - t)\right]^2. \tag{A.123}$$

Therefore,

$$K_r = \left[\frac{(p^2 - q^2)s - pq(r - t)}{(p^2 + q^2)(1 + p^2 + q^2)}\right]^2 \tag{A.124}$$

for N. Comparing this result and the rotor formula (Eq. A.101), one finds

$$K_r = \frac{\overline{G}^2}{(1 + \overline{G}^2)^2} rot^2 \tag{A.125}$$

for N, so total ring curvature is proportional to rot^2, with a nonnegative factor depending on gradient only. This means that K_r describes flow line twisting (i.e., one of the characteristics of terrain dissection), but unlike rotor, K_r does not take into account a direction of a flow line turn (Shary et al., 2002b).

THEOREM 6.

In any subset of N, where elevation may be expressed as a function of $\sqrt{(x - x_0)^2 + (y - y_0)^2}$, total ring curvature K_r is equal to zero (Shary, 1995).

THE PROOF.

Let ρ, ϕ be plane polar coordinates with the pole at a given point (x_0, y_0). Then elevation $z(x, y) = z(\rho)$. For this function

$$E = -\frac{R - \frac{P}{\rho}(1 + P^2)}{2\sqrt{(1 + P^2)^3}}, \qquad (A.126)$$

$$H = -\frac{\left|R - \frac{P}{\rho}(1 + P^2)\right|}{2\sqrt{(1 + P^2)^3}}, \qquad (A.127)$$

where $R = \frac{d^2 z}{d\rho^2}$, $P = \frac{dz}{d\rho}$. From here it follows that $M = |E|$. So, according to Eq. (A.120),

$$K_r = (|E| + E)(|E| - E) = 0 \qquad (A.128)$$

for N. The theorem is proven.

A.3.11 Total Accumulation Curvature

Total accumulation curvature, K_a, is defined as the product of vertical and horizontal curvatures, (Shary, 1995):

$$K_a = k_h k_v. \qquad (A.129)$$

Using formulas (Eqs. A.58 and A.61), one obtains:

$$K_a = k_h k_v = \frac{(q^2 r - 2pqs + p^2 t)(p^2 r + 2pqs + q^2 t)}{[(p^2 + q^2)(1 + p^2 + q^2)]^2} \qquad (A.130)$$

for N. Using formulas (Eqs. A.104 and A.105), one obtains for N:

$$K_a = H^2 - E^2. \qquad (A.131)$$

THEOREM 7.

In any subset of N, where elevation z may be expressed as a function of polar angle ϕ with a pole at some point (x_0, y_0), total accumulation curvature is equal to zero (Shary, 1995).

THE PROOF.

In this case, $z(x, y) = z(\phi)$. For this function

$$H = E = \frac{-T}{2\rho^2 \sqrt{\left(\frac{(1+Q^2)}{\rho^2}\right)^3}}, \quad (A.132)$$

where $T = \frac{d^2z}{d\phi^2}$, $Q = \frac{dz}{d\phi}$. It follows from Eq. (A.131) and from $H = E$ that

$$K_a = (H + E)(H - E) = 0 \quad (A.133)$$

for N. The theorem is proven.

THEOREM 8
(Theorem of total curvatures)

$K_a = K + K_r$ for N (Shary, 1995).

THE PROOF.

It was proved above that $K_a = H^2 - E^2$, $K = H^2 - M^2$, and $K_r = M^2 - E^2$ for N (Eqs. A.131, A.108, and A.120). The proof follows from these formulas. The theorem is proven.

The smooth surface S is two-dimensional; therefore it needs two curvatures for description of its local shape. However, an additional curvature is needed to describe gravitational effects (Koenderink and van Doorn, 1994). It follows from the above results that three curvatures—H, E, and M—are independent and can be chosen for description of S by curvatures (Shary, 1995). Most curvatures may be expressed by simple formulas through these independent curvatures, as summarized in Table A.1. Note that an algebraic sum of any two simple curvatures

TABLE A.1 Relationships between three independent curvatures (H, E, and M) and other curvatures (K, K_a, K_r, k_{max}, k_{min}, k_h, k_v, k_{he}, and k_{ve})

Simple Curvatures (units are m^{-1})		Total Curvatures (units are m^{-2})
$k_{max} = H + M$	$k_h = H - E$	$K = H^2 - M^2$
$k_{min} = H - M$	$k_{ve} = M + E$	$K_a = H^2 - E^2$
$k_v = H + E$	$k_{he} = M - E$	$K_r = M^2 - E^2$

From (Shary, 2006, Table IV).

results in a curvature of the same system. That is why this system of curvatures may by considered as complete (Shary, 1995).

A.4 GENERATING FUNCTION

The generating function of crest and thalwegs lines, T, can be defined as the derivative of k_h (Shary and Stepanov, 1991):

$$T = \frac{dk_h}{d\xi}, \qquad (A.134)$$

where ξ is a contour line length. From Eqs. (A.134) and (A.58) follows:

$$\frac{dk_h}{d\xi} = \frac{\partial k_h}{\partial p}\frac{dp}{d\xi} + \frac{\partial k_h}{\partial q}\frac{dq}{d\xi} + \frac{\partial k_h}{\partial r}\frac{dr}{d\xi} + \frac{\partial k_h}{\partial s}\frac{ds}{d\xi} + \frac{\partial k_h}{\partial t}\frac{dt}{d\xi}. \qquad (A.135)$$

See formulas of $\frac{dp}{d\xi}$ and $\frac{dq}{d\xi}$ in (Eqs. A.46). Similarly,

$$\frac{dr}{d\xi} = g\frac{dx}{d\xi} + k\frac{dy}{d\xi}, \quad \frac{ds}{d\xi} = k\frac{dx}{d\xi} + m\frac{dy}{d\xi}, \quad \frac{dt}{d\xi} = m\frac{dx}{d\xi} + h\frac{dy}{d\xi}, \qquad (A.136)$$

for N, where

$$g = \frac{\partial^3 z}{\partial x^3}, \quad h = \frac{\partial^3 z}{\partial y^3}, \quad k = \frac{\partial^3 z}{\partial x^2 \partial y}, \quad m = \frac{\partial^3 z}{\partial x \partial y^2}. \qquad (A.137)$$

See formulas of $\frac{dx}{d\xi}$ and $\frac{dy}{d\xi}$ in Eqs. (A.45).

After differentiation, the terms in Eq.(A.135) are as follows (the sign depends on the contour line orientation):

$$\frac{\partial k_h}{\partial p}\frac{dp}{d\xi} = \pm \frac{2(pt-qs)(ps-qr)}{\sqrt{(p^2+q^2)^3}\sqrt{1+p^2+q^2}} \pm \frac{(ps-qr)p[2+3(p^2+q^2)](q^2r-2pqs+p^2t)}{\sqrt{(p^2+q^2)^5}\sqrt{(1+p^2+q^2)^3}}, \qquad (A.138)$$

$$\frac{\partial k_h}{\partial q}\frac{dq}{d\xi} = \pm \frac{-2(pt-qs)(ps-qr)}{\sqrt{(p^2+q^2)^3}\sqrt{1+p^2+q^2}} \pm \frac{(pt-qs)q[2+3(p^2+q^2)](q^2r-2pqs+p^2t)}{\sqrt{(p^2+q^2)^5}\sqrt{(1+p^2+q^2)^3}}, \qquad (A.139)$$

$$\frac{\partial k_h}{\partial r}\frac{dr}{d\xi} = \pm \frac{q^2(qg-pk)}{\sqrt{(p^2+q^2)^3}\sqrt{1+p^2+q^2}}, \qquad (A.140)$$

$$\frac{\partial k_h}{\partial s}\frac{ds}{d\xi} = \pm \frac{2pq(pm-qk)}{\sqrt{(p^2+q^2)^3}\sqrt{1+p^2+q^2}}, \qquad (A.141)$$

$$\frac{\partial k_h}{\partial t}\frac{dt}{d\xi} = \pm \frac{-p^2(ph-qm)}{\sqrt{(p^2+q^2)^3}\sqrt{1+p^2+q^2}}. \qquad (A.142)$$

for N. The sum of these terms is:

$$T = \frac{dk_h}{d\xi} = \frac{\partial k_h}{\partial p}\frac{dp}{d\xi} + \frac{\partial k_h}{\partial q}\frac{dq}{d\xi} + \frac{\partial k_h}{\partial r}\frac{dr}{d\xi} + \frac{\partial k_h}{\partial s}\frac{ds}{d\xi} + \frac{\partial k_h}{\partial t}\frac{dt}{d\xi}$$

$$= \pm \frac{(ps-qr)p[2+3(p^2+q^2)](q^2r-2pqs+p^2t)}{\sqrt{(p^2+q^2)^5(1+p^2+q^2)^3}}$$

$$\pm \frac{(pt-qs)q[2+3(p^2+q^2)](q^2r-2pqs+p^2t)}{\sqrt{(p^2+q^2)^5(1+p^2+q^2)^3}}$$

$$\pm \frac{q^2(qg-pk)+2pq(pm-qk)-p^2(ph-qm)}{\sqrt{(p^2+q^2)^3(1+p^2+q^2)}} \quad \text{(A.143)}$$

for N. After simple algebraic operations, one obtains:

$$T = \pm \frac{q^3g - 3pq^2k + 3p^2qm - p^3h}{\sqrt{(p^2+q^2)^3(1+p^2+q^2)}}$$

$$\pm \frac{[2+3(p^2+q^2)](q^2r-2pqs+p^2t)}{\sqrt{(p^2+q^2)^5(1+p^2+q^2)^3}} \left[-pqr + (p^2-q^2)s + pqt\right] \quad \text{(A.144)}$$

for N. The sign of T is not important because T is zero in concave and convex spurs, where $|k_h|$ is maximal, so that T (or rather its zero values) might be considered as the generating function of crest and thalweg lines in a local approximation. The formula Eq.(A.144) may be rewritten for the sign " + " (Florinsky, 2009b):

$$T = \frac{q^3g - 3pq^2k + 3p^2qm - p^3h}{\sqrt{(p^2+q^2)^3(1+p^2+q^2)}}$$

$$+ \frac{[2+3(p^2+q^2)](q^2r-2pqs+p^2t)}{\sqrt{(p^2+q^2)^5(1+p^2+q^2)^3}} \left[(p^2-q^2)s - pq(r-t)\right] \quad \text{(A.145)}$$

for N. Considering the formulas for k_h and rot (Eqs. A.58 and A.101), one finally obtains:

$$T = \frac{q^3g - 3pq^2k + 3p^2qm - p^3h}{\sqrt{(p^2+q^2)^3(1+p^2+q^2)}} - k_h rot \frac{2+3(p^2+q^2)}{1+p^2+q^2} \quad \text{(A.146)}$$

for N.

Appendix B: LandLord—A Brief Description of the Software

Currently, there are more than 20 computer programs that have been developed or can be used for digital terrain modeling (Wood, 2009). For details, interested readers are invited to look up the second part of the book by Hengl and Reuter (2009).

The software LandLord is intended for research-oriented digital terrain modeling. LandLord is designed for use on 32-bit personal computers under Windows 95/98/2000/NT/XP. Minimal hardware requirements are: 24 MB RAM, a 256 color monitor, screen resolution of 800×600 pixels. The software size is about 760 KB. Currently, LandLord includes two modules: LandLord 4.0 calculation module and LandLord Viewer 1.3.

For LandLord 4.0, initial data are DEMs based on plane square grids or spheroidal equal angular grids. DEMs are represented as binary and ASCII files (LandLord formats *.REG and *.RGC, respectively). One can also import some types of DEMs:

- Binary DEMs generated by the software Surfer (© Golden Software Inc.)
- Global binary DEMs from ETOPO2, GTOPO30, GLOBO, and SRTM3 digital archives, as well as planetary DEMs from NASA digital archives

LandLord 4.0 allows the following operations:

1. Calculation of digital models of local morphometric variables. Four methods intended for plane square grids can be applied: the Evans–Young method (Section 4.1), the Zevenbergen–Thorne method, the Shary method, and the author's method (Section 4.2). One other author's method (Section 4.3) is utilized to derive digital models of local morphometric variables on spheroidal equal angular grids. The methods are used to calculate the following variables and models:
    a. All variables of the complete system of curvatures (Sections 2.2 and A.3).
    b. Some other local topographic attributes: slope gradient (Sections 2.2 and A.1.1), slope aspect (Sections 2.2 and A.2.1), rotor (Section A.3.7), plan curvature (Section A.3.1), generating function (Sections 4.5.2 and A.4), and so on.
    c. Insolation (Sections 2.5 and A.2.2).

d. Landscape segmentation models using the Gaussian classification (Section 2.7.1), the Efremov–Krcho classification, the concept of accumulation zones (Section 2.7.2), and the Shary classification (Section 2.7.3).
2. Calculation of digital models of nonlocal morphometric variables. On plane square grids, they are derived by the Martz–de Jong method (Section 4.4). This method, adapted to the geometry of spheroidal trapezoids (Section 4.3.3), is used to process DEMs on spheroidal equal angular grids. The following morphometric variables can be derived:
   a. Minimal and maximal (specific) catchment area (Sections 2.3 and 4.4)
   b. Minimal and maximal (specific) dispersive area (Sections 2.3 and 4.4)
   c. Minimal and maximal topographic index (Sections 2.6 and 4.4)
   d. Minimal and maximal stream power index (Sections 2.6 and 4.4)
3. DTM smoothing on plane square grids and spheroidal equal angular grids:
   a. Smoothing by a moving weighted average (Section 6.2.5)
   b. Depression filling by the Martz–de Jong method (Section 4.4)
4. Some simple mathematical operations: addition, subtraction, multiplication, division, taking logarithm, exponentiation, as well as area and volume estimation.
5. Statistical estimations:
   a. Calculation of predictive digital soil models using linear regression equations with DTMs as predictors (Section 10.4)
   b. Calculation of RMSE digital models for all variables of the complete system of curvatures, slope gradient, aspect, plan curvature, rotor, and generating function (Section 5.3). Calculation can be carried out on plane square grids and spheroidal equal angular grids.
6. Export of DEMs from LandLord formats *.REG and *.RGC to ASCII file, ArcInfo ASCII format *.ASC, and Surfer binary format *.GRD.

LandLord Viewer 1.3 allows visualizing DTMs from LandLord files *.REG using several color and gray scales. In the book, all author's maps were generated by this viewer.

Early versions (1993–2001) of LandLord calculation modules were written in C and Delphi by T.I. Grokhlina, N.L. Mikhailova, and P.V. Kozlov, under the supervision of the author (Florinsky et al., 1995). LandLord Viewer 1.3 was developed using C++ by G.L. Andrienko and N.V. Andrienko. Starting in the year 2002, the author has developed the software using Delphi.

References

Aandahl, A. R. (1948). The characterization of slope positions and their influence on the total nitrogen content of a few virgin soils of western Iowa. *Soil Science Society of America Proceedings, 13*, 449–454.

Abdelguerfi, M., Wynne, C., Cooper, E., Roy, L., & Shaw, K. (1998). Representation of 3-D elevation in terrain databases using hierarchical triangulated irregular networks: A comparative analysis. *International Journal of Geographical Information Science, 12*, 853–873.

Abelsky, A. M., & Lastochkin, A. N. (1969). Revealing and analysis of relief-forming wave-like deformations by a method of directional stacking with simultaneous frequency filtering (exemplified by the northeast of the Siberian Platform). *Uchenye Zapiski NII Geologii Arktiki, Regional Geology Series, 14*, 101–108 (in Russian).

Acker, W. L., III (1974). *Basic procedures for soil sampling and core drilling* (246 p.). Scranton, PA: Acker Drill Co.

Ackermann, F. (1978). Experimental investigation into the accuracy of contouring from DTM. *Photogrammetric Engineering and Remote Sensing, 44*, 1537–1548.

Adiyaman, Ö., Chorowicz, J., Arnaud, O. N., Gündogdu, M. N., & Gourgaud, A. (2001). Late Cenozoic tectonics and volcanism along the North Anatolian Fault: New structural and geochemical data. *Tectonophysics, 338*, 135–165.

Afanasiev, J. N. (1927). *The classification problem in Russian soil science* (51 p.). Leningrad: Academy of Sciences of the USSR.

Agishtein, M. E., & Migdal, A. A. (1991). Smooth surface reconstruction from scattered data points. *Computers and Graphics, 15*, 29–39.

Aguilar, F. J., Agüera, F., Aguilar, M. A., & Carvajal, F. (2005). Effects of terrain morphology, sampling density, and interpolation methods on grid DEM accuracy. *Photogrammetric Engineering and Remote Sensing, 71*, 805–816.

Aguilar, F. J., & Mills, J. P. (2008). Accuracy assessment of lidar-derived digital elevation models. *Photogrammetric Record, 23*, 148–169.

Ai, T., & Li, J. (2010). A DEM generalization by minor valley branch detection and grid filling. *ISPRS Journal of Photogrammetry and Remote Sensing, 65*, 198–207.

Akeno, K. (1996). DEM generation from multisensor stereopairs—AVHRR and MSS. *International Archives of Photogrammetry and Remote Sensing, 31*(B4), 36–40.

Akima, H. (1974). A method of bivariate interpolation and smooth surface fitting based on local procedures. *Communications of the ACM, 17*, 18–20.

Akovetsky, V. G. (1994). Efficiency improvement of stereoscopic measurements. *Geodezia i Cartografia*, No. 1, 29–33 (in Russian).

Albani, M., & Klinkenberg, B. (2003). A spatial filter for the removal of striping artifacts in digital elevation models. *Photogrammetric Engineering and Remote Sensing, 69*, 755–765.

Albani, M., Klinkenberg, B., Andison, D. W., & Kimmins, J. P. (2004). The choice of window size in approximating topographic surfaces from Digital Elevation Models. *International Journal of Geographical Information Science, 18*, 577–593.

Albee, A. L. (2000). Mars 2000. *Annual Review of Earth and Planetary Sciences, 28*, 281–304.

Al-Rousan, N., Cheng, P., Petrie, G., Toutin, T., & Valadan Zoej, M. J. (1997). Automated DEM extraction and orthoimage generation from SPOT Level 1B imagery. *Photogrammetric Engineering and Remote Sensing, 63*, 965–974.

Amante, C., & Eakins, B. W. (2009). *ETOPO1 1 arc-minute global relief model: Procedures, data sources and analysis* (19 p.). NOAA Technical Memorandum NESDIS NGDC-24.

Ames, W. F. (1977). *Numerical methods for partial differential equations* (2nd ed., 365 p.). New York: Academic Press.

Anderson, M. G., & Burt, T. P. (1978). The role of topography in controlling throughflow generation. *Earth Surface Processes, 3*, 331–344.

Anderson, M. G., & Burt, T. P. (1980). The role of topography in controlling throughflow generation: A reply. *Earth Surface Processes, 5*, 193–195.

Anisimov, I. G., Bordunov, A. A., Deeva, N. F., Kamalov, L. F., Lopachev, N. A., Muratova, V. S., et al. (1977). A technique for compilation of the series of medium-scale thematic maps "Natural and Reclamation Estimation of the Mid-Region of the USSR.". In I. N. Stepanov (Ed.), *Estimation of natural and reclamation conditions and prediction of their changes (Exemplified by Central Asia)* (pp. 23–93). Pushchino, USSR: Biological Research Centre Press (in Russian).

Anokhin, V. M., & Odesskii, I. A. (2001). Characteristics of the global pattern of planetary fracturing. *Geotectonics, 35*, 335–340.

Ansell, J. H., & Bannister, S. C. (1996). Shallow morphology of the subducted Pacific plate along the Hikurangi margin, New Zealand. *Physics of the Earth and Planetary Interiors, 93*, 3–20.

Arabelos, D. (2000). Intercomparisons of the global DTMs ETOPO5, TerrainBase and JGP95E. *Physics and Chemistry of the Earth (A), 25*, 89–93.

Armstrong, R. N., & Martz, L. W. (2003). Topographic parameterization in continental hydrology: A study in scale. *Hydrological Processes, 17*, 3763–3781.

Arnold, N. (2010). A new approach for dealing with depressions in digital elevation models when calculating flow accumulation values. *Progress in Physical Geography, 34*, 781–809.

Arrell, K., Wise, S., Wood, J., & Donoghue, D. (2008). Spectral filtering as a method of visualising and removing striped artefacts in digital elevation data. *Earth Surface Processes and Landforms, 33*, 943–961.

Arrouays, D., Daroussin, J., Kicin, J. L., & Hassika, P. (1998). Improving topsoil carbon storage prediction using a digital elevation model in temperate forest soils of France. *Soil Science, 163*, 103–108.

Audet, P. (2011). Directional wavelet analysis on the sphere: Application to gravity and topography of the terrestrial planets. *Journal of Geophysical Research, 116*, E01003, doi:10.1029/2010JE003710.

Australian Drilling Industry Training Committee (1997). *Drilling: The manual of methods, applications, and management* (4th ed., 615 p.). Boca Raton, FL: CRC Press.

Babiker, M., & Gudmundsson, A. (2004). The effects of dykes and faults on groundwater flow in an arid land: The Red Sea Hills, Sudan. *Journal of Hydrology, 297*, 256–273.

Baker, K. D. (1982). Basic image processing concepts. In N. B. Jones (Ed.), *Digital signal processing* (pp. 287–318). Stevenage, England: Peter Peregrinus.

Balassanian, S. Y. (2005). Sensitive energy active points (SEAP): A clue to better understanding of earthquake physics? *Russian Geology and Geophysics, 46*, 85–100.

Balce, A. E. (1987). Determination of optimum sampling interval in grid digital elevation models (DEM) data acquisition. *Photogrammetric Engineering and Remote Sensing, 53*, 323–330.

Baldi, P., Bonvalot, S., Briole, P., & Marsella, M. (2000). Digital photogrammetry and kinematic GPS applied to the monitoring of Vulcano Island, Aeolian Arc, Italy. *Geophysical Journal International, 142*, 801–811.

Ballabio, C. (2009). Spatial prediction of soil properties in temperate mountain regions using support vector regression. *Geoderma, 151*, 338–350.

Ballabio, C., & Comolli, R. (2010). Mapping heavy metal content in soils with multi-kernel SVR and LiDAR derived data. In J. L. Boettinger, D. W. Howell, A. C. Moore,

A. E. Hartemink, & S. Kienast-Brown (Eds.), *Digital soil mapping—Bridging research, environmental application, and operation* (pp. 205–216). Dordrecht: Springer.

Baltsavias, E. P. (1999). A comparison between photogrammetry and laser scanning. *ISPRS Journal of Photogrammetry and Remote Sensing, 54*, 83–94.

Bamber, J. L., Layberry, R. L., & Gogineni, S. P. (2001). A new ice thickness and bed data set for the Greenland ice sheet. 1: Measurement, data reduction, and errors. *Journal of Geophysical Research, 106*(D), 33773–33780.

Band, L. E. (1986). Topographic partition of watersheds with digital elevation models. *Water Resources Research, 22*, 15–24.

Band, L. E. (1993). Extraction of channel networks and topographic parameters from digital elevation data. In K. Beven, & M. J. Kirkby (Eds.), *Channel network hydrology* (pp. 13–42). Chichester, England: Wiley.

Band, L. E., & Moore, I. D. (1995). Scale: Landscape attributes and geographical information systems. *Hydrological Processes, 9*, 401–422.

Band, L. E., Patterson, P., Nemani, R., & Running, S. W. (1993). Forest ecosystem processes at the watershed scale: Incorporating hillslope hydrology. *Agricultural and Forest Meteorology, 63*, 93–126.

Baransky, N. N. (1946). Generalization in cartography and geographical textual description. *Uchenye Zapiski Moscovskogo Universiteta, 119*(2), 180–205 (in Russian).

Barringer, J. R. F., Hewitt, A. E., Lynn, I. H., & Schmidt, J. (2008). National mapping of landform elements in support of S-Map, a New Zealand soils database. In Q. Zhou, B. Lees, & G.-A. Tang (Eds.), *Advances in digital terrain analysis* (pp. 443–458). Berlin: Springer.

Barton, K. E., Howell, D. G., Vigil, J. F., Reed, J. C., & Wheeler, J. O. (2003). *The North America tapestry of time and terrain. Geologic investigations series map I-2781, scale 1 : 8,000,000.* U.S. Geological Survey.

Bastida, F., Aller, J., & Bobillo-Ares, N. C. (1999). Geometrical analysis of folded surfaces using simple functions. *Journal of Structural Geology, 21*, 729–742.

Bater, C. W., & Coops, N. C. (2009). Evaluating error associated with lidar-derived DEM interpolation. *Computers and Geosciences, 35*, 289–300.

Batschelet, E. (1981). *Circular statistics in biology* (371 p.). London: Academic Press.

Batson, R. M., Edwards, K., & Eliason, E. M. (1975). Computer-generated shaded-relief images. *Journal of Research of the US Geological Survey, 3*, 401–408.

Baxter, S. J., & Oliver, M. A. (2005). The spatial prediction of soil mineral N and potentially available N using elevation. *Geoderma, 128*, 325–339.

Beauchamp, E. G., & Bergstrom, D. W. (1993). Denitrification. In M. R. Carter (Ed.), *Soil sampling and methods of analysis* (pp. 351–357). Boca Raton, FL: Lewis.

Becker, J. J., Sandwell, D. T., Smith, W. H. F., Braud, J., Binder, B., Depner, J., et al. (2009). Global bathymetry and elevation data at 30 arc seconds resolution: SRTM30_PLUS. *Marine Geodesy, 32*, 355–371.

Beckett, P. H. T., & Webster, R. (1971). Soil variability: A review. *Soils and Fertilizers, 34*, 1–15.

Bedard-Haughn, A. K., & Pennock, D. J. (2002). Terrain controls on depressional soil distribution in a hummocky morainal landscape. *Geoderma, 110*, 169–190.

Behrens, T., Förster, H., Scholten, T., Steinrücken, U., Spies, E.-D., & Goldschmitt, M. (2005). Digital soil mapping using artificial neural networks. *Journal of Plant Nutrition and Soil Science, 168*, 21–33.

Behrens, T., & Scholten, T. (2007). A comparison of data-mining techniques in predictive soil mapping. In P. Lagacherie, A. B. McBratney, & M. Voltz (Eds.), *Digital soil mapping. An introductory perspective* (pp. 353–364). Amsterdam: Elsevier.

Behrens, T., Zhu, A-X., Schmidt, K., & Scholten, T. (2010). Multi-scale digital terrain analysis and feature selection for digital soil mapping. *Geoderma, 155*, 175–185.

Bell, J. C., Cunningham, R. L., & Havens, M. W. (1992). Calibration and validation of a soil-landscape model for predicting soil drainage class. *Soil Science Society of America Journal, 56*, 1860–1866.

Bell, J. C., Cunningham, R. L., & Havens, M. W. (1994a). Soil drainage class probability mapping using a soil-landscape model. *Soil Science Society of America Journal, 58*, 464–470.

Bell, J. C., Grigal, D. F., & Bates, P. C. (2000). A soil–terrain model for estimating spatial patterns of soil organic carbon. In J. Wilson, & J. Gallant (Eds.), *Terrain analysis: Principles and applications* (pp. 295–310). New York: Wiley.

Bell, J. C., Thompson, J. A., Butler, C. A., & McSweeney, K. (1994b). Modeling soil genesis from a landscape perspective. In B.J.D. Etchevers (Ed.), *Transactions of the 15th World congress of soil science, July 1994, Acapulco, Mexico* (Vol. 6a, pp. 179–195). Mexico: ISSS.

Bellian, J. A., Kerans, C., & Jennette, D. D. (2005). Digital outcrop models: Applications of terrestrial scanning lidar technology in stratigraphic modeling. *Journal of Sedimentary Research, 75*, 166–176.

Belonin, M. D., & Zhukov, I. M. (1968). Geometrical properties of surfaces of the Alexeevskoye elevated block, Kuibyshev Region. In M. A. Romanova (Ed.), *Problems of mathematical geology* (pp. 194–207). Leningrad: Nauka (in Russian).

Bergbauer, S. (2007). Testing the predictive capability of curvature analyses. *Geological Society of London, Special Publications, 292*, 185–202.

Bergbauer, S., Mukerji, T., & Hennings, P. (2003). Improving curvature analyses of deformed horizons using scale-dependent filtering techniques. *American Association of Petroleum Geologists Bulletin, 87*, 1255–1272.

Bergbauer, S., & Pollard, D. D. (2003). How to calculate normal curvatures of sampled geological surfaces. *Journal of Structural Geology, 25*, 277–289.

Bergstrom, D. W., Monreal, C. M., & St. Jacques, E. (2001a). Influence of tillage practice on carbon sequestration is scale-dependent. *Canadian Journal of Soil Science, 81*, 63–70.

Bergstrom, D. W., Monreal, C. M., & St. Jacques, E. (2001b). Spatial dependence of soil organic carbon mass and its relationship to soil series and topography. *Canadian Journal of Soil Science, 81*, 53–62.

Berlyant, A. M. (1966). Cartographic methods to study contemporary tectonics and their classification. *Izvestiya Academii Nauk SSSR, Geographical Series*, No. 2, 71–80 (in Russian).

Berlyant, A. M. (1986). *Image of the space: Map and information* (240 p.). Moscow: Mysl (in Russian).

Bernardin, T., Cowgill, E., Kreylos, O., Bowles, C., Gold, P., Hamann, B., et al. (2011). Crusta: A new virtual globe for real-time visualization of sub-meter digital topography at planetary scales. *Computers and Geosciences, 37*, 75–85.

Besprozvanny, P. A., Borodzich, E. V., & Bush, V. A. (1994). Numerical analysis of ordering relations in the global network of lineaments. *Physics of the Solid Earth, 30*, 150–159.

Beven, K. (1997). TOPMODEL: A critique. *Hydrological Processes, 11*, 1069–1085.

Beven, K., & Freer, J. (2001). A dynamic TOPMODEL. *Hydrological Processes, 15*, 1993–2011.

Beven, K. J., & Kirkby, M. J. (1979). A physically-based variable contributing area model of basin hydrology. *Hydrological Science Bulletin, 24*, 43–69.

Beven, K. J., Kirkby, M. J., Schoffield, N., & Tagg, A. F. (1984). Testing a physically-based flood forecasting model (TOPMODEL) for three UK catchments. *Journal of Hydrology, 69*, 119–143.

Bevis, M. (1986). The curvature of Wadati–Benioff Zones and the torsional rigidity of subducting plates. *Nature, 323*, 52–53.

Bhatia, S. C., Chetti, T. R. K., Filimonov, M. V., Gorshkov, A. I., Rantsman, E. Y., & Rao, M. N. (1992). Identification of potential areas for the occurrence of strong earthquakes in Himalayan arc region. *Proceedings of the Indian Academy of Sciences (Earth and Planetary Sciences), 101*, 369–385.

Bierkens, M. F. P., Finke, P. A., & de Willigen, P. (Eds.), (2000). *Upscaling and downscaling methods for environmental research* (204 p.). Dordrecht: Kluwer.
Bindlish, R., & Barros, A. P. (1996). Aggregation of digital terrain data using a modified fractal interpolation scheme. *Computers and Geosciences, 22*, 907−917.
Bishop, T. F. A., & McBratney, A. B. (2001). A comparison of prediction methods for the creation of field-extent soil property maps. *Geoderma, 103*, 149−160.
Bishop, T. F. A., Minasny, B., & McBratney, A. B. (2006). Uncertainty analysis for soil−terrain models. *International Journal of Geographical Information Science, 20*, 117−134.
Bjerhammar, A. (1973). *Theory of errors and generalized matrix inverses* (420 p.). Amsterdam: Elsevier.
Bjørke, J. T., & Nilsen, S. (2003). Wavelets applied to simplification of digital terrain models. *International Journal of Geographical Information Science, 17*, 601−621.
Bjørke, J. T., & Nilsen, S. (2004). Examination of a constant-area quadrilateral grid in representation of global digital elevation models. *International Journal of Geographical Information Science, 18*, 653−664.
Blazkova, S., Beven, K. J., & Kulasova, A. (2002). On constraining TOPMODEL hydrograph simulations using partial saturated area information. *Hydrological Processes, 16*, 441−458.
Blondel, P., & Murton, B. J. (1997). *Handbook of seafloor sonar imagery* (314 p.). Chichester, England: Wiley−Praxis.
Böhner, J., & Antonić, O. (2009). Land-surface parameters specific to topo-climatology. In T. Hengl, & H. I. Reuter (Eds.), *Geomorphometry: Concepts, software, applications* (pp. 195−226). Amsterdam: Elsevier.
Bolshakov, V. D., & Gaidaev, P. A. (1977). *Theory of mathematical processing of geodetic measurements* (2nd rev enl ed., 367 p.). Moscow: Nedra (in Russian).
Bolstad, P. V., & Stowe, T. (1994). An evaluation of DEM accuracy: Elevation, slope, and aspect. *Photogrammetric Engineering and Remote Sensing, 60*, 1327−1332.
Borisenko, L. S. (1986). Geological criteria of seismic activity in the Crimea. *Seismological Researches, 9*, 38−48 (in Russian, with English abstract).
Bourennane, H., King, D., & Couturier, A. (2000). Comparison of kriging with external drift and simple linear regression for predicting soil horizon thickness with different sample densities. *Geoderma, 97*, 255−271.
Bourillet, J. F., Edy, C., Rambert, F., Satra, C., & Loubrieu, B. (1996). Swath mapping system processing: Bathymetry and cartography. *Marine Geophysical Researches, 18*, 487−506.
Boussinesq, M. J. (1872). Sur les lignes de faîte et de thalweg. *Comptes Rendus Hebdomadaires des Séances de l'Académie des Sciences, 75*, 835−837.
Bowring, B. R. (1996). Total inverse solutions for the geodesic and great elliptic. *Survey Review, 33*, 461−476.
Brassel, K. E., & Weibel, R. (1988). A review and conceptual framework of automated map generalization. *International Journal of Geographical Information Systems, 2*, 229−244.
Brenner, A. C., DiMarzio, J. P., & Zwally, H. J. (2007). Precision and accuracy of satellite radar and laser altimeter data over the continental ice sheets. *IEEE Transactions on Geoscience and Remote Sensing, 45*, 321−331.
Bretar, F., & Chehata, N. (2010). Terrain modeling from lidar range data in natural landscapes: A predictive and Bayesian framework. *IEEE Transactions on Geoscience and Remote Sensing, 48*, 1568−1578.
Breton de Champ, M. (1877). Mémoire sur les lignes de faîte et de thalweg que l'on est conduit à considérer en topographie. *Journal de Mathématique Pure et Appliquées, Series 3, 3*, 99−114.
Breunig, M. (1996). *Integration of spatial information for geo-information systems* (171 p.). Berlin: Springer.

Brocklehurst, S. H. (2010). Tectonics and geomorphology. *Progress in Physical Geography, 34*, 357–383.

Brovelli, M. A., Cannata, M., & Longoni, U. M. (2004). LIDAR data filtering and DTM interpolation within GRASS. *Transactions in GIS, 8*, 155–174.

Brown, A. R. (2004). *Interpretation of three-dimensional seismic data* (6th ed., 541 p.). Tulsa, OK: American Association of Petroleum Geologists.

Brown, D. G., & Bara, T. J. (1994). Recognition and reduction of systematic error in elevation and derivative surfaces from 7.5-minute DEMs. *Photogrammetric Engineering and Remote Sensing, 60*, 189–194.

Bryukhanov, V. N., & Mezhelovsky, N. V. (Eds.), (1987). *Geology of the USSR from space* (240 p.). Moscow: Nedra (in Russian).

Buckley, S. J., Howell, J. A., Enge, H. D., & Kurz, T. H. (2008). Terrestrial laser scanning in geology: Data acquisition, processing and accuracy considerations. *Journal of the Geological Society, 165*, 625–638.

Bui, E. N., & Moran, C. J. (2001). Disaggregation of polygons of surficial geology and soil maps using spatial modelling and legacy data. *Geoderma, 103*, 79–94.

Bui, E. N., & Moran, C. J. (2003). A strategy to fill gaps in soil survey over large spatial extents: An example from the Murray–Darling basin of Australia. *Geoderma, 111*, 21–44.

Buivydaite, V. V., & Mozgeris, G. (2007). Comparison of automated landform classification and soil mapping units at a farm level. *International Journal of Ecology and Development, 8*(F07), 26–38.

Burba, G. A. (1984). *Application of cartographic method to study topography of the terrestrial planets*. Ph.D. Thesis. Lomonosov Moscow State University, Moscow, 290 p. (in Russian).

Burbank, D. W., & Anderson, R. S. (2001). *Tectonic geomorphology* (274 p.). Malden, MA: Blackwell Science.

Bürgmann, R., Rosen, P. A., & Fielding, E. J. (2000). Synthetic aperture radar interferometry to measure Earth's surface topography and its deformation. *Annual Review of Earth and Planetary Sciences, 28*, 169–209.

Burrough, P. A. (1986). *Principles of geographical information systems for land resources assessment* (193 p.). Oxford: Clarendon Press.

Burrough, P. A. (1993). Soil variability: A late 20th century view. *Soils and Fertilizers, 56*, 529–562.

Burt, T. P., & Butcher, D. P. (1985). Topographic controls of soil moisture distributions. *Journal of Soil Science, 36*, 469–486.

Burton, D. L., McMahon, S. K., & Chen, Y. (2000). Influence of method of manure application on greenhouse gas emissions. In *Proceedings of the 43rd Annual meeting of the Manitoba society of soil science, 25–26 January 2000, Winnipeg, Canada* (pp. 22–30). Winnipeg, MB: MSSS.

Byrne, C. J. (2007). A large basin on the near side of the Moon. *Earth, Moon, and Planets, 101*, 153–188.

Cahill, T., & Isacks, B. L. (1992). Seismicity and shape of the subducted Nazca Plate. *Journal of Geophysical Research, 97*(B), 17503–17529.

Callow, J. N., van Niel, K. P., & Boggs, G. S. (2007). How does modifying a DEM to reflect known hydrology affect subsequent terrain analysis? *Journal of Hydrology, 332*, 30–39.

Campagna, D. J., & Levandowski, D. W. (1991). The recognition of strike-slip fault systems using imagery, gravity, and topographic data sets. *Photogrammetric Engineering and Remote Sensing, 57*, 1195–1201.

Campbell, J. B. (1979). Spatial variability of soils. *Annals of the Association of American Geographers, 69*, 544–556.

Carey, S. W. (1988). *Theories of the Earth and Universe: A history of dogma in the earth sciences* (413 p.). Stanford, CA: Stanford University Press.

Carlisle, B. H. (2005). Modelling the spatial distribution of DEM error. *Transactions in GIS, 9*, 521–540.
Carrara, A., Bitelli, G., & Carla, R. (1997). Comparison of techniques for generating digital terrain models from contour lines. *International Journal of Geographical Information Science, 11*, 451–473.
Carré, F., & Girard, M. C. (2002). Quantitative mapping of soil types based on regression kriging of taxonomic distances with landform and land cover attributes. *Geoderma, 110*, 241–263.
Carré, F., & McBratney, A. B. (2005). Digital terron mapping. *Geoderma, 128*, 340–353.
Carré, F., Reuter, H. I., Daroussin, J., & Scheurer, O. (2008). From a large to a small scale soil map: Top–down against bottom–up approaches. Application to the Aisne soil map (France). In A. E. Hartemink, A. McBratney, & M. L. Mendonça–Santos (Eds.), *Digital soil mapping with limited data* (pp. 203–212). Dordrecht: Springer.
Carson, M. A., & Kirkby, M. J. (1972). *Hillslope form and process* (475 p.). Cambridge: Cambridge University Press.
Carter, J. R. (1988). Digital representations of topographic surfaces. *Photogrammetric Engineering and Remote Sensing, 54*, 1577–1580.
Carter, J. R. (1992). The effect of data precision on the calculation of slope and aspect using gridded DEMs. *Cartographica, 29*, 22–34.
Carter, M. R. (Ed.), (1993). *Soil sampling and methods of analysis* (823 p.). Boca Raton, FL: Lewis.
Caumon, G., Collon-Drouaillet, P., Le Carlier de Veslud, C., Viseur, S., & Sausse, J. (2009). Surface-based 3D modeling of geological structures. *Mathematical Geosciences, 41*, 927–945.
Cayley, A. (1859). On contour and slope lines. *The London, Edinburgh and Dublin Philosophical Magazine and Journal of Science, Series 4, 18*, 264–268.
Central Board of Geodesy and Cartography (1953). *Topographic map, scale 1 : 300,000. Sheets: VI-L-36 (Dzhankoi), VIII-L-36 (Sevastopol), IX-L-36 (Simferopol), V-L-36 (Kherson).* Moscow: Central Board of Geodesy and Cartography (in Russian).
Central Board of Geodesy and Cartography (1968). *Topographic map, scale 1 : 1,000,000. Sheet L-38 (Pyatigorsk).* Moscow: Central Board of Geodesy and Cartography (in Russian).
Chai, X., Shen, C., Yuan, X., & Huang, Y. (2008). Spatial prediction of soil organic matter in the presence of different external trends with REML-EBLUP. *Geoderma, 148*, 159–166.
Chang, K.-T., & Tsai, B.-W. (1991). The effect of DEM resolution on slope and aspect mapping. *Cartography and Geographic Information Systems, 18*, 69–77.
Chaplot, V., Bernoux, M., Walter, C., Curmi, P., & Herpin, U. (2001). Soil carbon storage prediction in temperate hydromorphic soils using a morphologic index and digital elevation model. *Soil Science, 166*, 48–60.
Chaplot, V., & Walter, C. (2003). Subsurface topography to enhance the prediction of the spatial distribution of soil wetness. *Hydrological Processes, 17*, 2567–2580.
Chaplot, V., Walter, C., & Curmi, P. (2000). Improving soil hydromorphy prediction according to DEM resolution and available pedological data. *Geoderma, 97*, 405–422.
Chebanenko, I. I. (1963). *Principal regularities of fault tectonics of the Earth's crust and its problems* (155 p.). Kiev: Ukrainian Academic Press (in Russian).
Chebanenko, I. I., & Fedorin, Y. V. (1983). On a new type of rotational tectonic lines in the Earth's lithosphere. *Doklady Akademii Nauk SSSR, 270*, 406–409 (in Russian).
Chekunov, A. V., Garkalenko, I. A., & Kharchenko, G. E. (1965). Ancient Precambrian faults of the southern Russian platform and their extension in the northern Black Sea Coast area. *Geophysical Communications, 14*, 24–34 (in Russian).
Chekunov, A. V., Gavrilenko, N. M., Shnyukov, E. F., Tikhonenkov, E. P., Baisarovich, M. N., Yakovlev, E. V., et al. (1990). Geological structure and geodynamics of the

region of the Crimean nuclear power plant. *Geophysical Journal, 12*(3), 3−27 (in Russian, with English abstract)
Chen, L. C., & Rau, J. Y. (1993). A unified solution for digital terrain model and ortho-image generation from SPOT stereopairs. *IEEE Transactions on Geoscience and Remote Sensing, 31*, 1243−1252.
Chen, T.-C., Tsay, J.-D., & Gutowski, W. J., Jr. (2008). A comparison study of three polar grids. *Journal of Applied Meteorology and Climatology, 47*, 2993−3007.
Chentsov, V. N. (1940). Morphometric indices of topography as applied to geomorphological maps. *Transactions of the Institute of Geography, 36*, 69−71 (in Russian).
Cherednichenko, A. I., Burmistenko, V. M., Tokovenko, V. S., & Chebanenko, I. I. (1966). An attempt of laboratory simulation of planetary faults (lineaments) of the Earth. *Dopovidi Academii Nauk Ukrainy*, No. 10, 1333−1336 (in Ukrainian, with English abstract).
Chigirev, A. A. (Ed.), (1976). *Digital filtering of airphotos, stereomodels, and maps* (200 p.). Leningrad: Nedra (in Russian).
Chopra, S., & Marfurt, K. J. (2005). Seismic attributes—A historical perspective. *Geophysics, 70*, 3SO−28SO.
Chopra, S., & Marfurt, K. J. (2007a). *Seismic attributes for prospect identification and reservoir characterization* (464 p.). Tulsa, OK: Society of Exploration Geophysicists.
Chopra, S., & Marfurt, K. J. (2007b). Volumetric curvature attributes add value to 3D seismic data interpretation. *Leading Edge, 26*, 856−867.
Chorley, R. J., & Huggett, P. (1965). Trend-surface mapping in geographical research. *Transactions of the Institute of British Geographers, 37*, 47−67.
Chorowicz, J., Bréard, J.-Y., Guillande, R., Morasse, C.-R., Prudon, D., & Rudant, J.-P. (1991). Dip and strike measured systematically on digitized three-dimensional geological maps. *Photogrammetric Engineering and Remote Sensing, 57*, 431−436.
Chorowicz, J., Collet, B., Bonavia, F. F., Mohr, P., Parrot, J. F., & Korme, T. (1998). The Tana basin, Ethiopia: Intra-plateau uplift, rifting and subsidence. *Tectonophysics, 295*, 351−367.
Chorowicz, J., Dhont, D., & Gündoğdu, N. (1999). Neotectonics in the eastern North Anatolian fault region (Turkey) advocates crustal extension: Mapping from SAR ERS imagery and digital elevation model. *Journal of Structural Geology, 21*, 511−532.
Chui, C. K. (1992). *An introduction to wavelets* (266 p.). San Diego: Academic Press.
Cialella, A. T., Dubayah, R., Lawrence, W., & Levine, E. (1997). Predicting soil drainage class using remotely sensed and digital elevation data. *Photogrammetric Engineering and Remote Sensing, 63*, 171−178.
Clark, L. A., & Pregibon, D. (1992). Tree-based models. In J. M. Chambers, & T. J. Hastie (Eds.), *Statistical models in S* (pp. 377−419). Pacific Grove, CA: Wadsworth and Brooks.
Clark, R. L., & Lee, R. (1998). Development of topographic maps for precision farming with kinematic GPS. *Transactions of the American Society of Agricultural Engineers, 41*, 909−916.
Clarke, D. D., & Phillips, C. C. (2003). Three-dimensional geologic modeling and horizontal drilling bring more oil out of the Wilmington oil field of southern California. *AAPG Methods in Exploration, 14*, 27−47.
Clarke, J. I. (1966). Morphometry from maps. In G. H. Dury (Ed.), *Essays in geomorphology* (pp. 235−274). London: Heinemann.
Clarke, K. C. (1988). Scale-based simulation of topographic relief. *American Cartographer, 15*, 173−181.
Claus, M. (1984). Digital terrain models through digital stereo correlation. *Photogrammetria, 39*, 183−192.

Clayton, J. S., Ehrlich, W. A., Cann, D. B., Day, J. H., & Marshall, I. B. (1977). *Soils of Canada. Vol. 1: Soil report* (243 p.). Ottawa: Research Branch, Canada Department of Agriculture.

Codilean, A. T., Bishop, P., & Hoey, T. B. (2006). Surface process models and the links between tectonics and topography. *Progress in Physical Geography, 30,* 307−333.

Cole, N. J., & Boettinger, J. L. (2007). Pedogenic understanding raster classification methodology for mapping soils, Powder River Basin, Wyoming, USA. In P. Lagacherie, A. B. McBratney, & M. Voltz (Eds.), *Digital soil mapping: An introductory perspective* (pp. 377−388). Amsterdam: Elsevier.

Collet, B., Parrot, J. F., & Taud, H. (2000a). Orientation of absolute African plate motion revealed by tomomorphometric analysis of the Ethiopian dome. *Geology, 28,* 1147−1149.

Collet, B., Taud, H., Parrot, J. F., Bonavia, F., & Chorowicz, J. (2000b). A new kinematic approach for the Danakil block using a digital elevation model representation. *Tectonophysics, 316,* 343−357.

Conrad, R. (1996). Soil microorganisms as controllers of atmospheric trace gases (H_2, CO, CH_4, OCS, N_2O, and NO). *Microbiological Reviews, 60,* 609−640.

Cook, S. E., Corner, R. J., Grealish, G., Gessler, P. E., & Chartres, C. J. (1996). A rule-based system to map soil properties. *Soil Science Society of America Journal, 60,* 1893−1900.

Cooley, J. W., & Tukey, J. W. (1965). An algorithm for the machine calculation of complex Fourier series. *Mathematics of Computation, 19,* 297−301.

Corre, M. D., van Kessel, C., & Pennock, D. J. (1996). Landscape and seasonal patterns of nitrous oxide emissions in a semiarid region. *Soil Science Society of America Journal, 60,* 1806−1815.

Da Ros, D., & Borga, M. (1997). Use of digital elevation model data for the derivation of the geomorphological instantaneous unit hydrograph. *Hydrological Processes, 11,* 13−33.

Daniel, W. W. (2000). *Applied nonparametric statistics* (2nd ed., 656 p.). Belmont, CA: Cengage Learning.

Danilov, D. L., & Zhigljavsky, A. A. (Eds.), (1997). *Principal components of time series: The "Caterpillar" method* (308 p.). St. Petersburg: St. Petersburg University Press (in Russian).

Daubechies, I. (1992). *Ten lectures on wavelets* (357 p.). Philadelphia: Society for Industrial and Applied Mathematics.

Davis, F. W., & Dozier, J. (1990). Information analysis of a spatial database for ecological land classification. *Photogrammetric Engineering and Remote Sensing, 56,* 605−613.

Davis, J. C. (1986). *Statistics and data analysis in geology* (2nd ed., 646 p.). New York: Wiley.

De Boer, D. H. (1992). Hierarchies and spatial scale in process geomorphology: A review. *Geomorphology, 4,* 303−318.

De Bruin, S., & Stein, A. (1998). Soil−landscape modelling using fuzzy c-means clustering of attribute data derived from a Digital Elevation Model (DEM). *Geoderma, 83,* 17−33.

De Donatis, M., Borraccini, F., & Susini, S. (2009). Sheet 280−Fossombrone 3D: A study project for a new geological map of Italy in three dimensions. *Computers and Geosciences, 35,* 19−32.

De Floriani, L., Magillo, P., & Puppo, E. (1997). Multiresolution representation and reconstruction of triangulated surfaces. In C. Arcelli, L. Cordella, & G. Sanniti di Baja (Eds.), *Advances in visual form analysis* (pp. 140−149). Singapore: World Scientific.

De Floriani, L., Magillo, P., & Puppo, E. (2000). VARIANT: A system for terrain modeling at variable resolution. *GeoInformatica, 4,* 287−315.

De Floriani, L., Mirra, D., & Puppo, E. (1993). Extracting contour lines from a hierarchical surface model. *Computer Graphics Forum, 12,* 249−260.

De Saint-Venant, M. (1852). Surfaces à plus grande pente constitutées sur des lignes courbes. *Extraits des Procès-Verbaux des Séances de la Société Philomatique de Paris, Series 5, 17,* 24−30.

Debella-Gilo, M., & Etzelmüller, B. (2009). Spatial prediction of soil classes using digital terrain analysis and multinomial logistic regression modeling integrated in GIS: Examples from Vestfold County, Norway. *Catena, 77*, 8–18.

Declercq, F. A. N. (1996). Interpolation methods for scattered sample data: Accuracy, spatial patterns, processing time. *Cartography and Geographic Information Systems, 23*, 128–144.

Deffontaines, B., Lacombe, O., Angelier, J., Chu, H. T., Mouthereau, F., Lee, C. T., et al. (1997). Quaternary transfer faulting in the Taiwan Foothills: Evidence from a multi-source approach. *Tectonophysics, 274*, 61–82.

Delarue, F., Cornu, S., Daroussin, J., Salvador–Blanes, S., Bourennane, H., Albéric, P., et al. (2009). 3D representation of soil distribution: An approach for understanding pedogenesis. *Comptes Rendus Geoscience, 341*, 486–494.

Delaunay, B. (1934). Sur la sphère vide (A la memoire de Georges Voronoi). *Bulletin de l'Academie des Sciences de l'USSR, Classes des Sciences Mathematiques et Naturelles, 7*, 793–800.

Deng, Y. (2007). New trends in digital terrain analysis: Landform definition, representation, and classification. *Progress in Physical Geography, 31*, 405–419.

Desmet, P. J. J. (1997). Effects of interpolation errors on the analysis of DEMs. *Earth Surface Processes and Landforms, 22*, 563–580.

Desmet, P. J. J., & Govers, G. (1996). Comparison of routing algorithms for digital elevation models and their implications for predicting ephemeral gullies. *International Journal of Geographical Information Science, 10*, 311–332.

Devdariani, A. S. (1967). *Mathematical analysis in geomorphology* (155 p.). Moscow: Nedra (in Russian).

Dhont, D., & Chorowicz, J. (2006). Review of the neotectonics of the Eastern Turkish–Armenian Plateau by geomorphic analysis of digital elevation model imagery. *International Journal of Earth Sciences, 95*, 34–49.

Dhont, D., Chorowicz, J., Yürür, T., Froger, J.-L., Köse, O., & Gündogdu, N. (1998a). Emplacement of volcanic vents and geodynamics of Central Anatolia, Turkey. *Journal of Volcanology and Geothermal Research, 85*, 33–54.

Dhont, D., Chorowicz, J., Yürür, T., & Köse, O. (1998b). Polyphased block tectonics along the North Anatolian Fault in the Tosya basin area (Turkey). *Tectonophysics, 299*, 213–227.

Dhont, D., Luxey, P., & Chorowicz, J. (2005). 3-D modeling of geologic maps from surface data. *American Association of Petroleum Geologists Bulletin, 89*, 1465–1474.

Dikau, R. (1988). Case studies in the development of derived geomorphic maps. *Geologisches Jahrbuch, A104*, 329–338.

Dimarzio, J. P., Brenner, A. C., Fricker, H. A., Schutz, B. E., Shuman, C. A., & Zwally, H. J. (2005). Digital elevation models of the Antarctic and Greenland ice sheets from ICESat. *American geophysical union, Fall meeting 2005, 5–9 December 2005, San Francisco, CA*. Abstract # C51B-0276.

Dixon, T. H., Naraghi, M., McNutt, M. K., & Smith, S. M. (1983). Bathymetric prediction from SEASAT altimeter data. *Journal of Geophysical Research, 88*(C), 1563–1571.

Dmitriev, E. A. (1998). The concept of relief flexure in relation to soil science. *Eurasian Soil Science, 31*, 338–346.

Dobos, E., Micheli, E., Baumgardner, M. F., Biehl, L., & Helt, T. (2000). Use of combined digital elevation model and satellite radiometric data for regional soil mapping. *Geoderma, 97*, 367–391.

Dobos, E., Micheli, E., & Montanarella, L. (2007). The population of a 500-m resolution soil organic matter spatial information system for Hungary. In P. Lagacherie, A. B. McBratney, & M. Voltz (Eds.), *Digital soil mapping. An introductory perspective* (pp. 487–495). Amsterdam: Elsevier.

Dokuchaev, V. V. (1883). *Russian Chernozem. Report to the Imperial Free Economic Society* (376 p.). St. Petersburg: Decleron and Evdokimov Press (in Russian).

Dokuchaev, V. V. (1886). In *Materials for land evaluation of the Nizhny Novgorod Governorate. Natural and historical part: Report to the Nizhny Novgorod Governorate zemstvo. Vol. 1: Key points in the history of land evaluation in the European Russia, with classification of Russian Soils* (391 p.). St. Petersburg: Evdokimov Press (in Russian).

Dokuchaev, V. V. (1891). On the relationships between, on the one hand, age and elevation of the terrain, and, on the other hand, character and distribution of Chernozems, forest soils, and Solonets. *Revue des Sciences Naturelles*, No. 1, 1–16; No. 2, 57–67; No. 3, 112–123 (in Russian, with French abstract).

Dokuchaev, V. V. (1899). *Report to the transcaucasian statistical committee on soil taxonomy in general and especially for the Transcaucasia. Horizontal and vertical soil zones* (19 p.). Tiflis, Russia: Office Press of the Civilian Affairs Commander-in-Chief in the Caucasus (in Russian).

Dolitsky, A. V., & Kiyko, I. A. (1963). On the causes of deformation of the Earth's crust. In D. V. Nalivkin, & N. V. Tupitsin (Eds.), *Problems of planetary geology* (pp. 291–312). Moscow: Gosgeoltekhizdat (in Russian).

Douglas, D. H. (1986). Experiments to locate ridges and channels to create a new type of digital elevation model. *Cartographica, 23*, 29–61.

Doyle, F. J. (1978). Digital terrain models: An overview. *Photogrammetric Engineering and Remote Sensing, 44*, 1481–1485.

Dubayah, R., & Rich, P. M. (1995). Topographic solar radiation models for GIS. *International Journal of Geographical Information Systems, 9*, 405–419.

Dubyansky, A. A., & Khakman, S. A. (Eds.), (1949). *Geology of the USSR, Vol. 6: Braynsk, Orel, Kursk, Voronezh, and Tambov Regions. Pt. 1: Geological description* (575 p.). Moscow: Geological Literature State Press (in Russian).

Durand, P., Robson, A., & Neal, C. (1992). Modelling the hydrology of submediterranean mountain catchments (Mont Lozere, France), using TOPMODEL: Initial results. *Journal of Hydrology, 139*, 1–14.

Durham, L. S. (1999). 3-D models are making cents. *AAPG Explorer*, No. 11, 8–9, 18.

Dury, G. H. (1962). *Map interpretation* (2nd ed., 209 p.). London: Pitman and Sons.

Dutton, G. (1999). *A hierarchical coordinate system for geoprocessing and cartography* (231 p.). Berlin: Springer.

Eberly, D., Gardner, R., Morse, B., Pizer, S., & Scharlach, C. (1994). Ridges for image analysis. *Journal of Mathematical Imaging and Vision, 4*, 351–371.

Eckert, M. (1908). On the nature of maps and map logic. *Bulletin of the American Geographical Society, 40*, 344–351.

Edwards, K., & Davis, P. A. (1994). The use of intensity–hue–saturation transformation for producing color shaded-relief images. *Photogrammetric Engineering and Remote Sensing, 60*, 1369–1374.

Efremov, Y. K. (1949). An experience on morphological classification of elements and simple forms of topography. *Voprosy Geografii, 11*, 109–136 (in Russian).

Efron, B. (1982). *The Jackknife, the Bootstrap, and other resampling plans* (92 p.). Philadelphia: Society for Industrial Mathematics.

Eguez, A., Alvarado, A., Yepes, H., Machette, M. N., Costa, C., & Dart, R. L. (2003). *Database and map of Quaternary faults and folds of Ecuador and its offshore regions. USGS Open-File Report 03-289* (71 p.). U.S. Geological Survey.

Ehsani, A. H., & Quiel, F. (2008). Application of Self Organizing Maps and SRTM data to characterize yardangs in the Lut Desert, Iran. *Remote Sensing of Environment, 112*, 3284–3294.

Eklundh, L., & Mårtensson, U. (1995). Rapid generation of digital elevation models from topographic maps. *International Journal of Geographical Information Science, 9*, 329–340.

Ekman, M. (1988). Gaussian curvature of postglacial rebound and the discovery of caves created by major earthquakes in Fennoscandia. *Geophysica, 24*, 47–56.

Eliason, J. R., & Eliason, V. L. C. (1987). *Process for structural geologic analysis of topography and point data*. US Patent No 698759, 107 p.

Elkins, T. A. (1951). The second derivative method of gravity interpretation. *Geophysics, 16*, 29–50.

El-Sheimy, N., Valeo, C., & Habib, A. (2005). *Digital terrain modeling: Acquisition, manipulation, and applications* (257 p.). Boston: Artech House.

Elsner, J. B., & Tsonis, A. A. (1996). *Singular spectrum analysis: A new tool in time series analysis* (164 p.). New York: Plenum Press.

Elsner, P., & Bonnici, M. (2007). Vertical accuracy of Shuttle Radar Topography Mission (SRTM) elevation and void-filled data in the Libyan desert. *International Journal of Ecology and Development, 8*(F07), 66–80.

Emeis, S., & Knoche, H. R. (2009). Applications in meteorology. In T. Hengl, & H. I. Reuter (Eds.), *Geomorphometry: Concepts, software, applications* (pp. 603–622). Amsterdam: Elsevier.

Endreny, T. A., Wood, E. F., & Lettenmaier, D. P. (2000). Satellite-derived digital elevation model accuracy: Hydrogeomorphological analysis requirements. *Hydrological Processes, 14*, 1–20.

Erskine, R. H., Green, T. R., Ramirez, J. A., & MacDonald, L. H. (2007). Digital elevation accuracy and grid cell size: Effects on estimated terrain attributes. *Soil Science Society of America Journal, 71*, 1371–1380.

Evans, I. S. (1972). General geomorphometry, derivations of altitude, and descriptive statistics. In R. J. Chorley (Ed.), *Spatial analysis in geomorphology* (pp. 17–90). London: Methuen.

Evans, I. S. (1979). *Statistical characterization of altitude matrices by computer. An integrated system of terrain analysis and slope mapping. The final report on grant DA-ERO-591-73-G0040* (192 p.). Durham, England: Department of Geography, University of Durham.

Evans, I. S. (1980). An integrated system of terrain analysis and slope mapping. *Zeitschrift für Geomorphologie, Suppl. 36*, 274–295.

Evans, I. S., & Minár, J. (2011). A classification of geomorphometric variables. In T. Hengl, I. S. Evans, J. P. Wilson, & M. Gould (Eds.), *Proceedings of Geomorphometry 2011, 7–11 September 2011, Redlands, CA.* (pp. 105–108).

Evans, J. S., & Hudak, A. T. (2007). A multiscale curvature algorithm for classifying discrete return LiDAR in forested environments. *IEEE Transactions on Geoscience and Remote Sensing, 45*, 1029–1038.

Eyton, J. R. (1986). Digital elevation model perspective plot overlays. *Annals of the Association of American Geographers, 76*, 570–576.

Falorni, G., Teles, V., Vivoni, E. R., Bras, R. L., & Amaratunga, K. S. (2005). Analysis and characterization of the vertical accuracy of digital elevation models from the Shuttle Radar Topography Mission. *Journal of Geophysical Research, 110*(F), F02005, doi:10.1029/2003JF000113.

Famiglietti, J. S., & Wood, E. F. (1995). Effects of spatial variability and scale on areally averaged evapotranspiration. *Water Resources Research, 31*, 699–712.

Farenhorst, A., Florinsky, I. V., Monreal, C. M., & Muc, D. (2003). Evaluating the use of digital terrain modelling for quantifying the spatial variability of 2,4-D sorption by soil within agricultural landscapes. *Canadian Journal of Soil Science, 83*, 557–563.

Farr, T. G., Rosen, P. A., Caro, E., Crippen, R., Duren, R., Hensley, S., et al. (2007). The Shuttle Radar Topography Mission. *Reviews of Geophysics, 45*, RG2004, doi:10.1029/2005RG000183.

Fedorov, A. E. (1991). *Hexagonal networks of linear heterogeneities of the Earth* (128 p.). Moscow: Nedra (in Russian).

Fedoseev, A. P. (1959). Soil moisture and terrain topography. In N. A. Konyukhov (Ed.), *Agricultural meteorology* (pp. 66–88). Moscow: Hydrometeorological Press (in Russian).

Felicísimo, A. M. (1994a). *Modelos digitales del terreno. Introducción y aplicaciones en las ciencias ambientales* (222 p.). Oviedo, Spain: Pentalfa Ediciones.

Felicísimo, A. M. (1994b). Parametric statistical method for error detection in digital elevation models. *ISPRS Journal of Photogrammetry and Remote Sensing, 49*, 29–33.

Felicísimo, A.M. (1995). Error propagation analysis in slope estimation by means of digital elevation models. In *Cartography crossing borders: Proceedings 1 of the 17th International cartographic conference and 10th General assembly of International cartographic association, 3–9 September 1995, Barcelona, Spain* (pp. 94–98). Barcelona: Institut Cartogràfic de Catalunya.

Feranec, J., Kolár, J., & Krcho, J. (1991). Mapping of the surface water logging intensity of the soils by applying Landsat TM data and complex digital terrain model. *Bulletin du Comité Français de Cartographie, 127–128*, 154–157.

Fernandez, O., Jones, S., Armstrong, N., Johnson, G., Ravaglia, A., & Muñoz, J. A. (2009). Automated tools within workflows for 3D structural construction from surface and subsurface data. *Geoinformatica, 13*, 291–304.

Fichtenholz, G. M. (1966a). *A course in differential and integral calculus* (6th ed., Vol. 1, 607 p.). Moscow: Nauka (in Russian).

Fichtenholz, G. M. (1966b). *A course in differential and integral calculus* (4th ed., Vol. 3, 656 p.). Moscow: Nauka (in Russian).

Filatov, M. M. (1927). *Diagrammatic meridional soil profile of the European part of USSR* (19 p.). Moscow: Mospoligraf.

Filippov, Y. V. (Ed.), (1955). *Principles of generalization of small-scale geographical maps* (336 p.). Moscow: Geodezizdat (in Russian).

Filosofov, V. P. (1960). *A brief guide of the morphometric method to search tectonic structures* (94 p.). Saratov, USSR: Saratov University Press (in Russian).

Filosofov, V. P. (1975). *Principles of the morphometric method to search tectonic structures* (232 p.). Saratov, USSR: Saratov University Press (in Russian).

Finkl, C. W., Benedet, L., & Andrews, J. L. (2005). Interpretation of seabed geomorphology based on spatial analysis of high-density airborne laser bathymetry. *Journal of Coastal Research, 21*, 501–514.

Fiore Allwardt, P. F., Bellahsen, N., & Pollard, D. D. (2007). Curvature and fracturing based on global positioning system data collected at Sheep Mountain anticline, Wyoming. *Geosphere, 3*, 408–421.

Fischer, M. P., & Keating, D. P. (2005). Photogrammetric techniques for analyzing displacement, strain, and structural geometry in physical models: Application to the growth of monoclinal basement uplifts. *Geological Society of America Bulletin, 117*, 369–382.

Fischer, M. P., & Wilkerson, M. S. (2000). Predicting the orientation of joints from fold shape: Results of pseudo-three-dimensional modeling and curvature analysis. *Geology, 28*, 15–18.

Fishbaugh, K. E., & Head, J. W., III (2001). Comparison of the north and south polar caps of Mars: New observations from MOLA data and discussion of some outstanding questions. *Icarus, 154*, 145–161.

Fisher, D. (1993). If Martian ice caps flow: Ablation mechanisms and appearance. *Icarus, 105*, 501–511.

Fisher, P. F., & Tate, N. J. (2006). Causes and consequences of error in digital elevation models. *Progress in Physical Geography, 30*, 467–489.

Fitzmaurice, J., Eilers, R. G., St., Jacques, E., & Waddell, A. (1999). *Soils of SE 32-14-25W—Miniota precision agriculture research site. Special report series 99-1* (47 p.). Winnipeg, MB: Land Resource Unit, Brandon Research Centre, Agriculture and Agri-Food Canada.

Florinsky, I. V. (1991). *Map generalization—A brief review of the problem* (55 p.). Pushchino, USSR: Pushchino Research Centre Press (in Russian).

Florinsky, I. V. (1992). *Recognition of lineaments and ring structures: Quantitative topographic techniques* (47 p.). Pushchino, Russia: Pushchino Research Centre Press (in Russian, with English abstract).

Florinsky, I. V. (1993). *Analyzing digital elevation models to reveal linear structures of the land surface*. Ph.D. Thesis. Institute of Soil Science and Photosynthesis, Russian Academy of Sciences, Pushchino, Russia, 133 p. (in Russian).

Florinsky, I. V. (1996). Quantitative topographic method of fault morphology recognition. *Geomorphology*, 16, 103−119.

Florinsky, I. V. (1998a). Accuracy of local topographic variables derived from digital elevation models. *International Journal of Geographical Information Science*, 12, 47−61.

Florinsky, I. V. (1998b). Combined analysis of digital terrain models and remotely sensed data in landscape investigations. *Progress in Physical Geography*, 22, 33−60.

Florinsky, I. V. (1998c). Derivation of topographic variables from a digital elevation model given by a spheroidal trapezoidal grid. *International Journal of Geographical Information Science*, 12, 829−852.

Florinsky, I. V. (2000). Relationships between topographically expressed zones of flow accumulation and sites of fault intersection: Analysis by means of digital terrain modelling. *Environmental Modelling and Software*, 15, 87−100.

Florinsky, I. V. (2002). Errors of signal processing in digital terrain modelling. *International Journal of Geographical Information Science*, 16, 475−501.

Florinsky, I. V. (2005). Artificial lineaments in digital terrain modelling: Can operators of topographic variables cause them? *Mathematical Geology*, 37, 357−372.

Florinsky, I. V. (2007a). Small-scale morphometric maps of the Northern Eurasia. *Geodezia i Cartografia*, No. 2, 15−21 (in Russian).

Florinsky, I. V. (2007b). Solving three problems of exploration and engineering geology by digital terrain analysis. *International Journal of Ecology and Development*, 8(F07), 52−65.

Florinsky, I. V. (2008a). Global lineaments: Application of digital terrain modelling. In Q. Zhou, B. Lees, & G.-A. Tang (Eds.), *Advances in digital terrain analysis* (pp. 365−382). Berlin: Springer.

Florinsky, I. V. (2008b). Global morphometric maps of Mars, Venus, and the Moon. In A. Moore, & I. Drecki (Eds.), *Geospatial vision: New dimensions in cartography* (pp. 171−192). Berlin: Springer.

Florinsky, I. V. (2008c). Morphometric maps of the world. *Geodezia i Cartografia*, No. 1, 24−27 (in Russian).

Florinsky, I. V. (2008d). On calculation accuracy in digital terrain modeling. *Geodezia i Cartografia*, No. 6, 28−32 (in Russian).

Florinsky, I. V. (2008e). 100 years anniversary of the Tunguska Event: Digital terrain modeling of the epicenter area. *Geodezia i Cartografia*, No. 8, 20−22 (in Russian).

Florinsky, I. V. (2009a). Accurate method for derivation of local topographic variables. *Geodezia i Cartografia*, No. 4, 19−23 (in Russian).

Florinsky, I. V. (2009b). Computation of the third-order partial derivatives from a digital elevation model. *International Journal of Geographical Information Science*, 23, 213−231.

Florinsky, I. V. (2010). *Theory and applications of mathematical cartographic modeling of topography*. D.Sc. Thesis. Institute of Mathematical Problems of Biology, Russian Academy of Sciences, Pushchino, Russia, 267 p. (in Russian).

Florinsky, I. V., & Arlashina, H. A. (1998). Quantitative topographic analysis of gilgai soil morphology. *Geoderma*, 82, 359−380.

Florinsky, I. V., & Eilers, R. G. (2002). Prediction of the soil carbon content at micro-, meso- and macroscales by digital terrain modelling. In *Transactions of the 17th World congress of soil science, 14−21 August 2002, Bangkok, Thailand, Symposium 52 (paper # 24)*. Bangkok: ISSS (CD ROM).

Florinsky, I. V., Eilers, R. G., Burton, D. L., McMahon, S. K., Monreal, C. M., & Farenhorst, A. (2009a). Predictive soil mapping based on digital terrain modeling. *Geoinformatika, No. 1*, 22−32 (in Russian, with English abstract).

Florinsky, I. V., Eilers, R. G., & Lelyk, G. W. (2000). Prediction of soil salinity risk by digital terrain modelling in the Canadian prairies. *Canadian Journal of Soil Science, 80*, 455−463.

Florinsky, I. V., Eilers, R. G., Manning, G., & Fuller, L. G. (1999). Application of digital terrain modelling to prediction of soil properties in the Prairie Ecozone. In *Papers of the 42nd Annual Manitoba society of soil science meeting, 2−3 February 1999, Winnipeg, Canada* (pp. 140−154). Winnipeg, MB: MSSS.

Florinsky, I. V., Eilers, R. G., Manning, G., & Fuller, L. G. (2002). Prediction of soil properties by digital terrain modelling. *Environmental Modelling and Software, 17*, 295−311.

Florinsky, I. V., Eilers, R. G., Wiebe, B. H., & Fitzgerald, M. M. (2009b). Dynamics of soil salinity in the Canadian prairies: Application of singular spectrum analysis. *Environmental Modelling and Software, 24*, 1182−1195.

Florinsky, I. V., Grokhlina, T. I., & Mikhailova, N. L. (1995). LANDLORD 2.0: The software for analysis and mapping of geometric characteristics of topography. *Geodezia i Cartografia, No. 5*, 46−51 (in Russian).

Florinsky, I. V., & Kuryakova, G. A. (1996). Influence of topography on some vegetation cover properties. *Catena, 27*, 123−141.

Florinsky, I. V., & Kuryakova, G. A. (2000). Determination of grid size for digital terrain modelling in landscape investigations—Exemplified by soil moisture distribution at a micro-scale. *International Journal of Geographical Information Science, 14*, 815−832.

Florinsky, I. V., McMahon, S., & Burton, D. L. (2004). Topographic control of soil microbial activity: A case study of denitrifiers. *Geoderma, 119*, 33−53.

Florinsky, I. V., McMahon, S., & Burton, D. L. (2009c). Topographic factors of nitrous oxide emission. In A. I. Sheldon, & E. P. Barnhart (Eds.), *Nitrous oxide emissions research progress* (pp. 105−126). New York: Nova Science Publishers.

Ford, P. G. (1992). *Magellan global topography, emissivity, reflectivity, and slope data, MGN-V-RDRS-5-GDR-TOPOGRAPHIC-V1.0, MGN-V-RDRS-5-GDR-EMISSIVITY-V1.0, MGN-V-RDRS-5-GDR-REFLECTIVITY-V1.0, and MGN-V-RDRS-5-GDR-SLOPE-V1.0*. NASA Planetary Data System. <http://pds-geosciences.wustl.edu/missions/magellan/gxdr/index.htm/>

Franke, R. (1982). Scattered data interpolation: Tests of some methods. *Mathematics of Computation, 38*, 181−200.

Franke, R. (1985). Thin plate spline with tension. *Computer Aided Geometric Design, 2*, 87−95.

Franklin, J. (1995). Predictive vegetation mapping: Geographic modelling of biospatial patterns in relation to environmental gradients. *Progress in Physical Geography, 19*, 474−499.

Frederiksen, P. (1981). Terrain analysis and accuracy prediction by means of the Fourier transformation. *Photogrammetria, 36*, 145−157.

Freeman, T. G. (1991). Calculating catchment area with divergent flow based on a regular grid. *Computers and Geosciences, 17*, 413−422.

French, J. R. (2003). Airborne LiDAR in support of geomorphological and hydraulic modelling. *Earth Surface Processes and Landforms, 28*, 321−335.

Fridland, V. M. (1976). *Pattern of the soil cover* (291 p.). Jerusalem: Israel Program for Scientific Translations.

Fridland, V. M. (Ed.), (1988). *Soil map of the Russian Soviet Federative Socialist Republic, scale 1 : 2,500,000* (16 p.). Moscow: Dokuchaev Soil Institute (in Russian).

Fukue, K., Shimoda, H., & Sakata, T. (1981). Complete lineament extraction with the aid of shadow-free Landsat image. In *Machine processing of remotely sensed data with special emphasis on range, forest, and wetlands assessment: Proceedings of the 7th International symposium, 23–26 June 1981, West Lafayette, IN* (pp. 94–102). West Lafayette, IN: Purdue University.

Gallant, J. C., & Hutchinson, M. F. (1996). Towards an understanding of landscape scale and structure. In *Proceedings of the 3rd International conference / workshop on integrating GIS and environmental modeling, 21–25 January 1996, Santa Fe, NM*. Santa Barbara, CA: National Center for Geographic Information and Analysis, <http://www.ncgia.ucsb.edu/conf/SANTA_FE_CD–ROM/sf_papers/gallant_john/paper.html/>.

Gallant, J. C., & Hutchinson, M. F. (1997). Scale dependence in terrain analysis. *Mathematics and Computers in Simulation, 43*, 313–321.

Gallant, J. C., & Read, A. (2009). Enhancing the SRTM data for Australia. In R. Purves, S. Gruber, T. Hengl, & R. Straumann (Eds.), *Proceedings of Geomorphometry 2009, 31 August–2 September 2009, Zurich, Switzerland* (pp. 149–154). Zurich: University of Zurich.

Ganas, A., Pavlides, S., & Karastathis, V. (2005). DEM-based morphometry of range-front escarpments in Attica, central Greece, and its relation to fault slip rates. *Geomorphology, 65*, 301–319.

Gao, J. (2009). Bathymetric mapping by means of remote sensing: Methods, accuracy and limitations. *Progress in Physical Geography, 33*, 103–116.

Gardiner, V., & Park, C. C. (1978). Drainage basin morphometry: Review and assessment. *Progress in Physical Geography, 2*, 1–35.

Garetsky, R. G., Karasev, O. I., Levkov, E. V., & Svyatogorov, A. A. (1983). The western part of the East European Platform. In V. G. Trifonov, V. I. Makarov, Y. G. Safonov, & P. V. Florensky (Eds.), *Space information in geology* (pp. 185–189). Moscow: Nauka (in Russian).

Gasperini, L., Alvisi, F., Biasini, G., Bonatti, E., Longo, G., Pipan, M., et al. (2007). A possible impact crater for the 1908 Tunguska Event. *Terra Nova, 19*, 245–251.

Gauch, J. M., & Pizer, S. M. (1993). Multiresolution analysis of ridges and valleys in grayscale images. *IEEE Transactions on Pattern Analysis and Machine Intelligence, 15*, 635–646.

Gauss, C. F. (1828). Disquisitiones generales circa superficies curvas. *Commentationes Societatis Regiae Scientiarum Gottingensis Recentiores, 6*, 99–146.

Gauss, C. F. (2009). General investigations of curved surfaces. In C. F. Gauss, *General investigations of curved surfaces of 1827 and 1825* (pp. 1–45). Alcester, England: Read Books.

Geiger, R. (1927). *Das Klima der bodennahen Luftschicht* (246 p.). Braunschweig: Vieweg.

Gelautz, M., Paillou, P., Chen, C. W., & Zebker, H. A. (2003). Radar stereo- and interferometry-derived digital elevation models: Comparison and combination using Radarsat and ERS-2 imagery. *International Journal of Remote Sensing, 24*, 5243–5264.

Gelfand, I. M., Guberman, S. I., Izvekova, M. L., Keilis-Borok, V. I., & Rantsman, E. Y. (1972). Criteria of high seismicity, determined by pattern recognition. *Tectonophysics, 13*, 415–422.

Gerasimov, I. P. (1959). *Structural features of the land surface topography of the USSR and their origin* (100 p.). Moscow: Soviet Academic Press (in Russian).

Gerasimov, I. P., & Korzhuev, S. S. (Eds.), (1979). *Morphostructural analysis of the drainage network of the USSR* (304 p.). Moscow: Nauka (in Russian).

Gerrard, A. J. (1981). *Soils and landforms. An integration of geomorphology and pedology* (219 p.). London: George Allen and Unwin.

Gessler, P. E., Chadwick, O. A., Chamran, F., Althouse, L., & Holmes, K. (2000). Modeling soil-landscape and ecosystem properties using terrain attributes. *Soil Science Society of America Journal, 64*, 2046–2056.

Gessler, P. E., Moore, I. D., McKenzie, N. J., & Ryan, P. J. (1995). Soil–landscape modelling and spatial prediction of soil attributes. *International Journal of Geographical Information Systems, 9*, 421–432.

Ghilani, C. D., & Wolf, P. R. (2008). *Elementary surveying: An introduction to geomatics* (12th ed., 931 p.). Upper Saddle River, NJ: Pearson/Prentice Hall.

Ghosh, J. K., Delampady, M., & Samanta, T. (2006). *An introduction to Bayesian analysis: Theory and methods* (352 p.). New York: Springer.

Giasson, E., Figueiredo, S. R., Tornquist, C. G., & Clarke, R. T. (2008). Digital soil mapping using logistic regression on terrain parameters for several ecological regions in southern Brazil. In A. E. Hartemink, A. McBratney, & M. L. Mendonça-Santos (Eds.), *Digital soil mapping with limited data* (pp. 225–232). Dordrecht: Springer.

Giles, P. T., & Franklin, S. E. (1996). Comparison of derivative topographic surfaces of a DEM generated from stereoscopic SPOT images with field measurements. *Photogrammetric Engineering and Remote Sensing, 62*, 1165–1171.

Glasko, M. P., & Rantsman, E. Y. (1996). Morphostructural groups as activization sites of natural processes. *Transactions (Doklady) of the Russian Academy of Sciences, Earth Science Section, 350*, 1171–1174.

GLOBE Task Team and others (1999). *Global land one-kilometer base elevation (GLOBE) digital elevation model, version 1.0*. Boulder, CO: National Oceanic and Atmospheric Administration, National Geophysical Data Center. <http://www.ngdc.noaa.gov/mgg/topo/globe.html/>.

Gogineni, S., Tammana, D., Braaten, D., Leuschen, C., Akins, T., Legarsky, J., et al. (2001). Coherent radar ice thickness measurements over the Greenland ice sheet. *Journal of Geophysical Research, 106*(D), 33761–33772.

Gold, C., & Mostafavi, M. A. (2000). Towards the global GIS. *ISPRS Journal of Photogrammetry and Remote Sensing, 55*, 150–163.

Golovchenko, A. V., & Polyanskaya, L. M. (1996). Seasonal dynamics of population and biomass of microorganisms in the soil profile. *Eurasian Soil Science, 29*, 1145–1150.

Golts, S., & Rosenthal, E. (1993). A morphotectonic map of the northern Arava in Israel, derived from isobase lines. *Geomorphology, 7*, 305–315.

Golyandina, N. E., Nekrutkin, V. V., & Zhigljavsky, A. A. (2001). *Analysis of time series structure: SSA and related techniques* (305 p.). London: Chapman and Hall/CRC.

Golyandina, N., & Stepanov, D. (2005). SSA-based approaches to analysis and forecast of multidimensional time series. In S. M. Ermakov, V. B. Melas, & A. N. Pepelyshev (Eds.), *Proceedings of the 5th St. Petersburg workshop on simulation, June 26–July 2 2005, St. Petersburg, Russia* (pp. 293–298). St. Petersburg: St. Petersburg State University.

Golyandina, N. E., & Usevich, K. D. (2010). 2D-extension of singular spectrum analysis: Algorithm and elements of theory. In V. Olshevsky, & E. Tyrtyshnikov (Eds.), *Matrix methods: Theory, algorithms, applications* (pp. 449–473). Singapore: World Scientific.

Golyandina, N. E., Usevich, K. D., & Florinsky, I. V. (2007). Filtering of digital terrain models by two-dimensional singular spectrum analysis. *International Journal of Ecology and Development, 8*(F07), 81–94.

Golyandina, N. E., Usevich, K. D., & Florinsky, I. V. (2008). Singular spectrum analysis for filtering of digital terrain models. *Geodezia i Cartografia, No. 5*, 21–28 (in Russian).

Gong, J., Li, Z., Zhu, Q., Sui, H., & Zhou, Y. (2000). Effects of various factors on the accuracy of DEMs: An intensive experimental investigation. *Photogrammetric Engineering and Remote Sensing, 66*, 1113–1117.

Goovaerts, P. (1997). *Geostatistics for natural resources evaluation* (483 p.). New York: Oxford University Press.

Gorshkov, A. I., & Solov'ev, A. A. (2009). Recognition of possible locations of future $M \geq 6.0$ earthquakes: The Mediterranean mountain belts. *Journal of Volcanology and Seismology, 3*, 210–219.

Gosteva, T. S., Patrakova, V. S., & Abramkina, V. A. (1983). Establishing the principles of spatial distribution of ring structures based on trend analysis of relief. *Soviet Geology and Geophysics, 24*, 62–67.

Grachev, A. F., Magnitsky, V. A., Mukhamediev, S. A., & Nikolaev, V. A. (2001). The effect of neotectonic movements on the gradients and curvatures of the Northern Eurasian lithosphere surface. *Physics of the Solid Earth, 37*, 89–106.

Graupe, D. (2007). *Principles of artificial neural networks* (2nd ed., 303 p.). Singapore: World Scientific.

Greve, M. H., Greve, M. B., Bou Kheir, R., Bøcher, P. K., Larsen, R., & McCloy, K. (2010). Comparing decision tree modeling and indicator kriging for mapping the extent of organic soils in Denmark. In J. L. Boettinger, D. W. Howell, A. C. Moore, A. E. Hartemink, & S. Kienast-Brown (Eds.), *Digital soil mapping—Bridging research, environmental application, and operation* (pp. 267–280). Dordrecht: Springer.

Grigorenko, A. M. (1998). *Some problems of technical information theory* (111 p.). Moscow: Ubex Press (in Russian).

Grinand, C., Arrouays, D., Laroche, B., & Martin, M. P. (2008). Extrapolating regional soil landscapes from an existing soil map: Sampling intensity, validation procedures, and integration of spatial context. *Geoderma, 143*, 180–190.

Groffman, P. M., & Hanson, G. C. (1997). Wetland denitrification: Influence of site quality and relationships with wetland delineation protocols. *Soil Science Society of America Journal, 61*, 323–329.

Groffman, P. M., & Tiedje, J. M. (1989). Denitrification in north temperate forest soils: Spatial and temporal patterns at the landscape and seasonal scales. *Soil Biology and Biochemistry, 21*, 613–620.

Grohmann, C. H. (2004a). Morphometric analysis in geographic information systems: Applications of free software GRASS and R. *Computers and Geosciences, 30*, 1055–1067.

Grohmann, C. H. (2004b). *Técnicas de geoprocessamento aplicadas a análise morfométrica*. M.Sc. Thesis. Institute of Geosciences, University of São Paulo, São Paulo, Brazil, 78 p.

Grohmann, C. H. (2005). Trend-surface analysis of morphometric parameters: A case study in southeastern Brazil. *Computers and Geosciences, 31*, 1007–1014.

Grohmann, C. H., Riccomini, C., & Alves, F. M. (2007). SRTM-based morphotectonic analysis of the Poços de Caldas Alkaline Massif, southeastern Brazil. *Computers and Geosciences, 33*, 10–19.

Grohmann, C. H., Riccomini, C., & Chamani, M. A. C. (2011). Regional scale analysis of landform configuration with base-level (isobase) maps. *Hydrology and Earth System Sciences, 15*, 1493–1504.

Groshong, R. H., Jr. (2006). *3-D structural geology. A practical guide to quantitative surface and subsurface map interpretation* (2nd ed., 400 p.). Berlin: Springer.

Grunwald, S. (2009). Multi-criteria characterization of recent digital soil mapping and modeling approaches. *Geoderma, 152*, 195–207.

Grunwald, S., Barak, P., McSweeney, K., & Lowery, B. (2000). Soil landscape models at different scales portrayed in Virtual Reality Modeling Language. *Soil Science, 165*, 598–615.

Grunwald, S., Ramasundaram, V., Comerford, N. B., & Bliss, C. M. (2007). Are current scientific visualisation and virtual reality techniques capable to represent real soil-landscapes? In P. Lagacherie, A. B. McBratney, & M. Voltz (Eds.), *Digital soil mapping. An introductory perspective* (pp. 571–580). Amsterdam: Elsevier.

Guberman, S., Pikovski, Y., & Rantsman, E. (1997). Methodology for prediction of the locations of giant oil and gas reservoirs: Field results. In *Proceedings of the SPE 67th Annual western regional meeting, 25–27 June 1997, Long Beach, CA* (pp. 321–330). Richardson, TX: Society of Petroleum Engineers.

Gudmundsson, A. (2000). Active fault zones and groundwater flow. *Geophysical Research Letters, 27*, 2993–2996.

Gugan, D. J., & Dowman, I. J. (1988). Topographic mapping from SPOT imagery. *Photogrammetric Engineering and Remote Sensing, 54*, 1409–1414.

Guillaume, B., Dhont, D., & Brusset, S. (2008). Three-dimensional geologic imaging and tectonic control on stratigraphic architecture: Upper Cretaceous of the Tremp Basin (south-central Pyrenees, Spain). *American Association of Petroleum Geologists Bulletin, 92*, 249–269.

Guth, P. (2007). Global SRTM geomorphometric atlas. In *Proceedings of the 9th International conference on geocomputation, 3–5 September 2007, Maynooth, Ireland.* GeoComputation CD-ROM, 6 p. (CD–ROM).

Gvin, V. Y. (1963). The use of morphometry in structural studies of the Upper and Mid Volga and Kama Regions. *Voprosy Geografii, 63*, 64–80 (in Russian).

Gzovsky, M. V. (1954). The main problems of fault classification. *Sovetskaya Geologia, 41*, 131–169 (in Russian).

Haggett, P., Chorley, R. J., & Stoddart, D. R. (1965). Scale standards in geographical research: A new measure of areal magnitude. *Nature, 205*, 844–847.

Hall, J. K. (1996). Digital topography and bathymetry of the area of the Dead Sea Depression. *Tectonophysics, 266*, 177–185.

Hancock, G. R. (2005). The use of digital elevation models in the identification and characterization of catchments over different grid scales. *Hydrological Processes, 19*, 1727–1749.

Hancock, G., & Willgoose, G. (2001). The production of digital elevation models for experimental model landscapes. *Earth Surface Processes and Landforms, 26*, 475–490.

Hannah, M. J. (1981). Error detection and correction in digital terrain models. *Photogrammetric Engineering and Remote Sensing, 47*, 63–69.

Hantke, R., & Scheidegger, A. E. (1999). Tectonic predesign in geomorphology. *Lecture Notes in Earth Sciences, 78*, 251–266.

Haralick, R. (1983). Ridges and valleys on digital images. *Computer Vision, Graphics, and Image Processing, 22*, 28–38.

Harrison, J. M., & Lo C.-P. (1996). PC-based two-dimensional discrete Fourier transform programs for terrain analysis. *Computers and Geosciences, 22*, 419–424.

Hart, B. S., & Sagan, J. A. (2007). Curvature for visualization of seismic geomorphology. *Geological Society of London, Special Publications, 277*, 139–149.

Hastie, T., & Tibshirani, R. (1990). *Generalized additive models* (335 p.). London: Chapman and Hall.

Hastings, D. A., & Dunbar, P. K. (1998). Development and assessment of the global land one-km base elevation digital elevation model (GLOBE). *ISPRS Archives, 32*, 218–221.

Hato, M., Tsu, H., Tachikawa, T., Abrams, M., & Bailey, B. (2009). The ASTER Global Digital Elevation Model (GDEM)—For societal benefit. In *American Geophysical Union, Fall Meeting 2009, 14–18 December 2009, San Francisco, CA.* Abstract # U33B–0065.

Haugerud, R. A., Harding, D. J., Johnson, S. Y., Harless, J. L., Weaver, C. S., & Sherrod, B. L. (2003). High-resolution lidar topography of the Puget Lowland, Washington—A bonanza for earth science. *GSA Today, 13*, 4–10.

Hayes, A. G., Wolf, A. S., Aharonso, O., Zebker, H., Lorenz, R., Kirk, R. L., et al. (2010). Bathymetry and absorptivity of Titan's Ontario Lacus. *Journal of Geophysical Research, 115*, E09009, doi:10.1029/2009JE003557.

Heddadj, D., & Gascuel-Odoux, C. (1999). Topographic and seasonal variations of unsaturated hydraulic conductivity as measured by tension disc infiltrometers at the field scale. *European Journal of Soil Science, 50*, 275–283.

Heller, M. (1990). Triangulation algorithms for adaptive terrain modeling. In K. Brassel, & H. Kisomoto (Eds.), *Proceedings of the 4th International symposium on spatial data handling, July 1990, Zurich, Switzerland* (Vol. 1, pp. 163–174). Zurich: University of Zurich.

Hengl, T. (2006). Finding the right pixel size. *Computers and Geosciences, 32*, 1283–1298.

Hengl, T., Heuvelink, G. B. M., & Rossiter, D. G. (2007a). About regression kriging: From equations to case studies. *Computers and Geosciences, 33*, 1301–1315.

Hengl, T., Heuvelink, G. B. M., & Stein, A. (2004). A generic framework for spatial prediction of soil variables based on regression-kriging. *Geoderma, 120*, 75–93.

Hengl, T., & Reuter, H. I. (Eds.), (2009). *Geomorphometry: Concepts, software, applications* (796 p.). Amsterdam: Elsevier.

Hengl, T., Toomanian, N., Reuter, H. I., & Malakouti, M. J. (2007b). Methods to interpolate soil categorical variables from profile observations: Lessons from Iran. *Geoderma, 140*, 417–427.

Herbst, M., Diekkrüger, B., & Vereecken, H. (2006). Geostatistical co-regionalization of soil hydraulic properties in a micro-scale catchment using terrain attributes. *Geoderma, 132*, 206–221.

Heuvelink, G. B. M., Burrough, P. A., & Stein, A. (1989). Propagation of errors in spatial modelling with GIS. *International Journal of Geographical Information Systems, 3*, 303–322.

Heuvelink, G. B. M., & Webster, R. (2001). Modelling soil variation: Past, present, and future. *Geoderma, 100*, 269–301.

Hewitt, E., & Hewitt, R. E. (1980). The Gibbs–Wilbraham phenomenon: An episode in Fourier analysis. *Archive for History of Exact Sciences, 21*, 129–160.

Hilley, G. E., Mynatt, I., & Pollard, D. D. (2010). Structural geometry of Raplee Ridge monocline and thrust fault imaged using inverse Boundary Element Modeling and ALSM data. *Journal of Structural Geology, 32*, 45–58.

Hills, E. S. (1956). A contribution to the morphotectonics of Australia. *Journal of Geological Society of Australia, 3*, 1–15.

Hirano, A., Welch, R., & Lang, H. (2003). Mapping from ASTER stereo image data: DEM validation and accuracy assessment. *ISPRS Journal of Photogrammetry and Remote Sensing, 57*, 356–370.

Hjelle, O., & Dæhlen, M. (2006). *Triangulations and applications* (234 p.). Berlin: Springer.

Hobbs, W. H. (1904). Lineaments of Atlantic Border region. *Geological Society of America Bulletin, 15*, 483–506.

Hobbs, W. H. (1911). Repeating patterns in the relief and in the structure of the land. *Geological Society of America Bulletin, 22*, 123–176.

Hodgson, M. E. (1995). What cell size does the computed slope/aspect angle represent? *Photogrammetric Engineering and Remote Sensing, 61*, 513–517.

Hodgson, R. A., Gay, S. P., & Benjamins, J. Y. (Eds.), (1976). *Proceedings of the 1st International conference on the new basement tectonics, 3–7 June 1974, Salt Lake City, UT*. Salt Lake City: Utah Geological Association.

Hollingsworth, I. D., Bui, E. N., Odeh, I. O. A., & McLeod, P. (2007). Rule-based land unit mapping of the Tiwi Islands, Northern Territory, Australia. In P. Lagacherie, A. B. McBratney, & M. Voltz (Eds.), *Digital soil mapping. An introductory perspective* (pp. 401–414). Amsterdam: Elsevier.

Holmes, K. W., Chadwick, O. A., & Kyriakidis, P. C. (2000). Error in a USGS 30-meter digital elevation model and its impact on terrain modeling. *Journal of Hydrology, 233*, 154–173.

Horn, B. K. P. (1981). Hill shading and the reflectance map. *Proceedings of the IEEE, 69*, 14–47.

Hortal, M., & Simmons, A. J. (1991). Use of reduced Gaussian grids in spectral models. *Monthly Weather Review, 119*, 1057–1074.

Horton, R. E. (1945). Erosional development of streams and their drainage basins, hydrophysical approach to quantitative morphology. *Geological Society of America Bulletin, 56*, 275–370.

Howard, A. D. (1978). Origin of the stepped topography of the Martian poles. *Icarus, 34*, 581–599.

Howell, D., Kim, Y., Haydu-Houdeshell, C., Clemmer, P., Almaraz, R., & Ballmer, M. (2007). Fitting soil property spatial distribution models in the Mojave Desert for digital soil mapping. In P. Lagacherie, A. B. McBratney, & M. Voltz (Eds.), *Digital soil mapping: An introductory perspective* (pp. 465–475). Amsterdam: Elsevier.

Huang, X., Senthilkumar, S., Kravchenko, A., Thelen, K., & Qi, J. (2007). Total carbon mapping in glacial till soils using Near Infrared Spectroscopy, Landsat Imagery, and topographical information. *Geoderma, 141*, 34–42.

Hudleston, P. J. (1973). Fold morphology and some geometrical implications of theories of fold development. *Tectonophysics, 16*, 1–46.

Huggett, R. J. (1975). Soil landscape systems: A model of soil genesis. *Geoderma, 13*, 1–22.

Huggett, R. J., & Cheesman, J. (2002). *Topography and the environment* (274 p.). Harlow, England: Pearson Education.

Huising, E. J., & Gomes-Pereira, L. M. (1998). Errors and accuracy estimates of laser data acquired by various laser scanning systems for topographic applications. *ISPRS Journal of Photogrammetry and Remote Sensing, 53*, 245–261.

Hunter, G. J., & Goodchild, M. F. (1995). Dealing with error in spatial databases: A simple case study. *Photogrammetric Engineering and Remote Sensing, 61*, 529–537.

Hutchinson, G. L., & Mosier, A. R. (1981). Improved soil cover method for field measurement of nitrous oxide fluxes. *Soil Science Society of America Journal, 45*, 311–316.

Ichoku, C., Chorowicz, J., & Parrot, J. F. (1994). Computerized construction of geological cross sections from digital maps. *Computers and Geosciences, 20*, 1321–1327.

Ioffe, A. I., & Kozhurin, A. I. (1997). Active tectonics and geoenvironmental zoning of the Moscow Region. *Bulletin of Moscow Society of Naturalists, Geological Series, 72*(5), 31–35 (in Russian, with English abstract).

Isaacson, D. L., & Ripple, W. J. (1990). Comparison of 7.5-minute and 1-degree digital elevation models. *Photogrammetric Engineering and Remote Sensing, 56*, 1523–1527.

Ivanov, M. A. (2008). Global geological map of Venus: Preliminary results. In *Abstracts of the 39th Lunar and planetary science conference, 10–14 March 2008, League City, TX*. Abstract # 1017.

Ivanov, S. S. (1994). Estimation of fractal dimension of the global relief. *Oceanology, 34*, 94–98.

Ivanov, V. I., & Kruzhkov, V. A. (1992). Evaluation of the optimal discretization step for a digital elevation model. *Geodezia i Cartografia, No. 5*, 47–50 (in Russian).

Jakobsson, M., Macnab, R., Mayer, L., Anderson, R., Edwards, M., Hatzky, J., et al. (2008). An improved bathymetric portrayal of the Arctic Ocean: Implications for ocean modeling and geological, geophysical and oceanographic analyses. *Geophysical Research Letters, 35*, L07602, doi:10.1029/2008GL033520.

Jancaitis, J. R. (1978). *Elevation data compaction by polynomial modeling. Report No. ETL-0140* (42 p.). Fort Belvoir, VA: U.S. Army Engineer Topographic Laboratories.

Jarvis, A., Rubiano, J., Nelson, A., Farrow, A., & Mulligan, M. (2004). *Practical use of SRTM data in the tropics—Comparisons with digital elevation models generated from cartographic data* (32 p.). Cali, Columbia: Centro Internacional de Agricultura Tropical.

Jenčo, M. (1992). The morphometric analysis of georelief in terms of a theoretical conception of the complex digital model of georelief. *Acta Facultatis Rerum Naturalium Universitatis Comenianae, Geographica, 33*, 133–151.

Jenks, G. F. (1963). Generalization in statistical mapping. *Annals of the Association of American Geographers, 53*, 15–26.

Jenny, H. (1941). *Factors of soil formation. A system of quantitative pedology* (281 p.). New York: McGraw Hill.

Jenny, H., Jenny, B., & Hurni, L. (2010). Interactive design of 3D maps with progressive projection. *Cartographic Journal, 47*, 211–221.

Jensen, J. R. (1995). Issues involving the creation of digital elevation models and terrain corrected orthoimagery using soft-copy photogrammetry. *Geocarto International, 10*, 5–21.

Jenson, S. K., & Domingue, J. Q. (1988). Extracting topographic structure from digital elevation data for geographic information system analysis. *Photogrammetric Engineering and Remote Sensing, 54,* 1593−1600.

Jerri, A. J. (1998). *The Gibbs phenomenon in Fourier analysis, splines and wavelet approximation* (336 p.). Boston: Kluwer.

Jet Propulsion Laboratory (2009). *ASTER global digital elevation model (GDEM).* Pasadena, CA: Jet Propulsion Laboratory, California Institute of Technology, NASA. <http://asterweb.jpl.nasa.gov/gdem.asp/>.

Ji, W., Civco, D. L., & Kennard, W. C. (1992). Satellite remote bathymetry—A new mechanisms for modeling. *Photogrammetric Engineering and Remote Sensing, 58,* 545−549.

Johansson, M. (1999). Analysis of digital elevation data for palaeosurfaces in south-western Sweden. *Geomorphology, 26,* 279−295.

Jones, R. (2002). Algorithms for using a DEM for mapping catchment areas of stream sediment samples. *Computers and Geosciences, 28,* 1051−1060.

Jones, R. R., McCaffrey, K. J. W., Clegg, P., Wilson, R. W., Holliman, N. S., Holdsworth, R. E., et al. (2009). Integration of regional to outcrop digital data: 3D visualisation of multi-scale geological models. *Computers and Geosciences, 35,* 4−18.

Jones, T. A., & Johnson, C. R. (1983). Stratigraphic relationships and geologic history depicted by computer mapping. *American Association of Petroleum Geologists Bulletin, 67,* 1415−1421.

Jordan, G. (2003). Morphometric analysis and tectonic interpretation of digital terrain data: A case study. *Earth Surface Processes and Landforms, 28,* 807−822.

Jordan, G. (2007). Adaptive smoothing of valleys in DEMs using TIN interpolation from ridgeline elevations: An application to morphotectonic aspect analysis. *Computers and Geosciences, 33,* 573−585.

Jordan, M. C. (1872). Sur les lignes de faîte et de thalweg. *Comptes Rendus Hebdomadaires des Séances de l'Académie des Sciences, 74,* 1457−1459.

Joughin, I., Winebrenner, D., Fahnestock, M., Kwok, R., & Krabill, W. (1996). Measurement of ice-sheet topography using satellite-radar interferometry. *Journal of Glaciology, 42,* 10−22.

Kachanoski, R. G., de Jong, E., & Rolston, D. E. (1985a). Spatial and spectral relationships of soil properties and microtopography: II. Density and thickness of B horizon. *Soil Science Society of America Journal, 49,* 812−816.

Kachanoski, R. G., Rolston, D. E., & de Jong, E. (1985b). Spatial and spectral relationships of soil properties and microtopography: I. Density and thickness of A horizon. *Soil Science Society of America Journal, 49,* 804−812.

Karakhanian, A. S. (1981). Recognition of large landslides by satellite image interpretation. *Izvestiya Vuzov, Geologia i Razvedka, No. 3,* 130−131 (in Russian).

Karkee, M., Steward, B. L., & Aziz, S. A. (2008). Improving quality of public domain digital elevation models through data fusion. *Biosystems Engineering, 101,* 293−305.

Kasimov, N. S. (1980). *Landscape geochemistry of fault zones (Exemplified by Kazakhstan)* (119 p.). Moscow: Moscow University Press (in Russian).

Kasimov, N. S., Kovin, M. I., Proskuryakov, Y. V., & Shmelkova, N. A. (1978). Geochemistry of the soils of fault zones (exemplified by Kazakhstan). *Soviet Soil Science, 10,* 397−406.

Kats, Y. G., Makarova, N. V., Kozlov, V. V., & Trofimov, D. M. (1981). Structural and geomorphological analysis of the Crimea by satellite image interpretation. *Izvestiya Vuzov, Geologia i Razvedka, No. 3,* 8−20 (in Russian).

Kats, Y. G., Poletaev, A. I., & Rumyantseva, E. F. (1986). *Principles of lineament tectonics* (140 p.). Moscow: Nedra (in Russian).

Katterfeld, G. N., & Charushin, G. V. (1973). General grid systems of planets. *Modern Geology, 4,* 253−287.

Kaufmann, O., & Martin, T. (2008). 3D geological modelling from boreholes, cross-sections and geological maps, application over former natural gas storages in coal mines. *Computers and Geosciences, 34*, 278–290.

Keating, D. P., & Fischer, M. P. (2008). An experimental evaluation of the curvature–strain relation in fault-related folds. *American Association of Petroleum Geologists Bulletin, 92*, 869–884.

Keller, E. A. (1986). Investigation of active tectonics: Use of surficial earth processes. In Geophysics Study Committee, Geophysics Research Forum, and National Research Council, *Active tectonics* (pp. 136–147). Washington, DC: National Academy Press.

Kerrich, R. (1986). Fluid transport in lineaments. *Philosophical Transactions of the Royal Society of London, Series A, 317*, 219–251.

Kershaw, K. A., & Looney, J. H. H. (1985). *Quantitative and dynamic plant ecology* (3rd ed., 282 p.). London: Arnold.

Khalil, M. A. K., & Rasmussen, R. A. (1992). The global sources of nitrous oxide. *Journal of Geophysical Research, 97*(D), 14651–14660.

King, D., Bourennane, H., Isambert, M., & Macaire, J. J. (1999). Relationship of the presence of a non-calcareous clay-loam horizon to DEM attributes in a gently sloping area. *Geoderma, 89*, 95–111.

King, P. B., & Beikman, H. M. (1974). *Geologic map of the United States (Exclusive of Alaska and Hawaii), scale 1 : 2,500,000* (3 p.). Reston, VA: U.S. Geological Survey.

Kirkby, M. J. (1997). TOPMODEL: A personal view. *Hydrological Processes, 11*, 1087–1097.

Kirkby, M. J., & Chorley, R. J. (1967). Throughflow, overland flow and erosion. *Bulletin of the International Association of Scientific Hydrology, 12*, 5–21.

Kleim, R. F., Skaugset, A. E., & Bateman, D. S. (1999). Digital terrain modeling of small stream channels with a total-station theodolite. *Advances in Water Resources, 23*, 41–48.

Kleinbaum, D. G., Kupper, L. L., Nizam, A., & Muller, K. E. (2008). *Applied regression analysis and other multivariable methods* (4th ed., 928 p.). Belmont, CA: Thomson Brooks/Cole.

Klemeš, V. (1983). Conceptualization and scale in hydrology. *Journal of Hydrology, 65*, 1–23.

Knetsch, G. (1965). Über ein Structur-Experiment an einer Kugel und Beziehungen zwischen Gross-Lineamenten und Pol-Lagen in der Erdeschichte. *Geologische Rundschau, 54*, 523–548.

Koenderink, J. J., & van Doorn, A. J. (1992). Surface shape and curvature scales. *Image and Vision Computing, 10*, 557–565.

Koenderink, J. J., & van Doorn, A. J. (1993). Local features of smooth shapes: Ridges and courses. *Proceedings of the SPIE, 2013*, 2–13.

Koenderink, J. J., & van Doorn, A. J. (1994). Two-plus-one-dimensional differential geometry. *Pattern Recognition Letters, 15*, 439–443.

Koike, K., Nagano, S., & Kawaba, K. (1998). Construction and analysis of interpreted fracture planes through combination of satellite-image derived lineaments and digital elevation model data. *Computers and Geosciences, 24*, 573–583.

Kondratyev, K. Y., Pivovarova, Z. I., & Fedorova, M. P. (1978). *Radiation regime of inclined surfaces* (215 p.). Leningrad: Hydrometeoizdat (in Russian, with English abstract).

Konovalov, N. E. (1960). *A project of a new railway line, with development of ideas on computers application to solve particular tasks*. M.Sc. Thesis. Kharkov Institute of Engineers for Railway Transport, Kharkov, USSR, 120 p. (in Russian).

Kopp, M. L. (2000). The recent deformations of the Scythian and Southern East European Platforms as a result of pressure form the Arabian Plate. *Geotectonics, 34*, 106–120.

Korobeynik, V. M., Komarova, M. V., & Shtengelov, Y. S. (1982). Permeable faults of the Earth crust in the Crimea and the north-western Black Sea area. *Doklady Academii Nauk Ukrainskoi SSR, Series B, No. 2*, 13–16 (in Russian, with English abstract).

Korsakova, O. P. (2002). Morphological analysis of the topography of the north-eastern part of the Baltic Shield. *Geomorfologia, No. 4*, 87–95 (in Russian, with English abstract).

Koshkarev, A. V. (1982). Topography as an input parameter for mathematical cartographic models of geosystems. In O. A. Evteev (Ed.), *Geographical cartography in scientific research and national economic practices* (pp. 117–131). Moscow: Moscow Branch, Soviet Geographical Society (in Russian).

Kotelnikov, V. A. (1933). On the transmission capacity of the "ether" and wire in electro-communications. In *Proceedings of the 1st All-union conference on the technological reconstruction of communications sector and the development of low-current engineering, January 1933, Moscow, USSR* (pp. 1–33). Moscow: Military Communication Department Press (in Russian).

Kotelnikov, V. A. (2001). On the transmission capacity of the "ether" and wire in electro-communications. In J. J. Benedetto, & P. J. S. G. Ferreira (Eds.), *Modern sampling theory: Mathematics and applications* (pp. 27–45). Boston: Birkhäuser.

Kovalevsky, S. A. (1965). The Mid deep fault of the Crimean Peninsula. *Doklady Academii Nauk SSSR, 162*, 887–890 (in Russian).

Kovda, V. A. (1946). *Origin of saline soils and their regime* (Vol. 1, 573 p.). Moscow: Soviet Academic Press (in Russian).

Kovda, V. A. (1971). *Origin of saline soils and their regime* (Vol. 1, 509 p.). Jerusalem: Israel Program for Scientific Translations.

Kovda, V. A. (1973). *The principles of pedology. General theory of soil formation* (Vol. 1, 447 p.). Moscow: Nauka (in Russian).

Kozlov, V. V. (1991). *Application of remotely sensed data in seismic studies* (32 p.). Moscow: All-Union Scientific Research Institute of Economics of Mineral Raw Material and Prospecting (in Russian).

Krasilnikov, N. N., Krasilnikova, O. I., & Shelepin, Y. E. (2000). Perception of achromatic, monochromatic, pure chromatic, and chromatic noisy images by real human–observer under threshold conditions. *Proceedings of the SPIE, 3981*, 78–85.

Kraus, K. (1994). Visualization of the quality of surfaces and their derivatives. *Photogrammetric Engineering and Remote Sensing, 60*, 457–462.

Kravchenko, A. N., Bollero, G. A., Omonode, R. A., & Bullock, D. G. (2002). Quantitative mapping of soil drainage classes using topographical data and soil electrical conductivity. *Soil Science Society of America Journal, 66*, 235–243.

Krcho, J. (1973). Morphometric analysis of relief on the basis of geometric aspect of field theory. *Acta Geographica Universitatis Comenianae, Geographico-Physica, 1*, 7–233.

Krcho, J. (1983). Teoretická koncepcia a interdisciplinárne aplikácie komplexného digitálneho modelu reliéfu pri modelovaní dvojdimenzionálnych polí. *Geografický Časopis, 35*, 265–291.

Kuhn, K., & Fedorko, E. (2006). Development of a 1/9th arc-second elevation dataset for West Virginia. In *Proceedings of MAPPS / ASPRS 2006 fall conference, 6–10 November 2006, San Antonio, TX* (8 p.). Bethesda, MD: American Society for Photogrammetry and Remote Sensing (CD–ROM).

Kühni, A., & Pfiffner, O. A. (2001a). Drainage patterns and tectonic forcing: A model study for the Swiss Alps. *Basin Research, 13*, 169–197.

Kühni, A., & Pfiffner, O. A. (2001b). The relief of the Swiss Alps and adjacent areas and its relation to lithology and structure: Topographic analysis from a 250-m DEM. *Geomorphology, 41*, 285–307.

Kukowski, N., Hampel, A., Hoth, S., & Bialas, J. (2008). Morphotectonic and morphometric analysis of the Nazca plate and the adjacent offshore Peruvian continental slope—Implications for submarine landscape evolution. *Marine Geology, 254*, 107–120.

Kukowski, N., Schillhorn, T., Huhn, K., von Rad, U., Husen, S., & Flueh, E. R. (2001). Morphotectonics and mechanics of the central Makran accretionary wedge off Pakistan. *Marine Geology, 173*, 1–19.

Kumler, M. P. (1994). An intensive comparison of triangulated irregular networks (TINs) and digital elevation models (DEMs). *Cartographica, 31*, 1−99.

Kuriakose, S. L., Devkota, S., Rossiter, D. G., & Jetten, V. G. (2009). Prediction of soil depth using environmental variables in an anthropogenic landscape, a case study in the Western Ghats of Kerala, India. *Catena, 79*, 27−38.

Kuryakova, G. A. (1996). *A methodology for studying and preparation of initial data for DTM-based mapping of biogeocenoses*. Ph.D. Thesis. Moscow State University of Geodesy and Cartography, Moscow, 136 p. (in Russian).

Kuryakova, G. A., & Florinsky, I. V. (1991). *The analysis of spatial relationships between ring structures, topography, and soil cover* (14 p.). Pushchino, USSR: Pushchino Research Centre Press (in Russian, with English abstract).

Kuryakova, G. A., Florinsky, I. V., & Shary, P. A. (1992). On the correlation between soil moisture and some topographic variables. In V. L. Andronikov (Ed.), *The modern problems of soil geography and mapping: Proceedings of the conference, 24−26 September 1991, Moscow, USSR* (pp. 70−71). Moscow: Dokuchaev Soil Institute (in Russian).

Kutina, J. (1974). Structural control of volcanic ore deposits in the context of global tectonics. *Bulletin of Volcanology, 38*, 1037−1069.

Kvietkauskas, V. (1963−1964). Keturspalvis morfograsfinis zemelapis. In S. Tarvydas (Ed.), *Geografinis metrastis VI-VII. Ledyninio reljefo morfogeneze ir dabartiniai egzogeniniai procesai* (pp. 87−107). Vilnius, USSR: Lietuvos TSR Geografine Draugija.

Kweon, I. S., & Kanade, T. (1994). Extracting topographic terrain features from elevation maps. *CVGIP: Image Understanding, 59*, 171−182.

Lagacherie, P. (2008). Digital soil mapping: A state of the art. In A. E. Hartemink, A. McBratney, & M. L. Mendonça-Santos (Eds.), *Digital soil mapping with limited data* (pp. 3−14). Dordrecht: Springer.

Lagacherie, P., & Holmes, S. (1997). Addressing geographical data errors in a classification tree for soil unit prediction. *International Journal of Geographical Information Science, 11*, 183−198.

Lagacherie, P., & McBratney, A. B. (2007). Spatial soil information systems and spatial soil inference systems: Perspectives for digital soil mapping. In P. Lagacherie, A. B. McBratney, & M. Voltz (Eds.), *Digital soil mapping: An introductory perspective* (pp. 3−22). Amsterdam: Elsevier.

Lagacherie, P., Robbez-Masson, J. M., Nguyen-The, N., & Barthès, J. P. (2001). Mapping of reference area representativity using a mathematical soilscape distance. *Geoderma, 101*, 105−118.

Lam, N. S. (1983). Spatial interpolation methods: A review. *American Cartographer, 10*, 129−149.

Lanyon, L. E., & Hall, G. F. (1983). Land surface morphology: 2. Predicting potential landscape instability in eastern Ohio. *Soil Science, 136*, 382−386.

Lark, R. M. (1999). Soil−landform relationships at within-field scales: An investigation using continuous classification. *Geoderma, 92*, 141−165.

Lark, R. M. (2007). Decomposing digital soil information by spatial scale. In P. Lagacherie, A. B. McBratney, & M. Voltz (Eds.), *Digital soil mapping: An introductory perspective* (pp. 301−326). Amsterdam: Elsevier.

Lark, R. M., Bishop, T. F. A., & Webster, R. (2007). Using expert knowledge with control of false discovery rate to select regressors for prediction of soil properties. *Geoderma, 138*, 65−78.

Lark, R. M., Cullis, B. R., & Welham, S. J. (2006). On spatial prediction of soil properties in the presence of a spatial trend: The empirical best linear unbiased predictor (E-BLUP) with REML. *European Journal of Soil Science, 57*, 787−799.

Lascelles, B., Favis-Mortlock, D., Parsons, T., & Boardman, J. (2002). Automated digital photogrammetry: A valuable tool for small-scale geomorphological research for the non-photogrammetrist? *Transactions in GIS, 6*, 5−15.

Lastochkin, A. N. (1987). *Morphodynamical analysis* (254 p.). Leningrad: Nedra (in Russian).
Lattman, L. H. (1958). Technique of mapping geologic fracture traces and lineaments on aerial photographs. *Photogrammetric Engineering, 24*, 568–576.
Lattman, L. H., & Parizek, R. R. (1964). Relationships between fracture traces and occurrence of ground water in carbonate rocks. *Journal of Hydrology, 2*, 73–91.
Lawrence, G. R. P. (1971). *Cartographic methods* (162 p.). London: Methuen.
Lea, N. L. (1992). An aspect driven kinematic routing algorithm. In A. J. Parsons, & A. D. Abrahams (Eds.), *Overland flow: Hydraulics and erosion mechanics* (pp. 393–407). New York: Chapman and Hall.
Lebedev, T. S., & Orovetsky, Y. P. (1966). Tectonic peculiarities of the Mountain Crimea (in the context of new geological and geophysical data). *Geophysical Communications, 18*, 34–41 (in Russian).
Leenaers, H., Okx, J. P., & Burrough, P. A. (1990). Employing elevation data for efficient mapping of soil pollution on floodplains. *Soil Use and Management, 6*, 105–114.
Legates, D. R., Mahmood, R., Levia, D. F., DeLiberty, T. L., Quiring, S. M., Houser, C., et al. (2011). Soil moisture: A central and unifying theme in physical geography. *Progress in Physical Geography, 35*, 65–86.
Lehman, R. M. (2007). Microbial distributions and their potential controlling factors in terrestrial subsurface environments. In R. B. Franklin, & A. L. Mills (Eds.), *The spatial distribution of microbes in the environment* (pp. 135–178). Dordrecht: Springer.
Lehmann, J. G. (1816). *Die Lehre der Situation-Zeichnung, oder Anweisung zum richtigen Erkennen und genauen Abbilden der Erdoberfläche in topographischen Karten und Situation-Planen* (2nd ed., 60 p.). Dresden: Arnoldische Buchhandlung.
Leigh, C. L., Kidner, D. B., & Thomas, M. C. (2009). Use of LiDAR in digital surface modelling: Issues and errors. *Transactions in GIS, 13*, 345–361.
Lemmens, M. J. P. M. (1999). Quality description problems of blindly sampled DEMs. In W. Shi, M. F. Goodchild, & P. F. Fisher (Eds.), *Proceedings of the International symposium on spatial data quality'99, 18–20 July 1999, Hong Kong, China* (pp. 210–218). Hong Kong: Hong Kong Polytechnic University.
Lemon, A. M., & Jones, N. L. (2003). Building solid models from boreholes and user-defined cross-sections. *Computers and Geosciences, 29*, 547–555.
Leonowicz, A. M., Jenny, B., & Hurni, L. (2009). Automatic generation of hypsometric layers for small-scale maps. *Computers and Geosciences, 35*, 2074–2083.
Levorsen, A. I. (1927). Convergence studies in the Mid-Continent region. *American Association of Petroleum Geologists Bulletin, 11*, 657–682.
Li, Z. (1994). A comparative study of the accuracy of digital terrain models (DTMs) based on various data models. *ISPRS Journal of Photogrammetry and Remote Sensing, 49*, 2–11.
Li, Z., Zhu, Q., & Gold, C. (2005). *Digital terrain modeling: Principles and methodology* (323 p.). New York: CRC Press.
Liang, C., & Mackay, D. S. (2000). A general model of watershed extraction and representation using globally optimal flow paths and up-slope contributing areas. *International Journal of Geographical Information Science, 14*, 337–358.
Lidov, V. P. (1949). From the experience in landscape mapping of the Prioksko-Terrasny State Reserve. *Voprosy Geografii, 16*, 179–190 (in Russian).
Linder, W. (2006). *Digital photogrammetry: A practical course* (2nd ed., 214 p.). Berlin: Springer.
Lindsay, J. B., & Creed, I. F. (2005). Removal of artifact depressions from digital elevation models: Towards a minimum impact approach. *Hydrological Processes, 19*, 3113–3126.
Liner, C. L. (2004). *Elements of 3D seismology* (2nd ed., 608 p.). Tulsa, OK: PennWell Books.
Lisle, R. J. (1994). Detection of zones of abnormal strains in structures using Gaussian curvature analysis. *American Association of Petroleum Geologists Bulletin, 78*, 1811–1819.

Lisle, R. J., & Robinson, J. M. (1995). The Mohr circle for curvature and its application to fold description. *Journal of Structural Geology, 17*, 739–750.

Lisle, R. J., & Toimil, N. C. (2007). Defining folds on three-dimensional surfaces. *Geology, 35*, 519–522.

Lisle, R. J., Toimil, N., Aller, J., Bobillo-Ares, N., & Bastida, F. (2010). The hinge lines of non-cylindrical folds. *Journal of Structural Geology, 32*, 166–171.

Liu, H., Jezek, K. C., & Li, B. (1999). Development of an Antarctic digital elevation model by integrating cartographic and remotely sensed data: A GIS based approach. *Journal of Geophysical Research, 104*(B), 23199–23213.

Liu, H., Jezek, K., Li, B., & Zhao, Z. (2001). *Radarsat Antarctic mapping project digital elevation model, version 2*. Boulder, CO: National Snow and Ice Data Center. <http://nsidc.org/data/nsidc-0082.html/>.

Liu, X. (2008). Airborne LiDAR for DEM generation: Some critical issues. *Progress in Physical Geography, 32*, 31–49.

Lloyd, C. D., & Atkinson, P. M. (2006). Deriving ground surface digital elevation models from LiDAR data with geostatistics. *International Journal of Geographical Information Science, 20*, 535–563.

Lopatin, D. V. (2002). Lineament tectonics and giant ore deposits of the Northern Eurasia. *Issledovanie Zemli iz Cosmosa*, No. 2, 77–91 (in Russian, with English abstract).

Lopatin, D. V. (2008). Cryptomorphic structures of the lithosphere: Their reflection on space images and the Earth's surface. *Doklady Earth Sciences, 421*, 983–986.

López, A. M., Lumbreras, F., & Serrat, J. (1998). Creaseness from level set extrinsic curvature. *Lecture Notes in Computer Science, 1407*, 156–169.

López, A. M., Lumbreras, F., Serrat, J., & Villanueva, J. J. (1999). Evaluation of methods for ridge and valley detection. *IEEE Transactions on Pattern Analysis and Machine Intelligence, 21*, 327–335.

López, A. M., & Serrat, J. (1996). Tracing crease curves by solving a system of differential equations. *Lecture Notes in Computer Science, 1064*, 241–250.

López, C. (1997). Locating some types of random errors in Digital Terrain Models. *International Journal of Geographical Information Science, 11*, 677–698.

Lukatela, H. (2000). A seamless global terrain model in the Hipparchus System. In M. Goodchild, & A. J. Kimerling (Eds.), *Discrete global grids*. Santa Barbara, CA: National Center for Geographic Information and Analysis. <http://www.ncgia.ucsb.edu/globalgrids-book/terra/>.

Lukin, A. A. (1987). *Development of a technique for morphostructural hydrogeological analysis* (111 p.). Novosibirsk: Nauka (in Russian).

Lythe, M. B., Vaughan, D. G., & the BEDMAP Consortium (2001). BEDMAP: A new ice thickness and subglacial topographic model of Antarctica. *Journal of Geophysical Research, 106*(B), 11335–11351.

Lyubashin, V. N., & Lisitsin, V. V. (1981). *A program of the engineering survey for justification of the technical project of anti-landslide actions and development at the slope of the Oka River valley in the city of Pushchino, Serpukhov District, Moscow Region. Progress report No. 3603* (40 p.). Moscow: Central Trust of Engineering Surveys (in Russian).

Lyubkov, A. N., & Martynenko, A. I. (1963). Producing topographic models by computer controlled milling machines. *Sbornik Nauchno-Tekhnicheskikh i Proizvodstvennykh Statei NII Voenno–Topograficheskoi Sluzhby, 12*, 102–109 (in Russian).

Mackaness, W. A., Ruas, A., & Sarjakoski, L. T. (Eds.), (2007). *Generalisation of geographic information: Cartographic modelling and applications* (370 p.). Amsterdam: Elsevier.

MacMillan, R. A. (2008). Experiences with applied DSM: Protocol, availability, quality and capacity building. In A. E. Hartemink, A. McBratney, & M. L. Mendonça–Santos (Eds.), *Digital soil mapping with limited data* (pp. 113–135). Dordrecht: Springer.

MacMillan, R. A., & Pettapiece, W. W. (2000). *Alberta landforms: Quantitative morphometric descriptions and classification of typical Alberta landforms. Technical bulletin No. 2000-2E* (118 p.). Swift Current, SK: Semiarid Prairie Agricultural Research Centre, Agriculture and Agri-Food Canada.

MacMillan, R. A., Pettapiece, W. W., Nolan, S. C., & Goddard, T. W. (2000). A generic procedure for automatically segmenting landforms into landform elements using DEMs, heuristic rules and fuzzy logic. *Fuzzy Sets and Systems, 113*, 81–109.

MacMillan, R. A., & Shary, P. A. (2009). Landforms and landform elements in geomorphometry. In T. Hengl, & H. I. Reuter (Eds.), *Geomorphometry: Concepts, software, applications* (pp. 227–254). Amsterdam: Elsevier.

Maerten, L., Pollard, D. D., & Maerten, F. (2001). Digital mapping of three-dimensional structures of the Chimney Rock fault system, central Utah. *Journal of Structural Geology, 23*, 585–592.

Magnus, J. R., & Neudecker, H. (2007). *Matrix differential calculus with applications in statistics and econometrics* (3rd ed., 450 p.). Chichester, England: Wiley.

Makarov, V. I. (1981). Lineaments: Problems and trends in airborne and spaceborne studies. *Soviet Journal of Remote Sensing, No. 4*, 646–658.

Makarovič, B. (1972). Information transfer in reconstruction of data from sampled points. *Photogrammetria, 28*, 111–130.

Makarovič, B. (1973). Progressive sampling for digital terrain models. *ITC Journal, No. 3*, 397–416.

Makarovič, B. (1977). Composite sampling for digital terrain models. *ITC Journal, No. 3*, 406–433.

Mallavan, B. P., Minasny, B., & McBratney, A. B. (2010). Homosoil, a methodology for quantitative extrapolation of soil information across the globe. In J. L. Boettinger, D. W. Howell, A. C. Moore, A. E. Hartemink, & S. Kienast-Brown (Eds.), *Digital soil mapping—Bridging research, environmental application, and operation* (pp. 137–150). Dordrecht: Springer.

Malone, B. P., McBratney, A. B., Minasny, B., & Laslett, G. M. (2009). Mapping continuous depth functions of soil carbon storage and available water capacity. *Geoderma, 154*, 138–152.

Manning, G. R. (1999). *Relations between spatial variability of soil properties and grain yield response to nitrogen fertilizer in a variable Manitoba soil-landscape*. M.Sc. Thesis. University of Manitoba, Winnipeg, MB, 280 p.

Marchetti, A., Piccini, C., Francaviglia, R., Santucci, S., & Chiuchiarelli, I. (2010). Estimating soil organic matter content by regression kriging. In J. L. Boettinger, D. W. Howell, A. C. Moore, A. E. Hartemink, & S. Kienast-Brown (Eds.), *Digital soil mapping—Bridging research, environmental application, and operation* (pp. 241–254). Dordrecht: Springer.

Mardia, K. V. (1972). *Statistics of directional data* (357 p.). London: Academic Press.

Mark, D. M. (1975a). Computer analysis of topography: A comparison of terrain storage methods. *Geografiska Annaler, Series A, 57*, 179–188.

Mark, D. M. (1975b). Geomorphometric parameters: A review and evaluation. *Geografiska Annaler, Series A, 57*, 165–177.

Mark, D. M. (1979). Phenomenon-based data-structuring and digital terrain modelling. *Geo-Processing, 1*, 27–36.

Mark, D. M. (1984). Automated detection of drainage networks from digital elevation models. *Cartographica, 21*, 168–178.

Martz, L. W., & de Jong, E. (1988). CATCH: A Fortran program for measuring catchment area from digital elevation models. *Computers and Geosciences, 14*, 627–640.

Maslov, L. A., & Anokhin, V. M. (2006). The Earth's decelerated rotation and regularities in orientation of its surface lineaments and faults. *Planetary and Space Science, 54*, 216–218.

Masoud, A., & Koike, K. (2006). Tectonic architecture through Landsat-7 ETM +/SRTM DEM-derived lineaments and relationship to the hydrogeologic setting in Siwa region, NW Egypt. *Journal of African Earth Sciences, 45*, 467–477.

Masumoto, S., Raghavan, V., Yonezawa, G., Nemoto, T., & Shiono, K. (2004). Construction and visualization of a three dimensional geologic model using GRASS GIS. *Transactions in GIS, 8*, 211–223.

Maxwell, J. C. (1870). On hills and dales. *The London, Edinburgh and Dublin Philosophical Magazine and Journal of Science, Series 4, 40*, 421–427.

McArthur, D. E., Fuentes, R. W., & Devarajan, V. (2000). Generation of hierarchical multi-resolution terrain databases using wavelet filtering. *Photogrammetric Engineering and Remote Sensing, 66*, 287–295.

McBratney, A. B., Mendonça Santos, M. L., & Minasny, B. (2003). On digital soil mapping. *Geoderma, 117*, 3–52.

McBratney, A. B., & Odeh, I. O. A. (1997). Application of fuzzy sets in soil science: Fuzzy logic, fuzzy measurement, and fuzzy decisions. *Geoderma, 77*, 85–113.

McBratney, A. B., Odeh, I. O. A., Bishop, T. F. A., Dunbar, M. S., & Shatar, T. M. (2000). An overview of pedometric techniques for use in soil survey. *Geoderma, 97*, 293–327.

McClean, C. J., & Evans, I. S. (2000). Apparent fractal dimensions from continental scale digital elevation models using variogram methods. *Transactions in GIS, 4*, 361–378.

McCullagh, M. J. (1981). Creation of smooth contours over irregularly distributed data using local surface patches. *Geographical Analysis, 13*, 51–63.

McCullagh, M. J. (1988). Terrain and surface modelling systems: Theory and practice. *Photogrammetric Record, 12*, 747–779.

McCullagh, P., & Nelder, J. (1999). *Generalized linear models* (2nd ed., 511 p.). Boca Raton, FL: Chapman and Hall/CRC.

McIntyre, M. L., Naar, D. F., Carder, K. L., Donahue, B. T., & Mallison, D. J. (2006). Coastal bathymetry from hyperspectral remote sensing data: Comparisons with high resolution multibeam bathymetry. *Marine Geophysical Researches, 27*, 128–136.

McKenzie, N. J., & Ryan, P. J. (1999). Spatial prediction of soil properties using environmental correlation. *Geoderma, 89*, 67–94.

McMahon, M. J., & North, C. P. (1993). Three-dimensional integration of remotely sensed geological data: A methodology for petroleum exploration. *Photogrammetric Engineering and Remote Sensing, 59*, 1251–1256.

McMahon, S. (2001). *Influence of long-term hydrologic regime, as indicated by landscape position, on soil denitrifier populations*. B.Sc. Thesis. University of Manitoba, Winnipeg, MB, 82 p.

McMaster, R. B., & Shea, K. S. (1992). *Generalization in digital cartography* (134 p.). Washington, DC: Association of American Geographers.

Mead, R. (1974). A test for spatial pattern at several scales using data from a grid of contiguous quadrates. *Biometrics, 30*, 295–307.

Meixner, F. X., & Eugster, W. (1999). Effects of landscape pattern and topography on emissions and transport. In J. D. Tenhunen, & P. Kabat (Eds.), *Integrating hydrology, ecosystem dynamics, and biogeochemistry in complex landscapes* (pp. 147–175). Chichester, England: Wiley.

Mendonça-Santos, M. L., Dart, R. O., Santos, H. G., Coelho, M. R., Barbara, R. L. L., & Lumbreras, J. F. (2010). Digital soil mapping of topsoil organic carbon content of Rio de Janeiro State, Brazil. In J. L. Boettinger, D. W. Howell, A. C. Moore, A. E. Hartemink, & S. Kienast-Brown (Eds.), *Digital soil mapping—Bridging research, environmental application, and operation* (pp. 255–266). Dordrecht: Springer.

Mendonça-Santos, M. L., Guenat, C., Bouzelboudjen, M., & Golay, F. (2000). Three-dimensional GIS cartography applied to the study of the spatial variation of soil horizons in a Swiss floodplain. *Geoderma, 97*, 351–366.

Mendonça-Santos, M. L., Santos, H. G., Dart, R. O., & Pares, J. G. (2008). Digital mapping of soil classes in Rio de Janeiro State, Brazil: Data, modelling and prediction. In A. E. Hartemink, A. McBratney, & M. L. Mendonça-Santos (Eds.), *Digital soil mapping with limited data* (pp. 381–396). Dordrecht: Springer.

Merot, P., Ezzahar, B., Walter, C., & Aurousseau, P. (1995). Mapping waterlogging of soils using digital terrain models. *Hydrological Processes, 9*, 27–34.

Merriman, M. (1899). *Elements of precise surveying and geodesy* (261 p.). New York: Wiley.

Meshcheryakov, Y. A. (1965). *Structural geomorphology of plain lands* (390 p.). Moscow: Nauka (in Russian).

Miller, C. L., & Leflamme, R. A. (1958). The digital terrain model—Theory and application. *Photogrammetric Engineering, 24*, 433–442.

Minár, J., & Evans, I. S. (2008). Elementary forms for land surface segmentation: The theoretical basis of terrain analysis and geomorphological mapping. *Geomorphology, 95*, 236–259.

Minasny, B., & McBratney, A. B. (2007). Spatial prediction of soil properties using EBLUP with the Matérn covariance function. *Geoderma, 140*, 324–336.

Mishra, D., Narumalani, S., Lawson, M., & Rundquist, D. (2004). Bathymetric mapping using IKONOS multispectral data. *GIScience and Remote Sensing, 41*, 301–321.

Mishustin, E. N., & Shilnikova, V. K. (1971). *Biological fixation of atmospheric nitrogen* (420 p.). London: Macmillan.

Mitas, L., & Mitasova, H. (1999). Spatial interpolation. In P. Longley, M. F. Goodchild, D. J. Maguire, & D. W. Rhind (Eds.), *Geographical information systems: Principles, techniques, management and applications* (2nd abr ed) (pp. 481–492). Hoboken, NJ: Wiley.

Mitášová, H., & Hofierka, J. (1993). Interpolation by regularized spline with tension: II. Application to terrain modeling and surface geometry analysis. *Mathematical Geology, 25*, 657–669.

Mitášová, H., & Mitáš, L. (1993). Interpolation by regularized spline with tension: I. Theory and implementation. *Mathematical Geology, 25*, 641–655.

Mitusov, A. V. (2001). *The role of topography in soil fertility development in biogeocenoses of the forest steppe zone of the European part of Russia*. Ph.D. Thesis. Kursk State Agricultural Academy, Kursk, Russia, 270 p. (in Russian).

Moličová, H., Grimaldi, M., Bonell, M., & Hubert, P. (1997). Using TOPMODEL towards identifying and modelling the hydrological patterns within a headwater, humid, tropical catchment. *Hydrological Processes, 11*, 1169–1196.

Moody, J. D. (1966). Crustal shear patterns and orogenesis. *Tectonophysics, 3*, 479–522.

Mooney, W., Laske, G., & Master, T. (1998). CRUST 5.1: A global crustal model at 5×5. *Journal of Geophysical Research, 103*(B), 727–747.

Moonjun, R., Farshad, A., Shrestha, D. P., & Vaiphasa, C. (2010). Artificial neural network and decision tree in predictive soil mapping of Hoi Num Rin sub-watershed, Thailand. In J. L. Boettinger, D. W. Howell, A. C. Moore, A. E. Hartemink, & S. Kienast–Brown (Eds.), *Digital soil mapping—Bridging research, environmental application, and operation* (pp. 151–164). Dordrecht: Springer.

Moore, I. D., Gessler, P. E., Nielsen, G. A., & Peterson, G. A. (1993). Soil attribute prediction using terrain analysis. *Soil Science Society of America Journal, 57*, 443–452.

Moore, I. D., Grayson, R. B., & Ladson, A. R. (1991). Digital terrain modelling: A review of hydrological, geomorphological and biological applications. *Hydrological Processes, 5*, 3–30.

Moore, I. D., Mackay, S. M., Wallbrink, P. J., Burch, G. J., & O'Loughlin, E. M. (1986). Hydrologic characteristics and modelling of a small forested catchment in southeastern New South Wales. Pre-logging condition. *Journal of Hydrology, 83*, 307–335.

Moore, R. F., & Simpson, C. J. (1983). Image analysis—A new aid in morphotectonic studies. In *Proceedings of the 17th International symposium on remote sensing of environment, 9–13 May 1983, Ann Arbor, MI* (Vol. 3, pp. 991–1002). Ann Arbor, MI: Environmental Research Institute of Michigan.

Mor, M., & Lamdan, T. (1972). A new approach to automatic scanning of contour maps. *Communications of the ACM, 15*, 809–812.

Mora-Vallejo, A., Claessens, L., Stoorvogel, J., & Heuvelink, G. B. M. (2008). Small scale digital soil mapping in Southeastern Kenya. *Catena, 76*, 44–53.

Moran, C. J., & Bui, E. (2002). Spatial data mining for enhanced soil map modelling. *International Journal of Geographical Information Science, 16*, 533–542.

Morozov, V. I., Kovalenko, A. P., & Pasynkov, A. A. (1988). Sites of abnormally high discharges of the Crimean Mountains. *Geologichesky Journal, No. 2*, 65–69 (in Russian, with English abstract)

Morozov, V. P. (1979). *A course in spheroidal geodesy* (2nd enl rev ed., 296 p.). Moscow: Nedra (in Russian).

Morris, K. (1991). Using knowledge-base rules to map the three-dimensional nature of geological features. *Photogrammetric Engineering and Remote Sensing, 57*, 1209–1216.

Mourier, B., Walter, C., & Merot, P. (2008). Soil distribution in valleys according to stream order. *Catena, 72*, 395–404.

Mueller-Dombois, D., & Ellenberg, H. (1974). *Aims and methods of vegetation ecology* (547 p.). New York: Wiley.

Mulder, V. L., de Bruin, S., Schaepman, M. E., & Mayr, T. R. (2011). The use of remote sensing in soil and terrain mapping—A review. *Geoderma, 162*, 1–19.

Muñoz-Espinoza, R. E. (1968). *Fracture finding by structural curvature mapping*. M.Sc. Thesis. University of Texas at Austin, Austin, TX, 44 p.

Muratov, M. V. (Ed.), (1969). *Geology of the USSR, Vol. 8: The Crimea. Pt. 1: Geological description* (575 p.). Moscow: Nedra (in Russian).

Murray, G. H., Jr. (1968). Quantitative fracture study—Sanish Pool, McKenzie County, North Dakota. *American Association of Petroleum Geologists Bulletin, 52*, 57–65.

Murray, R. E., Parsons, L. L, & Smith, M. S. (1989). Kinetics of nitrate utilization by mixed populations of denitrifying bacteria. *Applied and Environmental Microbiology, 55*, 717–721.

Mynatt, I., Bergbauer, S., & Pollard, D. D. (2007a). Using differential geometry to describe 3-D folds. *Journal of Structural Geology, 29*, 1256–1266.

Mynatt, I., Hilley, G. E., & Pollard, D. D. (2007b). Inferring fault characteristics using fold geometry constrained by Airborne Laser Swath Mapping at Raplee Ridge, Utah. *Geophysical Research Letters, 34*, L16315, doi:10.1029/2007GL030548.

Myrold, D. D., & Tiedje, J. M. (1985). Establishment of denitrification capacity in soil: Effects of carbon, nitrate and moisture. *Soil Biology and Biochemistry, 17*, 819–822.

Myrzin, Y. N., Furinevich, O. S., & Artemyeva, E. S. (1988). *Report on hydrogeological and engineering geological surveys (scale 1 : 200,000) for areas of topographic sheets M-36-V, -VI, and -XII, 1984–1988* (489 p.). Moscow: Centergeologia (in Russian).

Nalivkin, D. V. (Ed.), (1970). *Geological map of the Russian Platform and adjacent areas, scale 1 : 1,500,000* (16 p.). Moscow: All-Union Aerial Geological Trust (in Russian).

Nalivkin, D. V. (Ed.), (1974). *Basement structure of platform areas of the USSR* (400 p.). Moscow: Nedra (in Russian).

NASA (2003). *SRTM3 Version 2*, <ftp://e0srp01u.ecs.nasa.gov/srtm/version2/SRTM3/>

National Geophysical Data Center (2001). *ETOPO2, 2-minute gridded global relief data*. Boulder, CO: National Geophysical Data Center, National Oceanic and Atmospheric Administration, US Department of Commerce. <http://www.ngdc.noaa.gov/mgg/fliers/01mgg04.html/>.

National Geophysical Data Center (2006). *ETOPO2v2, global gridded 2-minute database*. Boulder, CO: National Geophysical Data Center, National Oceanic and Atmospheric Administration, US Department of Commerce. <http://www.ngdc.noaa.gov/mgg/global/etopo2.html/>.

Natural Resources Canada (1997). *Canadian digital elevation data: Standards and specifications* (11 p.). Sherbrooke, QC: Centre for Topographic Information, Natural Resources Canada.

Nazari, H., Ritz, J.-F., Salamati, R., Shafei, A., Ghassemi, A., Michelot, J.-L., et al. (2009). Morphological and palaeoseismological analysis along the Taleghan fault (Central Alborz, Iran). *Geophysical Journal International, 178*, 1028–1041.

Neustruev, S. S. (1915). On soil combinations in plain and mountainous terrains. *Pochvovedenie, 17*, 62–73 (in Russian).

Neustruev, S. S. (1927). *Genesis of soils* (98 p.). Leningrad: Academy of Sciences of the USSR.

Neustruev, S. S. (1930). *Elements of soil geography* (240 p.). Moscow: Selkhozgiz (in Russian, with English contents).

Nezametdinova, S. S. (1970). *Strike analysis of regional faults within oil and gas bearing areas (Exemplified by the Precaucasus)*. Ph.D. Thesis. Leningrad Mining Institute, Leningrad, 326 p. (in Russian).

Nico, G., Leva, D., Fortuny-Guasch, J., Antonello, G., & Tarchi, D. (2005a). Generation of digital terrain models with a ground-based SAR system. *IEEE Transactions on Geoscience and Remote Sensing, 43*, 45–49.

Nico, G., Rutigliano, P., Benedetto, C., & Vespe, R. (2005b). Terrain modelling by kinematical GPS survey. *Natural Hazards and Earth System Sciences, 5*, 293–299.

Nielson, G., & Franke, R. (1984). A method for construction of surfaces under tension. *Rocky Mountain Journal of Mathematics, 14*, 203–221.

Nikolsky, S. M. (1977a). *A course of mathematical analysis* (Vol. 1, 460 p.). Moscow: Mir Publishers.

Nikolsky, S. M. (1977b). *A course of mathematical analysis* (Vol. 2, 440 p.). Moscow: Mir Publishers.

Nikonov, A. A. (1995). Active faults: Definition and recognition problems. *Geoecologia, No. 4*, 16–27 (in Russian).

Nothard, S., McKenzie, D., Haines, J., & Jackson, J. (1996). Gaussian curvature and the relationship between the shape and the deformation of the Tonga slab. *Geophysical Journal International, 127*, 311–327.

Nozette, S., Rustan, P., Pleasance, L. P., Horan, D. M., Regeon, P., Shoemaker, E. M., et al. (1994). The Clementine mission to the Moon: Scientific overview. *Science, 266*, 1835–1839.

O'Callaghan, J. F., & Mark, D. M. (1984). The extraction of drainage networks from digital elevation data. *Computer Vision, Graphics, and Image Processing, 28*, 323–344.

O'Driscoll, E. S. T. (1980). The double helix in global tectonics. *Tectonophysics, 63*, 397–417.

O'Driscoll, E. S. T. (1986). Observations of the lineament–ore relation. *Philosophical Transactions of the Royal Society of London, Series A, 317*, 195–218.

O'Leary, D. W., Friedman, J. D., & Pohn, H. A. (1976). Lineament, linear, lineation: Some proposed new standards for old terms. *Geological Society of America Bulletin, 87*, 1463–1469.

O'Loughlin, E. M. (1981). Saturation regions in catchments and their relations to soil and topographic properties. *Journal of Hydrology, 53*, 229–246.

Odeh, I. O. A., Chittleborough, D. J., & McBratney, A. B. (1991). Elucidation of soil–landform interrelationships by canonical ordination analysis. *Geoderma, 49*, 1–32.

Odeh, I. O. A., Chittleborough, D. J., & McBratney, A. B. (1992). Soil pattern recognition with fuzzy-c-means: Application to classification and soil–landform interrelationships. *Soil Science Society America Journal, 56*, 505–516.

Odeh, I. A., Crawford, M., & McBratney, A. B. (2007). Digital mapping of soil attributes for regional and catchment modelling, using ancillary covariates, statistical and geostatistical techniques. In P. Lagacherie, A. B. McBratney, & M. Voltz (Eds.), *Digital soil mapping. An introductory perspective* (pp. 437–453). Amsterdam: Elsevier.

Odeh, I. O. A., McBratney, A. B., & Chittleborough, D. J. (1994). Spatial prediction of soil properties from landform attributes derived from a digital elevation model. *Geoderma, 63*, 197−214.

Odeh, I. O. A., McBratney, A. B., & Chittleborough, D. J. (1995). Further results on prediction of soil properties from terrain attributes: Heterotopic cokriging and regression-kriging. *Geoderma, 67*, 215−226.

Oksanen, J., & Sarjakoski, T. (2005). Error propagation of DEM-based surface derivatives. *Computers and Geosciences, 31*, 1015−1027.

Oliver, M. A., & Webster, R. (1986). Combining nested and linear sampling for determining the scale and form of spatial variation of regionalized variables. *Geographical Analysis, 18*, 227−242.

Ollier, C. (1981). *Tectonics and landforms* (324 p.). London: Longman.

Onorati, G., Poscolieri, M., Salvi, S., & Trigila, R. (1987). Use of TM Landsat data as a support to classical ground-based methodologies in the investigation of a volcanic site in Central Italy the Caldera of Latera. In *IGARSS'87. Remote sensing: Understanding the earth as a system: Proceedings of the IEEE international geoscience and remote sensing symposium, 18−21 May 1987, Ann Arbor, MI*, (Vol. 2, pp. 1173−1178). New York: IEEE.

Onorati, G., Poscolieri, M., Ventura, R., Chiarini, V., & Crucilla, U. (1992). The digital elevation model of Italy for geomorphology and structural geology. *Catena, 19*, 147−178.

Orlova, A. V. (1975). *Block structures and topography* (232 p.). Moscow: Nedra (in Russian).

Ototzky, P. V. (1901). On the relationship between elevation and the character of Chernozems in the Poltava Governorate. *Pochvovedenie, 3*, 197−206 (in Russian, with French abstract).

Pachepsky, Y. A., Timlin, D. J., & Rawls, W. J. (2001). Soil water retention as related to topographic variables. *Soil Science Society of America Journal, 65*, 1787−1795.

Pacina, J. (2010). Comparison of approximation methods for partial derivatives of surfaces on regular grids. *Geomorphologia Slovaca et Bohemica, 1*(10), 25−32.

Palyvos, N., Bantekas, I., & Kranis, H. (2006). Transverse fault zones of subtle geomorphic signature in northern Evia Island (central Greece extensional province): An introduction to the Quaternary Nileas graben. *Geomorphology, 76*, 363−374.

Panin, A. V., & Gelman, R. N. (1997). Experience of the GPS technique application to derive detailed digital terrain models. *Geodezia i Cartografia, No. 10*, 22−27 (in Russian).

Pannekoek, A. J. (1967). Generalized contour maps, summit level maps, and streamline surface maps as geomorphological tools. *Zeitschrift für Geomorphologie, 11*, 169−182.

Papo, H. B., & Gelbman, E. (1984). Digital terrain models for slopes and curvatures. *Photogrammetric Engineering and Remote Sensing, 50*, 695−701.

Park, S. J., McSweeney, K., & Lowery, B. (2001). Identification of the spatial distribution of soils using a process-based terrain characterization. *Geoderma, 103*, 249−272.

Park, S. J., & van de Giesen, N. (2004). Soil-landscape delineation to define spatial sampling domains for hillslope hydrology. *Journal of Hydrology, 295*, 28−46.

Parsons, L. L., Murray, R. E., & Smith, M. S. (1991). Soil denitrification dynamics: Spatial and temporal variations of enzyme activity, populations, and nitrogen gas loss. *Soil Science Society of America Journal, 55*, 90−95.

Patterson, T. (2001). DEM manipulation and 3-D terrain visualization: Techniques used by the US National Park Service. *Cartographica, 38*, 89−101.

Payne, W. J. (1981). *Denitrification* (214 p.). New York: Wiley.

Pearce, M. A., Jones, R. R., Smith, S. A. F., McCaffrey, K. J. W., & Clegg, P. (2006). Numerical analysis of fold curvature using data acquired by high-precision GPS. *Journal of Structural Geology, 28*, 1640−1646.

Pe'eri, S., Gardner, J. V., Ward, L. G., & Morrison, J. R. (2011). The seafloor: A key factor in lidar bottom detection. *IEEE Transactions on Geoscience and Remote Sensing, 49*, 1150−1157.

Pei, T., Qin, C.-Z., Zhu, A-X., Yang, L., Luo, M., Li, B., et al. (2010). Mapping soil organic matter using the topographic wetness index: A comparative study based on different flow-direction algorithms and kriging methods. *Ecological Indicators, 10*, 610–619.

Peive, A. V. (1956a). The main types of deep faults. 1. General characteristics, classification, and spatial distribution of deep faults. *Izvestiya Academii Nauk SSSR, Geological Series, No. 1*, 90–105 (in Russian).

Peive, A. V. (1956b). The main types of deep faults. 2. Relationships of sedimentation, folding, magmatism, and ore deposits with deep faults. *Izvestiya Academii Nauk SSSR, Geological Series, No. 3*, 57–71 (in Russian).

Penck, W. (1924). *Die Morphologische Analyse. Ein Kapital der Physikalischen Geologie* (283 p.). Stuttgart: J. Engelhorns Nachfolger.

Penizek, V., & Boruvka, L. (2008). The digital terrain model as a tool for improved delineation of alluvial soils. In A. E. Hartemink, A. McBratney, & M. L. Mendonça-Santos (Eds.), *Digital soil mapping with limited data* (pp. 319–326). Dordrecht: Springer.

Pennock, D. J., & Corre, M. D. (2001). Development and application of landform segmentation procedures. *Soil and Tillage Research, 58*, 151–162.

Pennock, D. J., van Kessel, C., Farrel, R. E., & Sutherland, R. A. (1992). Landscape-scale variations in denitrification. *Soil Science Society of America Journal, 56*, 770–776.

Pennock, D. J., Zebarth, B. J., & de Jong, E. (1987). Landform classification and soil distribution in hummocky terrain, Saskatchewan, Canada. *Geoderma, 40*, 297–315.

Pereberin, A. V. (2001). Systematization of wavelet transforms. *Numerical Methods and Programming, 2*(3), 15–40 (in Russian, with English abstract).

Pereira, V., & FitzPatrick, E. A. (1998). Three-dimensional representation of tubular horizons in sandy soils. *Geoderma, 81*, 295–303.

Pettengill, G. H., Ford, P. G., Johnson, W. T. K., Raney, R. K., & Soderblom, L. A. (1991). Magellan: Radar performance and data products. *Science, 252*, 260–265.

Peucker, T. K. (1980a). The impact of different mathematical approaches to contouring. *Cartographica, 17*, 73–95.

Peucker, T. K. (1980b). The use of computer graphics for displaying data in three dimensions. *Cartographica, 17*, 59–72.

Peucker, T. K., Fowler, R. J., Little, J. J., & Mark, D. M. (1976). *Digital representation of three-dimensional surfaces by triangulated irregular networks (TIN). Technical report No. 10, Office of naval research, USA Contract # N00014-75-C-0886* (63 p.). Burnaby, BC: Department of Geography, Simon Fraser University.

Pflug, R., & Harbaugh, J. W. (Eds.), (1992). *Computer graphics in geology: Three-dimensional computer graphics in modeling geologic structures and simulating geologic processes* (298 p.). Berlin: Springer.

Phillips, J. D. (1988). The role of spatial scale in geomorphic systems. *Geographical Analysis, 20*, 308–317.

Phillips, J. D. (1995). Biogeomorphology and landscape evolution: The problem of scale. *Geomorphology, 13*, 337–347.

Pike, R. J. (1995). Geomorphometry—Progress, practice, and prospect. *Zeitschrift für Geomorphologie, Suppl. 101*, 221–238.

Pike, R. J. (2000). Geomorphometry—Diversity in quantitative surface analysis. *Progress in Physical Geography, 24*, 1–20.

Pike, R. J. (2001). Digital terrain modelling and industrial surface metrology—Converging crafts. *International Journal of Machine Tools and Manufacture, 41*, 1881–1888.

Pike, R. J. (2002). *A bibliography of terrain modeling (Geomorphometry), the quantitative representation of topography, Suppl. 4.0. USGS open-file report 02-465* (157 p.). Menlo Park, CA: U.S. Geological Survey.

Pike, R. J., Evans, I. S., & Hengl, T. (2009). Geomorphometry: A brief guide. In T. Hengl, & H. I. Reuter (Eds.), *Geomorphometry: Concepts, software, applications* (pp. 3–30). Amsterdam: Elsevier.

Piñol, J., Beven, K., & Freer, J. (1997). Modelling the hydrological response of Mediterranean catchments, Prades, Catalonia: The use of distributed models as aids to hypothesis formulation. *Hydrological Processes, 11*, 1287–1306.

Pirson, S. J. (1970). *Geologic well log analysis* (370 p.). Houston, TX: Gulf Publishing.

Pogorelov, A. V. (1957). *Differential geometry* (171 p.). Groningen, the Netherlands: Noordhoff.

Poletaev, A. I. (1992). *Fault intersections of the Earth's crust* (50 p.). Moscow: Geoinformmark (in Russian).

Poletaev, A. I., Kats, Y. G., & Leonov, N. N. (1991). Revealing of active fault and lineament structures within the area of the Smolensk nuclear power plant by visual and automated analysis. In B. N. Mozhaev (Ed.), *Digital processing of video information in structural geological and seismotectonic investigations* (pp. 42–55). Leningrad: Aerogeologia (in Russian).

Poletaev, A. I., Tevelev, A. V., Bryantseva, G. V., & Blyumkina, N. V. (1992). *Study of fault and lineament structure of the area of the Kursk nuclear power plant. Report No. 271, Pt. 1* (204 p.). Moscow: Lomonosov Moscow State University (in Russian).

Polidori, L., Chorowicz, J., & Guillande, R. (1991). Description of terrain as a fractal surface, and application to digital elevation model quality assessment. *Photogrammetric Engineering and Remote Sensing, 57*, 1329–1332.

Pollard, D. D., Bergbauer, S., & Mynatt, I. (2004). Using differential geometry to characterize and analyse the morphology of joints. *Geological Society of London, Special Publications, 231*, 153–182.

Pollard, D. D., & Fletcher, R. C. (2005). *Fundamentals of structural geology* (500 p.). Cambridge: Cambridge University Press.

Polyanskaya, L. M., & Zvyagintsev, D. G. (1995). Microbial succession in soil. *Physiology and General Biology Reviews, 9*, 1–68.

Ponagaibo, N. D. (1915). *Influence of microtopography on soil character, its temperature, moisture, and productivity* (96 p.). Poltava, Russia: Frishberg Press (in Russian).

Pustovitenko, B. G., & Trostnikov, V. N. (1977). Towards a relationship between seismic processes and tectonics in the Crimea. *Geophysical Communications, 77*, 13–23 (in Russian, with English abstract).

Puzachenko, M. Y., Puzachenko, Y. G., Kozlov, D. N., & Fedyaeva, M. V. (2006). Mapping of the thickness of organogenic and humic horizons of forest soils and bogs in the southern taiga (south-west of the Valdai Hills) using three-dimensional topographic model and remotely sensed data (Landsat 7). *Issledovanie Zemli iz Cosmosa, No. 4*, 1–9 (in Russian, with English abstract).

Puzachenko, Y. G., Onufrenya, I. A., & Aleshchenko, G. M. (2002). Analysis of hierarchical organization of topography. *Izvestiya Academii Nauk, Geographical Series, No. 4*, 29–38 (in Russian, with English abstract).

Qi, F., & Zhu, A. (2006). Fuzzy soil mapping based on prototype category theory. *Geoderma, 136*, 774–787.

Quinn, P. F., & Beven, K. J. (1993). Spatial and temporal prediction of soil moisture dynamics, runoff, variable source areas and evapotranspiration for Plynlimon, Mid-Wales. *Hydrological Processes, 7*, 425–448.

Quinn, P. F., Beven, K. J., Chevallier, P., & Planchon, O. (1991). The prediction of hillslope flowpaths for distributed modelling using digital terrain models. *Hydrological Processes, 5*, 59–80.

Quinn, P., Beven, K., & Lamb, R. (1995). The $\ln(a/\tan\beta)$ index: How to calculate it and how to use it within the TOPMODEL framework. *Hydrological Processes, 9*, 161–182.

Rabus, B., Eineder, M., Roth, A., & Bamler, R. (2003). The Shuttle Radar Topography Mission—A new class of digital elevation models acquired by spaceborne radar. *ISPRS Journal of Photogrammetry and Remote Sensing, 57*, 241–262.
Ramasundaram, V., Grunwald, S., Mangeot, A., Comerford, N. B., & Bliss, C. M. (2005). Development of an environmental virtual field laboratory. *Computers and Education, 45*, 21–34.
Ramsay, J. G. (1967). *Folding and fracturing of rocks* (568 p.). New York: McGraw-Hill.
Rance, H. (1967). Major lineaments and torsional deformation of the Earth. *Journal of Geophysical Research, 72*, 2213–2217.
Rance, H. (1968). Plastic flow and fracture in a torsionally stressed planetary sphere. *Journal of Mathematics and Mechanics, 17*, 953–974.
Rance, H. (1969). Lineaments and torsional deformation of the Earth: Indian Ocean. *Journal of Geophysical Research, 74*, 3271–3272.
Rappaport, N. J., Konopliv, A. S., Kucinskas, A. B., & Ford, P. G. (1999). An improved 360 degree and order model of Venus topography. *Icarus, 139*, 19–31.
Rastsvetaev, L. M. (1977). The Crimean Mountains and the Northern Black Sea area. In A. I. Suvorov (Ed.), *Faults and horizontal movements in mountain regions of the USSR* (pp. 95–113). Moscow: Nauka (in Russian).
Raupach, M. R., & Finnigan, J. J. (1997). The influence of topography on meteorological variables and surface–atmosphere interactions. *Journal of Hydrology, 190*, 182–213.
Rhind, D. W. (1971). Automated contouring—An empirical evaluation of some differing techniques. *Cartographic Journal, 8*, 145–158.
Richards, F. B. (1991). A Gibbs phenomenon for spline functions. *Journal of Approximation Theory, 66*, 334–351.
Rieger, J. (1997). Topographical properties of generic images. *International Journal of Computer Vision, 23*, 79–92.
Rieger, W. (1996). Accuracy of slope information derived from DEM-data. *International Archives of Photogrammetry and Remote Sensing, 31*(B4), 690–695.
Rieger, W. (1998). A phenomenon-based approach to upslope contributing area and depressions in DEMs. *Hydrological Processes, 12*, 857–872.
Roberts, A. (2001). Curvature attributes and their application to 3D interpreted horizons. *First Break, 19*(2), 85–100.
Roberts, A. M., Corfield, R. I., Kusznir, N. J., Matthews, S. J., Hansen, E.-K., & Hooper, R. J. (2009). Mapping palaeostructure and palaeobathymetry along the Norwegian Atlantic continental margin: Møre and Vøring basins. *Petroleum Geoscience, 15*, 27–43.
Robeson, S. M. (1997). Spherical methods for spatial interpolation: Review and evaluation. *Cartography and Geographic Information Systems, 24*, 3–20.
Robin, G. de Q., Evans, S., & Bailey, J. T. (1969). Interpretation of radio echo sounding in polar ice sheets. *Philosophical Transactions of the Royal Society of London, Series A, 265*, 437–505.
Robinson, J. E., Charlesworth, H. A. K., & Ellis, M. J. (1969). Structural analysis using spatial filtering in Interior Plains of south-central Alberta. *American Association of Petroleum Geologists Bulletin, 53*, 2341–2367.
Rode, A. A. (1953). Origin of microtopography of the Northern Caspian Lowland. *Voprosy Geografii, 33*, 249–260 (in Russian).
Rodríguez, E., Morris, C. S., & Belz, J. E. (2006). A global assessment of the SRTM performance. *Photogrammetric Engineering and Remote Sensing, 72*, 249–260.
Roecker, S. M., & Thompson, J. A. (2010). Scale effects on terrain attribute calculation and their use as environmental covariates for digital soil mapping. In J. L. Boettinger, D. W. Howell, A. C. Moore, A. E. Hartemink, & S. Kienast-Brown (Eds.), *Digital soil mapping—Bridging research, environmental application, and operation* (pp. 55–66). Dordrecht: Springer.

Romanova, E. N. (1963). Some regularities of water redistribution on slopes. In I. A. Goltsberg (Ed.), *Problems of microclimatology* (pp. 66–82). Leningrad: Hydrometeoizdat (in Russian).

Romanova, E. N. (1970). Seasonal humidification of soils in contrast geomorphic conditions. In I. A. Goltsberg, & E. N. Romanova (Eds.), *Microclimatology* (pp. 23–43). Leningrad: Hydrometeoizdat (in Russian).

Romanova, E. N. (1971). An approach of measurement and mapping of soil moisture using morphometric data. In I. A. Goltsberg, & F. F. Davitaya (Eds.), *Soil climate* (pp. 39–51). Leningrad: Hydrometeoizdat (in Russian).

Romanova, E. N. (1977). *Microclimatic variability of the main elements of climate* (279 p.). Leningrad: Hydrometeoizdat (in Russian, with English abstract).

Romashkevich, A. I., Rantsman, E. Y., & Mikheev, G. A. (1997). The anomalies in soil properties and soil mantle composition in mountainous regions and seismotectonics. *Eurasian Soil Science, 30*, 472–482.

Rosenberg, P. (1955). Information theory and electronic photogrammetry. *Photogrammetric Engineering, 21*, 543–555.

Rosenfeld, A., & Kak, A. C. (1982). *Digital picture processing* (2nd ed., Vol. 1, 435 p). New York: Academic Press.

Rothe, R. (1915). Zum Problem des Talwegs. *Sitzungsberichte der Berliner Mathematischen Gesellschaft, 14*, 51–68.

Rudy, R. M. (1999). *Methods to study relief of the land surface*. D.Sc. Thesis. Ivano-Frankovsk State Technical University of Oil and Gas, Ivano-Frankovsk, Ukraine, 375 p. (in Ukrainian).

Ryakhovsky, V., Rundquist, D., Gatinsky, Y., & Chesalova, E. (2003). GIS-project: Geodynamic globe for global monitoring of geological processes. *Geophysical Research Abstracts, 5*, 11645.

Ryan, C., & Boyd, M. (2003). CatchmentSIM: A new GIS tool for topographic geocomputation and hydrologic modelling. In M. Boyd, J. Ball, M. Babister, & J. Green (Eds.), *Proceedings of the 28th International hydrology and water resources symposium, 10–14 November 2003, Wollongong, Australia* (Vol. 1, pp. 35–42). Barton, Australia: Institution of Engineers.

Rybakov, M., Fleischer, L., & ten Brink, U. (2003). The Hula Valley subsurface structure inferred from gravity data. *Israel Journal of Earth Sciences, 52*, 113–122.

Saadi, N. M., Aboud, E., & Watanabe, K. (2009). Integration of DEM, ETM+, geologic, and magnetic data for geological investigations in the Jifara Plain, Libya. *IEEE Transactions on Geoscience and Remote Sensing, 47*, 3389–3398.

Sahr, K., White, D., & Kimerling, A. J. (2003). Geodesic discrete global grid systems. *Cartography and Geographic Information Science, 30*, 121–134.

Salishchev, K. A. (1955). On cartographic method of investigation. *Vestnik Moscovskogo Universiteta. Physical, Mathematical and Natural Sciences Series, No. 10*, 161–170 (in Russian).

Samson, P. P., & Mallet, J. L. (1997). Curvature analysis of triangulated surfaces in structural geology. *Mathematical Geology, 29*, 391–412.

Samsonov, T. E. (2010). *Multiscale mapping of topography using geodatabases*. Ph.D. Thesis Summary. Lomonosov Moscow State University, Moscow, 25 p. (in Russian).

Samsonova, V. P., Pozdnyakov, A. I., & Meshalkina, J. L. (2007). Study of disturbed soil cover using soil electrical resistivity and topographic data. *International Journal of Ecology and Development, 8*(F07), 39–51.

Sandwell, D. T., & Smith, W. H. F. (2001). Bathymetric estimation. In L.-L. Fu, & A. Cazenave (Eds.), *Satellite altimetry and earth sciences* (pp. 441–458). San Diego: Academic Press.

Sard, A. (1942). The measure of the critical values of differentiable maps. *Bulletin of American Mathematical Society, 48*, 883–890.

Sasowsky, K. C., Petersen, G. W., & Evans, B. M. (1992). Accuracy of SPOT digital elevation model and derivatives: Utility for Alaska's North Slope. *Photogrammetric Engineering and Remote Sensing, 58*, 815–824.

Satalkin, A. I. (Ed.), (1988). *Soil map of the Moscow Region, scale 1 : 300,000* (1 p.). Moscow: Central Board of Geodesy and Cartography (in Russian).

Saul, J. M. (1978). Circular structures of large scale and great age on the Earth's surface. *Nature, 271*, 345–349.

Saunders, A. M., & Boettinger, J. L. (2007). Incorporating classification trees into a pedogenic understanding raster classification methodology, Green River Basin, Wyoming, USA. In P. Lagacherie, A. B. McBratney, & M. Voltz (Eds.), *Digital soil mapping. An introductory perspective* (pp. 389–399). Amsterdam: Elsevier.

Saunders, R. S., & Pettengill, G. H. (1991). Magellan: Mission summary. *Science, 252*, 247–249.

Schaetzl, R. J., & Anderson, S. (2005). *Soils: Genesis and geomorphology* (817 p.). Cambridge: Cambridge University Press.

Scheidegger, A. E. (2004). *Morphotectonics* (197 p.). Berlin: Springer.

Schmid-McGibbon, G. (1995). Generalization of digital terrain models for use in landform mapping. *Cartographica, 32*, 26–38.

Schmidt, J., & Andrew, R. (2005). Multi-scale landform characterization. *Area, 37*, 341–350.

Schmidt, J., Evans, I. S., & Brinkmann, J. (2003). Comparison of polynomial models for land surface curvature calculation. *International Journal of Geographical Information Science, 17*, 797–814.

Schmidt, J., & Hewitt, A. (2004). Fuzzy land element classification from DTMs based on geometry and terrain position. *Geoderma, 121*, 243–256.

Schmidt, J. P., Taylor, R. K., & Gehl, R. J. (2003). Developing topographic maps using a sub-meter accuracy global positioning receiver. *Applied Engineering in Agriculture, 19*, 291–302.

Schowengerdt, R. A., & Glass, C. E. (1983). Digitally processed topographic data for regional tectonic evaluations. *Geological Society of America Bulletin, 94*, 549–556.

Schröder, P., & Sweldens, W. (2000). Spherical wavelets: Efficiently representing functions on a sphere. *Lecture Notes in Earth Sciences, 90*, 158–188.

Schuler, D. L., Ainsworth, T. L., Lee, J. S., & de Grandi, G. F. (1998). Topographic mapping using polarimetric SAR data. *International Journal of Remote Sensing, 19*, 141–160.

Schuler, D. L., Lee, J. S., & de Grandi, G. (1996). Measurement of topography using polarimetric SAR images. *IEEE Transactions on Geoscience and Remote Sensing, 34*, 1266–1277.

Schultz-Ela, D. D., & Yeh, J. (1992). Predicting fracture permeability from bed curvature. In J. R. Tillerson, & W. R. Wawersik (Eds.), *Proceedings of the 33rd US symposium on rock mechanics, 3–5 June 1992, Santa Fe, NM* (pp. 579–589). Rotterdam: Balkema.

Schumm, S. A., & Lichty, R. W. (1965). Time, space, and causality in geomorphology. *American Journal of Science, 263*, 110–119.

Schut, G. H. (1976). Review of interpolation methods for digital terrain models. *Canadian Surveyor, 30*, 389–412.

Schutz, B. E., Zwally, H. J., Shuman, C. A., Hancock, D., & DiMarzio, J. P. (2005). Overview of the ICESat Mission. *Geophysical Research Letters, 32*, L21S01, doi:10.1029/2005GL024009.

Scott, D. H., & Carr, M. H. (1978). *Geologic map of Mars, I-1083, scale 1 : 25,000,000*. Reston, VA: U.S. Geological Survey.

Scott, M. L., & Needelman, B. A. (2007). Utilizing water well logs for soil parent material mapping in the Mid-Atlantic coastal plain. *Soil Science, 172*, 701–720.

Scull, P., Franklin, J., & Chadwick, O. A. (2005). The application of classification tree analysis to soil type prediction in a desert landscape. *Ecological Modelling, 181*, 1–15.

Scull, P., Franklin, J., Chadwick, O. A., & McArthur, D. (2003). Predictive soil mapping: A review. *Progress in Physical Geography, 27*, 171–197.

Seifert, T., Tauber, F., & Kayser, B. (2001). A high-resolution spherical grid topography of the Baltic Sea—2nd edition. In U. Brenner (Ed.), *Abstracts, Baltic Sea science congress 2001: Past, present and future—A joint venture, 25—29 November 2001, Stockholm, Sweden* (p. 298). Stockholm: Stockholm Marine Research Centre, Stockholm University.

Semenov, A. M., van Bruggen, A. H. C., & Zelenev, V. V. (1999). Moving waves of bacterial populations and total organic carbon along roots of wheat. *Microbial Ecology, 37,* 116—128.

Senthilkumar, S., Kravchenko, A. N., & Robertson, G. P. (2009). Topography influences management system effects on total soil carbon and nitrogen. *Soil Science Society of America Journal, 73,* 2059—2067.

Serbenyuk, S. N. (1990). *Cartography and geoinformatics—Their interaction* (160 p.). Moscow: Moscow University Press (in Russian).

Shalimov, A. I. (1966). A new tectonic scheme of the Crimea and the connection between fold structures in the Mountain Crimea and the Northwest Caucasus. In V. A. Magnitsky, Y. D. Bulange, & Y. A. Meshcheryakov (Eds.), *The structure of the Black Sea depression* (pp. 49—58). Moscow: Nedra (in Russian, with English abstract).

Shannon, C. E. (1949). Communication in the presence of noise. *Proceedings of the IRE, 37,* 10—21.

Shary, P. A. (1991). The second derivative topographic method. In I. N. Stepanov (Ed.), *The geometry of the Earth surface structures.* Pushchino, USSR: Pushchino Research Centre Press (in Russian).

Shary, P. A. (1995). Land surface in gravity points classification by complete system of curvatures. *Mathematical Geology, 27,* 373—390.

Shary, P. A. (2005). *Estimating interrelations between topography, soil, and plants with new approaches of geomorphometry (Exemplified by an agrolandscape and a forest ecosystem in the southern Moscow Region).* Ph.D. Thesis. Institute of Physical, Chemical, and Biological Problems of Soil Science, Russian Academy of Sciences, Pushchino, Russia, 224 p. (in Russian).

Shary, P. A. (2006). Variables of geomorphometry: The current state-of-art. In X. Liu, & Y. Wang (Eds.), *Proceedings of the International symposium on terrain analysis and digital terrain modelling (TADTM 2006), 23—25 November 2006, Nanjing, China* (17 p.). Nanjing: Nanjing Normal University (CD–ROM).

Shary, P. A. (2008). Models of topography. In Q. Zhou, B. Lees, & G.-A. Tang (Eds.), *Advances in digital terrain analysis* (pp. 29—57). Berlin: Springer.

Shary, P. A., Kuryakova, G. A., & Florinsky, I. V. (1991). On the international experience of the use of topographic techniques in landscape investigations (a brief review). In I. N. Stepanov (Ed.), *The geometry of the Earth surface structures* (pp. 15—29). Pushchino, USSR: Pushchino Research Centre Press (in Russian).

Shary, P. A., Rukhovich, O. V., Sharaya, L. S., & Mitusov, A. V. (2002a). Soils and topography: Accumulation zones and non-local approaches. In *Transactions of the 17th World congress of soil science, 14—21 August 2002, Bangkok, Thailand, Symposium 48 (paper # 2310).* Bangkok: ISSS (CD ROM).

Shary, P. A., Sharaya, L. S., & Mitusov, A. V. (2002b). Fundamental quantitative methods of land surface analysis. *Geoderma, 107,* 1—32.

Shary, P. A., Sharaya, L. S., & Mitusov, A. V. (2005). The problem of scale-specific and scale-free approaches in geomorphometry. *Geografia Fisica e Dinamica Quaternaria, 28,* 81—101.

Shary, P. A., & Stepanov, I. N. (1991). Application of the method of second derivatives in geology. *Transactions (Doklady) of the USSR Academy of Sciences, Earth Science Sections, 320*(7), 87—91.

Sherstyankin, P. P., Alekseev, S. P., Abramov, A. M., Stavrov, K. G., de Batist, M., Hus, R., et al. (2006). Computer-based bathymetric map of Lake Baikal. *Doklady Earth Sciences, 408*, 564–569.

Shi, X., Zhu, A-X., Burt, J. E., Qi, F., & Simonson, D. (2004). A case-based approach to fuzzy soil mapping. *Soil Science Society of America Journal, 68*, 885–894.

Shiryaev, E. E. (1977). *New methods of cartographic representation and analysis of geoinformation by computers* (182 p.). Moscow: Nedra (in Russian).

Shtengelov, Y. S. (1978). On the fan-like recent spreading of the Earth crust, and on the nature of Benioff zones. *Doklady Academii Nauk SSSR, 240*, 922–925 (in Russian).

Shtengelov, Y. S. (1982). Zones of recent and modern spreading of continental crust. *International Geology Review, 24*, 759–770.

Shults, S. S. (1971). Planetary fractures and tectonic deformations. *Geotectonics, 5*, 203–207.

Sibirtsev, N. M. (1899). *Soil science. Pts. II and III* (360 p.). St. Petersburg: Skorokhodov Press (in Russian).

Sinai, G., Zaslavsky, D., & Golany, P. (1981). The effect of soil surface curvature on moisture and yield—Beer Sheba observation. *Soil Science, 132*, 367–375.

Sithole, G., & Vosselman, G. (2004). Experimental comparison of filter algorithms for bare-Earth extraction from airborne laser scanning point clouds. *ISPRS Journal of Photogrammetry and Remote Sensing, 59*, 85–101.

Sitnikov, A. B. (1978). *Water dynamics in unsaturated and saturated grounds of the aeration zone* (155 p.). Kiev: Naukova Dumka (in Russian).

Sitnikov, A. B. (1980). A concept of competent volumes. In A. E. Babinets (Ed.), *Introduction to modeling of hydrogeological processes* (pp. 20–25). Kiev: Naukova Dumka (in Russian).

Sjogren, W. L. (1997). *Magellan spherical harmonic and gravity map data V1.0, MGN-V-RSS-5-GRAVITY-L2-V1.0*. NASA Planetary Data System. <http://pds-geosciences.wustl.edu/missions/magellan/shadr_topo_grav/index.htm/>.

Skaryatin, V. D. (1973). Study of fault tectonics using a set of multiscale satellite images of the Earth (a method of multistep generalization). *Izvestiya Vuzov, Geologia i Razvedka, No. 7*, 34–50 (in Russian).

Skidmore, A. K. (1989). A comparison of techniques for calculation gradient and aspect from a gridded digital elevation model. *International Journal of Geographical Information Systems, 3*, 323–334.

Skidmore, A. K. (1990). Terrain position as mapped from a gridded digital elevation model. *International Journal of Geographical Information Systems, 4*, 33–49.

Skidmore, A. K., Ryan, P. J., Dawes, W., Short, D., & O'Loughlin, E. (1991). Use of an expert system to map forest soils from a geographical information system. *International Journal of Geographical Information Systems, 5*, 431–445.

Slemmons, D. B., & Depolo, C. M. (1986). Evaluation of active faulting and associated hazards. In Geophysics Study Committee, Geophysics Research Forum, and National Research Council, *Active Tectonics* (pp. 45–62). Washington, DC: National Academy Press.

Slyuta, E. N., Kudrin, L. V., & Sinilo, V. P. (1989). Preliminary data on the nature of a planetary system of lineaments observed in radar images of Venus (data from Venera-15 and -16). *Cosmic Research, 27*, 786–797.

Smith, B., Kessler, H., Scheib, A. J., Brown, S. E., Palmer, R. C., Kuras, O., et al. (2008). 3D modelling of geology and soils—A case study from the UK. In A. E. Hartemink, A. McBratney, & M. L. Mendonça-Santos (Eds.), *Digital soil mapping with limited data* (pp. 183–191). Dordrecht: Springer.

Smith, B., & Sandwell, D. (2003). Accuracy and resolution of Shuttle Radar Topography Mission data. *Geophysical Research Letters, 30*, 1467, doi:10.1029/2002GL016643.

Smith, D. E. (2003). *MGS MOLA mission experiment gridded data record, MGS-M-MOLA-5-MEGDR-L3-V1.0*. NASA Planetary Data System. <http://pds-geosciences.wustl.edu/missions/mgs/megdr.html/>.

Smith, D. E., Zuber, M. T., Neumann, G. A., & Lemoine, F. G. (1997). Topography of the Moon from the Clementine lidar. *Journal of Geophysical Research, 102*(E), 1591−1611.

Smith, D. E., Zuber, M. T., Solomon, S. C., Phillips, R. J., Head, J. W., Garvin, J. B., et al. (1999). The global topography of Mars and implications for surface evolution. *Science, 284,* 1495−1503.

Smith, D. E., Zuber, M. T., Neumann, G. A., Lemoine, F. G., Mazarico, E., Torrence, M. H., et al. (2010). Initial observations from the Lunar Orbiter Laser Altimeter. *Geophysical Research Letters, 37,* L18204, doi:10.1029/2010GL043751.

Smith, M. P., Zhu, A-X., Burt, J. E., & Stiles, C. (2006). The effects of DEM resolution and neighborhood size on digital soil survey. *Geoderma, 137,* 58−69.

Smith, S., Bulmer, C., Flager, E., Frank, G., & Filatow, D. (2010). Digital soil mapping at multiple scales in British Columbia, Canada. In *Proceedings of the 4th International workshop on digital soil mapping, 24−26 May 2010, Rome, Italy,* 8 p. (CD−ROM).

Smith, W. H. F. (1993). On the accuracy of digital bathymetric data. *Journal of Geophysical Research, 98*(B), 9591−9603.

Smith, W. H. F., & Sandwell, D. T. (1997). Global sea floor topography from satellite altimetry and ship depth soundings. *Science, 277,* 1956−1962.

Sobolevsky, P. K. (1932). The modern mining geometry. *Socialisticheskaya Reconstructsiya i Nauka, No. 7,* 42−78 (in Russian).

Sodano, E. M. (1965). General non-iterative solution of the inverse and direct geodetic problems. *Bulletin Géodésique, 75,* 69−89.

Soil Classification Working Group (1998). *The Canadian system of soil classification* (3rd ed., 187 p.). Ottawa: NRC Research Press.

Solomon, S. C., Head, J. W., Kaula, W. M., McKenzie, D., Parsons, B., Phillips, R. J., et al. (1991). Venus tectonics: Initial analysis from Magellan. *Science, 252,* 297−312.

Sonder, R. A. (1938). Die Lineamenttektonik und ihre Probleme. *Eclogae Geologicae Helvetiae, 31,* 199−238.

Sørensen, R., & Seibert, J. (2007). Effects of DEM resolution on the calculation of topographical indices: TWI and its components. *Journal of Hydrology, 347,* 79−89.

Sorokina, N. P., & Kozlov, D. N. (2009). Experience in digital mapping of soil cover patterns. *Eurasian Soil Science, 42,* 182−193.

Speight, J. G. (1974). A parametric approach to landform regions. In E. H. Brown, & R. S. Waters (Eds.), *Progress in geomorphology: Papers in honour of D.L. Linton* (pp. 213−230). London: Institute of British Geographers.

Speight, J. G. (1980). The role of topography in controlling throughflow generation: A discussion. *Earth Surface Processes, 5,* 187−191.

Spooner, C. S., Jr., Dossi, S. W., & Misulia, M. G. (1957). Let's go over the hill—Potential benefits of profile scanning the stereo-model. *Photogrammetric Engineering, 23,* 909−920.

Spot Image (2008). SPOT DEM, <http://www.spotimage.com/web/772-spot-dem.php/>.

Steen, Ø., Sverdrup, E., & Hanssen, T. H. (1998). Predicting the distribution of small faults in a hydrocarbon reservoir by combining outcrop, seismic and well data. *Geological Society of London, Special Publications, 147,* 299−312.

Stefanovic, P., Radwan, M. M., & Tempfli, K. (1977). Digital terrain models: Data acquisition, processing and applications. *ITC Journal, No. 1,* 61−76.

Steger, C. (1999). Extraction of watersheds from DTM and images with subpixel precision. *International Archives of Photogrammetry and Remote Sensing, 32*(3-2W5), 55−60.

Stepanov, I. N. (Ed.), (1984). *Map of soil cover systems of the Turkmen Soviet Socialist Republic, scale 1 : 1,500,000* (1 p.). Moscow: Central Board of Geodesy and Cartography (in Russian).

Stepanov, I. N. (Ed.), (1989). *Map of land surface systems and soil cover of the part of Central Asia, scale 1 : 1,500,000* (2 p.). Moscow: Central Board of Geodesy and Cartography (in Russian).

Stepanov, I. N. (1996). Introduction of soil geological forms of flow structures to cartography. *Eurasian Soil Science, 28*(9), 17−33.

Stepanov, I. N., Abdunazarov, U. K., Brynskikh, M. N., Deeva, N. F., Ilyina, A. A., Peido, L. P., et al. (1984). *Temporal guide for compilation of large- and medium-scale maps of Relief Plasticity* (20 p.). Pushchino, USSR: Biological Research Centre Press (in Russian).

Stepanov, I. N., Florinsky, I. V., & Shary, P. A. (1991). On the conceptual scheme of landscape investigations. In I. N. Stepanov (Ed.), *The geometry of the Earth surface structures* (pp. 9−15). Pushchino, USSR: Pushchino Research Centre Press (in Russian)

Stepanov, I. N., & Loshakova, N. A. (1998). On three types of soil delineations on the soil maps. *Eurasian Soil Science, 31*, 328−337.

Stepanov, I. N., Loshakova, N. A., Kovaleva, A. E., & Stepanova, V. I. (1998). On the publication of regional soil maps of Russia: The problem of borders. *Geodezia i Cartografia, No. 12*, 29−36 (in Russian).

Stepanov, I. N., Loshakova, N. A., Satalkin, A. I., & Andronova, M. I. (1987). Compilation of soil maps using the Relief Plasticity cartographic method. In V. A. Kovda (Ed.), *The method of Relief Plasticity in thematic mapping* (pp. 7−22). Pushchino, USSR: Biological Research Centre Press (in Russian).

Stepanov, I. N., & Sabitova, N. I. (1983). *Detection of natural soil geomorphic bodies on topographic maps and aerial photographs: Classification of structures of the land surface* (56 p.). Pushchino, USSR: Biological Research Centre Press (in Russian).

Sterner, R. (1995). *Color landform atlas of the United States*. Laurel, MD: The Johns Hopkins University Applied Physics Laboratory. <http://fermi.jhuapl.edu/states/>.

Steward, H. J. (1974). Cartographic generalisation: Some concepts and explanation. *Cartographica, 11*, 1−78.

Stewart, S. A., & Podolski, R. (1998). Curvature analysis of gridded geological surfaces. *Geological Society of London, Special Publications, 127*, 133−147.

Strahler, A. N. (1952). Hypsometric (area-altitude) analysis of erosional topography. *Geological Society of America Bulletin, 63*, 1117−1141.

Strahler, A. N. (1957). Quantitative analysis of watershed geomorphology. *Transactions of the American Geophysical Union, 38*, 913−920.

Strakhov, V. N. (2007). Change of epochs in Earth sciences. *Russian Journal of Earth Sciences, 9*, ES1001, doi:10.2205/2007ES000217.

Sulebak, J. R., & Hjelle, Ø. (2003). Multiresolution spline models and their applications in geomorphology. In I. S. Evans, R. Dikau, E. Tokunaga, H. Ohmori, & M. Hirano (Eds.), *Concepts and modelling in geomorphology: International perspectives* (pp. 221−237). Tokyo: Terrapub.

Sulebak, J. R., Tallaksen, L. M., & Erichsen, B. (2000). Estimation of areal soil moisture by use of terrain data. *Geografiska Annaler, Series A, 82*, 89−105.

Sumfleth, K., & Duttmann, R. (2008). Prediction of soil property distribution in paddy soil landscapes using terrain data and satellite information as indicators. *Ecological Indicators, 8*, 485−501.

Sundararajan, D. (2001). *The discrete Fourier transform: Theory, algorithms and applications* (374 p.). Singapore: World Scientific.

Svetlitchnyi, A. A., Plotnitskiy, S. V., & Stepovaya, O. Y. (2003). Spatial distribution of soil moisture content within catchments and its modelling on the basis of topographic data. *Journal of Hydrology, 277*, 50−60.

Swets, J. A. (1961). Detection theory and psychophysics: A review. *Psychometrika, 26*, 49−63.

Szynkaruk, E., Graduno-Monroy, V. H., & Bocco, G. (2004). Active fault systems and tectono-topographic configuration of the central Trans-Mexican Volcanic Belt. *Geomorphology, 61*, 111−126.

Tabor, R. W. (1986). Changing concepts of geologic structure and the problem of siting nuclear reactors: Examples from Washington State. *Geology, 14*, 738−742.

Takahashi, H. (1981). A lineament enhancement technique for active fault analysis. In *Machine processing of remotely sensed data with special emphasis on range, forest, and wetlands assessment: Proceedings of the 7th International symposium, 23−26 June 1981, West Lafayette, IN* (pp. 103−112). West Lafayette, IN: Purdue University.

Targulian, V. O., & Krasilnikov, P. V. (2007). Soil system and pedogenic processes: Self-organization, time scales, and environmental significance. *Catena, 71*, 373−381.

Targulian, V. O., & Sokolova, T. A. (1996). Soil as a biotic/abiotic natural system: A reactor, memory, and regulator of biospheric interactions. *Eurasian Soil Science, 29*, 30−41.

Taychinov, S. N, & Fayzullin, M. M. (1958). Dynamics of soil moisture in relation to topography. *Soviet Soil Science, No. 10*, 1121−1126.

Taylor, J. A., & Odeh, I. O. A. (2007). Comparing discriminant analysis with binomial logistic regression, regression kriging and multi-indicator kriging for mapping salinity risk in Northwest New South Wales, Australia. In P. Lagacherie, A. B. McBratney, & M. Voltz (Eds.), *Digital soil mapping. An introductory perspective* (pp. 455−464). Amsterdam: Elsevier.

Tempfli, K. (1980). Spectral analysis of terrain relief for the accuracy estimation of digital terrain models. *ITC Journal, No. 3*, 478−510.

Terra, J. A., Shaw, J. N., Reeves, D. W., Raper, R. L., van Santen, E., & Mask, P. L. (2004). Soil carbon relationships with terrain attributes, electrical conductivity, and a soil survey in a coastal plain landscape. *Soil Science, 169*, 819−831.

Tesfa, T. K., Tarboton, D. G., Chandler, D. G., & McNamara, J. P. (2010). A generalized additive soil depth model for a mountainous semi-arid watershed based upon topographic and land cover attributes. In J. L. Boettinger, D. W. Howell, A. C. Moore, A. E. Hartemink, & S. Kienast-Brown (Eds.), *Digital soil mapping—Bridging research, environmental application, and operation* (pp. 29−41). Dordrecht: Springer.

Thelin, G. P., & Pike, R. J. (1991a). *Landforms of the conterminous United States − A digital shaded-relief portrayal. Miscellaneous investigations map I-2206, scale 1 : 3,500,000*. U.S. Geological Survey.

Thelin, G. P., & Pike, R. J. (1991b). *Landforms of the conterminous United States—A digital shaded-relief portrayal. Text to accompany map I-2206* (14 p.). U.S. Geological Survey.

Thienssen, R. L., Soofi, K., & Sheline, H. (1994). A new expandable detector applied to digital topography and TM image data in support of petroleum exploration. *Photogrammetric Engineering and Remote Sensing, 60*, 77−85.

Thomas, A. L., King, D., Dambrine, E., Couturier, A., & Roque, J. (1999). Predicting soil classes with parameters derived from relief and geologic materials in a sandstone region of the Vosges mountains (Northeastern France). *Geoderma, 90*, 291−305.

Thompson, J. A., Bell, J. C., & Butler, C. A. (2001). Digital elevation model resolution: Effects on terrain attribute calculation and quantitative soil−landscape modeling. *Geoderma, 100*, 67−89.

Thompson, J. A., & Kolka, R. K. (2005). Soil carbon storage estimation in a forested watershed using quantitative soil-landscape modeling. *Soil Science Society of America Journal, 69*, 1086−1093.

Thompson, J. C., & Moore, R. D. (1996). Relations between topography and water table depth in a shallow forest soil. *Hydrological Processes, 10*, 1513−1525.

Tiedje, J. M. (1994). Denitrifiers. In R. W. Weaver, S. Angle, P. Bottomley, D. Bezdicek, S. Smith, A. Tabatabai, & A. Wollum (Eds.), *Methods of soil analysis. Pt. 2: Microbiological and biochemical properties* (pp. 245−267). Madison, WI: Soil Science Society of America.

Tillo, A. A. (1890). Orography of the European Russia from the hypsometric map. *Proceedings of the Imperial Russian Geographical Society, 26*, 8−32 (in Russian).

Timoshenko, S., & Goodier, J. N. (1951). *Theory of elasticity* (2nd ed., 506 p.). New York: McGraw-Hill.

Tobler, W. R. (1966). Numerical map generalization. *Michigan Inter-University Community of Mathematical Geographers, Discussion Paper, 8*, 1−27.
Tobler, W. R. (1969). Geographical filters and their inverses. *Geographical Analysis, 1*, 234−253.
Tomer, M. D., & Anderson, J. L. (1995). Variation of soil water storage across a sand plain hillslope. *Soil Science Society of America Journal, 59*, 1091−1100.
Tooth, S. (2006). Virtual globes: A catalyst for the re-enchantment of geomorphology? *Earth Surface Processes and Landforms, 31*, 1192−1194.
Toutin, T. (2004a). DSM generation and evaluation from QuickBird stereo imagery with 3D physical modelling. *International Journal of Remote Sensing, 25*, 5181−5193.
Toutin, T. (2004b). DTM generation from Ikonos in-track stereo images using a 3D physical model. *Photogrammetric Engineering and Remote Sensing, 70*, 695−702.
Toutin, T. (2006). Generation of DSMs from SPOT-5 in-track HRS and across-track HRG stereo data using spatiotriangulation and autocalibration. *ISPRS Journal of Photogrammetry and Remote Sensing, 60*, 170−181.
Toutin, T. (2008). ASTER DEMs for geomatic and geoscientific applications: A review. *International Journal of Remote Sensing, 29*, 1855−1875.
Toutin, T. (2010). Impact of Radarsat-2 SAR ultrafine-mode parameters on stereo-radargrammetric DEMs. *IEEE Transactions on Geoscience and Remote Sensing, 48*, 3816−3823.
Toutin, T., & Gray, A. L. (2000). State-of-the-art of extraction of elevation data using satellite SAR data. *ISPRS Journal of Photogrammetry and Remote Sensing, 5*, 13−33.
Trangmar, B. B., Yost, R. S., Wade, M. K., Uehara, G., & Sudjadi, M. (1987). Spatial variation of soil properties and rice yield on recently cleared land. *Soil Science Society of America Journal, 51*, 668−674.
Tribe, A. (1992). Automated recognition of valley lines and drainage networks from grid digital elevation models: A review and a new method. *Journal of Hydrology, 139*, 263−293.
Trifonov, V. G. (1983). *Late Quaternary tectonics* (224 p.). Moscow: Nauka (in Russian, with English contents).
Trifonov, V. G. (1995). World map of active faults (preliminary results of studies). *Quaternary International, 25*, 3−12.
Trifonov, V. G., Makarov, V. I., Safonov, Y. G., & Florensky, P. V. (Eds.), (1983). *Space information in geology* (535 p.). Moscow: Nauka (in Russian, with English contents).
Troeh, F. R. (1964). Landform parameters correlated to soil drainage. *Soil Science Society of America Proceedings, 28*, 808−812.
Twidale, C. R. (2004). River patterns and their meaning. *Earth-Science Reviews, 67*, 159−218.
Twidale, C. R., & Bourne, J. A. (2007). Lineaments and cryptostructural effects. In J. A. Bourne, & C. R. Twidale (Eds.), *Crustal structures and mineral deposits: E.S.T. O'Driscoll's contribution to mineral exploration* (pp. 153−164). Kenthurst, Australia: Rosenberg Publishing.
Ufimtsev, G. F. (1988). Topography and geological structure. In N. A. Logachev, D. A. Timofeev, & G. F. Ufimtsev (Eds.), *Problems of theoretical geomorphology* (pp. 59−75). Moscow: Nauka (in Russian).
Umscheid, L., & Bannon, P. R. (1977). A comparison of three global grids used in numerical prediction models. *Monthly Weather Review, 105*, 618−635.
US DMA (1991). *Digital chart of the world, scale 1 : 1,000,000*. Fairfax, VA: U.S. Defense Mapping Agency. 2 CD-ROMS.
USGS (1972). *Geologic atlas of the Moon, scale 1 : 1,000,000*. Washington, DC: U.S. Geological Survey.
USGS (1993). *Digital elevation models: Data users guide 5* (48 p.). Reston, VA: U.S. Geological Survey.

USGS (1996). *GTOPO30, A 30-arc seconds global digital elevation model*. Sioux Falls, ID: Earth Resources Observation and Science Center, U.S. Geological Survey. <http://eros.usgs.gov/#/Find_Data/Products_and_Data_Available/gtopo30_info/>.

USSR General Headquarters (1981). *Topographic map, scale 1 : 200,000. Sheet 13-36-06 (M-36-VI), Lgov.* Moscow: General Headquarters (in Russian).

USSR General Headquarters (1986). *Topographic map, scale 1 : 500,000. Sheet 12-36-4 (L-36-G), Simferopol*. Moscow: General Headquarters (in Russian).

Ustinova, V. N., & Ustinov, V. G. (2004). Seismic morphological analysis in mapping of high-capacity collector layers: I. Morphological classification. Manifestation of facies type of a collector layer in paleotopography. *Izvestiya Tomskogo Politekhnicheskogo Universiteta, 307*(3), 49–53 (in Russian).

Vakhtin, B. (1930). On the determination of mathematical characteristics of topography. *Geodezist, No. 2–3*, 7–16 (in Russian).

Van Kessel, C., Pennock, D. J., & Farrel, R. E. (1993). Seasonal variations in denitrification and nitrous oxide evolution. *Soil Science Society of America Journal, 57*, 988–995.

Van Kreveld, M. (1997). Digital elevation models and TIN algorithms. *Lecture Notes in Computer Science, 1340*, 37–78.

Van Niel, K. P., Laffan, S. W., & Lees, B. G. (2004). Effect of error in the DEM on environment variables for predictive vegetation modelling. *Journal of Vegetation Science, 15*, 747–756.

Vapnik, V. N. (2000). *The nature of statistical learning theory* (2nd ed., 314 p.). New York: Springer.

Vening Meinesz, F. A. (1947). Shear patterns of the Earth's crust. *Transactions of the American Geophysical Union, 28*, 1–61.

Venteris, E. R., McCarty, G. W., Ritchie, J. C., & Gish, T. (2004). Influence of management history and landscape variables on soil organic carbon and soil redistribution. *Soil Science, 169*, 787–795.

Veregin, H. (1997). The effects of vertical error in digital elevation models on the determination of flow-path direction. *Cartography and Geographic Information Systems, 24*, 67–79.

Verkhovtsev, V. G. (2008). *Contemporary platform geostructures of the Ukraine and dynamics of their development*. D.Sc. Thesis. Institute of Geological Sciences, Ukrainian National Academy of Sciences, Kiev, 423 p. (in Russian).

Vigil, J. F., Pike, R. J., & Howell, D. G. (2000a). *A tapestry of time and terrain, Geologic investigations series map I-2720, scale 1 : 3,500,000.* U.S. Geological Survey.

Vigil, J. F., Pike, R. J., & Howell, D. G. (2000b). *A tapestry of time and terrain. Pamphlet to accompany Geologic investigations series I-2720* (15 p.). Denver, CO: U.S. Geological Survey.

Vincenty, T. (1975). Direct and inverse solutions of geodesics on the ellipsoid with application of nested equations. *Survey Review, 23*, 88–93.

Vinogradova, A. I., & Eremin, V. K. (Eds.), (1971). *Aerial methods for geological investigations* Leningrad: Nedra (in Russian).

Volchanskaya, I. K. (1981). *Morphostructural regularities in location of endogenous mineralization* (239 p.). Moscow: Nauka (in Russian).

Volkov, N. M. (1950). *Principles and methods of cartometry* (327 p.). Moscow: Soviet Academic Press (in Russian).

Volkov, Y. V. (1995). Loxodromy and minerageny (Influence of astronomic resonances in the Earth–Moon system on the origin of ore deposits in the Earth's crust). *Bulletin of Moscow Society of Naturalists, Geological Series, 70*(6), 90–94 (in Russian, with English abstract).

Volkova, N. I., & Zhuchkova, V. K. (2000). Critical analysis of the morphoisographic method as applied to soil mapping. *Eurasian Soil Science, 33*, 1025–1037.

Vörösmarty, C. J., Fekete, B. M., Meybeck, M., & Lammers, R. B. (2000). Geomorphometric attributes of the global system of rivers at 30-minute spatial resolution. *Journal of Hydrology*, 237, 17–39.

Vysotsky, G. N. (1906). On oro-climatic bases of soil classification. *Pochvovedenie*, 8, 1–18 (in Russian).

Wahba, G. (1981). Spline interpolation and smoothing on the sphere. *SIAM Journal on Scientific and Statistical Computing*, 2, 5–16.

Waksman, S. A., & Deemer, R. B. (1928). Participants in the official tour of the 1st International Congress of Soil Science through the United States and Canada. In S. A. Waksman, & R. B. Deemer (Eds.), *Proceedings and papers of the 1st International congress of soil science, 13–22 June 1927, Washington, DC. Transcontinental excursion and impressions of the congress and of America* (pp. 22–39). Washington, DC: American Organizing Committee of the 1st International Congress of Soil Science.

Walker, P. H., Hall, G. F., & Protz, R. (1968). Relation between landform parameters and soil properties. *Soil Science Society of America Proceedings*, 32, 101–104.

Walter, C., Lagacherie, P., & Follain, S. (2007). Integrating pedological knowledge into digital soil mapping. In P. Lagacherie, A. B. McBratney, & M. Voltz (Eds.), *Digital soil mapping. An introductory perspective* (pp. 281–300). Amsterdam: Elsevier.

Warner, W. S. (1995). Mapping a three-dimensional soil surface with hand-held 35 mm photography. *Soil and Tillage Research*, 34, 187–197.

Watson, D. (1992). *Contouring: A guide to the analysis and display of spatial data* (340 p.). Oxford: Pergamon.

Webster, E. A., & Hopkins, D. W. (1996). Contributions from different microbial processes to N_2O emission from soil under different moisture regimes. *Biology and Fertility of Soils*, 22, 331–335.

Webster, R., & Burrough, P. A. (1974). Multiple discriminant analysis in soil survey. *Journal of Soil Science*, 25, 121–134.

Webster, R., & Oliver, M. A. (1992). Sample adequately to estimate variograms of soil properties. *Journal of Soil Science*, 43, 177–192.

Webster, R., & Oliver, M. A. (2007). *Geostatistics for environmental scientists* (2nd ed., 315 p.). Chichester, England: Wiley.

Wegener, A. (1915). *Die Entstehung der Kontinente und Ozeane* (94 p.). Braunschweig, Germany: Sammlung Vieweg.

Wehr, A., & Lohr, U. (1999). Airborne laser scanning—An introduction and overview. *ISPRS Journal of Photogrammetry and Remote Sensing*, 54, 68–82.

Weibel, R. (1992). Models and experiments for adaptive computer-assisted terrain generalization. *Cartography and Geographic Information Systems*, 19, 133–153.

Weibel, R. (1997). Generalization of spatial data: Principles and selected algorithms. *Lecture Notes in Computer Science*, 1340, 99–152.

Weibel, R., & Brändli, M. (1995). Adaptive methods for the refinement of digital terrain models for geomorphometric applications. *Zeitschrift für Geomorphologie, Suppl. 101*, 13–30.

Weibel, R., & Heller, M. (1991). Digital terrain modelling. In D. J. Maguire, M. F. Goodchild, & D. Rhind (Eds.), *Geographical information systems: Principles and applications, Vol. 1: Principles* (pp. 269–297). Harlow, England: Longman.

Weijermars, R. (1985/86). The polar spirals of Mars may be due to glacier surges deflected by Coriolis forces. *Earth and Planetary Science Letters*, 76, 227–240.

Weinberg, B. P. (1934a). Experience of mathematical definition of geomorphological concepts and mathematical characterization of geomorphic peculiarities. In *Proceedings of the 1st All-union geographical congress, 11–18 April 1933, Leningrad, USSR* (Vol. 3, pp. 126–135). Leningrad: State Geographical Society (in Russian).

Weinberg, B. P. (1934b). Some peculiarities of the Earth's crust folding. In *Proceedings of the 1st All-union geographical congress, 11–18 April 1933, Leningrad, USSR* (Vol. 3, pp. 123–126). Leningrad: State Geographical Society (in Russian).
Welch, R., Jordan, T., Lang, H., & Murakami, H. (1998). ASTER as a source for topographic data in the late 1990's. *IEEE Transactions on Geoscience and Remote Sensing, 36*, 1282–1289.
Wertz, J. B. (1974). Intersections linéamentaires et minéralisation dans les Montagnes Rocheuses de l'Amérique du Nord. *Geologische Rundschau, 63*, 708–754.
Western, A. W., Grayson, R. B., Blöschl, G., Willgoose, G. R., & McMahon, T. A. (1999). Observed spatial organization of soil moisture and its relation to terrain indices. *Water Resources Research, 35*, 797–810.
Whelan, M. J., & Gandolfi, C. (2002). Modelling of spatial controls on denitrification at the landscape scale. *Hydrological Processes, 16*, 1437–1450.
White, D. (2000). Global grids from recursive diamond subdivisions of the surface of an octahedron or icosahedron. *Environmental Monitoring and Assessment, 64*, 93–103.
Wieczorek, M. A. (2007). Gravity and topography of the terrestrial planets. In T. Spohn (Ed.), *Treatise on geophysics* (Vol. 10, pp. 165–206). Amsterdam: Elsevier.
Willmott, C. J., Rowe, C. M., & Philpot, W. D. (1985). Small-scale climate maps: A sensitivity analysis of some common assumptions associated with grid-point interpolation and contouring. *American Cartographer, 12*, 5–16.
Wilson, J. P., Aggett, G., Deng, Y., & Lam, C. S. (2008). Water in the landscape: A review of contemporary flow routing algorithms. In Q. Zhou, B. Lees, & G.-A. Tang (Eds.), *Advances in digital terrain analysis* (pp. 213–236). Berlin: Springer.
Wilson, J. P., & Gallant, J. C. (Eds.), (2000). *Terrain analysis: Principles and applications* New York: Wiley.
Wilson, J. P., Spangrud, D. J., Nielsen, G. A., Jacobsen, J. S., & Tyler, D. A. (1998). Global positioning system sampling intensity and pattern effects on computed topographic attributes. *Soil Science Society of America Journal, 62*, 1410–1417.
Wilson, J. T. (1941). Structural features in the Northwest Territories. *American Journal of Science, 239*, 493–502.
Wise, D. U. (1969). Regional and sub-continental sized fracture systems detectable by topographic shadow techniques. *Geological Survey of Canada Paper, 68–52*, 175–198.
Wise, D. U., Funiciello, R., Parotto, M., & Salvini, F. (1985). Topographic lineament swarms: Clues to their origin from domain analysis of Italy. *Geological Society of America Bulletin, 96*, 952–967.
Wise, S. (2000). Assessing the quality for hydrological applications of digital elevation models derived from contours. *Hydrological Processes, 14*, 1909–1929.
Wise, S. M. (2007). Effect of differing DEM creation methods on the results from a hydrological model. *Computers and Geosciences, 33*, 1351–1365.
Wladis, D. (1999). Automatic lineament detection using digital elevation models with second derivative filters. *Photogrammetric Engineering and Remote Sensing, 65*, 453–458.
Wolf, P. R., & Dewitt, B. A. (2000). *Elements of photogrammetry with applications in GIS* (3rd ed., 624 p.). Boston: McGraw-Hill.
Wolfman, Y. M., Gintov, O. B., Ostanin, A. M., Kolesnikova, E. Y., & Murovskaya, A. V. (2008). On the role of structural kinematic identification of tectonic disruptive faults in formation of the ideas on the structure and geodynamics of the Crimean region. *Geophysical Journal, 30*(1), 49–61 (in Russian, with English abstract).
Wolock, D. M., & Price, C. V. (1994). Effects of digital elevation model map scale and data resolution on a topography-based watershed model. *Water Resources Research, 30*, 3041–3052.

Wood, E. F., Sivapalan, M., & Beven, K. J. (1990). Similarity and scale in catchment storm response. *Reviews of Geophysics, 28*, 1–18.

Wood, E. F., Sivapalan, M., Beven, K. J., & Band, L. (1988). Effects of spatial variability and scale with implications to hydrologic modeling. *Journal of Hydrology, 102*, 29–47.

Wood, J. D. (1996). *The geomorphological characterisation of digital elevation models*. Ph.D. Thesis. University of Leicester, Leicester, England, 193 p.

Wood, J. (2009). Overview of software packages used in geomorphometry. In T. Hengl, & H. I. Reuter (Eds.), *Geomorphometry: Concepts, software, applications* (pp. 257–268). Amsterdam: Elsevier.

Wood, J. D., & Fisher, P. F. (1993). Assessing interpolation accuracy in elevation models. *IEEE Computer Graphics and Applications, 13*, 48–56.

World Data Center for Geophysics (1993). *5-minute gridded global relief data collection (ETOPO5)*. Boulder, CO: World Data Center for Geophysics & Marine Geology, National Oceanic and Atmospheric Administration. <http://www.ngdc.noaa.gov/mgg/fliers/93mgg01.html/>.

Wu, F. (2003). Scale-dependent representations of relief based on wavelet analysis. *Geo-Spatial Information Science, 6*, 66–69.

Wynn, T. J., & Stewart, S. A. (2003). The role of spectral curvature mapping in characterizing subsurface strain distributions. *Geological Society of London, Special Publications, 209*, 127–143.

Xu, T., Moore, I. D., & Gallant, J. C. (1993). Fractals, fractal dimensions and landscapes—A review. *Geomorphology, 8*, 245–262.

Yabrova, L. A. (1981). Concentrically radial drainage networks as the indicator of ring structures in the Caucasus and other mountain terrains. In G. K. Belyaev (Ed.), *Nature and natural resources of the central and eastern parts of the North Caucasus* (pp. 16–25). Ordzhonikidze, USSR: Khetagurov North-Ossetian State University (in Russian).

Yaroshenko, P. D. (1961). *Geobotany. The main concepts and methods* (474 p.). Moscow: Soviet Academic Press (in Russian)

Yoeli, P. (1975). Compilation of data for computer-assisted relief cartography. In J. C. Davis, & M. J. McCullagh (Eds.), *Display and analysis of spatial data* (pp. 352–367). London: Wiley.

Young, A. (1972). *Slopes* (288 p.). Edinburgh: Oliver and Boyd.

Young, M. (1978). *Statistical characterization of altitude matrices by computer. Terrain analysis: Program documentation. Report 5 on grant DA-ERO-591-73-G0040* (18 p.). Durham, England: Department of Geography, University of Durham.

Young, T. (1805). An essay on the cohesion of fluids. *Philosophical Transactions of Royal Society of London, Pt. I, 95*, 65–87.

Yuan, W., Zhuang, D., Yuan, W., & Qiu, D. (2010). Equal arc ratio projection and a new spherical triangle quadtree model. *International Journal of Geographical Information Science, 24*, 1703–1723.

Yudin, V. V., & Gerasimov, M. E. (2001). Thrusts of the Crimean Mountains. *Geophysical Journal, 23*(2), 121–129 (in Russian, with English abstract).

Zadeh, L. A. (1965). Fuzzy sets. *Information and Control, 8*, 338–353.

Zádorová, T., Penížek, V., Šefrna, L., Rohošková, M., & Borůvka, L. (2011). Spatial delineation of organic carbon-rich Colluvial soils in Chernozem regions by terrain analysis and fuzzy classification. *Catena, 85*, 22–33.

Zakharchenko, A. V., & Zakharchenko, N. V. (2006). Three-dimensional surface morphometry of soil horizons in field studies. *Eurasian Soil Science, 39*, 134–140.

Zakharov, S. A. (1911). On the role of micro- and macrotopography in the Podzol region. *Pochvovedenie, 13*, 49–72 (in Russian, with German abstract).

Zakharov, S. A. (1913). On characteristics of soils in mountain regions. *Izvestiya Konstantinovskogo Mezhevogo Instituta, 4,* 1−93 (in Russian).
Zakharov, S. A. (1927). *A course of soil science* (440 p.). Moscow: Gosizdat (in Russian).
Zakharov, S. A. (1940). Importance of slope aspect and gradient for soil and vegetation distribution in the Great Caucasus. *Journal Botanique de l'URSS, 25,* 378−405 (in Russian).
Zamani, A., & Hashemi, N. (2000). A comparison between seismicity, topographic relief, and gravity anomalies of the Iranian Plateau. *Tectonophysics, 327,* 25−36.
Zanchi, A., Francesca, S., Stefano, Z., Simone, S., & Graziano, G. (2009). 3D reconstruction of complex geological bodies: Examples from the Alps. *Computers and Geosciences, 35,* 49−69.
Zaslavsky, D., & Rogowski, A. S. (1969). Hydrologic and morphologic implications of anisotropy and infiltration in soil profile development. *Soil Science Society of America Proceedings, 33,* 594−599.
Zebker, H. A., Farr, T. G., Salazar, R. P., & Dixon, T. H. (1994a). Mapping the world's topography using radar interferometry: The TOPSAT mission. *Proceedings of the IEEE, 82,* 1774−1786.
Zebker, H., Werner, C., Rosen, P., & Hensley, S. (1994b). Accuracy of topographic maps derived from ERS-1 interferometric radar. *IEEE Transactions on Geoscience and Remote Sensing, 32,* 823−836.
Zeilik, B. S., Perfil'ev, L. G., Vasilenko, A. N., & Seitmuratova, E. Y. (1989). New techniques for geological interpretation of annular structures in semi-covered and covered areas. *Soviet Journal of Remote Sensing, 5,* 606−621.
Zernitz, E. R. (1931). Drainage patterns and their significance. *Journal of Geology, 40,* 498−521.
Zevenbergen, L. W., & Thorne, C. R. (1987). Quantitative analysis of land surface topography. *Earth Surface Processes and Landforms, 12,* 47−56.
Zhang, H.-P., Liu, S.-F., Yang, N., Zhang, Y.-Q., & Zhang, G.-W. (2006). Geomorphic characteristics of the Minjiang drainage basin (eastern Tibetan Plateau) and its tectonic implications: New insights from a digital elevation model study. *Island Arc, 15,* 239−250.
Zhang, W., & Montgomery, D. R. (1994). Digital elevation model grid size, landscape representation, and hydrologic simulation. *Water Resources Research, 30,* 1019−1028.
Zhao, X., Bai, J., & Chen, J. (2008). A seamless and adaptive LOD model of the global terrain based on the QTM. In Q. Zhou, B. Lees, & G.-A. Tang (Eds.), *Advances in digital terrain analysis* (pp. 85−103). Berlin: Springer.
Zhou, B., Zhang, X.-G., & Wang, R.-C. (2004). Automated soil resources mapping based on decision tree and Bayesian predictive modeling. *Journal of Zhejiang University Science, 5,* 782−795.
Zhou, Q., & Chen, Y. (2011). Generalization of DEM for terrain analysis using a compound method. *ISPRS Journal of Photogrammetry and Remote Sensing, 66,* 38−45.
Zhou, Q., & Liu, X. (2004). Error analysis on grid-based slope and aspect algorithms. *Photogrammetric Engineering and Remote Sensing, 70,* 957−962.
Zhu, A-X. (2000). Mapping soil landscape as spatial continua: The neural network approach. *Water Resources Research, 36,* 663−677.
Zhu, A-X., Band, L. E., Dutton, B., & Nimlos, T. J. (1996). Automated soil inference under fuzzy logic. *Ecological Modelling, 90,* 123−145.
Zhu, A-X., Burt, J. E., Smith, M., Wang, R., & Gao, J. (2008). The impact of neighbourhood size on terrain derivatives and digital soil mapping. In Q. Zhou, B. Lees, & G.-A. Tang (Eds.), *Advances in digital terrain analysis* (pp. 333−348). Berlin: Springer.
Zhu, A-X., Hudson, B., Burt, J., Lubich, K., & Simonson, D. (2001). Soil mapping using GIS, expert knowledge, and fuzzy logic. *Soil Science Society America Journal, 65,* 1463−1472.

Zhu, Q., & Lin, H. S. (2010). Comparing ordinary kriging and regression kriging for soil properties in contrasting landscapes. *Pedosphere, 20*, 594–606.

Ziadat, F. M., Taylor, J. C., & Brewer, T. R. (2003). Merging Landsat TM imagery with topographic data to aid soil mapping in the Badia region of Jordan. *Journal of Arid Environments, 54*, 527–541.

Zibilske, L. M. (1994). Carbon mineralization. In R. W. Weaver, S. Angle, P. Bottomley, D. Bezdicek, S. Smith, A. Tabatabai, & A. Wollum (Eds.), *Methods of soil analysis. Pt. 2: Microbiological and biochemical properties* (pp. 835–863). Madison, WI: Soil Science Society of America.

Zlatopolsky, A. A. (1992). Program LESSA (Lineament Extraction and Stripe Statistical Analysis): Automated linear image features analysis—Experimental results. *Computers and Geosciences, 18*, 1121–1126.

Zuber, M. T. (1996). *Clementine lunar topography V1.0, CLEM1-L-LIDAR-5-TOPO-V1.0.* NASA Planetary Data System. <http://pds-geosciences.wustl.edu/missions/clementine/gravtopo.html/>.

Zuber, M., Smith, D. E., Lemoine, F. G., & Neumann, G. (1994). The shape and internal structure of the Moon from the Clementine mission. *Science, 266*, 1839–1843.

Zuber, M. T., Solomon, S. C., Phillips, R. J., Smith, D. E., Tyler, G. L., Aharonson, O., et al. (2000). Internal structure and early thermal evolution of Mars from Mars Global Surveyor topography and gravity. *Science, 287*, 1788–1793.

Zuchiewicz, W. (1989). Selected geomorphic method applied to neotectonic studies in the Northern Carpathians. *Bulletin de l'Association Française pour l'Étude du Quaternaire, 26*, 131–136.

Zverev, A. T., & Strykov, A. I. (1985). Transformation of geological, geophysical, and topographic maps to reveal pseudo-round anomalies (forms) for automated recognition of ring structures on satellite images. *Izvestiya Vuzov, Geodesia i Aerofotosyemka, No. 5*, 46–54 (in Russian).

Zvyagintsev, D. G. (1994). Vertical distribution of microbial communities in soils. In K. Ritz, J. Dighton, & K. E. Giller (Eds.), *Beyond the biomass: Compositional and functional analysis of soil microbial communities* (pp. 29–37). Chichester, England: Wiley.

Zwally, H. J., Bindschadler, R. A., Brenner, A. C., Martin, T. V., & Thomas, R. H. (1983). Surface elevation contours of Greenland and Antarctic ice sheets. *Journal of Geophysical Research, 88*(C), 1589–1596.

Zwally, H. J., Bindschadler, R. A., Major, J. A., & Brenner, A. C. (1987). Ice measurements by Geosat radar altimetry. *Johns Hopkins APL Technical Digest, 8*, 251–254.

Index

A

A. *See* Slope aspect
Accumulation curvature (K), 10, 310–311
  calculation of, 15
  definitions and interpretations of, 13t, 27
Accumulation zones, 25–28, 26f, 256
  data processing for, 183, 258–261, 258f, 259f, 260t
  discussion of, 261–262, 261f
  Efremov-Krcho classification using, 25–28, 27f, 29f
  gravity-driven overland transport with, 25, 256
  intrasoil transport with, 256
  mapping of, 27, 29f, 184f, 258f
  materials for, 183, 258–261, 258f, 259f, 260t
  motivation for, 255–257
  results for, 183, 261–262, 261f
  study area for, 257
Accuracy, 65–102
  estimation, 70–71
  Gibbs phenomenon, 88–93
    data processing with, 90
    discussion of, 90–93, 91f
    materials in, 90
    motivation for, 88–90, 88f
    results with, 90–93, 91f
  grid displacement, 93–97
    data processing with, 94–96, 95f, 96f
    discussion of, 96–97, 97f, 98t
    Kolmogorov-Smirnov test with, 97, 97f, 98t
    materials in, 94–96, 95f, 96f
    motivation for, 88f, 93–94
    results with, 96–97, 97f, 98t
  linear artifacts, 98–101
    isotropy of local morphometric variables with, 100–101
    motivation for, 98–100, 99f
  local morphometric variables calculation, 71–81
    motivation, 71–73
    plane square grid calculation, 75–76, 76t
    RMSE formulas for local morphometric variables, 73–75
    RMSE formulas for partial derivatives, 75–78
    RMSE mapping, 78–81, 79f, 80f
    spheroidal equal angular grid calculation, 77–78
  sampling theorem, errors from ignoring, 81–88
    data processing with, 82–86, 84f, 85f
    discussion of, 86–88
    materials in, 82–86, 84f, 85f
    motivation for, 81–82, 82f
    results with, 86–88
  soil predictive mapping, 185–187
  sources of errors, 66–70, 66f, 68f
Aim-oriented generalization, 106
Airborne ice-penetrating radar techniques, 35
Airborne optical sensing, 34
Approximation
  Gibbs phenomenon as, 88
  partial derivatives, 43–44, 54–55, 286
  square wave function, 88f
Artifacts, linear, 98–101
  isotropy of local morphometric variables with, 100–101
  motivation for, 98–100, 99f
Artificial neural networks, soil predictive mapping with, 181
Aspect. *See* Slope aspect
ASTER GDEM, 32, 38
Augering, 34
AVHRR, 32
Azimuth. *See* Slope aspect

B

Basins. *See* Elliptical fold
Bathymetry, 32–34, 36, 119, 242f, 266–267
Bayesian analysis of evidence, soil predictive mapping with, 181

368 INDEX

Biscay-Santa Cruz helical structures, 280, 281f
Bottom-up approach, soil predictive mapping with, 183–185
Bulk density, 197, 206t, 210t, 217

C

CA. *See* Catchment area
Calculation methods, 43–64
  local topographic variables, 43–59
    Evans-Young method, 43–45, 44f
    plane square grid, 45–54
      accuracy with, 75–76, 76t
      data processing for validation of, 49–50
      Delaunay triangulation for validation of, 49
      discussion of method for, 50–54, 51f, 52f, 53f
      formulas, 46–49
      materials for validation of, 49–50
      method validation, 49–54
      motivation, 45
      results of, 50–54, 51f, 52f, 53f
      RMSE for validation of, 49, 50
      Taylor formula for, 46
    spheroidal equal angular grid, 54–59
      accuracy with, 77–78
      discussion of method for, 58–59
      formulas, 55–57
      linear sizes of window of, 57–58
      motivation, 54–55
  nonlocal topographic variables, 59–61, 60f
  structural lines, 61–64
    conventional algorithms, 61–62
    generating function, 62–64, 63f
Canadian prairies, 183, 184f, 192–220
  Black Soil Zone, 183, 184f
  Manitoba, 184f, 192–220
  Miniota site, 184f, 192–216, 193f, 194f
    materials and methods for, 196–213
    results and discussion for, 214–216, 213f
  Minnedosa site, 192–220, 193f, 195f
    materials and methods for, 196–213
    results and discussion for, 216–220, 213f
Cartographic generalization, 105–106
Catalan solid, 38
Catchment area (CA), 16–18, 19f
  definitions and interpretations of, 13t
  Earth, 272, 274f, 278f
  illustration of, 11f
  Mars, 272, 275f, 279f
  Moon, 272, 275f, 279f
  Venus, 275f, 279f
CatchmentSIM, 62
Cation exchange capacity, 175, 176
Caucasus-Clipperton double helix, 280, 281f
Cell-to-cell flow routing, structural line calculation with, 61
Classification
  heuristic, 174
  methods of soil mapping, 173–174
    heuristic classification approaches, 174
    methods to predict categorical variables, 174
    methods to predict quantitative soil properties, 173
    supervised classification approaches, 174
    unsupervised classification approaches, 174
  supervised, 174
  unsupervised, 174
Clementine Gravity and Topography Data, 267, 274
Clementine mission, 267
Cokriging, 176
Combined morphometric variable, 23, 24f
  steam power index, 14t, 23, 24f
  topographic index, 14t, 23, 24f
Concave break lines, 20
Contour lines, 19–20, 291–294, 297, 299, 312
  digitizing of, 35
  soil moisture distribution along, 160f
Contributing area, 13t
Convex break lines, 20
Crests, 20, 20f, 312
  calculation methods for, 62, 63f
Crimean peninsula, 241–248, 241f, 242f
  geological setting for, 242–245, 243f
  materials and data processing for, 82–83, 245, 258, 259, 258f, 259f, 260t
  results and discussion for, 83f, 84f, 86, 87, 245–248, 246f, 247f, 260t, 261, 261f, 262
Cross sections, 135–136, 137f
Curvature, 10, 10f, 297
  accumulation, 10, 13t, 15, 18f, 310–311

INDEX 369

difference, 10, 13t, 14, 17f, 27f, 306–308, 310–311
  Gaussian, 10, 12t, 14, 18f, 23–25, 26f, 27f, 121, 225–229, 232, 301–302, 311
  horizontal, 10, 11f, 13t, 14, 17f, 25–27, 80f, 85f, 96f, 118f, 126f, 128f, 129f, 131f, 147, 161f, 163t, 164f, 195f, 229f, 235–237, 238f, 246f, 251f, 269, 270f, 271f, 299, 308, 310–312
  horizontal excess, 10, 13t, 15, 308
  maximal, 10, 12t, 15, 17f, 195f, 224f, 227, 232, 301–302, 307–308, 311
  mean, 10, 12t, 14, 17f, 24–25, 26f, 27f, 79f, 161f, 163t, 225, 227, 299–302, 307, 311
  minimal, 10, 12t, 15, 17f, 194f, 224f, 227, 270f, 301–302, 307–308, 311
  plan, 10, 172, 297–299
  ring, 10, 13t, 15, 18f, 308–311
  unsphericity, 10, 12t, 15, 18f, 302–304, 311
  vertical, 10, 11f, 12t, 14, 17f, 84f, 96f, 118f, 120f, 147, 161f, 163t, 194f, 229f, 235–237, 238f, 246f, 251f, 270f, 272, 273f, 299–301, 306, 308, 311
  vertical excess, 10, 13t, 15, 18f, 308
Cutting method, 119–122, 121f
Cylindrical fold, 225

D

DA. *See* Dispersive area
Dakar helical structure, 280, 281f
D_d. *See* Drainage density
Decision trees, soil predictive mapping with, 180
Delaunay triangulation, 37, 40, 49, 83, 90, 94, 160, 198, 199, 239, 245, 250
DEM. *See* Digital elevation models
Denitrification
  study of, 192
  topography and, 192, 216–220
Denitrification rate, 197t, 198, 210t
Denitrifier, 197t, 198
Denitrifier enzyme activity, 197t, 210t
Denoising, 104–105
Depth to calcium carbonate, 148, 173, 175, 256
Difference curvature (E), 10, 17f, 306–308, 310–311
  calculation of, 14
  definitions and interpretations of, 13t

Differentiation, 58–59, 76, 81, 82f, 92
  spatial, 146, 147, 217, 218, 220
Digital elevation models (DEM), 31–42
  generation, 31–36
    airborne ice-penetrating radar techniques for, 35
    airborne optical sensing for, 34
    digitizing of contours for, 35
    geological boring for, 34
    kinematic GPS surveys for, 32
    LiDAR and laser surveys for, 33
    photogrammetric approaches for, 32
    radar techniques for, 32–33
    radio-echo sounding surveys for, 35
    satellite radar altimetry for, 34
    shipboard echo sounding for, 33–34
    soil augering for, 34
    three-dimensional seismic survey for, 34–35
    topographic surveys for, 31–32
  grid types, 36–38
  interpolation, 40–41
  resolution, 38–40
  topography information from, 2
Digital soil mapping (DSM), 169–170
  six basic steps of, 173
Digital terrain modeling
  filtering methods of, 109–122
    cutting method, 119–122, 121f
    Filosofov method, 109–111, 110f
    row and column elimination, 117–119, 120f
    smoothing, 115–117, 118f
    trend-surface analysis, 109
    two-dimensional discrete Fourier transform, 112–113, 114f
    two-dimensional discrete wavelet transform, 113–115, 116f
  filtering tasks of, 103–109
    decomposition of topographic surface, 104
    denoising, 104–105
    generalization, 105–109
  historical overview of, 1–4
  morphometric variables in, 7–30
  photogrammetry in history of, 2
  progress made in, 285
  research trends for, 4
  two research avenues of, 3
Digital terrain models (DTM), 2, 31
Digitizing of contours, 35
Dip, 240–241, 244f

Dip angle. *See* Slope gradient
Dip-slip fault, 235, 236, 236f, 237, 238f, 239, 246, 247f
Discretization, 39, 93–94, 95f, 98t
Discriminant analysis, soil predictive mapping with, 180
Dispersive area (*DA*), 16, 18, 19f
   definitions and interpretations of, 13t
   Earth, 272, 274f
   illustration of, 11f
   Mars, 272, 276f
   Moon, 276f
   Venus, 276f
Displacement, 240–241, 244f
Dissipation zones, 25–28, 26f, 183, 256, 260t
   mapping of, 27, 29f, 184f, 258f
Dokuchaev equation, 168
Dokuchaev hypothesis, 167–168
Domes. *See* Elliptical fold
Drainage density (D_d), 16–18, 19f
   definitions and interpretations of, 14t
DSM. *See* Digital soil mapping
DTM. *See* Digital terrain models

E

E. *See* Difference curvature
Earth
   catchment area, 272, 274f
   dispersive area, 272, 274f
   elevation of, 266f
   global helical structures of, 265f
   global morphometric maps, 270f
   horizontal curvature of, 269, 270f
   minimal curvature of, 270f
   vertical curvature of, 270f, 272
Earthquake hypocenter, 36, 227
Efremov-Krcho classification, 25–28, 27f, 29f
Eigentriple (ET), 124, 126–127, 127f, 130
Elevation. *See also* Digital elevation models
   Earth, 266f
   Mars, 268f
   Moon, 268f
   Venus, 268f
Elliptical fold (Domes, Basins), 225
Empirical best linear unbiased predictor, soil predictive mapping with, 182
Errors, 65–102
   accuracy in calculation, local morphometric variables, 71–81
      motivation, 71–73

plane square grid, 75–76, 76t
RMSE formulas for local morphometric variables, 73–75
RMSE formulas for partial derivatives, 75–78
RMSE mapping, 78–81, 79f, 80f
spheroidal equal angular grid, 77–78
accuracy in estimation for, 70–71
Gibbs phenomenon, 88–93
   data processing with, 90
   discussion of, 90–93, 91f
   materials in, 90
   motivation for, 88–90, 88f
   results with, 90–93, 91f
grid displacement, 93–97
   data processing with, 94–96, 95f, 96f
   discussion of, 96–97, 97f, 98t
   Kolmogorov-Smirnov test with, 97, 97f, 98t
   materials in, 94–96, 95f, 96f
   motivation for, 88f, 93–94
   results with, 96–97, 97f, 98t
linear artifacts, 98–101
   isotropy of local morphometric variables with, 100–101
   motivation for, 98–100, 99f
sampling theorem, from ignoring, 81–88
   data processing with, 82–86, 84f, 85f
   discussion of, 86–88
   materials in, 82–86, 84f, 85f
   motivation for, 81–82, 82f
   results with, 86–88
sources of, 66–70, 66f, 68f
ERS, 32–33, 34
ET. *See* Eigentriple
ETOPO1, 34, 35, 38, 54, 70
ETOPO2, 34, 35, 38, 54, 70, 119, 266
ETOPO5, 34, 35, 38, 54, 70
Euler theorem, 299, 302, 307
Evans-Young method, 43–45, 44f
   grid spacing in, 44–45
   polynomial for, 44–45

F

Factor field, 124
Fault dip, 240–241, 244f
Fault intersections, 255–263
   data processing for, 258–261, 258f, 259f, 260t
   discussion of, 261–262, 261f
   gravity-driven overland transport with, 256

intrasoil transport with, 256
materials for, 258–261, 258f, 259f, 260t
motivation for, 255–257
results for, 261–262, 261f
study area for, 257
Fault strike, 240–241, 244f
Fault zone, 232, 242, 243f, 244
Faults, 231–254
 block diagrams illustrating, 236f
 case studies, 241–253
  Crimean peninsula, 241–248, 241f, 242f
  Kursk nuclear power plant area, 248–253
 classification of, 232
 dip estimation for, 240–241, 244f
 dip-slip, 235, 236, 236f, 237, 238f, 239, 246, 247f
 dislocations from vertical *v.* horizontal tectonic motions differ, 232
 displacement estimation for, 240–241, 244f
 gaping, 237, 239
 hill shading for recognition of, 233, 234f
 materials and data processing for, 237–239
 method validation for, 237–241, 238f
 motivation for, 231–235, 233f
 oblique-slip, 237, 238f, 247, 247f
 results and discussion on, 238f, 239
  oblique-slip fault model, 238f, 239
  strike-slip fault model, 238f, 239
  thrust model, 238f, 239
 strike estimation for, 240–241, 244f
 strike-slip, 236–237, 236f, 238f, 244, 247
 theory of, 235–237
Filosofov method, 109–111, 110f
Filtering, 103–132
 global, 106–107
 methods of DTM, 109–122
  cutting method, 119–122, 121f
  Filosofov method, 109–111, 110f
  row and column elimination, 117–119, 120f
  smoothing, 115–117, 118f
  trend-surface analysis, 109
  two-dimensional discrete Fourier transform, 112–113, 114f
  two-dimensional discrete wavelet transform, 113–115, 116f
 selective, 107

 tasks of DTM, 103–109
  decomposition of topographic surface, 104
  denoising, 104–105
  generalization, 105–109
 two-dimensional singular spectrum analysis, 122–132
  algorithm for, 122–125
  materials and data processing for, 125–129, 125f, 126f, 127f, 128f, 129f
  results and discussion of, 126f, 127f, 128f, 129f, 130–132, 131f
Flexure, 111, 237, 251–252, 252f
Flow attributes, 9–11, 12t
Flow line, 12t, 60f, 135, 240–241, 244f, 293–294
Flow routing algorithm, 59, 60
Folding, 223–230
Folds, 223–230
 classification, 223–225, 224f
 cylindrical, 225
 elliptical (domes or basins), 225
 folding models, 226–229, 228f, 229f
 geometry, 223–225, 224f
 hyperbolical (saddles), 225
 predicting fold deformation degree, 225–226
 predicting fracturing of, 225–226
 Theorema Egregium and, 226–229, 229f
 three problems, 223
 visualization, 225
Form attributes, 9–11, 12t
Fourier transform, 112–113
 two-dimensional discrete, 112–113, 114f
Fractal topographic models, 8
Fracturing, 225–226
 rock, 258–262, 259f, 260t, 261f
Fuzzy logic, soil predictive mapping with, 178–180, 179f

G

G. *See* Slope gradient
Gaping fault, 237, 239
Gauss-Ostrogradsky theorem, 298
Gaussian classification, 23–25
 continual form of, 25
 illustration using terms of structural geology, 27f
 map of topographic segmentation using, 29f
Gaussian curvature (K), 10, 18f, 225, 301–302, 311

Gaussian curvature (K) (*Continued*)
  calculation of, 14
  definitions and interpretations of, 12t
General geomorphometry, 8–9
Generalization, 105–109
  cartographic, 105–106
  classes of DTM, 106–108
    global filtering, 106–107
    heuristic generalization, 107–108
    selective filtering, 107
  DTM, 106–109, 107f
Generalized additive models, soil predictive mapping with, 182
Generalized linear models, soil predictive mapping with, 182
Generating function (T), 10, 312–313
  structural line calculation with, 62–64, 63f
Geological boring, 34
Geomorphometry, 1–3, 23
  general, 8–9
  specific, 8–9
Geosat, 34
Gibbs phenomenon, errors and accuracy from, 88–93
  data processing with, 90
  discussion of, 90–93, 91f
  materials in, 90
  motivation for, 88–90, 88f
  results with, 90–93, 91f
Global filtering, 106–107
Global helical structures, 265f, 277–284, 280f, 281f, 282f, 282t, 283f
  Biscay-Santa Cruz, 280, 281f, 282t
  Caucasus-Clipperton double helix, 280, 281f
  Dakar, 280, 281f, 282t
  Marcus, 281f, 282t
  Palawan, 280, 281f, 282t
Global topography, 263–284
  data processing for, 266–269, 266f, 268f
  discussion of, 269–284
  materials for, 266–269, 266f, 268f
  motivation for, 263–266, 265f
  results for, 269–284
    general interpretation, 269–277, 270f, 271f, 273f, 274f, 275f, 276f
    global helical structures, 265f, 277–284, 278f, 279f, 280f, 281f, 282f, 282t, 283f
Global Topography Data Record, 267

GLOBE DEM, 35, 38, 54, 70, 78, 119, 184f, 266
GPS, 196. *See also* Kinematic GPS surveys
Gradient. *See* Slope gradient
Gravitational acceleration vector, 9–11, 289
Gravity, topographic surface, 8, 289
Grid
  calculation of local variables on
    spheroidal equal angular grid, 54–59, 77–78
    plane square grid, 45–54, 75–76, 76t
  calculation of nonlocal variables, 59–61, 60f
  displacement errors, 93–97
    data processing with, 94–96, 95f, 96f
    discussion of, 96–97, 97f, 98t
    Kolmogorov-Smirnov test with, 97, 97f, 98t
    materials in, 94–96, 95f, 96f
    motivation for, 88f, 93–94
    results with, 96–97, 97f, 98t
  irregular, 36–37, 37f
  regular, 37, 37f
  spheroidal, 38
  square, 37, 37f
  triangle, 37f, 37f
  types, 36–38
Grid spacing, 38–40
  adequate, 152
  determining, 153
  empirical approaches to selecting, 38–39
  Evans-Young method, 44–45
  morphometric variables with, 134t
  sampling theorem with, 39
  soil moisture correlated with, 163t
Groundwater, 27–28, 158, 255–256, 262
GTOPO30, 35, 36, 38, 54, 70, 107f, 125, 126, 132, 178

H

H. *See* Mean curvature
Hachures, 135, 136f
Helical structures, 265f, 277–284, 280f, 281f, 282f, 282t, 283f
  Biscay-Santa Cruz, 280, 281f, 282t
  Caucasus-Clipperton, 280, 281f
  Dakar, 280, 281f, 282t
  Marcus, 281f, 282t
  Palawan, 280, 281f, 282t
Heuristic classification, 174
Heuristic generalization, 107–108

Hierarchical level, 119–120, 132, 216, 223, 286
High-frequency noise, 8, 40, 45, 52, 58, 64, 71, 81, 104, 117, 267, 285
Hill shading
 fault recognition with, 233, 234f
 soil and geological data combined with, 141–142, 141f, 179f, 181f
Homosoil method, soil predictive mapping with, 182–183
Horizontal curvature (k_h), 10, 17f, 299, 308, 310–312
 calculation of, 14
 definitions and interpretations of, 13t
 Earth, 269, 270f
 illustration of, 11f
 Mars, 269, 271f
 Moon, 269, 271f
 soil moisture correlated with, 163t
 soil properties influenced by, 147
 Venus, 269, 271f
Horizontal excess curvature (k_{he}), 10, 18f, 308
 calculation of, 15
 definitions and interpretations of, 13t
Hybrid geostatistical methods, soil predictive mapping with, 176
Hyperbolical fold (Saddles), 225

I

IBCAO, 36
Ice sheet, 8, 33, 34
ICESat, 266–267
Ikonos, 32
InSAR. See Interferometric synthetic aperture radar
Insolation, 21–22, 21f, 22f, 296–297
Interferometric synthetic aperture radar (InSAR), 32–33
Interferometry, 32–33
Interpolation, 40–41
 formulation of, 40
 inverse distance-weighted, 40
 methods of, 40–41
 thin-plate spline, 41
 triangulation-based, 40–41
Inverse distance-weighted interpolation, 40
IS. See Shape index
Isotropy, 100–101

J

Jenny equation, 169

K

K. See Gaussian curvature
K_a. See Accumulation curvature
k_h. See Horizontal curvature
k_{he}. See Horizontal excess curvature
Kinematic GPS surveys, 3–4, 32, 66
 predictive soil mapping quantified by, 169
k_{max}. See Maximal curvature
k_{min}. See Minimal curvature
Kolmogorov-Smirnov test, grid displacement errors with, 97, 97f, 98t
Kotelnikov theorem. See Sampling theorem
K_r. See Ring curvature
Krasovsky ellipsoid, 267
Kriging, 176
Kriging with external drift, 176
Kursk nuclear power plant area, 79f, 248–253
 geological setting for, 248–249, 248f, 249f
 materials and data processing for, 249–250, 250f, 251f, 252f
 results and discussion for, 251–253
k_v. See Vertical curvature
k_{ve}. See Vertical excess curvature

L

Landforms
 classification of, 23–30
  Efremov-Krcho classification, 25–28, 27f, 29f
  Gaussian classification, 23–25, 27f, 29f
  Shary classification, 27f, 28–30, 29f
LandLord, 50, 80, 86, 90, 95, 128, 162, 199, 212, 229, 239, 245, 261, 269, 315, 316
Landsat MSS, 32
Laplacian, 147, 226, 301
Laser surveys, 33
Leveling, 31–32
LiDAR (Light Detection And Ranging), 33
 predictive soil mapping quantified by, 169
Lineaments, 231–254
 block diagrams illustrating, 236f
 case studies, 241–253
  Crimean peninsula, 241–248, 241f, 242f
  Kursk nuclear power plant area, 248–253
 classification of, 232

Lineaments (*Continued*)
 dip estimation for, 240–241, 244f
 dislocations from vertical *v.* horizontal tectonic motions differ, 232
 displacement estimation for, 240–241, 244f
 hill shading for recognition of, 233, 234f
 materials and data processing for, 237–239
 method validation for, 237–241, 238f
 motivation for, 231–235, 233f
 results and discussion on, 238f, 239
  oblique-slip fault model, 238f, 239
  strike-slip fault model, 238f, 239
  thrust model, 238f, 239
 strike estimation for, 240–241, 244f
 theory of, 235–237
Linear artifacts, 98–101
 isotropy of local morphometric variables with, 100–101
 motivation for, 98–100, 99f
Linear regression, soil predictive mapping with, 182
Local morphometric variable, 9–15, 10f, 11f, 12t, 15f, 16f
 calculation methods for
  spheroidal equal angular grid, 54–59
  plane square grid, 45–54
 flow attributes, 9–11, 12t
 form attributes, 9–11, 12t
 linear artifacts from isotropy of, 100–101
 soil properties influenced by, 146–148
 two classes of, 9–11
Lunar Orbiter Laser Altimeter mission, 274

M

M. See Unsphericity curvature
Magellan mission, 267
Manual geomorphometric techniques, 1, 2, 170–172, 185, 186f
Maple, 49, 57, 77, 88
Mapping, 133–142
 combined visualization of morphometric variables in, 135, 136f
 cross sections in, 135–136, 137f
 hill-shading maps combined with soil and geological data in, 141–142, 141f, 179f, 181f
 three-dimensional topographic modeling in, 136–141, 138f, 139f, 179f, 233f, 250f
 peculiarities of, 133–135

Marcus helical structure, 281f
Mars
 catchment area of, 272, 275f, 279f
 dispersive area of, 272, 276f
 elevation of, 268f
 global helical structures of, 282f, 283f
 horizontal curvature of, 269, 271f
 vertical curvature of, 272, 273f
Mars Global Surveyor mission, 267
Mars Orbiter Laser Altimeter mission, 267, 283f
Maximal curvature (k_{max}), 10, 17f, 301–302, 307–308, 311
 calculation of, 15
 definitions and interpretations of, 12t
Mean curvature (H), 10, 17f, 299–302, 307, 311
 calculation of, 14
 definitions and interpretations of, 12t
 soil moisture correlated with, 163t
Meusnier theorem, 299
Mega-scarps, 272
Microbial biomass carbon, 197t, 198, 210t, 213t
Microbial respiration rate, 197t, 198, 210t
Minimal curvature (k_{min}), 10, 17f, 301–302, 307–308, 311
 calculation of, 15
 definitions and interpretations of, 12t
 Earth, 270f
Moderate Resolution Imaging Spectroradiometer (MODIS), 178f
Moon
 catchment area of, 275f, 279f
 dispersive area of, 276f
 elevation of, 268f
 global helical structures of, 282f
 horizontal curvature of, 269, 271f
 vertical curvature of, 272, 273f
Morphoisographs, 172
Morphometric map, 1–2, 70, 211–212, 252–253
 peculiarities of, 133–135
Morphometric variable, 7–30
 combined, 23, 24f
  steam power index, 14t, 23, 24f
  topographic index, 14t, 23, 24f
 defined, 8–9
 landform classifications based on, 23–30
  Efremov-Krcho classification, 25–28, 27f, 29f

Gaussian classification, 23–25, 27f, 29f
Shary classification, 27f, 28–30, 29f
local, 9–15, 10f, 11f, 12t, 15f, 16f
  flow attributes, 9–11, 12t
  form attributes, 9–11, 12t
  linear artifacts from isotropy of, 100–101
  two classes of, 9–11
  main groups of, 9
  nonlocal, 13t, 16–19, 18f, 19f
  solar, 21–22, 21f
  structural lines, 19–20
  visualization and mapping with, 135, 136f
Moscow Region, 78, 80f
  Moscow, city, 117, 118f
  Severny Gully, Pushchino, 82–86, 85f, 157–165, 158f
    materials and methods for, 158–162, 159f
    results and discussion for, 162–165, 161f
Moving window, 43–44, 52, 55, 58, 61, 72–73, 75, 78, 267
  plane square, 43–44, 44f
  smoothing with, 115, 117
  spheroidal equal angular, 44f, 54, 55, 57, 58
  types of, 44f
Multinomial logistic regression, soil predictive mapping with, 180, 181f
Multiple flow direction algorithm, 59
Multiple regression analysis, soil predictive mapping with, 175–176, 188
Multiscale support vector regression, soil predictive mapping with, 181–182

N
Nitrous oxide flux (N_2O flux), 197t, 210t
Nonlocal morphometric variable, 13t, 16, 18, 19f
  calculation methods for, 59–61, 60f
  soil properties influenced by, 148–149
Normal section, 9, 10f, 12t, 236f, 299, 307
Northern Andes, 125–132, 125f
  materials for, 125–126, 126f
  data processing for, 126–129, 127f
  results and discussion for, 130–132, 126f, 128f, 129f

O
Oblique-slip fault, 237, 238f, 247, 247f
Oblique-slip fault model, 238f, 239
Overthrust, 239

P
Palawan helical structure, 280, 281f
Paleotopography, 227, 229, 252–253
Partial derivatives, 11, 290, 312
  approximation with, 43–44, 54–55, 286
  RMSE formulas for, 75–78
Pedometric techniques, 185–186
Photogrammetric approaches, 32
Photogrammetry, digital terrain models from, 2
Plan curvature, 10, 297–299. *See also* Horizontal curvature and Relief
Plasticity method
Plane square grid, 37, 37f
  calculation methods for, 45–54
    accuracy with, 75–76, 76t
    data processing for validation of, 49–50
    discussion of method for, 50–54, 51f, 52f, 53f
    formulas, 46–49
    materials for validation of, 49–50
    method validation, 49–54
    motivation, 45
    results of, 50–54, 51f, 52f, 53f
    RMSE for validation of, 49, 50
    Taylor formula for, 46
Plane square moving window, 43–44, 44f
Plasticity, 223, 227
Plate Carrée projection, 281f, 282f
Platonic solid, 38
Polarimetry, 32–33
Polynomial, 44, 46, 55–56
  Evans-Young method, 44
  quadric, 94–95, 160, 198–199, 239, 245
  trigonometric, 88–89, 88f, 109
Principal curvature, 3, 10, 119, 120–121, 224f. *See also* Maximal curvature and Minimal curvature
Profile curvature. *See* Vertical curvature

Q
QuickBird, 32

R
Radar techniques, 32–33
Radarsat, 32–33
Radio-echo sounding surveys, 35
Rare forms, 30, 121, 225

Regression kriging, soil predictive mapping with, 176–177, 177f, 178f
Relative accumulation zones. *See* Accumulation zones
Relative dissipation zones. *See* Dissipation zones
Relief Plasticity method, soil predictive mapping with, 172, 185, 186f
Resolution, 38–40
   field study on modeling, 157–165
      materials and methods, 158–162, 159f, 160f, 161f
      results and discussion, 162–165, 163t
      study site, 157–158, 158f, 159f
   modeling with adequate, 151–166
      elementary volume in, 154, 154f, 155
      motivation for, 151–153
      technical aspect of, 151–152
      theory of, 153–157, 154f
   sampling theorem with, 39
Ridge lines. *See* Crests
Ring curvature (K_r), 10, 18f, 308–311
   calculation of, 15
   definitions and interpretations of, 13t
RMSE. *See* Root mean square error
Rock fracturing, 259f, 261f, 260t, 258–262
Romanova methods, 171
Root mean square error (RMSE), 49, 50
   accuracy with formulas for local morphometric variables, 73–75
   accuracy with formulas for partial derivatives, 75–78
   mapping, 78–81, 79f, 80f
Rotational stress, 284, 287
Rotor, 304–306, 309
Row and column elimination, 117–119, 120f

S

Saddles. *See* Hyperbolical fold
Sampling theorem, 39
   errors from ignoring, 81–88
      data processing with, 82–86, 84f, 85f
      discussion of, 86–88
      materials in, 82–86, 84f, 85f
      motivation for, 81–82, 82f
      results with, 86–88
   three main sequences of, 39
SAR. *See* Synthetic aperture radars
Satellite laser altimetry, 33
Satellite radar altimetry, 34
SCA. *See* Specific catchment area
Scale
   topography, 1, 8
   upscaling of soil predictive mapping on small, 182–185
Scale-oriented generalization, 105–106
SCORPAN model, 169
SDA. *See* Specific dispersive area
Seasat, 34
Selective filtering, 107
Shape index (IS), Gaussian classification using, 25
Shary classification, 27f, 28–30, 29f
   five curvatures as basis of, 29–30
   flexibility of, 30
Shipboard echo sounding, 33–34
SI. *See* Stream power index
Single flow direction algorithm, 59, 277
Singular spectrum analysis (SSA), two-dimensional, 122–132
Slope aspect (A), 10, 16f, 294–296
   calculation of, 14
   definitions and interpretations of, 12t
   illustration of, 11f
Slope gradient (G), 10, 16f, 289–291
   calculation of, 14
   definitions and interpretations of, 12t
   illustration of, 11f
   soil moisture correlated with, 163t
Slope lines, 293–294
Smoothing, 115–117, 118f
Software, 315
   CatchmentSIM, 62
   LandLord, 50, 80, 86, 90, 95, 128, 162, 199, 212, 229, 239, 245, 261, 269, 315, 316
   Maple, 49, 57, 77, 88
   Statgraphics Plus, 50, 96, 128, 162, 212
   2D-SSA, 127
Soil, topography's influence on, 145–149
   discussion of, 149
   gravity-driven migration of water with, 146
   horizontal curvatures in, 147
   local morphometric variables in, 146–148
   nonlocal morphometric variables in, 148–149
   temperature differentiated by regime of slopes in, 146
   tillage practice in, 149
   topographic index in, 148–149
   vertical curvatures in, 147

Soil augering, 34
Soil forming factors, 1, 145–146, 167–169, 182–183, 286
Soil horizons thickness, 141, 176
Soil mapping *See* Digital soil mapping; Soil predictive mapping
Soil moisture
  distribution along contour of, 160f
  estimates in modeling resolution field study, 158–159
  grid spacing correlated with, 163t
  horizontal curvature correlated with, 163t
  importance of, 146
  mean curvature correlated with, 163t
  relative slope position in relation to, 148
  slope gradient correlated with, 163t
  TOPMODEL assumption regarding, 191–192
  vertical curvature correlated with, 163t
Soil moisture content, topography-soil system with, 204t, 208t
Soil organic carbon, 173, 175, 176, 217
Soil organic matter, predictive mapping of, 176–177, 177f, 178f
Soil predictive mapping, 167–190
  classification of methods for, 173–174
    heuristic classification approaches, 174
    methods to predict categorical variables, 174
    methods to predict quantitative soil properties, 173
    supervised classification approaches, 174
    unsupervised classification approaches, 174
  current predictive methods of, 172–187
  Dokuchaev hypothesis, 167–170
  early models of, 170–172
  GPS kinematic survey for quantifying, 169
  LiDAR for quantifying, 169
  manual geomorphometric techniques for, 170–172
  mathematical approaches to, 174–182
    artificial neural networks, 181
    Bayesian analysis of evidence, 181
    decision trees, 180
    discriminant analysis, 180
    empirical best linear unbiased predictor, 182
    fuzzy logic, 178–180, 179f
    generalized additive models, 182
    generalized linear models, 182
    hybrid geostatistical methods, 176
    linear regression, 182
    multinomial logistic regression, 180, 181f
    multiple regression analysis, 175–176
    multiscale support vector regression, 181–182
    regression kriging, 176–177, 177f, 178f
  prediction accuracy with, 185–187
  Romanova methods for, 171
  SCORPAN model for, 169
  small-scale predictive models and upscaling, 182–185
    bottom-up approach for, 183–185
    Homosoil method for, 182–183
    Relief Plasticity method for, 185, 186f
    top-down approach for, 183–185
    upscaling procedures for, 182–183, 184f, 185f
  TOPMODEL for, 171, 172
  topographic multivariable approach to, 187–189
Solar azimuth angle, 110f
Solar morphometric variable, 21–22, 21f
Solar zenith angle, 110f
Special points, 15, 19–20, 64, 291–293
Specific catchment area (*SCA*), 16, 18, 19f
  definitions and interpretations of, 13t
Specific dispersive area (*SDA*), 16, 18, 19f
  definitions and interpretations of, 13t
  Earth, 272, 274f
Specific geomorphometry, 8–9
Spheroidal equal angular grid, calculation methods for, 54–59
  accuracy with, 77–78
  discussion of method for, 58–59
  formulas, 55–57
  linear sizes of window of, 57–58
  motivation, 54–55
Spheroidal equal angular moving window, 44f, 54, 55, 57, 58
SPOT, 32
SRTM, 67–68, 99, 116f, 177f, 178f, 285
SRTM3, 32–33, 38, 54, 55, 117
SRTM30_PLUS, 33, 34, 36, 38, 54
SSA. *See* Singular spectrum analysis
Statgraphics Plus, 50, 96, 128, 162, 212
Stavropol Upland, 15f, 16f, 17f, 18f, 19f, 20f, 22f, 24f, 28f, 29f, 49, 51f, 53f, 95f, 96f

Stereo images, 66f, 67, 98
Stereo radargrammetry, 32–33
Stratigraphic horizon, 8, 34–35, 135–136, 250f
Stratigraphic surface, 36, 69–70, 112–113, 223, 226, 232, 235, 250
Stream power index (*SI*)
  calculation of, 23, 24f
  definitions and interpretations of, 14t
Strike, 240–241, 244f
Strike-slip fault, 236–237, 236f, 238f, 244, 247
Strike-slip fault model, 238f, 239
Structural lines, 19–20
  calculation methods for, 61–64
    cell-to-cell flow routing, 61
    conventional algorithms, 61–62
    generating function, 62–64, 63f, 312–313
Subduction zone, 36, 227
Supervised classification, 174
Surface
  stratigraphic, 36, 69–70, 112–113, 223, 226, 232, 235, 250
  three-dimensional seismic survey with, 34–35
  topographic, 7–8
    defined, 8
    filtering for decomposition of, 104
    restrictions for, 7–8
Synthetic aperture radars (SAR), 32–33

T

T. *See* Generating function
Tacheometer, 31–32, 158
Tangential curvature. *See* Horizontal curvature
Taylor formula, 46
Tectonic structures, 263–284
  data processing for, 266–269, 266f, 268f
  discussion of, 269–284
  materials for, 266–269, 266f, 268f
  motivation for, 263–266, 265f
  results for, 269–284
    general interpretation, 269–277, 270f, 271f, 273f, 274f, 275f, 276f
    global helical structures, 265f, 277–284, 278f, 279f, 280f, 281f, 282f, 282t, 283f
Thalweg, 20, 20f, 312
  calculation methods for, 61, 62, 63f
Theorema Egregium, 226–229, 229f

Thickness of soil horizons, 141, 176
Thin-plate spline interpolation, 41
Three-dimensional geological modeling, 33, 138f, 139f, 140
Three-dimensional seismic survey, 34–35
Three-dimensional topographic modeling, 136–141, 138f
Thrust, 236, 237, 238f, 239, 246
TI. *See* Topographic index
TIN. *See* Triangulated irregular networks
Top-down approach, soil predictive mapping with, 183–185
TOPMODEL, 171, 172
  soil moisture decreases with the soil depth assumption of, 191–192
Topographic index (*TI*)
  calculation of, 23, 24f
  definitions and interpretations of, 14t
  soil properties influenced by, 148–149
Topographic surface, 7–8
  defined, 8
  filtering for decomposition of, 104
  restrictions for, 7–8
Topographic surveys, 31–32
Topography. *See also* Global topography
  digital elevation models for, 2
  maps, 1–2
  near-surface layer of planet controlled by, 1
  soil properties influenced by, 145–149
    discussion of, 149
    gravity-driven migration of water with, 146
    horizontal curvatures in, 147
    local morphometric variables in, 146–148
    nonlocal morphometric variables in, 148–149
    temperature differentiated by regime of slopes in, 146
    tillage practice in, 149
    topographic index in, 148–149
    vertical curvatures in, 147
Topography-soil system, 191–221
  data processing for, 198–213
    statistical analysis in, 199–213, 213f, 200t, 202t, 204t, 206t, 208t, 210t, 211t, 212t, 213t
    topographic modeling in, 194f, 195f, 198–199
  field work, 196–197, 197t
  laboratory analyses, 197–198

materials and methods for, 195–213
motivation for, 191–192
results and discussion, 214–220
  depth variability, 215–216
  drier conditions, 218–219
  interpretations, 219–220
  relationship variability with soil and morphometric variables, 214–216
  temporal variability, 208t, 211t, 212t, 214–215
  topography and denitrification, 216–220
  wetter conditions, 217–218
study sites for, 192–195, 193f, 194f, 195f
  density traces and statistics, 200t, 202t
  microbial biomass carbon content, 213t
  parameters and statistics of regression equation, 211t, 212t
  soil attributes statistics, 206t
  soil moisture content, 204t
  soil moisture content correlations, 208t
  soil properties in wetter and drier conditions, 210t
Torsional deformation, 264, 265f, 280, 281
Transit zones, 25–28, 26f, 183, 256, 260t
  mapping of, 27, 29f, 184f, 258f
Trend-surface analysis, 109
Triangulated irregular networks (TIN), 37
Triangulation-based interpolation, 40–41
Two-dimensional discrete Fourier transform, 112–113, 114f
Two-dimensional discrete wavelet transform, 113–115, 116f
Two-dimensional singular spectrum analysis, 122–132
  algorithm for, 122–125
  materials and data processing for, 125–129, 125f, 126f, 127f, 128f, 129f
  results and discussion of, 126f, 127f, 128f, 129f, 130–132, 131f

U

Unsphericity curvature (M), 10, 18f, 302–304, 311
  calculation of, 15

definitions and interpretations of, 12t
Unsupervised classification, 174
Upscaling, soil predictive mapping with, 182–185, 184f, 185f

V

Valley lines. *See* Thalweg
Venus
  catchment area of, 275f, 279f
  dispersive area of, 276f
  elevation of, 268f
  global helical structures of, 282f, 283f
  horizontal curvature of, 269, 271f
  vertical curvature of, 272, 273f
Vertical curvature (k_v), 10, 17f, 299–301, 306, 308, 311
  calculation of, 14
  definitions and interpretations of, 12t
  Earth, 270f, 272
  illustration of, 11f
  Mars, 272, 273f
  Moon, 272, 273f
  soil moisture correlated with, 163t
  soil properties influenced by, 147
  Venus, 272, 273f
Vertical excess curvature (k_{ve}), 10, 18f, 308
  calculation of, 15
  definitions and interpretations of, 13t
Visualization, 133–142
  combination of morphometric variables, 135, 136f
  cross sections in, 135–136, 137f
  fold, 225
  hill-shading maps combined with soil and geological data in, 141–142, 141f, 179f, 181f
  three-dimensional topographic modeling in, 136–141, 138f, 139f
  peculiarities of, 133–135

W

Wavelet transform, 113–115, 116f

Z

Zakharov equation, 168

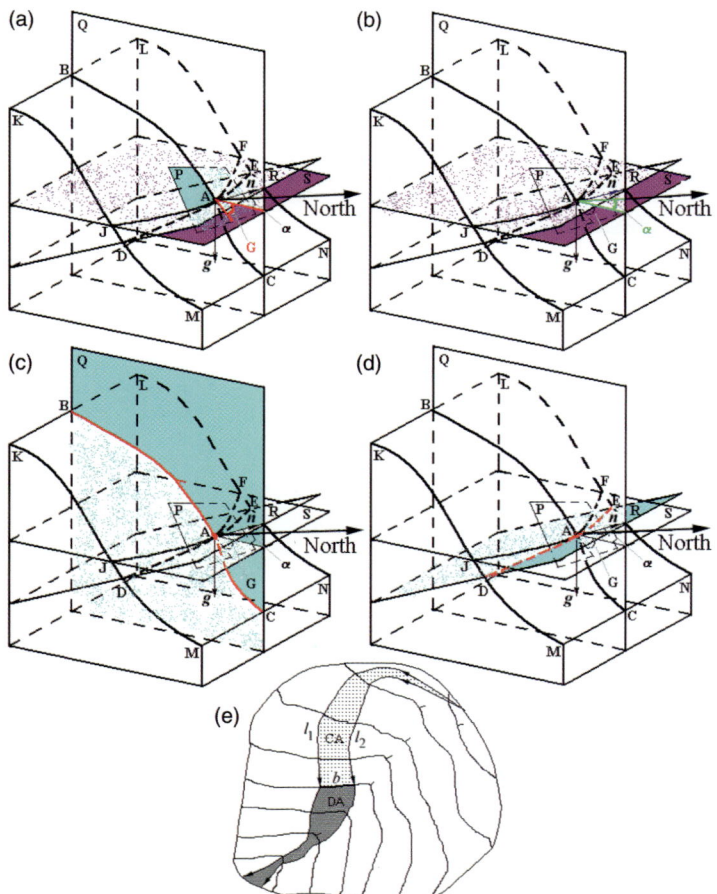

FIGURE 2.2 Illustrations for the definition of some morphometric variables: (a) slope gradient, (b) slope aspect, (c) vertical curvature, (d) horizontal curvature, (e) catchment and dispersive areas. For an explanation, see Table 2.1. *From (Florinsky, 2010, Fig. 1.2).*

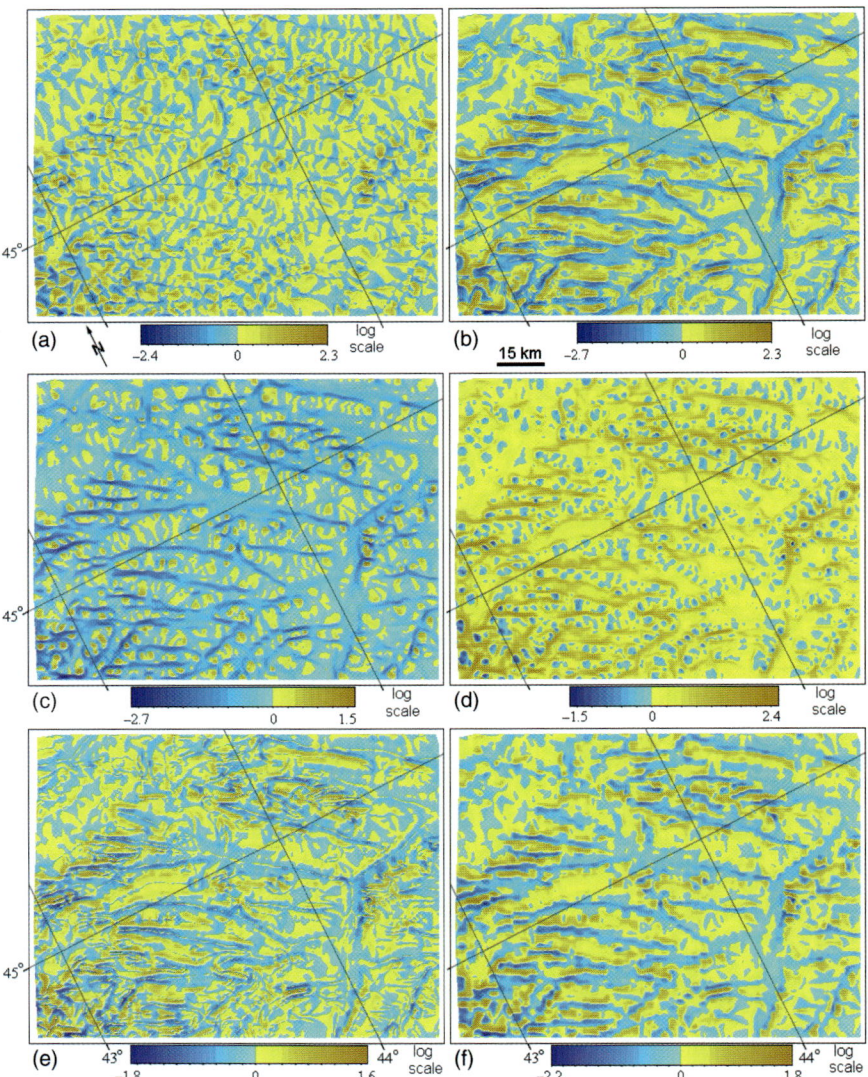

FIGURE 2.5 The Stavropol Upland: (a) horizontal curvature, (b) vertical curvature, (c) minimal curvature, (d) maximal curvature, (e) difference curvature, (f) mean curvature, (g) horizontal excess curvature, (h) vertical excess curvature, (i) the Gaussian curvature, (j) accumulation curvature, (k) ring curvature, and (l) unsphericity curvature. For details of calculation, see Section 4.2.3.1; for the elevation map, see Fig. 2.3. *From (Florinsky, 2009a, Fig. 2).*

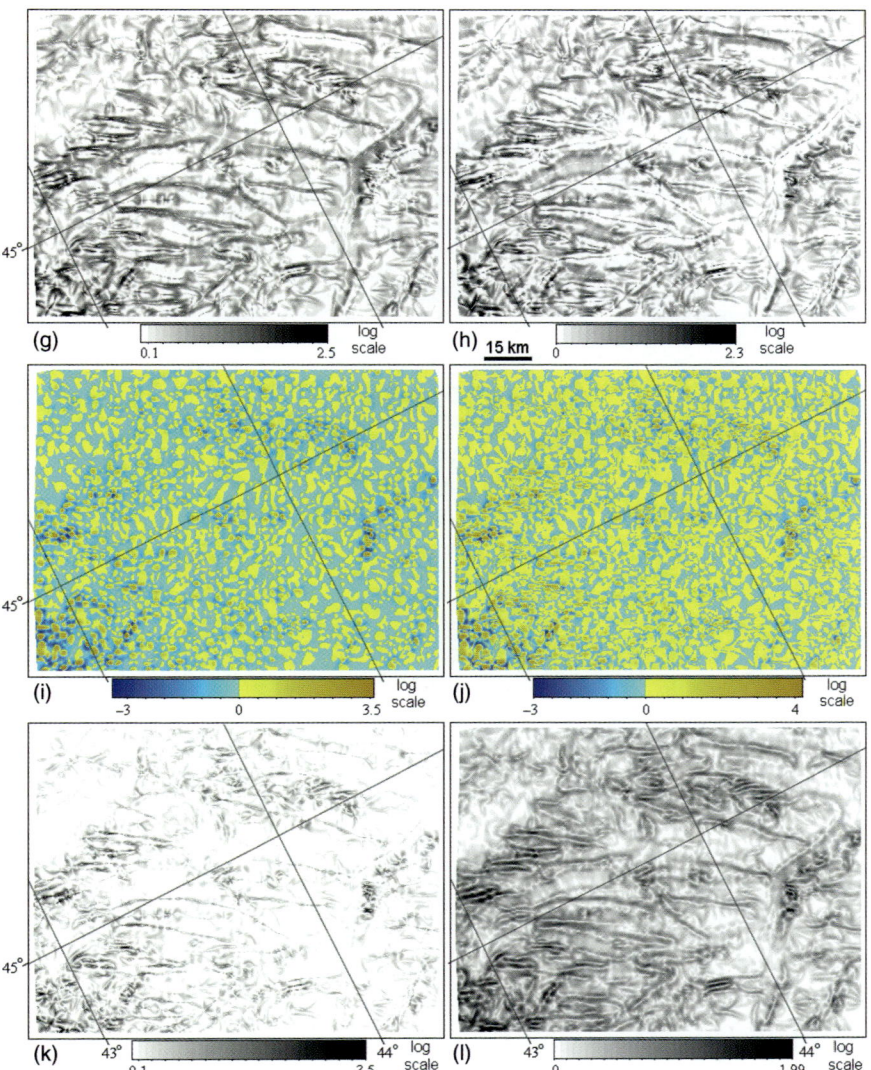

FIGURE 2.5 (Continued)

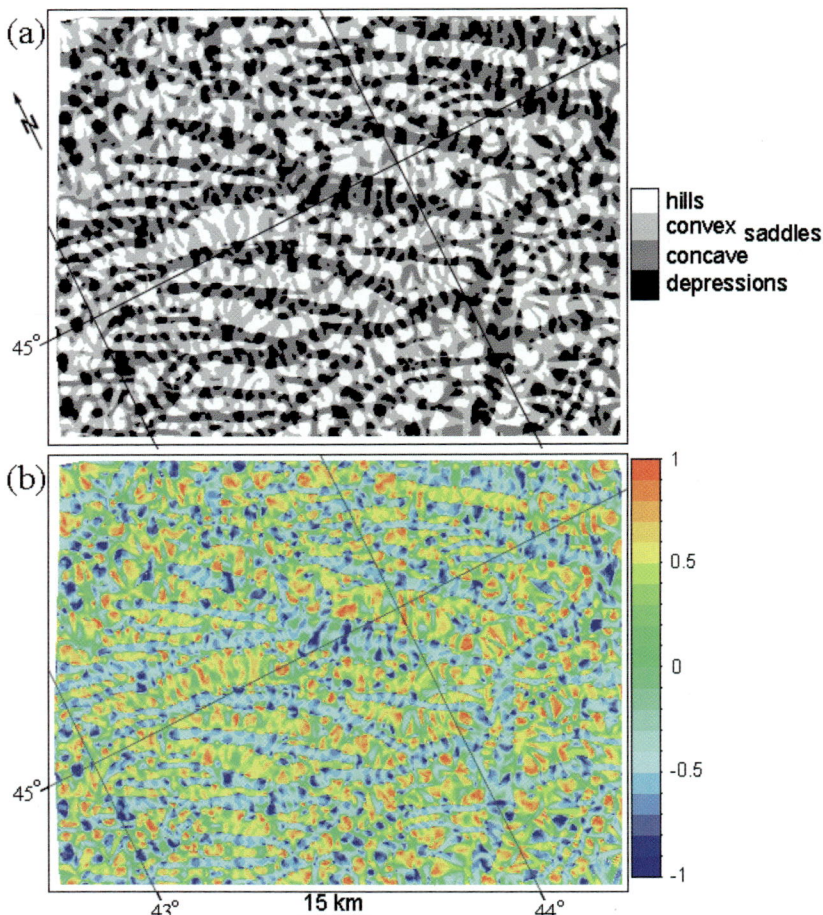

FIGURE 2.12 The Stavropol Upland: topographic segmentation performed using (a) the Gaussian classification; (b) shape index; (c) the Efremov–Krcho classification (zones of relative accumulation, transit, and dissipation); (d) the Shary classification. For the elevation map, see Fig. 2.3.

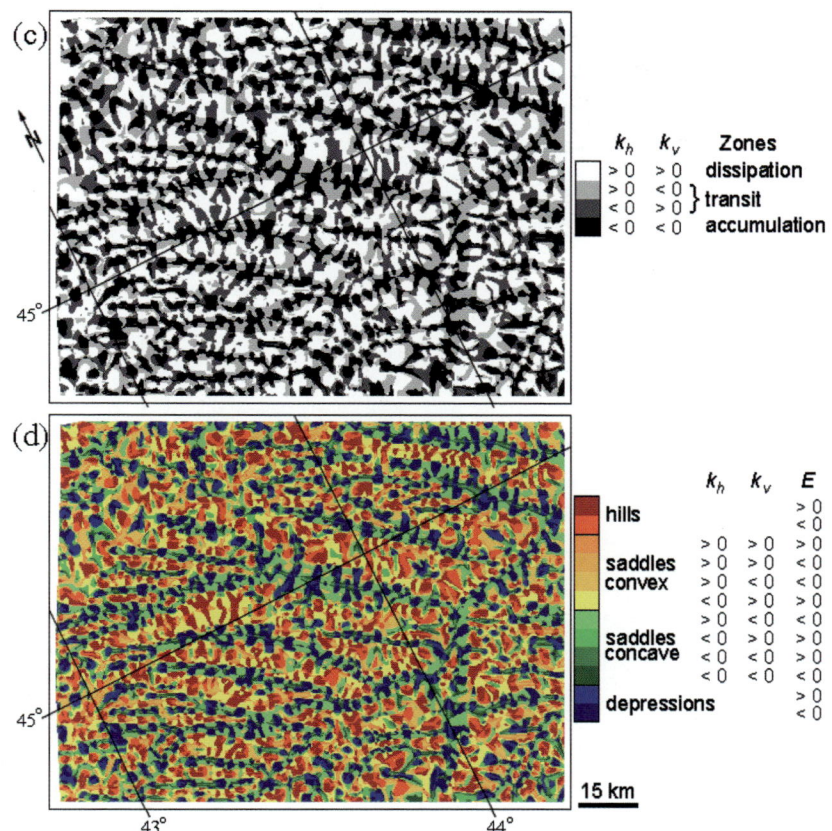

FIGURE 2.12 (Continued)

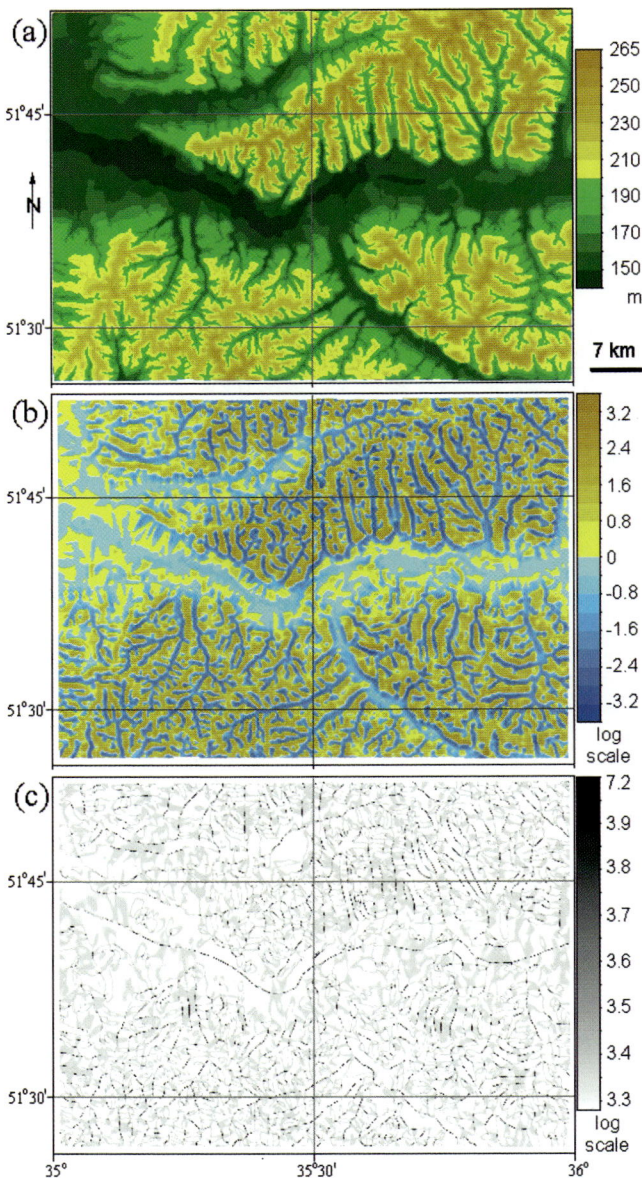

FIGURE 5.3 The Kursk Region near the Kursk nuclear power plant: (a) elevation, (b) mean curvature, (c) RMSE of the mean curvature calculation. *From (Florinsky, 2010, Fig. 2.6).*

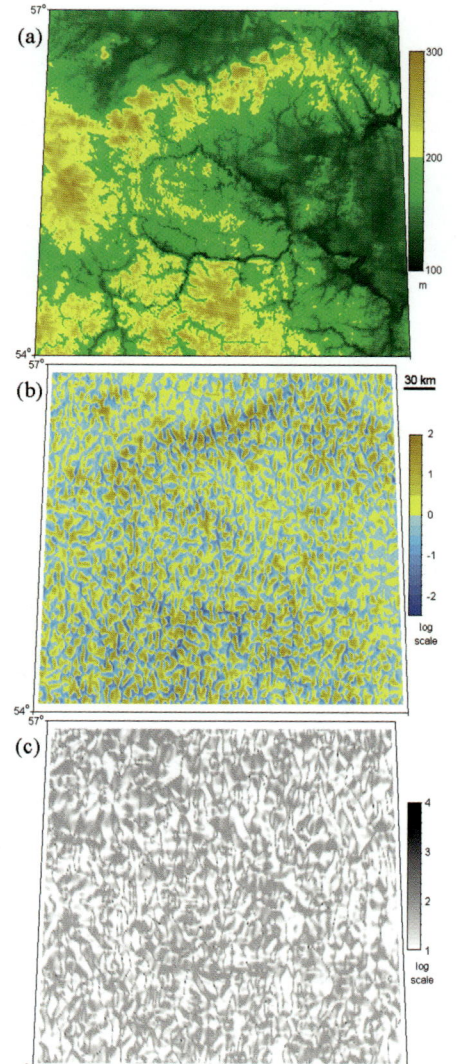

FIGURE 5.4 The Moscow Region and adjacent territories: (a) elevation, (b) horizontal curvature, (c) RMSE of the horizontal curvature calculation. *From (Florinsky, 2008d, Figs. 1 and 2a, c).*

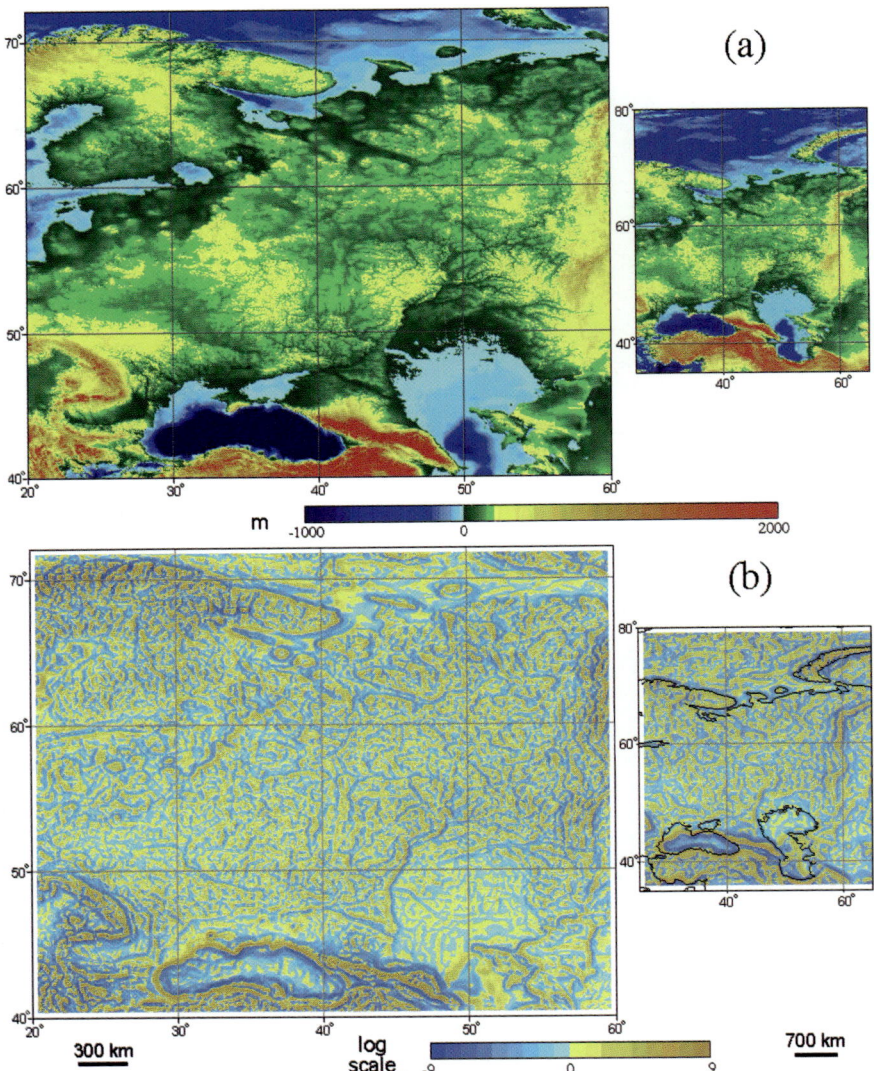

FIGURE 6.6 The East European Plain and adjacent territories: (a) elevation, (b) vertical curvature. Grid spacings are 4′ (left) and 10′ (right). DEMs include 289,081 points (the matrix 601 × 481) and 64,530 points (the matrix 239 × 270), respectively. *From (Florinsky, 2010, Fig. 1.23).*

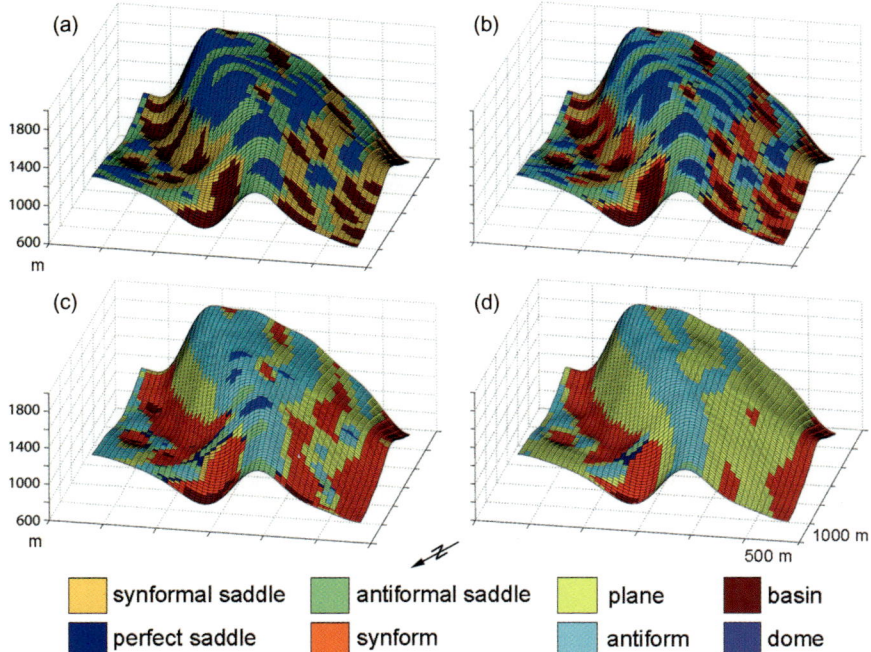

FIGURE 6.7 DTM generalization by the cutting method. Classification of the Sheep Mountain Anticline, Wyoming, USA, by signs of the Gaussian and mean curvatures with varying values of the curvature threshold: (a) $k_t = 0 \text{ m}^{-1}$, (b) $k_t = 10^{-4} \text{ m}^{-1}$, (c) $k_t = 5 \times 10^{-4} \text{ m}^{-1}$, (d) $k_t = 10^{-3} \text{ m}^{-1}$. A DEM was created by digitizing a structure contour map. The regular DEM includes 4641 points; the grid spacing is 50 m. *From (Mynatt et al., 2007a, Fig. 5), reproduced with permission.*

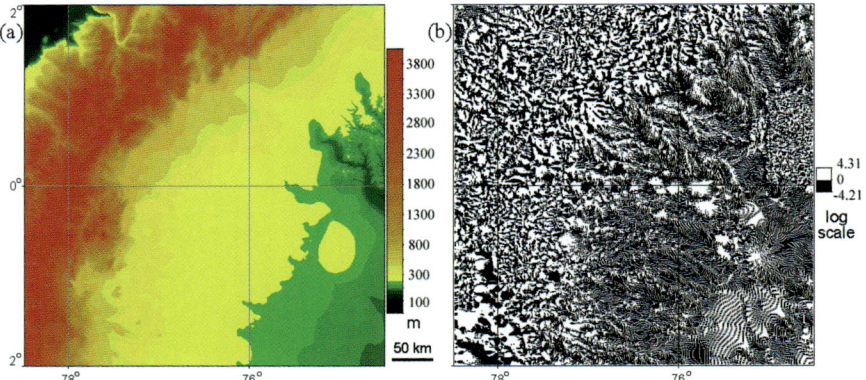

FIGURE 6.9 The Northern Andes: (a) elevation, (b) horizontal curvature derived from the initial DEM. *From (Golyandina et al., 2008, Fig. 2a, d).*

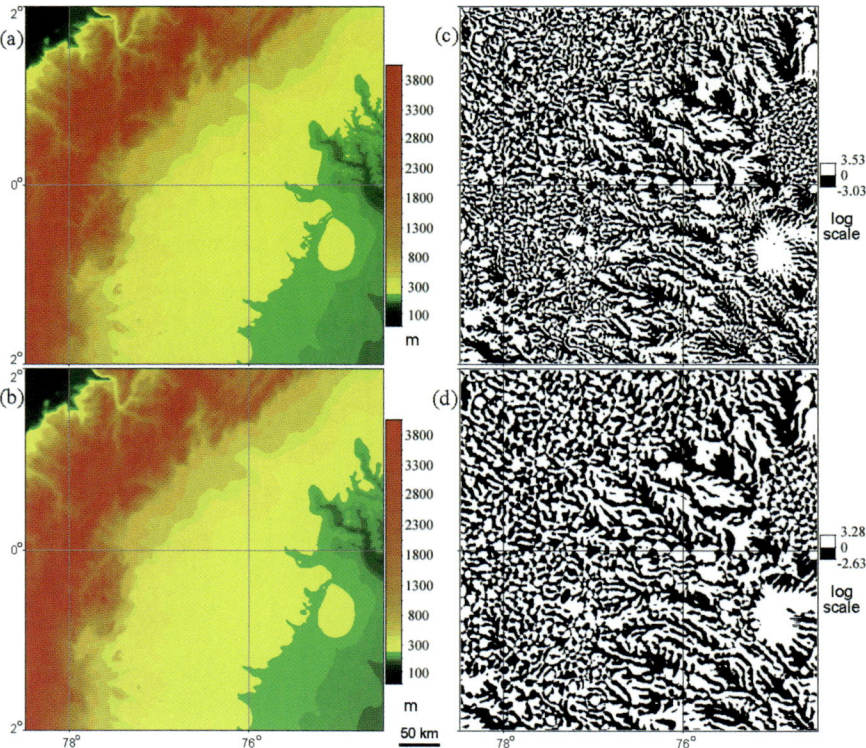

FIGURE 6.11 The Northern Andes, DTM denoising. Elevation reconstructed from: (a) the ET 1–100, (b) the ET 1–50. Horizontal curvature derived from: (c) the ET 1–100 DEM, (d) the ET 1–50 DEM. *From (Golyandina et al., 2008, Fig. 2b, c, e, f).*

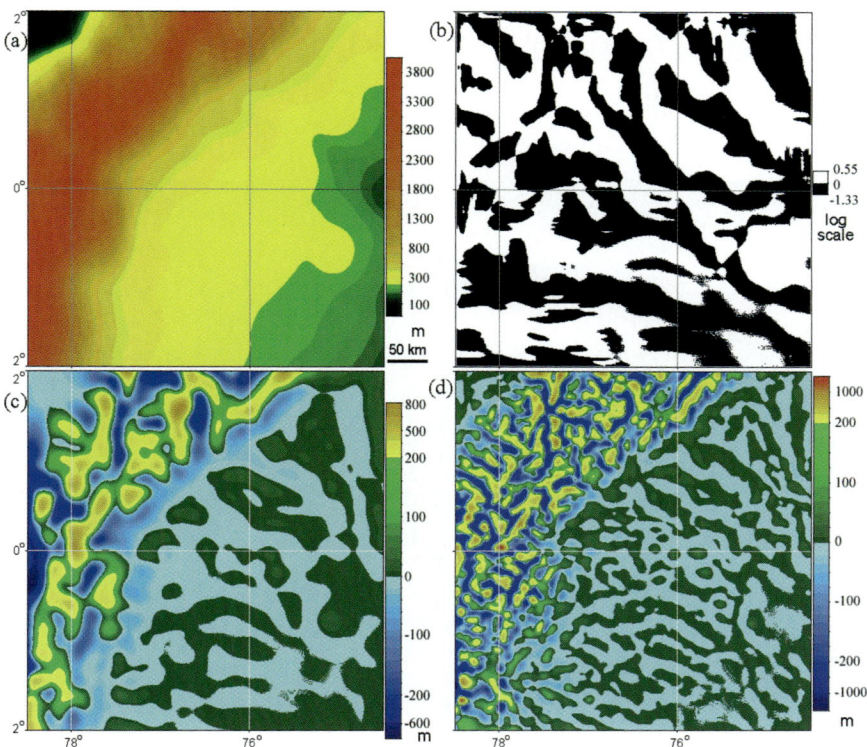

FIGURE 6.12 The Northern Andes, low-frequency components. (a) Elevation reconstructed from the ET 1. (b) Horizontal curvature derived from the ET 1 DEM. Elevation reconstructed from: (c) the ET 2–3, (d) the ET 4–25. *From (Golyandina et al., 2008, Fig. 3a–d).*

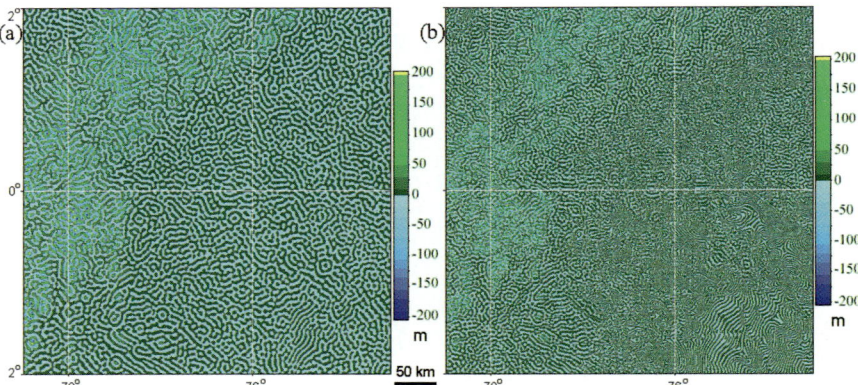

FIGURE 6.13 The Northern Andes, high-frequency components. Elevation reconstructed from: (a) the ET 51–100, (b) the ET 101–900. *From (Golyandina et al., 2008, Fig. 3e, f).*

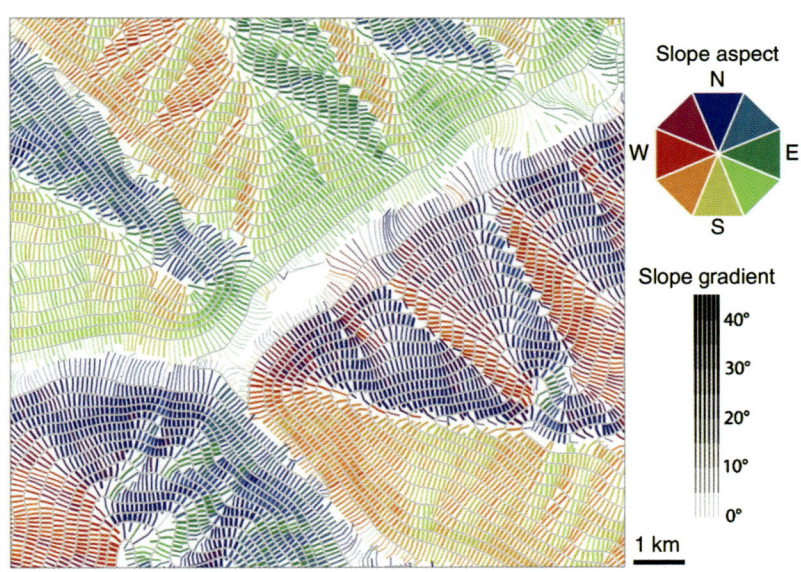

FIGURE 7.1 Combined visualization of slope gradient and aspect by colored hachures. *From (Samsonov, 2010, Fig. 9), reproduced with permission.*

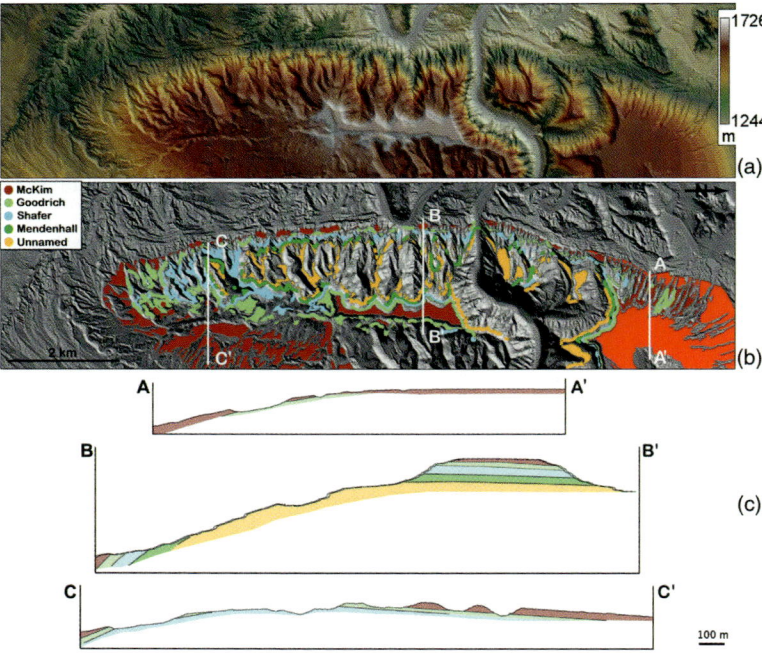

FIGURE 7.2 The use of DEM-derived cross sections in geological studies. The Raplee Ridge monocline, Utah, USA: (a) Elevation. (b) Hill shading with the spatial distribution of the five bedding-plane surfaces exhumed within the fold. (c) Cross sections (no vertical exaggeration). A DEM was derived from the Airborne Laser Swath Mapping data. The grid spacing is 1 m. *From (Hilley et al., 2010, Figs. 2a and 3), reproduced with permission.*

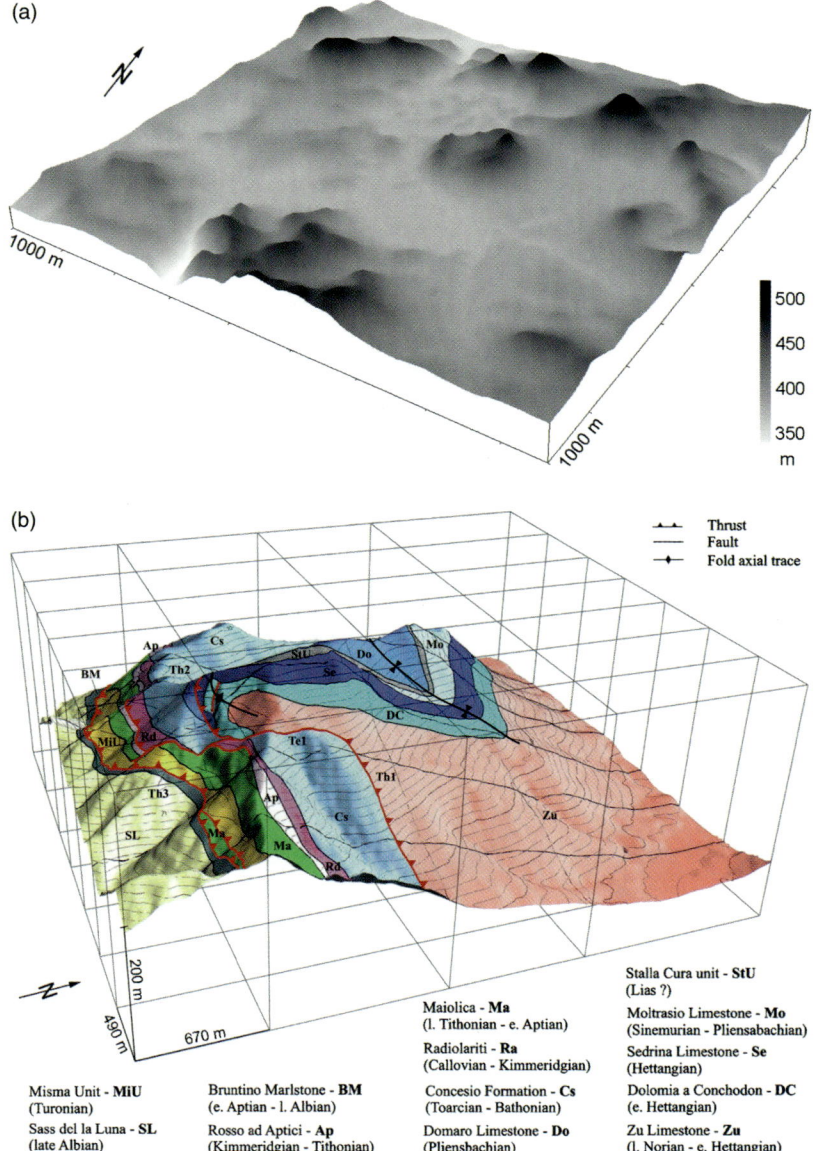

FIGURE 7.3 Examples of 3D topographic models: (a) 3D model of the Tunguska event area (60°53′09″N, 101°53′40″E). A DEM was derived from a topographic map, scale 1:100,000. The DEM includes 182,213 points (the matrix 431 × 423); the grid spacing is 20 m. The lower Triassic Kulikovsky paleovolcanic complex is visible due to the 5^x vertical exaggeration. *From (Florinsky, 2008e, Fig. 1)*. (b) 3D geological map of Mount Misma, Italian Alps. A DEM was derived from 10-m contour interval topographic maps. The grid spacing is 2.5 m. Geological information was derived from a 1:5000 scale geological map. *From (Zanchi et al., 2009, Fig. 10), reproduced with permission.*

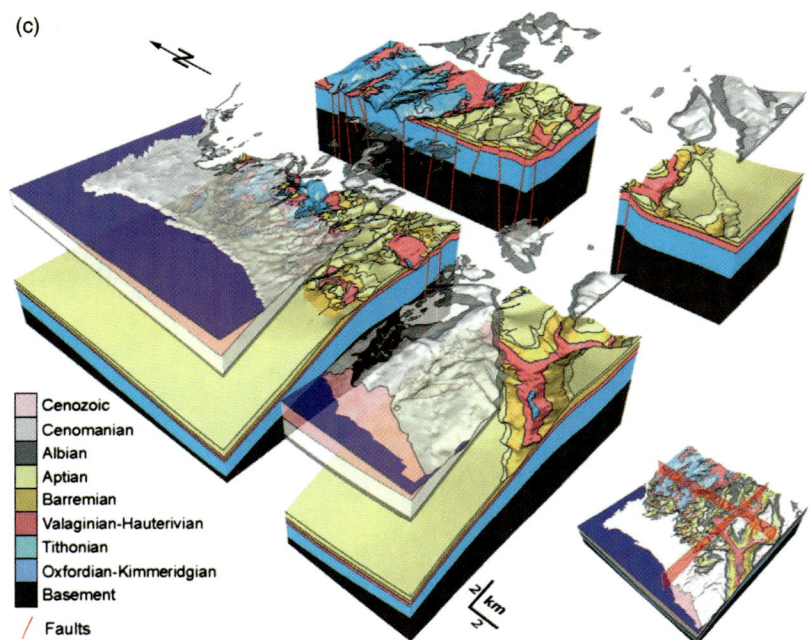

FIGURE 7.3 (Continued). (c) 3D geological model of the Beirut watershed, Lebanon. The model was cut along two cross sections (see right bottom corner), and each side of the cross section was pushed away from the other. The layers younger than the Aptian horizon are shown floating. The topographic surface was derived from a 30-m DEM obtained from the Terra Aster Product. Geological surfaces were derived from a 1:50,000 scale geological map. *From (Dhont et al., 2005, Fig. 4), reproduced with permission.*

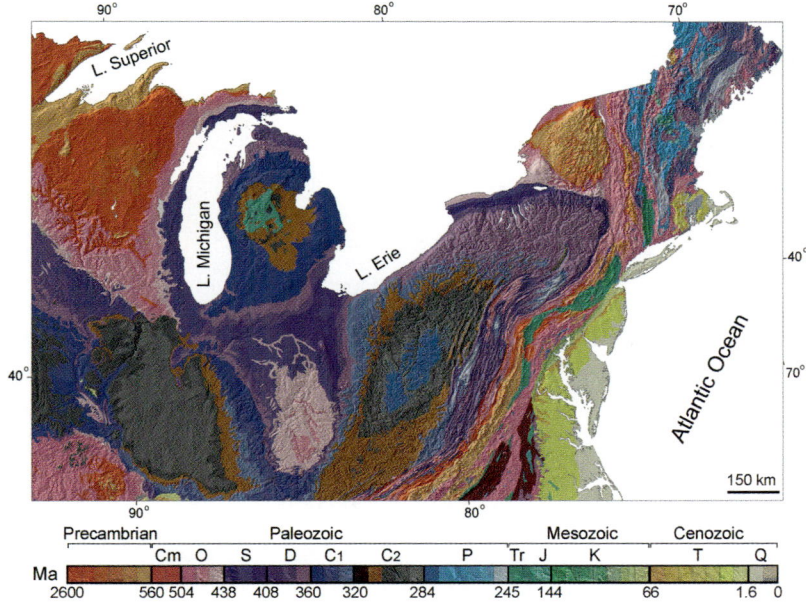

FIGURE 7.4 Combining geological information and hill-shading maps: a northeastern portion of the geological map of the United States. *From (Vigil et al., 2000a), courtesy of the U.S. Geological Survey.*

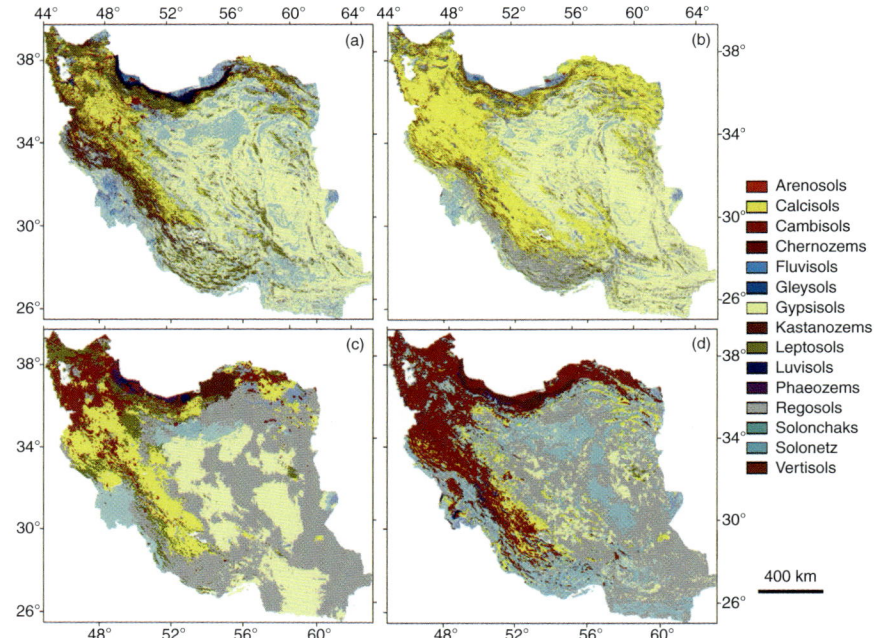

FIGURE 10.2 Small-scale predictive mapping of World Reference Base soil groups for the territory of Iran. Soil groups were interpolated by (a) supervised classification using maximum likelihood; (b) multinomial logistic regression; (c) regression kriging on memberships; and (d) per-pixel classification of interpolated taxonomic distances. A DEM was produced combining SRTM data (NASA, 2003) and GTOPO30 (USGS, 1996). The DEM grid spacing is 1 km. *G, H, TI,* and some other topographic attributes were used as predictors. *From (Hengl et al., 2007b, Fig. 5), reproduced with permission.*

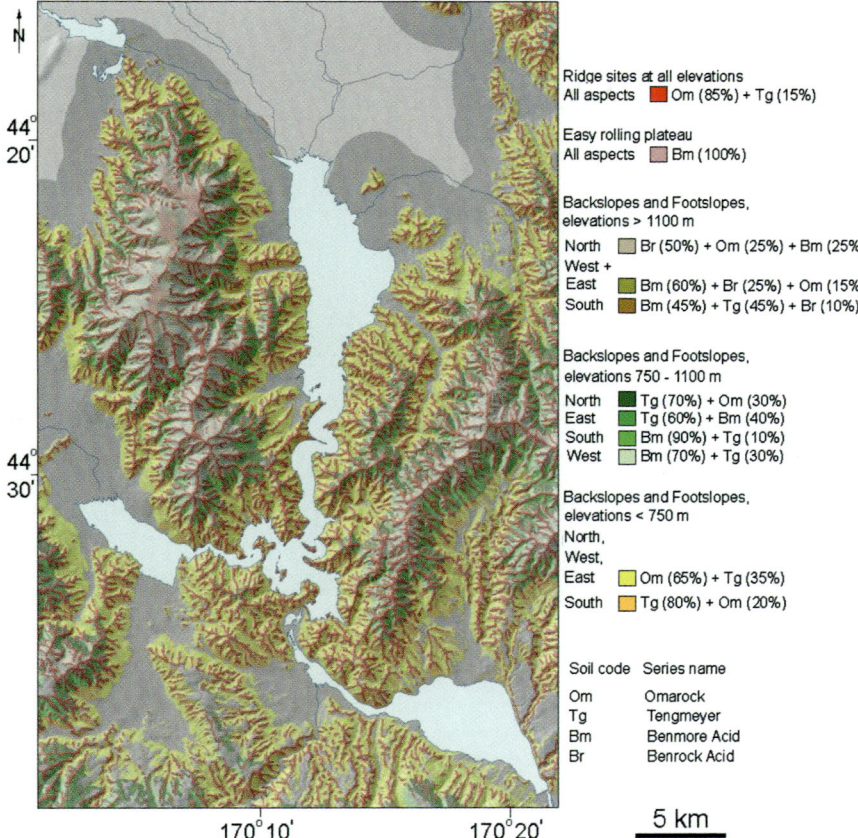

FIGURE 10.3 Large-scale predictive soil mapping by means of landscape segmentation and fuzzy logic (exemplified by the Benmore Range, South Island, New Zealand). The national DEM of New Zealand was used; the DEM grid spacing is 25 m. *From (Barringer et al., 2008, Fig. 3), © 2008 Springer-Verlag Berlin Heidelberg, reproduced with kind permission from Springer Science + Business Media.*

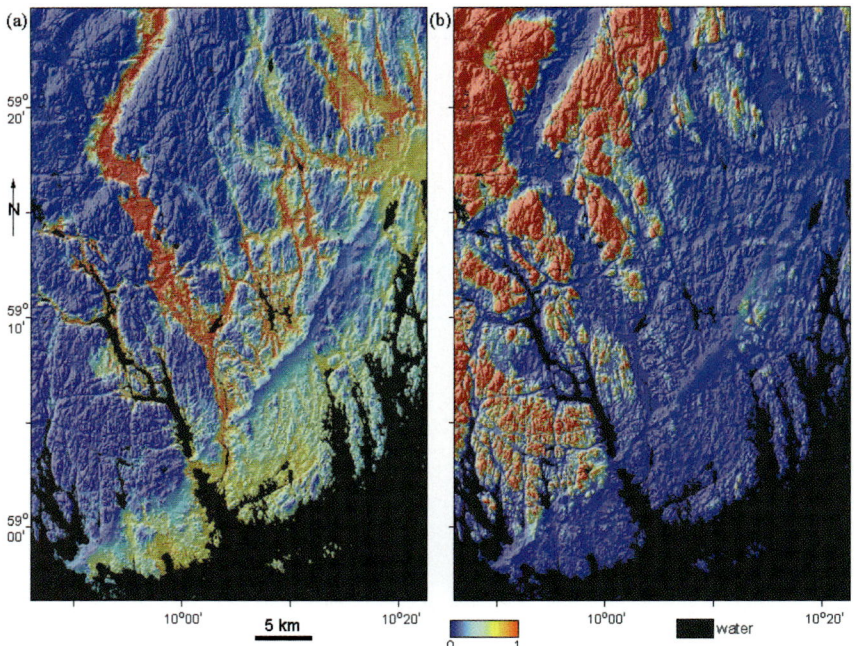

FIGURE 10.5 Large-scale mapping of the probability distribution, p-value, of (a) Cambisols and (b) Leptosols in the Vestfold County, southeastern Norway. Prediction was carried out by multinomial logistic regression. Leptosols occupy hills and rocky areas, and Cambisols occupy valleys. A DEM was created using 20-m contour topographic maps; the DEM grid spacing is 25 m. *From (Debella-Gilo and Etzelmüller, 2009, Fig. 5), reproduced with permission.*

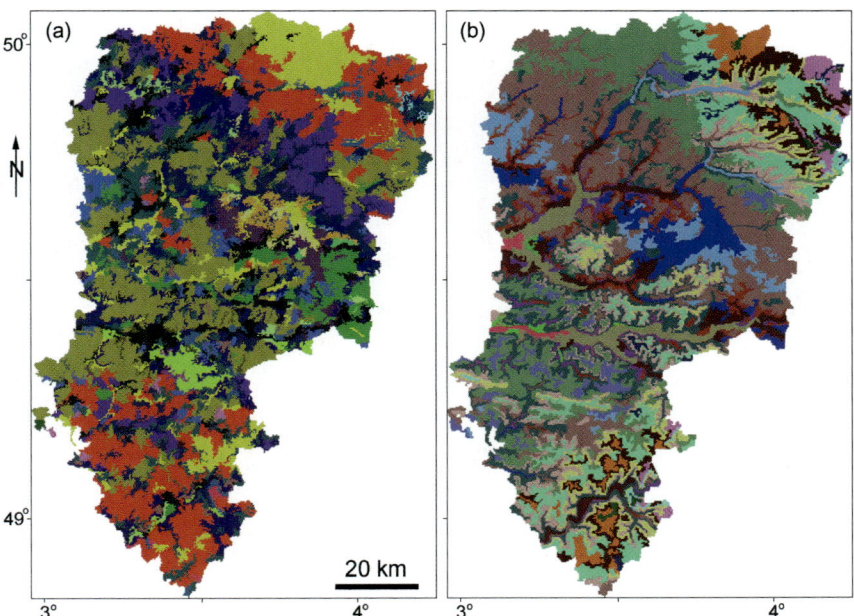

FIGURE 10.7 Upscaling soil contours for medium-scale soil mapping (exemplified by the Aisne Department, France): pedolandscape units resulting from the "bottom–up" (a) and "top-down" (b) approaches. The DEM grid spacing is 50 m. Map legends are not displayed as we discuss the geometry of soil contours only. *Modified after (Carré et al., 2008, Fig. 17.2), © 2008 Springer Science + Business Media B.V., reproduced with kind permission from Springer Science + Business Media.*

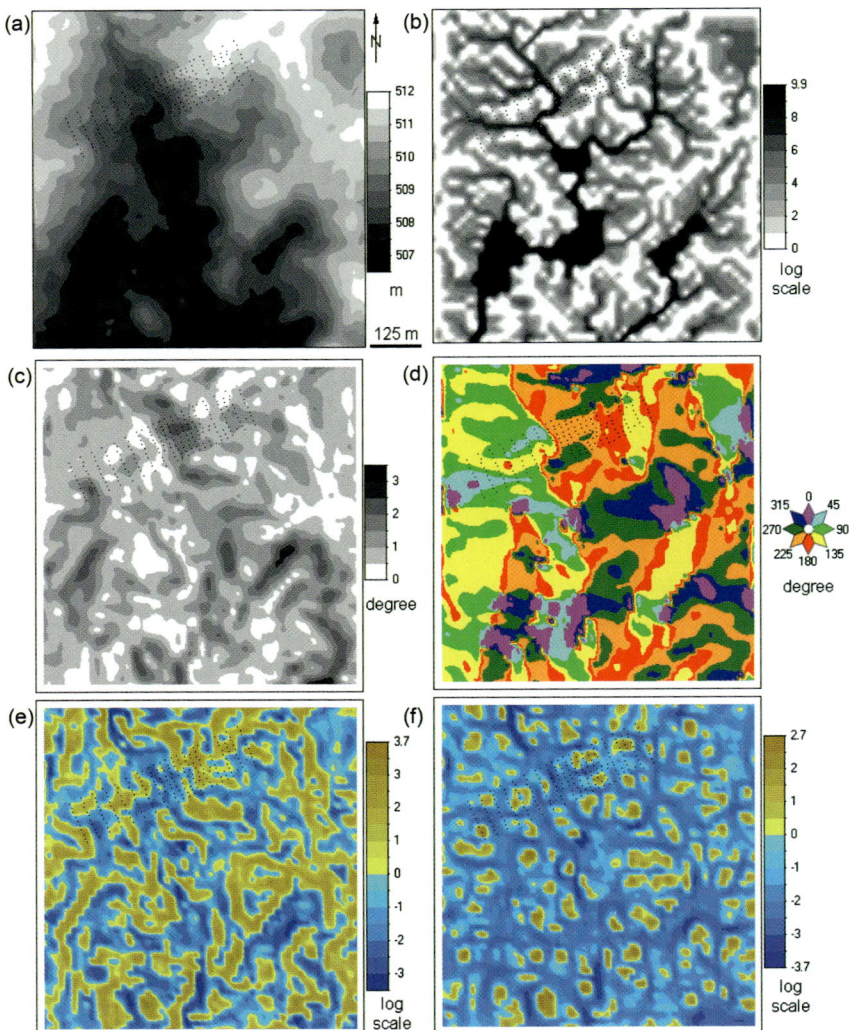

FIGURE 11.2 The Miniota site, some morphometric variables: (a) elevation, (b) specific catchment area, (c) slope gradient, (d) slope aspect, (e) vertical curvature, (f) minimal curvature. Dots indicate soil-sampling sites. *From (Florinsky et al., 2009a, Fig. 1).*

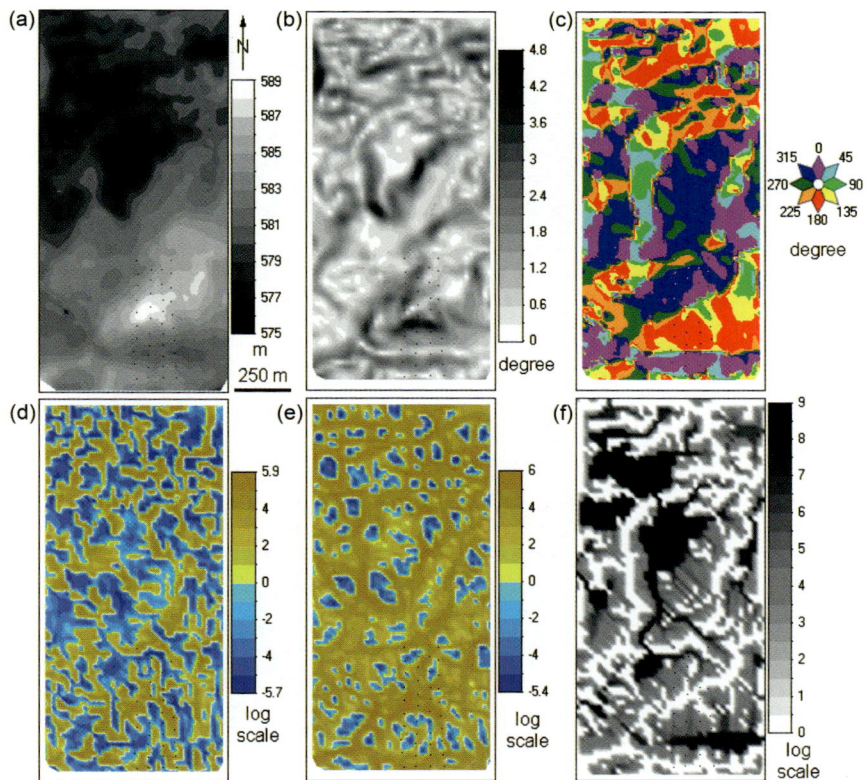

FIGURE 11.3 The Minnedosa site, some morphometric variables: (a) elevation, (b) slope gradient, (c) slope aspect, (d) horizontal curvature, (e) maximal curvature, (f) specific catchment area. Dots indicate soil-sampling sites. *From (Florinsky et al., 2009a, Fig. 2)*.

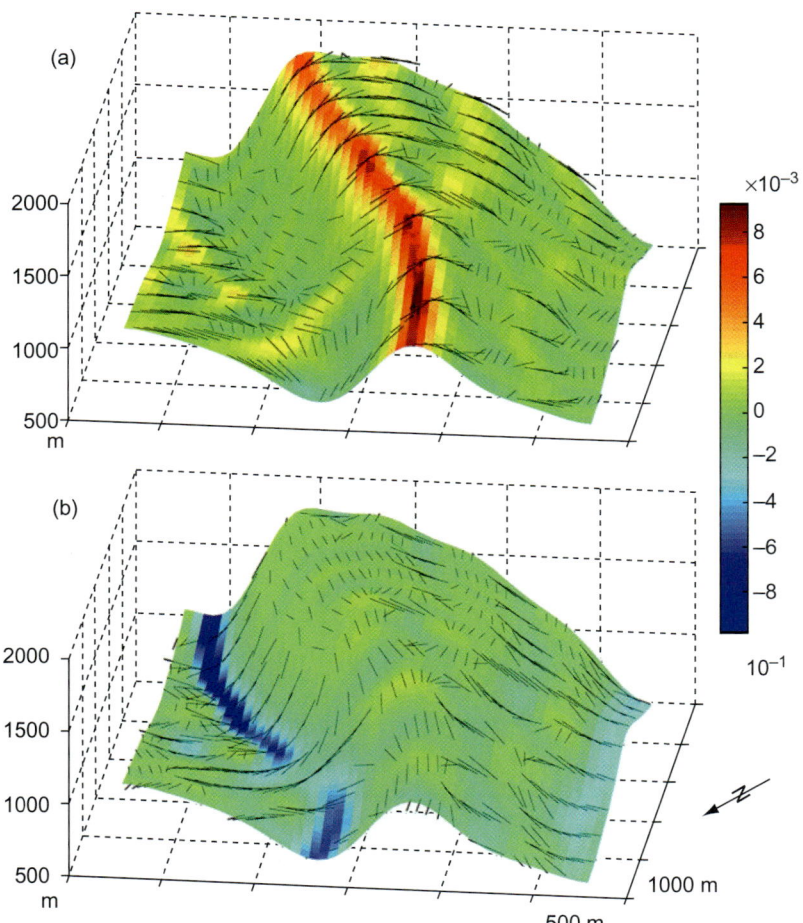

FIGURE 12.1 Magnitudes (colors) and axis directions (black tic marks) of the principal curvatures draped on the 3D model of the Sheep Mountain Anticline, Wyoming, USA: (a) maximal curvature, (b) minimal curvature. For the DEM description, see Fig. 6.7. *From (Mynatt et al., 2007a, Fig. 4), reproduced with permission.*

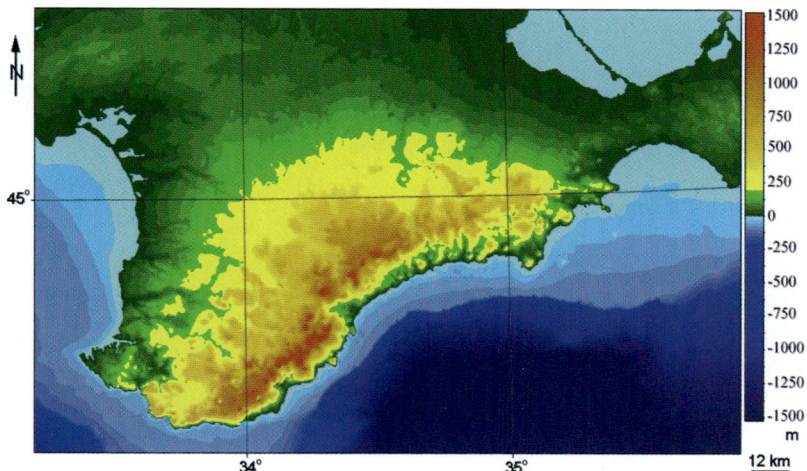

FIGURE 13.7 The Crimea and the adjacent sea bottom: elevation. *From (Florinsky, 2010, Fig. 4.5).*

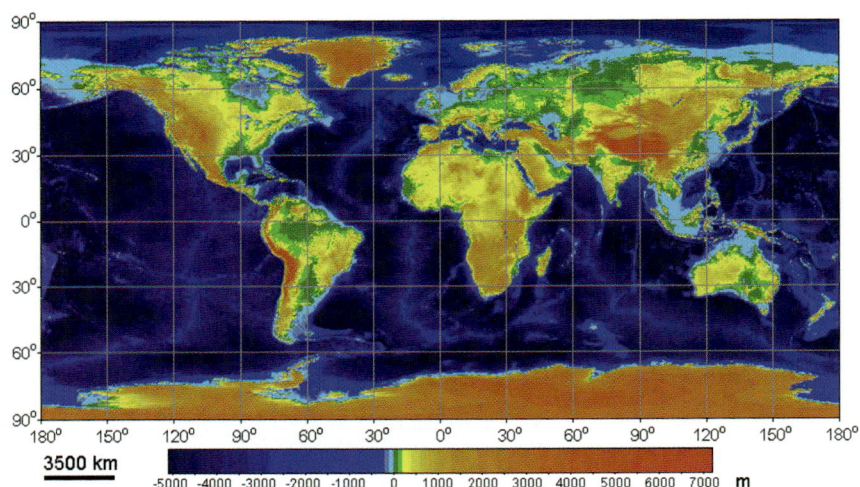

FIGURE 15.2 The Earth, elevation. *From (Florinsky, 2008c, Fig. 1).*

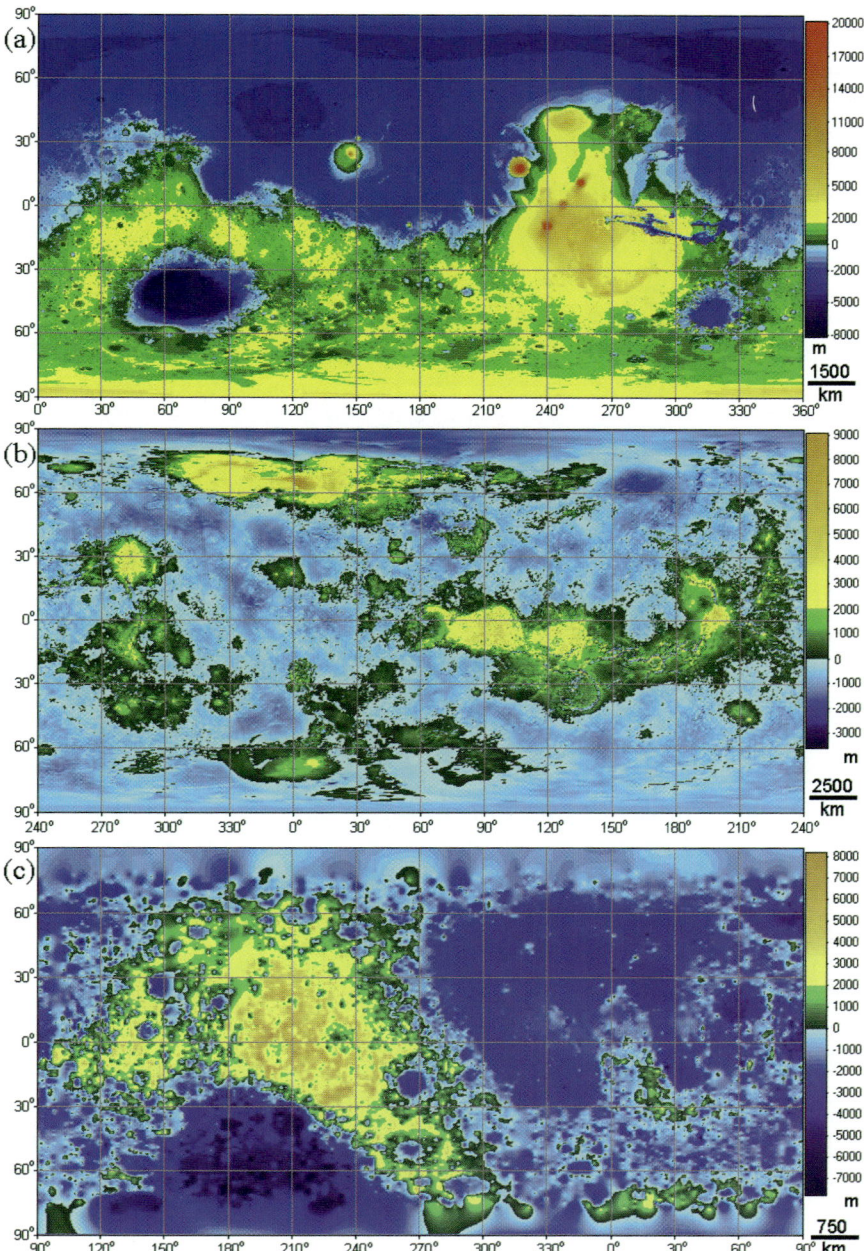

FIGURE 15.3 Elevation: (a) Mars, (b) Venus, (c) the Moon. *From (Florinsky, 2010, Fig. 4.19).*

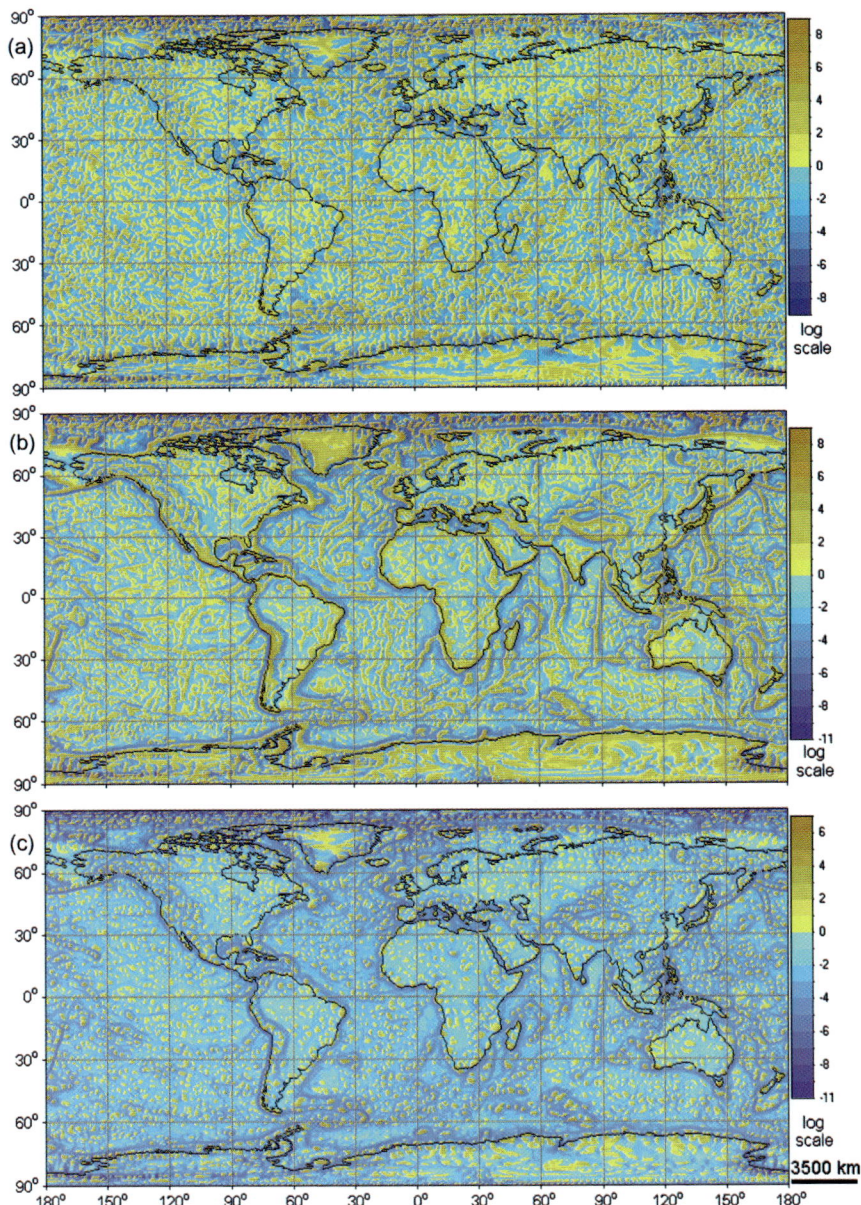

FIGURE 15.4 The Earth, global morphometric maps derived from the three times smoothed DEM: (a) horizontal curvature, (b) vertical curvature, (c) minimal curvature. *From (Florinsky, 2008c, Fig. 2a, b, c).*

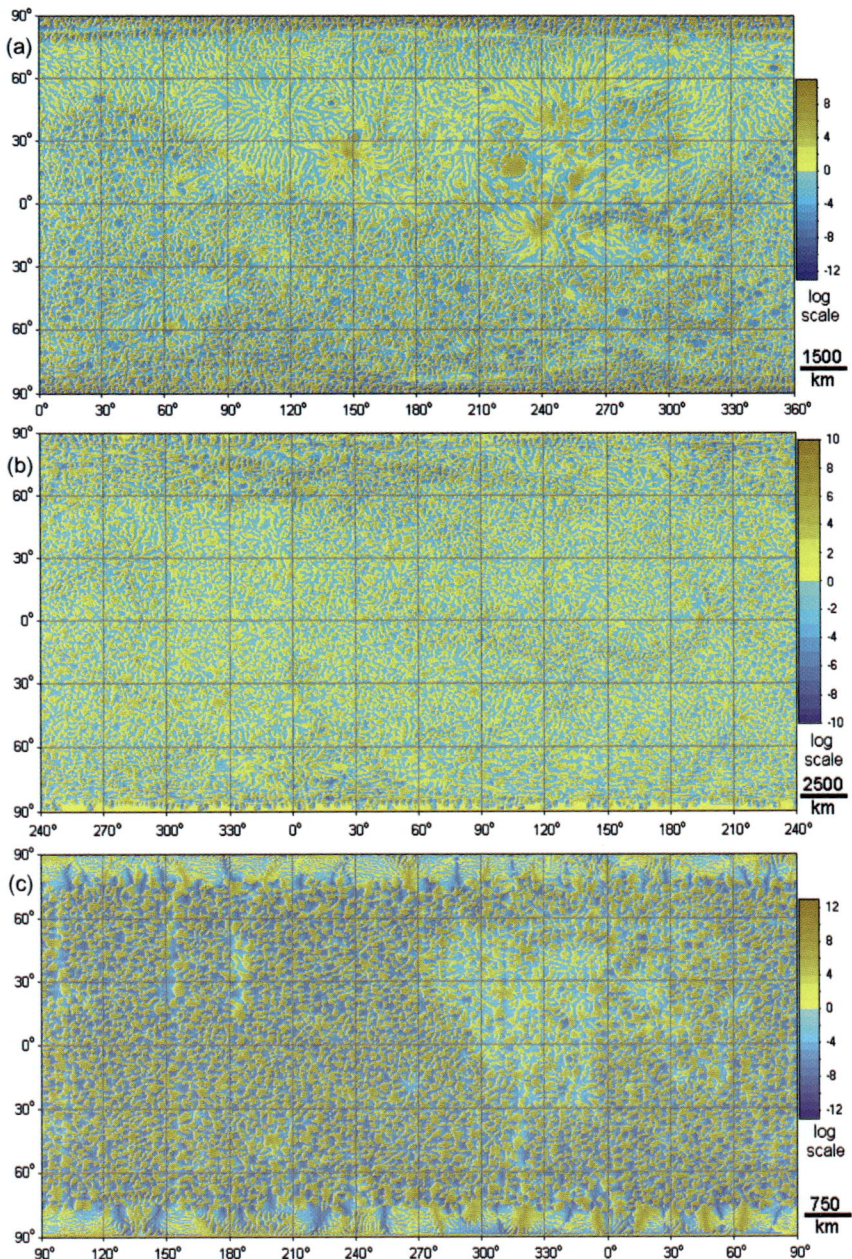

FIGURE 15.5 Horizontal curvature derived from the two times smoothed DEMs: (a) Mars, (b) Venus, (c) the Moon. *From (Florinsky, 2010, Fig. 4.21).*

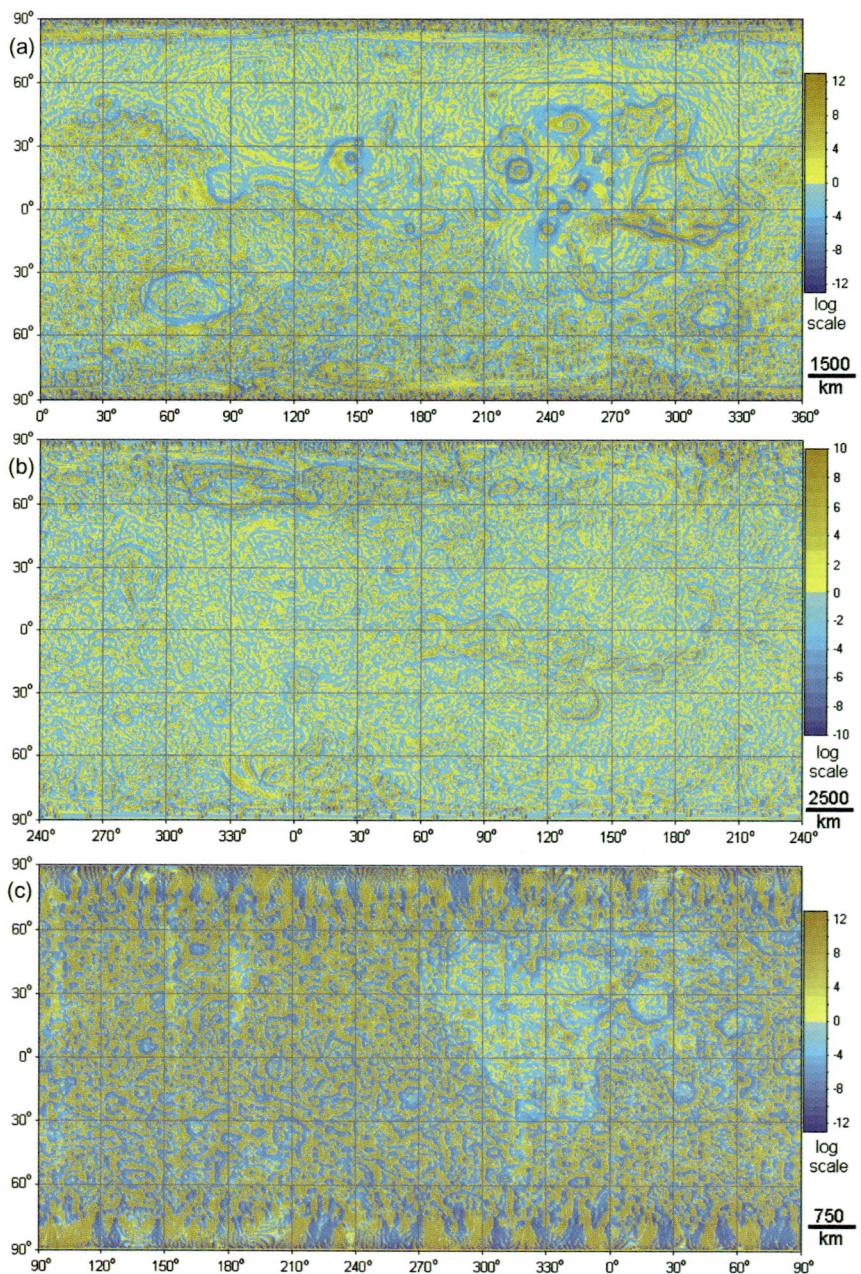

FIGURE 15.6 Vertical curvature derived from the two times smoothed DEMs: (a) Mars, (b) Venus, (c) the Moon. *From (Florinsky, 2010, Fig. 4.22).*

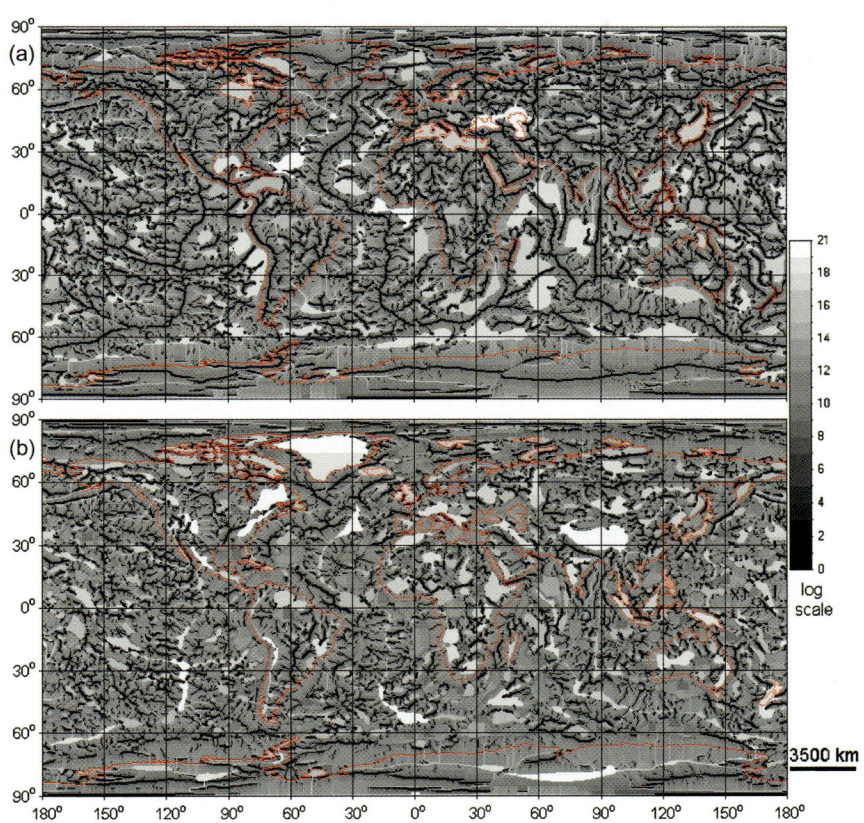

FIGURE 15.7 The Earth, global morphometric maps derived from the three times smoothed DEM: (a) specific catchment area, (b) specific dispersive area. *From (Florinsky, 2008c, Fig. 3).*

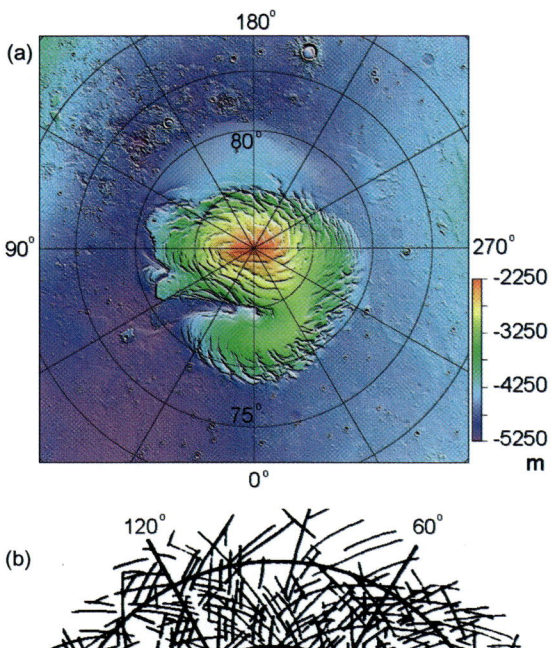
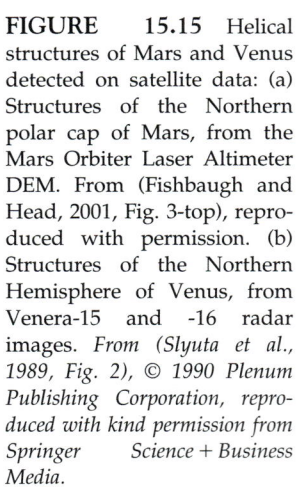
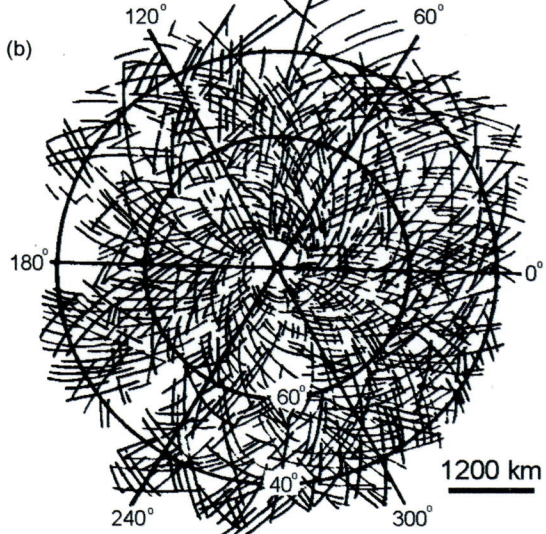

FIGURE 15.15 Helical structures of Mars and Venus detected on satellite data: (a) Structures of the Northern polar cap of Mars, from the Mars Orbiter Laser Altimeter DEM. From (Fishbaugh and Head, 2001, Fig. 3-top), reproduced with permission. (b) Structures of the Northern Hemisphere of Venus, from Venera-15 and -16 radar images. From (Slyuta et al., 1989, Fig. 2), © 1990 Plenum Publishing Corporation, reproduced with kind permission from Springer Science + Business Media.